Phase Transitions and Renormalization Group

Jean Zinn-Justin
Dapnia, CEA/Saclay, France

and

*Institut de Mathématiques de Jussieu-Chevaleret,
Université de Paris VII*

OXFORD
UNIVERSITY PRESS

Great Clarendon Street, Oxford OX2, 6DP,
United Kingdom

Oxford University Press is a department of the University of Oxford.
It furthers the Universitys objective of excellence in research, scholarship,
and education by publishing worldwide. Oxford is a registered trade mark of
Oxford University Press in the UK and in certain other countries

© Jean Zinn-Justin 2007

The moral rights of the author have been asserted

First published 2007
First published in paperback 2013

Impression: 1

All rights reserved. No part of this publication may be reproduced, stored in
a retrieval system, or transmitted, in any form or by any means, without the
prior permission in writing of Oxford University Press, or as expressly permitted
by law, by licence or under terms agreed with the appropriate reprographics
rights organization. Enquiries concerning reproduction outside the scope of the
above should be sent to the Rights Department, Oxford University Press, at the
address above

You must not circulate this work in any other form
and you must impose this same condition on any acquirer

British Library Cataloguing in Publication Data

Data available

ISBN 978-0-19-922719-8 (hbk.)
ISBN 978-0-19-966516-7 (pbk.)

Printed in Great Britain by
CPI Group (UK) Ltd, Croydon, CR0 4YY

Phase Transitions and Renormalization Group

Preface

Quantum field theory is at the basis of a notable part of the theoretical developments of twentieth century physics. The model that describes all fundamental interactions, apart from gravitation, at the microscopic scale, is a quantum field theory. Perhaps more surprisingly, quantum field theory has also led to a complete understanding of the singular macroscopic properties of a wide class of phase transitions near the transition point as well as statistical properties of some geometrical models.

However, unlike Newtonian or non-relativistic quantum mechanics, a quantum field theory in its most direct formulation leads to severe conceptual difficulties due to the appearance of infinities in the calculation of physical observables. Eventually, the problem of infinities was solved empirically by a method called *renormalization*. Only later did the method find a satisfactory interpretation, in the framework of the *renormalization group*. The problem of infinities is related to an unexpected phenomenon, the non-decoupling of very different length-scales in some physical situations.

It is within the framework of statistical physics and continuous phase transitions that the discussion of these conceptual problems is the simplest. This work thus tries to provide an elementary introduction to the notions of continuum limit and universality in physical systems with a large number of degrees of freedom. We will emphasize the role of Gaussian distributions and their relations with the mean-field approximation and Landau's theory of critical phenomena. We will show that quasi-Gaussian or mean-field approximations cannot describe correctly phase transitions in two and three space dimensions. We will assign this difficulty to the coupling of very different physical length-scales, even though the systems we will consider have only local, that is, short-range interactions. To analyse the problem, a new concept is required: the renormalization group, whose fixed points allow understanding the universality of physical properties at large distance, beyond the quasi-Gaussian or mean-field approximations.

Renormalization group arguments then lead to the idea that, in critical systems, correlations at large distance near the transition temperature can be described by local statistical field theories, formally quantum field theories in imaginary time.

This work corresponds to a course delivered, in various forms, for three years at the University of Paris 7 and first published in French [1]. It is organized in the following way.

Chapter 1 contains a short, semi-historical, introduction that tries to describe the evolution of ideas from the first works in quantum field theory [2–5] to the application of renormalization group methods to phase transitions.

In Chapter 2, we have collected a number of technical results concerning generating functions, Gaussian measures and the steepest descent method, which are indispensable for the understanding of the work.

Chapter 3 introduces several basic topics of the work: the notions of continuum limit and universality, through the examples of the central limit theorem and the random walk. We show that universality originates from the small probability of large deviations from the expectation value in probability distributions, which translates into an hypothesis of locality in random walks. In both examples, universality is related to the appearance of asymptotic Gaussian distributions. We then show that, beyond a direct calculation, universality can also be understood as resulting from the existence of fixed points of transformations acting on the space of probability distributions. These very simple examples will allow us to introduce immediately the renormalization group terminology. Finally, the existence of continuum limits leads naturally to a description in terms of path integrals.

In Chapter 4, we begin the study of classical statistical systems, the central topic of this work, with the example of one-dimensional models. This enables us to introduce the terminology of statistical physics, like correlation functions, thermodynamic limit, correlation length, and so on. Even if one-dimensional systems with short-range interactions do not exhibit phase transitions, it is nevertheless possible to define a continuum limit near zero temperature. Moreover, in the case of short-range interactions, these systems can be solved exactly by the transfer matrix method, and thus provide interesting pedagogical examples.

The continuum limit of one-dimensional models again leads to path integrals. We describe some of their properties in Chapter 5 (for a more systematic discussion see, for example, Ref. [6]).

In Chapter 6, we define more general statistical systems, in an arbitrary number of space dimensions. For convenience, we use the ferromagnetic language, even though, as a consequence of universality, the results that are derived in this work apply to much more general statistical systems. In addition to complete and connected correlation functions (whose decay properties at large distance, called cluster properties, are recalled), which we have already defined in the preceding chapters, we introduce vertex functions, which are related to the thermodynamic potential. The free energy and thermodynamic potential, like connected correlation functions and vertex functions, are related by a Legendre transformation of which we discuss a few properties.

Chapter 7 is devoted to the concept of phase transition, a concept that is far from being trivial in the sense that a phase transition requires the interaction of an infinite number of degrees of freedom. We first solve exactly a particular model in the limit in which the number of space dimensions becomes infinite. In this limit, the model exhibits a behaviour that the analysis presented in the following chapters will identify as quasi-Gaussian or mean-field like. Then, we discuss, in general terms, the existence of phase transitions as a function of the space dimension. We emphasize the difference between models with discrete and continuous symmetries in dimension two.

In Chapter 8, we examine the universal properties of phase transitions in the quasi-Gaussian or mean-field approximations. We study the singularities of thermodynamic functions at the transition point as well as the large-distance behaviour of the two-point correlation function. We summarize the universal properties in the form of Landau's theory [7]. We stress the peculiarities of models with continuous symmetries at low temperature due to the appearance of Goldstone modes. Finally, we evaluate corrections to the quasi-Gaussian approximation and show that the approximation is only consistent in space dimension larger than 4 (following the lines of Ref. [8]). We mention the possible existence of tricritical points.

In Chapter 9, we introduce the general concept of renormalization group [5] in the spirit of the work [9]. We study the role of fixed points and their stability properties. We exhibit a particular fixed point, the Gaussian fixed point, which is stable in dimension larger than 4. We identify the leading perturbation to the Gaussian fixed point in dimension ≤ 4. We discuss the possible existence of a non-Gaussian fixed point near dimension 4.

In Chapter 10, using the assumptions introduced in Chapter 9, we show that it is indeed possible to find a non-Gaussian fixed point in dimension $d = 4 - \varepsilon$ [10], both in models with reflection and rotation symmetries. We briefly introduce the field theory methods [11, 12] that we will describe more thoroughly in the following chapters. Finally, we present a selection of numerical results concerning critical exponents and some universal amplitude ratios [13–17], obtained by field theory methods using both the Callan–Symanzik formalism in three dimensions and the ε-expansion extrapolated to $\varepsilon = 1$.

Chapter 11 contains a general discussion of renormalization group equations and the properties of the corresponding fixed points, for a whole class of models that possess more general symmetries than the reflection and rotation groups considered so far, generalizing somewhat the results presented in [8, 18]. In particular, the analysis leads to an interesting conjecture, relating decay of correlation functions and stability of fixed points [8, 19].

With Chapter 12, we begin a more systematic presentation of field theory methods. Beyond a simple generalization of the perturbative methods already presented in the preceding chapters, several new concepts are introduced like the loop expansion, dimensional continuation and regularization [20].

With these technical tools, we can then justify, in Chapter 13, asymptotic renormalization group equations obtained by varying the cut-off, as they appear in field theory [4, 21]. General universality properties follow, as well as methods of calculating universal quantities as an expansion in powers of the deviation $\varepsilon = 4 - d$ from dimension 4. We conclude the chapter by a short presentation of the alternative formalism of renormalization group equations in renormalized form [22–25], in particular Callan–Symanzik equations [22] directly relevant to the numerical results reported in Chapter 10.

A class of field theories with an $O(N)$ orthogonal symmetry can be solved in the $N \to \infty$ limit, as we show in Chapter 14. All universal properties derived within the framework of the ε-expansion can also be proved at fixed dimension, within the framework of an expansion in powers of $1/N$ [26–36].

In models with continuous symmetries, phase transitions are dominated, at low temperature and large distance, by the interaction between Goldstone (massless) modes. The interaction can be described by the non-linear σ-model. Its study, using the renormalization group, allows generalizing the scaling properties of the critical theory at the transition to the whole low-temperature phase and studying properties of the phase transition near dimension 2 [37–40].

The renormalization group of quantum field theory has an interpretation as an asymptotic renormalization group when the relevant fixed point is close to the Gaussian fixed point. Quite early, more general formulations of the renormalization group have been proposed, which do not rely on such an assumption [41–42]. They lead to functional renormalization group (FRG) equations that describe the evolution of the effective interaction, but which are much more difficult to handle than the equations arising in field theory. They have been used to prove renormalizability without relying on a direct analysis of Feynman diagrams, unlike more traditional methods [43]. Moreover, more recently, they have inspired a number of new approximation schemes, different from the perturbative scheme of field theory [44]. Thus, for both pedagogical and practical reasons, we have decided to describe then in this work.

Finally, in the appendix, we have collected various technical considerations useful for a better understanding of the material presented in the work, and a few additional results concerning the FRG and functional flow equations based on partial integration over low-momentum (IR) modes [45].

Contents

1 Quantum field theory and the renormalization group 1
   1.1 Quantum electrodynamics: A quantum field theory 3
   1.2 Quantum electrodynamics: The problem of infinities 4
   1.3 Renormalization 7
   1.4 Quantum field theory and the renormalization group 9
   1.5 A triumph of QFT: The Standard Model 10
   1.6 Critical phenomena: Other infinities 12
   1.7 Kadanoff and Wilson's renormalization group 14
   1.8 Effective quantum field theories 16

2 Gaussian expectation values. Steepest descent method 19
   2.1 Generating functions 19
   2.2 Gaussian expectation values. Wick's theorem 20
   2.3 Perturbed Gaussian measure. Connected contributions 24
   2.4 Feynman diagrams. Connected contributions 25
   2.5 Expectation values. Generating function. Cumulants 28
   2.6 Steepest descent method 31
   2.7 Steepest descent method: Several variables, generating functions . . . 37
   Exercises 40

3 Universality and the continuum limit 45
   3.1 Central limit theorem of probabilities 45
   3.2 Universality and fixed points of transformations 54
   3.3 Random walk and Brownian motion 59
   3.4 Random walk: Additional remarks 71
   3.5 Brownian motion and path integrals 72
   Exercises 75

4 Classical statistical physics: One dimension 79
   4.1 Nearest-neighbour interactions. Transfer matrix 80
   4.2 Correlation functions 83
   4.3 Thermodynamic limit 85
   4.4 Connected functions and cluster properties 88
   4.5 Statistical models: Simple examples 90
   4.6 The Gaussian model 92

- 4.7 Gaussian model: The continuum limit 98
- 4.8 More general models: The continuum limit 102
- Exercises 104

5 Continuum limit and path integrals 111
- 5.1 Gaussian path integrals 111
- 5.2 Gaussian correlations. Wick's theorem 118
- 5.3 Perturbed Gaussian measure 118
- 5.4 Perturbative calculations: Examples 120
- Exercises 124

6 Ferromagnetic systems. Correlation functions 127
- 6.1 Ferromagnetic systems: Definition 127
- 6.2 Correlation functions. Fourier representation 133
- 6.3 Legendre transformation and vertex functions 137
- 6.4 Legendre transformation and steepest descent method 142
- 6.5 Two- and four-point vertex functions 143
- Exercises 145

7 Phase transitions: Generalities and examples 147
- 7.1 Infinite temperature or independent spins 150
- 7.2 Phase transitions in infinite dimension 153
- 7.3 Universality in infinite space dimension 158
- 7.4 Transformations, fixed points and universality 161
- 7.5 Finite-range interactions in finite dimension 163
- 7.6 Ising model: Transfer matrix 166
- 7.7 Continuous symmetries and transfer matrix 171
- 7.8 Continuous symmetries and Goldstone modes 173
- Exercises 175

8 Quasi-Gaussian approximation: Universality, critical dimension 179
- 8.1 Short-range two-spin interactions 181
- 8.2 The Gaussian model: Two-point function 183
- 8.3 Gaussian model and random walk 188
- 8.4 Gaussian model and field integral 190
- 8.5 Quasi-Gaussian approximation 194
- 8.6 The two-point function: Universality 196
- 8.7 Quasi-Gaussian approximation and Landau's theory 199
- 8.8 Continuous symmetries and Goldstone modes 200
- 8.9 Corrections to the quasi-Gaussian approximation 202
- 8.10 Mean-field approximation and corrections 207
- 8.11 Tricritical points 211
- Exercises 212

9 Renormalization group: General formulation 217
- 9.1 Statistical field theory. Landau's Hamiltonian 218
- 9.2 Connected correlation functions. Vertex functions 220
- 9.3 Renormalization group (RG): General idea 222
- 9.4 Hamiltonian flow: Fixed points, stability 226
- 9.5 The Gaussian fixed point 231

9.6 Eigen-perturbations: General analysis 234
9.7 A non-Gaussian fixed point: The ε-expansion 237
9.8 Eigenvalues and dimensions of local polynomials 241
10 Perturbative renormalization group: Explicit calculations 243
   10.1 Critical Hamiltonian and perturbative expansion 243
   10.2 Feynman diagrams at one-loop order 246
   10.3 Fixed point and critical behaviour 248
   10.4 Critical domain 254
   10.5 Models with $O(N)$ orthogonal symmetry 258
   10.6 RG near dimension 4 259
   10.7 Universal quantities: Numerical results 262
11 Renormalization group: N-component fields 267
   11.1 RG: General remarks 268
   11.2 Gradient flow 269
   11.3 Model with cubic anisotropy 272
   11.4 Explicit general expressions: RG analysis 276
   11.5 Exercise: General model with two parameters 281
   Exercises 284
12 Statistical field theory: Perturbative expansion 285
   12.1 Generating functionals 285
   12.2 Gaussian field theory. Wick's theorem 287
   12.3 Perturbative expansion 289
   12.4 Loop expansion 296
   12.5 Dimensional continuation and dimensional regularization . . . 299
   Exercises 306
13 The σ^4 field theory near dimension 4 307
   13.1 Effective Hamiltonian. Renormalization 308
   13.2 RG equations 313
   13.3 Solution of RG equations: The ε-expansion 316
   13.4 The critical domain above T_c 322
   13.5 RG equations for renormalized vertex functions 326
   13.6 Effective and renormalized interactions 328
14 The $O(N)$ symmetric $(\phi^2)^2$ field theory in the large N limit . . . 331
   14.1 Algebraic preliminaries 332
   14.2 Integration over the field ϕ: The determinant 333
   14.3 The limit $N \to \infty$: The critical domain 337
   14.4 The $(\phi^2)^2$ field theory for $N \to \infty$ 339
   14.5 Singular part of the free energy and equation of state 342
   14.6 The $\langle\lambda\lambda\rangle$ and $\langle\phi^2\phi^2\rangle$ two-point functions 345
   14.7 RG and corrections to scaling 347
   14.8 The $1/N$ expansion 350
   14.9 The exponent η at order $1/N$ 352
   14.10 The non-linear σ-model 353
15 The non-linear σ-model 355
   15.1 The non-linear σ-model on the lattice 355

15.2 Low-temperature expansion 357
15.3 Formal continuum limit 362
15.4 Regularization 363
15.5 Zero-momentum or IR divergences 364
15.6 Renormalization group 365
15.7 Solution of the RGE. Fixed points 370
15.8 Correlation functions: Scaling form 372
15.9 The critical domain: Critical exponents 374
15.10 Dimension 2 375
15.11 The $(\phi^2)^2$ field theory at low temperature 379

16 Functional renormalization group 383
16.1 Partial field integration and effective Hamiltonian 383
16.2 High-momentum mode integration and RG equations . . . 392
16.3 Perturbative solution: ϕ^4 theory 398
16.4 RG equations: Standard form 401
16.5 Dimension 4 404
16.6 Fixed point: ε-expansion 411
16.7 Local stability of the fixed point 413

Appendix 419
A1 Technical results 419
A2 Fourier transformation: Decay and regularity 423
A3 Phase transitions: General remarks 428
A4 $1/N$ expansion: Calculations 433
A5 Functional flow equations: Additional considerations . . . 435

Bibliography 443
Index 449

1 Quantum field theory and the renormalization group

Without a minimal understanding of quantum or statistical field theories (formally related by continuation to imaginary time), the theoretical basis of a notable part of twentieth century physics remains incomprehensible.

Indeed, field theory, in its various incarnations, describes fundamental interactions at the microscopic scale, singular properties of phase transitions (like liquid–vapour, ferromagnetic, superfluid, separation of binary mixture,...) at the transition point, properties of diluted quantum gases beyond the model of Bose–Einstein condensation, statistical properties of long polymeric chains (as well as self-avoiding random walks), or percolation, and so on.

In fact, quantum field theory offers at present the most comprehensive framework to discuss physical systems that are characterized by a large number of strongly interacting local degrees of freedom.

However, at its birth, quantum field theory was confronted with a somewhat unexpected problem, the problem of *infinities*. The calculation of most physical processes led to infinite results. An empirical recipe, *renormalization*, was eventually discovered that allowed extracting from divergent expressions finite predictions. The procedure would hardly have been convincing if the predictions were not confirmed with increasing precision by experiment. A new concept, the *renormalization group* related in some way to the renormalization procedure, but whose meaning was only fully appreciated in the more general framework of the theory of phase transitions, has led, later, to a satisfactory interpretation of the origin and role of renormalizable quantum field theories and of the renormalization process.

This first chapter tries to present a brief history of the origin and the development of quantum field theory, and of the evolution of our interpretation of renormalization and the renormalization group, which has led to our present understanding.

This history has two aspects, one directly related to the theory of fundamental interactions that describes physics at the microscopic scale, and another one related to the theory of phase transitions in macroscopic physics and their universal properties. That two so vastly different domains of physics have required the development of the same theoretical framework, is extremely surprising. It is one of the attractions of theoretical physics that such relations can sometimes be found.

A few useful dates:

1925 Heisenberg proposes a quantum mechanics, under the form of a mechanics of matrices.

1926 Schrödinger publishes his famous equation that bases quantum mechanics on the solution of a non-relativistic wave equation. Since relativity theory was already well established when quantum mechanics was formulated, this may surprise.

In fact, for accidental reasons, the spectrum of the hydrogen atom is better described by a non-relativistic wave equation than by a relativistic equation without spin,* the Klein–Gordon equation (1926).

1928 Dirac introduces a relativistic wave equation that incorporates the spin 1/2 of the electron, which describes much better the spectrum of the hydrogen atom, and opens the way for the construction of a relativistic quantum theory. In the two following years, Heisenberg and Pauli lay out, in a series of articles, the general principles of quantum field theory.

1934 First correct calculation in quantum electrodynamics (Weisskopf) and confirmation of the existence of divergences, called ultraviolet (UV) since they are due, in this calculation, to the short-wavelength photons.

1937 Landau publishes his general theory of phase transitions.

1944 Exact solution of the two-dimensional Ising model by Onsager.

1947 Measurement of the so-called Lamb shift by Lamb and Retherford, which agrees well with the prediction of quantum electrodynamics (QED) after cancellation between infinities.

1947–1949 Construction of an empirical general method to eliminate divergences called *renormalization* (Feynman, Schwinger, Tomonaga, Dyson, *et al*).

1954 Yang and Mills propose a non-Abelian generalization of Maxwell's equations based on non-Abelian gauge symmetries (associated to non-commutative groups).

1954–1956 Discovery of a formal property of quantum field theory characterized by the existence of a *renormalization group* whose deep meaning is not fully appreciated (Peterman–Stückelberg, Gell-Mann–Low, Bogoliubov–Shirkov).

1967–1975 The Standard Model, a renormalizable quantum field theory based on the notions of non-Abelian gauge symmetry and spontaneous symmetry breaking, is proposed, which provides a complete description of all fundamental interactions, but gravitation.

1971–1972 After the initial work of Kadanoff (1966), Wilson, Wegner, *et al*, develop a more general concept of renormalization group, which includes the field theory renormalization group as a limit, and which explains universality properties of continuous phase transitions (liquid–vapour, superfluidity, ferromagnetism) and later of geometrical models like self-avoiding random walks or percolation.

1972–1975 Several groups, in particular Brézin, Le Guillou and Zinn-Justin, develop powerful quantum field theory techniques that allow a proof of universality properties of critical phenomena and calculating universal quantities.

1973 Using renormalization group arguments, Politzer and Gross–Wilczek establish the property of *asymptotic freedom* of a class of non-Abelian gauge theories, which allows explaining the free-particle behaviour of quarks within nucleons.

1975–1976 Additional information about universal properties of phase transitions are derived from the study of the non-linear σ model and the corresponding $d - 2$ expansion (Polyakov, Brézin–Zinn-Justin).

---

* intrinsic angular momentum of particles, that takes half-integer (fermions) or integer (bosons) values in units of $\hbar$.

1977–1980 Following a suggestion of Parisi, Nickel calculates several successive terms of the series expansion of renormalization group functions, by field theory methods. Applying summation methods to these series, Le Guillou and Zinn-Justin obtain the first precise estimates of critical exponents in three-dimensional phase transitions from renormalization group methods.

1.1 Quantum electrodynamics: A quantum field theory

Quantum electrodynamics (QED) describes, in a quantum and relativistic framework, interactions between all electrically charged particles and the electromagnetic field. QED is not a theory of individual particles, as in non-relativistic quantum mechanics, but a *quantum field theory* (QFT). Indeed, it is also a quantum extension of a classical relativistic field theory: Maxwell electromagnetism in which the dynamical quantities are fields, the electric and magnetic fields. Moreover, the discovery that to the electromagnetic field were associated quanta, the photons, which are massless spin one particles, has naturally led to the postulate that all particles were also manifestations of quantum fields.

But such a theory differs radically from a particle theory in the sense that fields have an *infinite number of degrees of freedom*. Indeed, a point particle in classical mechanics has three degrees of freedom; it is characterized by its three Cartesian coordinates. By contrast, a field is characterized by its values at all space points, which thus constitutes an infinite number of data. The non-conservation of the number of particles in high-energy collisions is a manifestation of such a property.

Moreover, the field theories that describe microscopic physics have a *locality* property, a notion that generalizes the notion of point-like particle: they display no short-distance structure.

The combination of an infinite number of degrees of freedom and locality explain why QFT has somewhat 'exotic' properties.

Gauge symmetries. In what follows, we mention gauge symmetry and gauge theories, the simplest example being provided by QED. In non-relativistic quantum mechanics, in the presence of a magnetic field, gauge invariance corresponds simply to the possibility of adding a gradient term to the vector potential without affecting the equations of motion. In non-relativistic quantum mechanics, physics is not changed if one multiplies the wave function by a phase factor $e^{i\theta}$ (corresponding to a transformation of the Abelian, or commutative, group $U(1)$). In the case of a charged particle, in the presence of a magnetic field, one discovers a much larger symmetry, a gauge symmetry: it is possible to change the phase of the wave function at each point in space independently,

$$\psi(x) \mapsto e^{i\theta(x)} \psi(x),$$

by modifying in a correlated way the vector potential.

Unlike an ordinary symmetry that corresponds to transforming in a global way all dynamical variables, a gauge symmetry corresponds to independent transformations at each point in space (or space-time). Gauge symmetry is a dynamical principle: it generates interactions instead of simply relating them as an ordinary or global symmetry. To render a theory gauge invariant, it is necessary to introduce a vector potential coupled in a universal way to all charged particles. In a relativistic quantum theory, this vector potential takes the form of a gauge field corresponding to a spin-one particle, the photon in the case of QED.

Units in relativistic quantum theory. The phenomena that we evoke below are characteristic of a relativistic and very quantum limit. It is thus physically meaningful to take the speed of light, c, and Planck's constant, $\hbar$, as units. As a consequence, in a relativistic theory mass-scales M, momenta p and energies E can be related by the speed of light c:

$$E = pc = Mc^2$$

and, thus, expressed in a common unit like the electron volt (eV). It is then equivalent to talk about large momenta or large energies.

Moreover, in a quantum theory momentum-scales p can be related to length-scales ℓ by Planck's constant,

$$p\ell = \hbar.$$

As a consequence, high-energy experiments probe properties of matter at short distance.

1.2 Quantum electrodynamics: The problem of infinities

The discovery of the Dirac equation in 1928 opened the way to the construction of a quantum and relativistic theory, allowing a more precise description of the electromagnetic interactions between protons and electrons. This theory, whose principles were established by Heisenberg and Pauli (1929–1930), was a QFT and not a theory of individual particles, because the discovery that the electromagnetic field was associated to quanta suggested that, conversely, all particles could be manifestations of the existence of underlying fields.

After the articles of Heisenberg and Pauli, Oppenheimer and Waller (1930) published independent calculations of the effect of the electromagnetic field on the electron propagation, at first-order in the fine structure constant, a *dimensionless* constant that characterizes the intensity of the electromagnetic interactions,

$$\alpha = \frac{e^2}{4\pi \hbar c} \approx 1/137, \tag{1.1}$$

where e is the electron charge, defined in terms of the Coulomb potential parametrized as a function of the separation R as $e^2/4\pi R$. Since this constant α is numerically small (a meaningful statement because it is a dimensionless quantity), a first-order calculation is reasonable.

The physical process responsible of this contribution is a process typical of a QFT, the emission and absorption by an electron of energy-momentum (or quadri-momentum) q of a virtual photon of quadri-momentum k, as represented by the

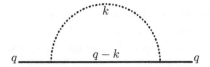

Fig. 1.1 Feynman diagram: A contribution to electron propagation, dotted line for photon, full line for electron.

Feynman diagram in Figure 1.1 (a representation proposed by Feynman several years later).

One possible motivation for undertaking such a calculation was a determination of the first corrections to the electron mass in QED and the solution of the puzzle of the 'classical model' of the electron. In a relativistic theory, the mass of a particle is proportional to its rest energy. This thus includes the potential self-energy. But it was well-known that the classical model of the electron as a charged sphere of radius R led to a result that diverged as e^2/R in the zero R limit. It could have been hoped that quantum mechanics, which is a theory of wave functions, would solve the problem generated by the point-like nature of the electron.

However, the first results were paradoxical. Not only was the contribution to the mass still infinite, but it diverged even more strongly than in the classical model: introducing a bound Λc^2 on the photon energy (this is equivalent to modifying the theory at a short-distance $R = \hbar/c\Lambda$), one found a quadratic divergence $\Lambda^2 \propto 1/R^2$. In fact, it was soon discovered that these results were wrong. Indeed, perturbative calculations with the technical tools of the time were laborious. The formalism was not explicitly relativistic; the role of the 'holes' in Dirac's theory (predicted to be anti-electrons or positrons in 1931 and experimentally discovered in 1932) was unclear, and gauge invariance was at the origin of additional difficulties. Only in 1934 was the correct result published by Weisskopf (after a last error had been pointed out by Furry). It confirmed that the contribution was definitively infinite, even though it diverged less strongly than in the classical model. The quadratic divergence was replaced by a less severe logarithmic divergence and the contribution to the electron mass at order α was found to be given by

$$\delta m_{\rm QED} = -3\frac{\alpha}{2\pi} m \ln(mRc/\hbar),$$

where m is the electron mass for $\alpha = 0$.

The divergence is generated by the summation over virtual photons of arbitrary high momenta (due to the absence of a short-distance structure), which explains the denomination of ultraviolet (UV) divergences. Moreover, conservation of probabilities implies that all processes contribute additively.

The conclusion was that QFT was less singular than the classical model. Nevertheless, the problem of infinities was not solved and no straightforward modification could be found to save QFT.

These divergences were understood to have a profound meaning, seeming to be an unavoidable consequence of unitarity (conservation of probabilities) and locality (point-like particles with contact interactions). Moreover, it appeared extremely difficult to conceive of a consistent relativistic theory with extended particles.

The problem thus was very deep and touched at the essence of the theory itself. QED was an incomplete theory, but it seemed difficult to modify it without sacrificing some fundamental physical principle. It was possible to render the theory finite by abandoning unitarity and, thus, conservation of probabilities (as proposed by Dirac (1942)), but the physics consequences seemed hardly acceptable. What is often called Pauli–Villars' regularization, a somewhat *ad hoc* and temporary procedure to render the perturbative expansion finite before *renormalization* (see below), has this nature. It seemed even more difficult to incorporate the theory into a relativistic, non-local extension (which would correspond to giving an internal structure to particles), though Heisenberg proposed in 1938 the introduction of a fundamental length. In fact, it is only in the 1980s that possible candidates for a non-local extension of QFT were proposed in the form of superstring theories.

The crisis was so severe that Wheeler (1937) and Heisenberg (1943) proposed abandoning QFT altogether in favour of a theory of physical observables, in fact scattering data between particles: the S-matrix theory, an idea that became quite fashionable in the 1960s in the theory of *strong interactions* (those that generate nuclear forces).

Infinities and the problem of charged scalar bosons. After the first QED calculations, some pragmatic physicists started calculating other physical quantities, exploring the form and nature of infinities. Let me mention here another important work of Weisskopf (1939) in which the author shows that logarithmic divergences persist to all orders in the perturbative expansion, that is, in the expansion in powers of α. But he also notes that in the case of scalar (i.e., spinless) charged particles the situation is much worse: the divergences are quadratic, which is disastrous. Indeed, if the divergences are suppressed by some momentum cut-off $\Lambda = \hbar/Rc$ related to some new, unknown, physics and if Λ/m is not too large (and for some time 100 MeV, which is the range of nuclear forces, seemed a plausible candidate), then the product $\alpha \ln(\Lambda/m)$ remains small: a logarithmic divergence produces undetermined but nevertheless small corrections, but this is no longer the case for quadratic divergences. This result could have been understood as an indication that scalar particles cannot be considered as fundamental.

Note that the problem is more relevant than ever since the Standard Model that describes all experimental results up to the highest available energies in colliders, contains a scalar particle, the Higgs boson, and is now called the *fine tuning* or *hierarchy* problem. Indeed, to cancel infinities, it is necessary to adjust one parameter of the initial theory with a precision related to the ratio of the physical mass and the large-momentum cut-off, something that is not *natural*. The problem has now become specially severe since physicists have realized that mass-scales as large as 10^{15} (the so-called unification mass) or 10^{19} GeV (Planck's mass) can be involved. It is one of the main motivations for the introduction of *supersymmetry* (a symmetry that, surprisingly, relates bosons to fermions). Recent experimental results at the new CERN proton accelerator, the Large Hadron Collider or LHC, indicate that the Higgs boson has presumably been detected (2012), but the elucidation of its properties remains now a major goal of the ongoing and future experiments.

1.3 Renormalization

Calculating a number of different physical quantities, physicists noticed that, although many physical quantities were divergent, it was always the same kind of divergent contributions that appeared. One could thus find combinations that were finite (Weisskopf 1936). However, the physical meaning of such a property, the cancellation of infinities, was somewhat obscure. In fact, in the absence of any deeper understanding of the problem, little progress could be expected.

Each time physicists are confronted with such conceptual difficulties, some clue must eventually come from experiment.

Indeed, in 1947 Lamb and Retheford measured precisely the separation between the $2s_{1/2}2p_{1/2}$ levels of the hydrogen atom, while Rabi's group at Columbia measured the anomalous moment of the electron. Remarkably enough, it was possible to organize the calculation of the Lamb shift in such the way that infinities cancel (first approximate calculation by Bethe) and the result was found to be in very good agreement with experiment. Shortly after, Schwinger obtained the leading contribution to the anomalous magnetic moment of the electron.

These results initiated extraordinary theoretical developments (earlier work of Kramers concerning the mass renormalization of the extended classical electron proved to be important to generalize the idea of cancellation of infinities by subtraction, to the idea of renormalization), and in 1949 Dyson, relying in particular on the work of Feynman, Schwinger and Tomonaga, gave a proof of the cancellation of infinities to all orders in the perturbative expansion. What became known as the *renormalization theory* led in QED to finite results for all physical observables.

The general idea is the following: one begins with an initial theory called *bare*, which depends on parameters like the *bare mass* m_0 and the *bare charge* e_0 of the electron (mass and charge in the absence of interactions), or equivalently bare fine structure constant $\alpha_0 = e_0^2/4\pi\hbar c$. Moreover, one introduces a large-momentum cut-off $c\Lambda$ (which corresponds to modifying in a somewhat arbitrary and unphysical way the theory at a very short distance of order $\hbar/c\Lambda$). One then calculates the physical values (i.e., those one measures), called *renormalized*, of the same quantities (as the observed charge e and, thus, the fine structure constant α, and the physical mass m) as functions of the bare parameters and the cut-off:

$$\alpha = \alpha_0 - \beta_2 \alpha_0^2 \ln(\Lambda/m_0) + \cdots,$$
$$m = m_0 - \gamma_1 \, m_0 \alpha \ln(\Lambda C_1/m_0) + \cdots.$$

(β_2, γ_1 and C_1 are three numerical constants.) One inverts these relations, now expressing bare quantities as functions of the renormalized one. In this substitution, one exchanges, for example, the bare constant α_0 with the physical or renormalized constant α as the expansion parameter:

$$\alpha_0 = \alpha + \beta_2 \alpha^2 \ln(\Lambda/m) + \cdots,$$
$$m_0 = m + \gamma_1 \, m\alpha \ln(\Lambda C_1/m) + \cdots.$$

One then expresses any other observable (i.e., any other measurable quantity), initially calculated in terms of bare parameters, in terms of these physical or renormalized quantities. Most surprisingly, in the limit of infinite cut-off Λ, all physical observables then have a finite limit.

This *a priori* somewhat strange procedure, renormalization, has allowed and still allows calculations of increasing precision in QED. The remarkable agreement between predictions and experiment convincingly demonstrates that renormalized QFT provides a suitable formalism to describe electrodynamics in the quantum regime (it is even the domain in physics where agreement between theory and experiment is verified with the highest precision).

Moreover, renormalization theory has led to the very important concept of *renormalizable theories*. Only a limited number of field theories lead to finite results by this procedure. This severely restricts the structure of possible theories.

Finally, let us point out that for more than 15 years theoretical progress had been stopped by the problem of divergences in QFT. However, once experiment started producing decisive information, in two years a complete and consistent framework for perturbative calculations was set up.

The mystery of renormalization. Though it was now obvious that QED was the correct theory to describe electromagnetic interactions, the renormalization procedure itself, allowing the extraction of finite results from initial infinite quantities, had remained a matter of some concern for theorists: the meaning of the renormalization 'recipe' and, thus, of the bare parameters remained obscure. Much effort was devoted to try to overcome this initial conceptual weakness of the theory. Several types of solutions were proposed:

(i) The problem came from the use of an unjustified perturbative expansion and a correct summation of the expansion in powers of α would solve it. Somewhat related, in spirit, was the development of the so-called *axiomatic QFT*, which tried to derive rigorous, non-perturbative, results from the general principles on which QFT was based.

(ii) The principle of QFT had to be modified: only renormalized perturbation theory was meaningful. The initial bare theory with its bare parameters had no physical meaning. This line of thought led to the BPHZ (Bogoliubov, Parasiuk, Hepp, Zimmerman) formalism and, finally, to the work of Epstein and Glaser, where the problem of divergences in position space (instead of momentum space) was reduced to a mathematical problem of a correct definition of singular products of distributions. The corresponding efforts much clarified the principles of perturbation theory, but disguised the problem of divergences in such a way that it seemed never having existed in the first place.

(iii) Finally, the cut-off had a physical meaning and was generated by additional interactions, non-describable by QFT. In the 1960s some physicists thought that *strong interactions* could play this role (the cut-off then being provided by the range of nuclear forces). Renormalizable theories could then be thought as theories somewhat insensitive to this additional unknown short-distance structure, a property that obviously required some better understanding.

This latter point of view is in some sense close to our modern understanding, even though the cut-off is no longer provided by strong interactions.

1.4 Quantum field theory and the renormalization group

In the mid 1950s, several groups, most notably Peterman and Stückelberg (1953), Gell-Mann and Low (1954) and Bogoliubov and Shirkov (1955–1956), noticed that in the limit of a QED of photons and massless electrons, the renormalized perturbative expansion has a peculiar formal property, a direct consequence of the renormalization process itself.

In a massive theory, the renormalized charge can be defined through the electric interaction of particles at rest (Coulomb force). This definition is no longer applicable to massless particles, which always travel at the speed of light. It becomes then necessary to introduce some arbitrary mass (or energy or momentum) scale μ to define the renormalized charge e: it is related to the observed strength of the electromagnetic interaction in particle collisions at momenta of order μ. One can then call the renormalized charge the *effective* charge at scale μ. However, since this mass-scale is arbitrary, one can find other couples $\{e', \mu'\}$ which give the same physical results. The set of transformations of the physical parameters associated with the change in scale μ and necessary to keep the physics constant was called the *renormalization group* (RG). Making an infinitesimal change of scale, one can describe the variation of the effective charge by a differential (flow) equation

$$\mu \frac{\mathrm{d}\alpha(\mu)}{\mathrm{d}\mu} = \beta\bigl(\alpha(\mu)\bigr), \quad \beta(\alpha) = \beta_2 \alpha^2 + O(\alpha^3), \tag{1.2}$$

where the function $\beta(\alpha)$ can be calculated as a series expansion in powers of α.

Actually, even in a massive theory one can introduce this definition of the renormalized charge. This effective charge then has the following physical interpretation. At large distance, the intensity of the electromagnetic force does not vary and the charge has the value measured through the Coulomb force. However, at distances much shorter than the wavelength $\hbar/mc$ associated with a particle (one explores in some sense the 'interior' of the particle), one observes screening effects. What is remarkable is that these short-distance effects have a direct relation with renormalization.

Since one main concern was the large-momentum divergences in QFT, Gell-Mann and Low tried to use the RG to study the large-momentum behaviour of the electron propagator, beyond perturbation theory, in relation with the large cut-off behaviour of the bare charge. The bare charge can indeed be considered as an effective charge at the cut-off scale. If the function $\beta(\alpha)$ were to have a zero with a negative slope, the zero would have been the finite limit at infinite cut-off of the bare charge, beyond perturbation theory.

Unfortunately QED is a so-called IR-free theory ($\beta_2 > 0$), which means that the effective charge decreases at low momentum, and conversely at large momentum increases until the perturbative expansion of the β-function is no longer reliable.

(This variation is observed in experiments since the effective value of α measured near the Z vector boson, i.e., at about 100 GeV, is about 4% larger than its low--energy value.)

It is quite striking that if they were to have turned the argument around, they would have found that, at fixed bare charge, the effective charge goes to zero as $1/\ln(\Lambda/m_{\mathrm{el.}})$, which is acceptable for any reasonable value of cut-off and may even account for the small value of α, but their hope of course was to get rid of the cut-off.

Note some related speculations: Landau and Pomeranchuk (1955) noticed that if, in the calculation of the electron propagation in an electromagnetic field, one sums the leading terms at large momentum at each order, one predicts the existence of a particle of mass $M \propto m\, \mathrm{e}^{1/\beta_2 \alpha}$. This could have corresponded to a boundstate, but unfortunately this particle has unphysical properties leading to non-conservation of probabilities, and thus was called the Landau 'ghost'. For Landau, this was obviously the sign of some inconsistency of QED, though of no immediate physical consequence, because α is so small that this mass is of the order of 10^{30} GeV. Bogoliubov and Shirkov correctly pointed out that this result amounted to solving the RG equation (1.2) at leading order, that is, for small effective charge. Since the effective charge becomes large at large momenta, perturbation theory can eventually no longer be trusted. It is amusing to note that, in the modern point of view, we believe that Landau's intuition was basically correct, even though the argument, as initially formulated, was somewhat too naive.

1.5 A triumph of QFT: The Standard Model

QFT in the 1960s. After the triumph of QED, the 1960s were a time of doubt and uncertainty for QFT. Three outstanding problems remained, related to the three other known interactions:

(i) *Weak interactions* (related to the weak nuclear force) were described by the non-renormalizable four-fermion Fermi (–Feynman–Gell-Mann) interaction. Since the coupling was weak and the interaction was of a current–current type, as in QED when the photon is not quantized, it was conceivable that the theory was, in some way, the leading approximation to a QED-like theory, but with at least two very heavy (of the order of 100 GeV) photons, because the interaction was essential point-like. Classical field theories containing several photons, called non-Abelian gauge theories, that is, theories in which interactions are generated by a generalized symmetry principle called gauge symmetry, had been proposed by Yang and Mills (1954). However, their quantization led to new and difficult technical problems. Moreover, gauge theories, like QED, have a strong tendency to produce massless vector fields. So a few theorists were trying both to quantize the so-called Yang–Mills fields and to find ways to generate mass terms for them, within the framework of renormalizable QFTs.

(ii) On the other hand, many thought that, in the theory of strong interactions, the case for QFT was desperate: because the interactions were so strong, no perturbative expansion could make sense. Only a theory of physical observables, called

S-matrix theory, could provide the right theoretical framework, and strict locality had to be abandoned. One can note the first appearance of string models in this context.

(iii) Finally, since *gravitational forces* are extremely small at short distance, there was no immediate urgency to deal with quantum gravity, and the solution to this problem of uncertain experimental relevance could be postponed.

The triumph of renormalizable QFT. Toward the end of the 1960s, the situation changed quite rapidly. At last, methods to quantize non-Abelian gauge theories were found (Faddeev–Popov, De Witt 1967). These new theories could be proved to be renormalizable ('t Hooft, 't Hooft–Veltman, Slavnov, Taylor, Lee–Zinn-Justin, Becchi–Rouet–Stora, Zinn-Justin, 1971–1975) even in the broken symmetry phase in which masses could be generated for vector bosons (the Higgs mechanism, Higgs, Brout–Englert, Guralnik–Hagen–Kibble 1964) and fermions. These developments allowed constructing a quantum version of a model for combined weak and electromagnetic interactions proposed earlier by Weinberg (1967) and Salam (1968). Its predictions were soon confirmed by experiment.

In the rather confusing situation of strong interactions, the solution came as often in such circumstances from experiments: *deep inelastic scattering* experiments at SLAC, probing the interior of protons or neutrons, revealed that hadrons were composed of almost free point-like objects, initially called *partons* and eventually identified with the quarks that had been used as mathematical entities to provide a simple description of the symmetries of the hadron spectrum.

To understand this peculiar phenomenon, RG ideas were recalled in the modernized version of Callan and Symanzik (1970), valid also for massive theories, but the phenomenon remained a puzzle for some time until field theories could be found in which interactions became weak at short distance (unlike QED) such as to explain SLAC results. Finally, the same theoretical advances in the quantization of non-Abelian gauge theories which had provided a solution to the problem of weak interactions, allowed constructing a theory of strong interactions: *quantum chromodynamics* (QCD). Indeed, it was found that non-Abelian gauge theories, with a limited number of fermions, were *asymptotically free* (Gross–Wilczek, Politzer 1973). Unlike QED, the first coefficient β_2 of the β-function there is negative. The weakness of the interactions between quarks at short distance becomes a consequence of the decrease of the effective strong charge.

Therefore, around 1973–1974, a complete QFT model for all fundamental interactions but gravity was proposed, now called the Standard Model, which has successfully survived all experimental tests up to now, more than 30 years later, except for some minor recent modifications to take into account the non-vanishing neutrino masses. This was the triumph of all ideas based on renormalizable QFT.

At this point it was tempting to conclude that some kind of new law of nature has been discovered: all interactions can be described by a *renormalizable* QFT and renormalized perturbation theory. The divergence problem had by then been so well hidden that many physicists no longer worried about it.

A remaining potential issue was what Weinberg called the *asymptotic safety* condition: the consistency of QFT on all length-scales seemed to require the existence of a UV fixed point, in the formalism of equation (1.2) a solution of

$$\beta(\alpha) = 0 \quad \text{with} \quad \beta'(\alpha) < 0,$$

(one of the options already considered by Gell-Mann and Low). Asymptotically free field theories share of course this property, but scalar fields (as required by the Higgs mechanism) have a tendency to destroy asymptotic freedom. Finally, it still remained to cast quantum gravity into the renormalizable framework and this became the goal of many theorists in the following years. The failure of this programme eventually led to the introduction of string theories.

1.6 Critical phenomena: Other infinities

The theory of *critical phenomena* deals with continuous, or second-order, phase transitions in macroscopic systems. Simple examples are liquid–vapour, binary mixtures, superfluid He and magnetic transitions. The simplest lattice model that exhibits such a phase transition is the famous Ising model.

These transitions are characterized by a collective behaviour on large scales near the transition temperature (the critical temperature T_c). For example, the correlation length, which characterizes the scale of distance on which a collective behaviour is observed, becomes infinite at the transition. Near T_c, these systems thus depend on two very different length-scales, a microscopic scale given by the size of atoms, the lattice spacing or the range of forces, and another scale dynamically generated, given by the correlation length. To the latter scale are associated non-trivial large-distance or macroscopic phenomena.

It could then have been expected that physics near the critical temperature could be described, in some leading approximation, by a few effective macroscopic parameters, without explicit reference to the initial degrees of freedom. This idea leads to *mean field theory* (MFT) and in its more general form to Landau's theory of critical phenomena (1937). Such a theory can be called *quasi-Gaussian*, in the sense that it assumes implicitly that the remaining correlations between stochastic variables at the microscopic scale can be treated perturbatively and, thus, that macroscopic expectation values are given by quasi-Gaussian distributions, in the spirit of the central limit theorem.

Among the simplest and most robust predictions of such a theory, one finds the *universality* of the singular behaviour of thermodynamic functions at the critical temperature T_c: for instance, the correlation length ξ always diverges as $(T - T_c)^{-1/2}$, the spontaneous magnetization vanishes like $(T_c - T)^{1/2}$, and so on, these properties being independent of the dimension of space, the symmetry of the system, and of course the detailed microscopic dynamics.

Therefore, physicists were surprised when some experiments as well as numerical calculations in simple lattice models started questioning MFT predictions. An additional blow to MFT came from Onsager's (1944) exact solution of the 2D Ising model which confirmed the corresponding lattice calculations. In the following years, empirical evidence accumulated that critical phenomena in two and three space dimensions could not be described quantitatively by MFT. In fact, the critical behaviour was found to depend on space dimensions, symmetries and some general properties of models. Nevertheless, there were also some indications that some universality survived, but in a more limited sense. Some specific properties were important, but not all details of the microscopic dynamics.

The non-decoupling of scales. To understand how deep the problem was, one has to realize that such a situation had never been met before (except, perhaps, in turbulence): indeed, the main ingredient in Landau's theory is the hypothesis that, as usual, physical phenomena on too different length-scales decouple. Let us illustrate this idea with a simple example. Naively, one derives the period τ of the pendulum, up to a numerical factor, from dimensional analysis,

$$\tau \propto \sqrt{\ell/g},$$

where ℓ is the length of the pendulum and g is the gravitational acceleration. But in this argument is hidden a deeper and essential hypothesis: the internal atomic structure of the pendulum, the size of the earth or the distance between earth and sun are irrelevant because these length-scales are either much too small or much too large compared to the size of the pendulum. One expects that they lead to corrections of order the ratio $\lambda =$ (small scale/large scale), which, thus, are totally negligible. Of course, one can find mathematical functions of the ratio λ that decrease only slowly with λ like, for example, $1/\ln \lambda$. But these functions are singular and we have reasons to believe that, in general, nature is not perverse and does not introduce singularities where they are not absolutely needed.

In the same way, in Newtonian mechanics, to describe the motion of planets one can forget, to a very good approximation, the existence of other stars, the size of the sun and the planets, which can be replaced by point-like objects. Again in the same way, in non-relativistic quantum mechanics, one can ignore the internal structure of the proton to calculate with a very good precision the energy levels of the hydrogen atom.

The failure of MFT has demonstrated, on the contrary, that the decoupling of scales is not always true in the theory of critical phenomena, a new and totally unexpected situation. In fact, if one tries to calculate corrections to MFT, one finds divergences at the critical temperature. These divergences are generated by contributions that depend on the ratio of the correlation length and the microscopic scale. This situation is reminiscent of particle physics, except that in the early interpretation of QFT, it is the microscopic scale that goes to zero, while here it is the macroscopic scale (the correlation length) that becomes infinite.

The divergences met in QFT, as describing microscopic physics, and in the theory of critical phenomena, have actually a common origin: the non-decoupling of very different physical scales. The infinities appear when one tries to ignore, as one does usually and is generally justified, the existence of other length-scales where relevant physical laws may be very different.

Therefore, one could have feared that macroscopic physics would be sensitive to the details of the short-distance structure, that large-scale phenomena would depend on the detailed microscopic dynamics and, thus, would essentially be unpredictable. The emergence of a surviving universality, even more limited, was therefore even more surprising. To understand these observations, a new conceptual framework obviously had to be invented.

1.7 Kadanoff and Wilson's renormalization group

In 1966, Kadanoff proposed a strategy to deal with the problem: calculate physical observables by summing recursively over short-distance degrees of freedom. One then obtains a sequence of effective models that have all the same large-distance properties. Following Kadanoff, we use the Ising model to illustrate the idea, but take a more general viewpoint.

Example: The Ising model. The Ising model is a statistical lattice model. To each lattice site i is associated a random variable S_i that takes only two values ± 1, a 'classical' spin. Thermodynamic quantities are calculated by averaging over all spin configurations with a Boltzmann weight $e^{-\mathcal{H}_a(S)/T}/\mathcal{Z}$, where T is the temperature, a the lattice spacing, $\mathcal{H}_a(S)$ a configuration energy corresponding to some short-range interactions (for example, only nearest-neighbour spins on the lattice are coupled) and $\mathcal{Z}$ is a normalization factor of the probability distribution called the *partition function*. Note that phase transitions can only occur in the infinite-volume limit, called the thermodynamic limit.

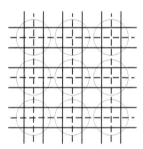

Fig. 1.2 Initial lattice (full line) and lattice with doubled spacing – – –.

Of course, in general expectation values cannot be calculated exactly. But, one would like, at least, to understand the origin of universality. The idea then is to sum over spins S_i at fixed average on a lattice of spacing $2a$. For example, on a square lattice one groups spins on a set of disjoint squares (inside the circles in Figure 1.2), and fixes the average over each square. After summation, the statistical sum is given by a summation over configurations of these average spins (which now take more than two values) belonging to a lattice of double spacing. To these spins corresponds a new configuration energy $\mathcal{H}_{2a}(S)$ called the *effective* interaction at scale $2a$.

This transformation can then be iterated,

$$\mathcal{H}_{2^n a}(S) = \mathcal{T}\left[\mathcal{H}_{2^{n-1}a}(S)\right], \tag{1.3}$$

as long as the lattice spacing remains small compared to the correlation length, that is, the scale of the macroscopic phenomena of interest. If the repeated application of the transformation produces an effective interaction whose asymptotic form is, to a large extent, independent of the initial interaction, one has found a mechanism that explains the remaining universality. Such asymptotic interactions will be *fixed points* or belong to *fixed surfaces* of the transformation $\mathcal{T}$:

$$\mathcal{H}^*(S) \underset{n\to\infty}{\to} \mathcal{H}_{2^n a}(S), \quad \mathcal{H}^*(S) = \mathcal{T}\left[\mathcal{H}^*(S)\right].$$

Eventually, Wilson (1971) transformed this initial, somewhat vague, idea into a precise operational scheme, unifying finally Kadanoff's renormalization group (RG) idea with the RG of QFT. This led to an understanding of universality, as being a consequence of the existence of large-distance (IR) fixed points of a general RG. It even became possible to develop systematic methods to calculate universal quantities, with the help of partially preexisting QFT techniques (Brézin–Le Guillou–Zinn-Justin 1973).

Continuum limit and QFT. A first step consists in understanding that the iteration (1.3) leads asymptotically to a field theory in continuum space, even if the initial model is a lattice model and the dynamical variables take only discrete values.

For example, in the Ising model it is clear that, after many iterations, the effective spin variable, which is a local average of a large number of spins, takes a dense discrete set of values and can be replaced by a continuum variable. Similarly, the spacing of the initial lattice becomes arbitrarily smaller than the spacing of the iterated lattice. One can thus replace the effective spin variable by a field $S(x)$ in continuum space. The sum over spins becomes an integral over fields (generalization of Feynman's path integral), formally analogous to the field integrals that allow calculating physical observables in QFT.

The Gaussian fixed point. On can verify that one RG fixed point has the form of a Gaussian distribution (a property in direct relation with the central limit theorem of probabilities). At the critical temperature, in the large-distance limit, the Gaussian fixed point takes the form of a free (scalar) QFT (particles do not interact). Moreover, the theory is massless (the correlation length plays the role of an inverse mass) in the language of microscopic physics. The weakly perturbed Gaussian model (quasi-Gaussian approximation) reproduces all universal results predicted by MFT.

A finer analysis shows, however, that a small perturbation of the Gaussian fixed point generates infinite contributions, at least in space dimensions smaller than, or equal to 4. Below four dimensions, the Gaussian fixed point then corresponds to an unstable point fixed of the RG.

Moreover, the most singular terms are generated by a renormalizable QFT whose large-distance properties can be studied. This is a very striking result since it indicates that renormalizable QFTs can emerge as effective theories describing the large-distance properties of critical phenomena. The RG of QFT then appears as an asymptotic form of the general Wilson–Kadanoff RG.

Conversely, one is then strongly tempted to apply the ideas that have emerged in the theory of phase transitions to the QFT that describes the physics of fundamental microscopic interactions.

1.8 Effective quantum field theories

The condition that fundamental microscopic interactions should be described by a renormalizable QFT has been used as a basic principle for constructing the Standard Model. From the success of the programme, it could have been concluded that the principle of renormalizability was a new law of nature. This would have implied that all interactions including gravity should be describable by such theories. The failure, so far, to exhibit a renormalizable version of quantum gravity has shed doubts on the whole programme itself. Indeed, if the Standard Model and its natural possible extensions are only approximate theories, it becomes difficult to understand why it should obey such an abstract principle.

The theory of critical phenomena, with the natural appearance of renormalizable QFTs, has led to another simpler and more plausible explanation. One can now imagine that fundamental interactions are described at distances much shorter than those presently accessible to experiment (unification, Planck's scales?), and thus at much higher energies, by a finite theory that does not have the form of a local QFT. Although this theory is entirely defined at the microscopic scale, for reasons that can be better understood only when we get a more precise idea about this more fundamental theory, it generates, by a cooperative effect of a large number of degrees of freedom, a non-trivial large-distance physics with effective interactions between very light particles. In phase transitions, it is the experimentalist that adjusts the temperature at its critical value to make the correlation length diverge (i.e., the mass vanish). In particle physics, this must happen automatically, otherwise one is confronted with the famous *fine tuning* problem. Since Planck's mass, for example, is at least of the order of 10^{13} times larger than the mass of the Higgs particle, whose existence is conjectured by the Standard Model, this would imply that one parameter of the theory is accidentally close to some critical value with a precision of 10^{-13}.

A few possible mechanisms are known, that generate massless particles, spontaneously broken continuous symmetries that generate massless scalar bosons (Goldstone bosons), gauge symmetries that generate massless vector bosons like the photon, chiral symmetry that produces massless fermions. But none of these mechanisms solves the problem of the Higgs particle.

Assuming this problem has been solved, then one can imagine that, as a consequence of the existence of a large-distance fixed point, low-energy or large-distance physics is described by an effective QFT. This field theory comes naturally equipped with a cut-off, a reflection of the initial microscopic structure, and contains all local interactions allowed by the field content and symmetries. If the free or Gaussian field theory is not too bad an approximation (at least in some energy range), which implies that the fixed point is close enough to the Gaussian fixed point, the interactions can be classified according to the dimensions of their coefficients. Then, interactions of non-renormalizable type, which have to be excluded in the traditional viewpoint, are automatically suppressed by powers of the cut-off (Einstein's gravitation corresponds presumably already to this class). Renormalizable interactions, which are dimensionless, evolve only slowly with the scale (logarithmically) and, thus, survive at large distance. They determine low-energy physics. The super-renormalizable interactions (this include possible mass terms), which are considered innocuous in the traditional presentation of QFT because they generate only a finite number of divergences, must be naturally absent or much suppressed because they grow as a power of the cut-off. The bare theory is then also a version of the effective theory in which all non-renormalizable interactions have already been omitted. QFT is not required to be physically consistent at very short distance where it is no longer a valid approximation and where it can be rendered finite by a modification that is, to a large extent, arbitrary.

Of course, such an interpretation has no immediate influence on perturbative calculations and one could thus consider the whole issue as somewhat philosophical. But this is not completely true!

We have mentioned above that taking the bare theory seriously, leads, in particular, to a confrontation with the fine tuning problem in the case of masses of scalar particles (and thus the Higgs boson) and, thus, forces us to look for solutions (supersymmetry, bound state of more fundamental fermions?).

This interpretation also solves the problem of *triviality*: renormalized interactions decreasing logarithmically with the cut-off are acceptable because the physical cut-off is finite. For example, let us consider QED and reverse Gell-Mann and Low's argument. At fixed bare charge, the effective charge at a mass-scale μ decreases like $1/\ln(\Lambda/\mu)$, which is acceptable for any sensible value of the cut-off Λ if, for instance, μ is of the order of the electron mass. This decrease may even explain the small value of the fine structure constant.

This interpretation also suggests that quantized Einstein gravitation is a surviving non-renormalizable interaction. Moreover, other non-renormalizable interactions could also be detected through very small symmetry violations. Indeed, the theory of critical phenomena provides examples of a possible mechanism. There, one finds situations in which the theory reduced to renormalizable interactions has more symmetry that the complete initial theory (cubic symmetry on the lattice leads to rotation symmetry at large distance).

This modern viewpoint, deeply based on RG ideas and the notion of scale-dependent effective interactions, not only provides a more consistent picture of QFT, but also a framework in which new physics phenomena can be discussed.

It implies that QFTs are somewhat temporary constructions. Due to an essential coupling of very different physical scales, *renormalizable* QFTs have a consistency limited to low-energy (or large-distance) physics. One uses the terminology of *effective QFT*, approximations of an as yet unknown more fundamental theory of a radically different nature.

2 Gaussian expectation values. Steepest descent method

In this work, Gaussian integrals and Gaussian expectation values play a major role. They appear naturally in probability theory because the average of a large number of independent random variables often has a Gaussian distribution. In statistical physics, in the theory of phase transitions and critical phenomena, somewhat similar considerations lead to the quasi-Gaussian model, which reproduces all the results coming from the mean-field approximation or Landau's theory. Thus, we recall in this chapter a few basic mathematical results that we will use, in a suitably generalized form, in this work.

It is first convenient, for technical reasons, to introduce the notion of generating function of the moments of a probability distribution. We then calculate Gaussian integrals and prove Wick's theorem for Gaussian expectation values, a result that is simple but of major practical importance.

The steepest descent method provides asymptotic evaluations, in some limits, of real or complex integrals. It leads to calculations of Gaussian expectation values, which explains its presence in this chapter. Moreover, the steepest descent method will be directly useful in this work

Finally, let us emphasize that the results we derive in this chapter are valid for integrals involving an arbitrary number of variables and can, thus, be generalized to the limit where this number becomes infinite, as in path or field integrals.

Notation. In what follows, we use systematically boldface characters to indicate matrices or vectors, and the corresponding italic characters, with indices, to indicate the corresponding matrix elements or vector components.

2.1 Generating functions

Let $\Omega(x_1, x_2, \ldots, x_n)$ be a positive measure or a normalized probability distribution, defined on $\mathbb{R}^n$. In what follows, we use the notation

$$\langle F \rangle \equiv \int \mathrm{d}^n x \, F(\mathbf{x}) \Omega(\mathbf{x}),$$

where $\mathrm{d}^n x \equiv \prod_{i=1}^n \mathrm{d}x_i$, for the expectation value of a function $F(\mathbf{x})$. By definition, $\langle 1 \rangle = 1$.

It is convenient to introduce the Fourier transform of the distribution, which is also the generating function of its moments. Here, we consider a special class of distributions whose Fourier transforms are analytic functions that can be defined even for imaginary arguments. We then define

$$\mathcal{Z}(\mathbf{b}) = \langle \mathrm{e}^{\mathbf{b} \cdot \mathbf{x}} \rangle = \int \mathrm{d}^n x \, \Omega(\mathbf{x}) \, \mathrm{e}^{\mathbf{b} \cdot \mathbf{x}} \quad \text{where} \quad \mathbf{b} \cdot \mathbf{x} = \sum_{i=1}^n b_i x_i \qquad (2.1)$$

with **b** real. The benefit of considering this particular function, rather than the Fourier transform, is that the integrand is still a positive measure.

The function $\mathcal{Z}(\mathbf{b})$ then is a *generating function* of the moments of the distribution, that is, of expectation values of monomials. Indeed, one recognizes, expanding the integrand in powers of the variables b_k, the series

$$\mathcal{Z}(\mathbf{b}) = \sum_{\ell=0}^{\infty} \frac{1}{\ell!} \sum_{k_1,k_2,\ldots,k_\ell=1}^{n} b_{k_1} b_{k_2} \ldots b_{k_\ell} \langle x_{k_1} x_{k_2} \ldots x_{k_\ell} \rangle.$$

Expectation values can thus be obtained by differentiating the function $\mathcal{Z}(\mathbf{b})$ with respect to its arguments. Differentiating both sides of equation (2.1) with respect to b_k, one obtains

$$\frac{\partial}{\partial b_k} \mathcal{Z}(\mathbf{b}) = \int d^n x \, x_k \, e^{\mathbf{b} \cdot \mathbf{x}} \Omega(\mathbf{x}). \tag{2.2}$$

Differentiating repeatedly and taking the limit $\mathbf{b} = 0$, one thus finds

$$\langle x_{k_1} x_{k_2} \ldots x_{k_\ell} \rangle = \left[\frac{\partial}{\partial b_{k_1}} \frac{\partial}{\partial b_{k_2}} \ldots \frac{\partial}{\partial b_{k_\ell}} \mathcal{Z}(\mathbf{b}) \right]_{\mathbf{b}=0}. \tag{2.3}$$

This notion of generating function is quite useful and in Section 5.1.1 will be extended to the limit where the number of variables becomes infinite.

2.2 Gaussian expectation values. Wick's theorem

For many reasons, among them the central limit theorem, Gaussian probability distributions play an important role in probability theory and in physics. Moreover, they have remarkable algebraic properties that we now recall. Although we derive all properties only for real integrals, most results generalize to complex Gaussian integrals.

2.2.1 Even Gaussian integrals

We consider the Gaussian integral

$$\mathcal{Z}(\mathbf{A}) = \int d^n x \, e^{-A(\mathbf{x})}, \tag{2.4}$$

where A is the real quadratic form

$$A(\mathbf{x}) = \frac{1}{2} \sum_{i,j=1}^{n} x_i A_{ij} x_j. \tag{2.5}$$

The integral converges only if the real symmetric matrix $\mathbf{A}$ with elements A_{ij} is positive (and thus all its eigenvalues are strictly positive). Then, various methods allow proving

$$\mathcal{Z}(\mathbf{A}) = (2\pi)^{n/2} (\det \mathbf{A})^{-1/2}. \tag{2.6}$$

Here, we prove the result only for real matrices, but since the expression (2.6) is an algebraic function of all matrix elements, the result also extends to complex integrals, with an appropriate determination of the square root.

Proof. The simple integral can easily be calculated by considering its square. For $a > 0$,
$$\int_{-\infty}^{+\infty} dx\, e^{-ax^2/2} = \sqrt{2\pi/a}. \tag{2.7}$$

More generally, the real symmetric matrix $\mathbf{A}$ in integral (2.4) can be diagonalized by an orthogonal transformation and, thus, can be written as
$$\mathbf{A} = \mathbf{O}\mathbf{D}\mathbf{O}^T, \tag{2.8}$$

where $\mathbf{O}$ is an orthogonal matrix and $\mathbf{D}$ a diagonal matrix with elements D_{ij}:
$$\mathbf{O}^T\mathbf{O} = \mathbf{1}, \quad D_{ij} = a_i \delta_{ij}, \quad a_i > 0.$$

Then, one changes variables, $\mathbf{x} \mapsto \mathbf{y}$, in the integral (2.4):
$$x_i = \sum_{j=1}^{n} O_{ij} y_j \quad \Rightarrow \quad \sum_{i,j} x_i A_{ij} x_j = \sum_{i,j,k} x_i O_{ik} a_k O_{jk} x_j = \sum_i a_i y_i^2.$$

The Jacobian of the transformation is $|\det \mathbf{O}| = 1$. The integral then factorizes:
$$\mathcal{Z}(\mathbf{A}) = \prod_{i=1}^{n} \int dy_i\, e^{-a_i y_i^2/2}.$$

Since all eigenvalues a_i of the matrix $\mathbf{A}$ are positive, each integral converges and is given by the result (2.7). It follows that
$$\mathcal{Z}(\mathbf{A}) = (2\pi)^{n/2}(a_1 a_2 \ldots a_n)^{-1/2} = (2\pi)^{n/2}(\det \mathbf{A})^{-1/2}.$$

The proof based on diagonalization, used for real matrices, has a complex generalization. Complex symmetric matrices $\mathbf{A}$ have a decomposition of the form
$$\mathbf{A} = \mathbf{U}\mathbf{D}\mathbf{U}^T,$$

where $\mathbf{U}$ is a unitary matrix and $\mathbf{D}$ a diagonal positive matrix.

2.2.2 General Gaussian integral

We now consider the general Gaussian integral

$$\mathcal{Z}(\mathbf{A},\mathbf{b}) = \int d^n x \, e^{-A(\mathbf{x})+\mathbf{b}\cdot\mathbf{x}}, \qquad (2.9)$$

where $A(\mathbf{x})$ is the quadratic form (2.5).

To calculate $\mathcal{Z}(\mathbf{A},\mathbf{b})$, one looks for the minimum of $A(\mathbf{x}) - \mathbf{b}\cdot\mathbf{x}$:

$$\frac{\partial}{\partial x_k}\left(\sum_{i,j=1}^n \tfrac{1}{2} x_i A_{ij} x_j - \sum_{i=1}^n b_i x_i\right) = \sum_{j=1}^n A_{kj} x_j - b_k = 0.$$

Introducing the inverse matrix

$$\mathbf{\Delta} = \mathbf{A}^{-1},$$

one can write the solution as

$$x_i = \sum_{j=1}^n \Delta_{ij} b_j. \qquad (2.10)$$

After the change of variables $x_i \mapsto y_i$ where

$$x_i = \sum_{j=1}^n \Delta_{ij} b_j + y_i, \qquad (2.11)$$

the integral becomes

$$\mathcal{Z}(\mathbf{A},\mathbf{b}) = e^{\Delta(\mathbf{b})} \int d^n y \, e^{-A(\mathbf{y})}, \qquad (2.12)$$

where we have set

$$\Delta(\mathbf{b}) = \frac{1}{2} \sum_{i,j=1}^n b_i \Delta_{ij} b_j. \qquad (2.13)$$

The change of variables has reduced the calculation to the integral (2.4). It follows that

$$\mathcal{Z}(\mathbf{A},\mathbf{b}) = (2\pi)^{n/2} (\det \mathbf{A})^{-1/2} e^{\Delta(\mathbf{b})}. \qquad (2.14)$$

Remark. The Gaussian integral has a remarkable property: if one integrates over a subset of variables, one still obtains a Gaussian integral. This structural stability explains the special role of Gaussian probability distributions.

2.2.3 Gaussian expectation values and Wick's theorem

When the matrix $\mathbf{A}$ is real and positive, the Gaussian function can be considered as a positive measure over $\mathbb{R}^n$ or a probability distribution. The expectation value of a function of the variables x_i is given by

$$\langle F(\mathbf{x})\rangle \equiv \mathcal{N}(\mathbf{A}) \int d^n x\, F(\mathbf{x})\, e^{-A(\mathbf{x})}, \tag{2.15}$$

where the normalization $\mathcal{N}$ is chosen such that $\langle 1 \rangle = 1$ and thus

$$\mathcal{N}(\mathbf{A}) = \mathcal{Z}^{-1}(\mathbf{A}, 0) = (2\pi)^{-n/2} (\det \mathbf{A})^{1/2} .$$

The function

$$\langle e^{\mathbf{b}\cdot\mathbf{x}} \rangle = \mathcal{Z}(\mathbf{A}, \mathbf{b})/\mathcal{Z}(\mathbf{A}, 0), \tag{2.16}$$

where $\mathcal{Z}(\mathbf{A}, \mathbf{b})$ is the function (2.9), thus is a *generating function* of the moments of the Gaussian distribution, which are expectation values of monomials (see Section 2.1). The expectation values can be obtained by differentiating expression (2.16) with respect to the variables b_i,

$$\langle x_{k_1} x_{k_2} \ldots x_{k_\ell} \rangle = (2\pi)^{-n/2} (\det \mathbf{A})^{1/2} \left(\frac{\partial}{\partial b_{k_1}} \frac{\partial}{\partial b_{k_2}} \cdots \frac{\partial}{\partial b_{k_\ell}} \mathcal{Z}(\mathbf{A}, \mathbf{b}) \right)\bigg|_{\mathbf{b}=0},$$

and replacing $\mathcal{Z}(\mathbf{A}, \mathbf{b})$ by the explicit expression (2.14),

$$\langle x_{k_1} \ldots x_{k_\ell} \rangle = \left(\frac{\partial}{\partial b_{k_1}} \cdots \frac{\partial}{\partial b_{k_\ell}}\, e^{\Delta(\mathbf{b})} \right)\bigg|_{\mathbf{b}=0}. \tag{2.17}$$

More generally, if $F(\mathbf{x})$ has a series expansion in the set of variables x_i, its expectation value is formally given by the identity

$$\langle F(\mathbf{x})\rangle = \left(F(\partial/\partial \mathbf{b})\, e^{\Delta(\mathbf{b})} \right)\bigg|_{\mathbf{b}=0}. \tag{2.18}$$

Wick's theorem. Identity (2.17) leads to Wick's theorem. Each time a derivative acts on the exponential in the right-hand side, it generates a factor b:

$$\frac{\partial}{\partial b_k} e^{\Delta(\mathbf{b})} = \sum_{k'} \Delta_{kk'} b_{k'}\, e^{\Delta(\mathbf{b})} .$$

Another derivative has to act later on the same factor, otherwise the corresponding contribution vanishes in the $\mathbf{b} = 0$ limit. One concludes that the expectation value of the product $x_{k_1} \ldots x_{k_\ell}$ with the Gaussian measure $e^{-A(\mathbf{x})}/\mathcal{Z}(\mathbf{A}, 0)$ is obtained in the following way: one considers all possible pairings of the indices $k_1, \ldots, k_\ell$ (ℓ

must thus be even). To each pair $k_p k_q$, one associates the element $\Delta_{k_p k_q}$ of the matrix $\boldsymbol{\Delta} = \mathbf{A}^{-1}$. Then,

$$\langle x_{k_1} \ldots x_{k_\ell} \rangle = \sum_{\substack{\text{over all possible} \\ \text{pairings } P \text{ of } \{k_1 \ldots k_\ell\}}} \Delta_{k_{P_1} k_{P_2}} \ldots \Delta_{k_{P_{\ell-1}} k_{P_\ell}}, \qquad (2.19)$$

$$= \sum_{\substack{\text{over all possible} \\ \text{pairings } P \text{ of } \{k_1 \ldots k_\ell\}}} \langle x_{k_{P_1}} x_{k_{P_2}} \rangle \ldots \langle x_{k_{P_{\ell-1}}} x_{k_{P_\ell}} \rangle. \qquad (2.20)$$

Equations (2.19) and (2.20) are characteristic properties of all centred ($\langle x_i \rangle = 0$) Gaussian measures. They are known under the denomination of Wick's theorem and are, in a form adapted to quantum mechanics or to quantum field theory, the basis of perturbation theory. The simplicity of the result should not, however, hide its *major practical relevance*. Note also that the proof is completely algebraic and, thus, extends to complex matrices $\mathbf{A}$ with non-vanishing determinant. Only the interpretation of Gaussian functions as measures or probability distributions then disappears.

Examples. One finds successively

$$\langle x_{i_1} x_{i_2} \rangle = \Delta_{i_1 i_2},$$
$$\langle x_{i_1} x_{i_2} x_{i_3} x_{i_4} \rangle = \Delta_{i_1 i_2} \Delta_{i_3 i_4} + \Delta_{i_1 i_3} \Delta_{i_2 i_4} + \Delta_{i_1 i_4} \Delta_{i_3 i_2}.$$

More generally, the expectation value of a product of $2p$ variables is the sum of $(2p-1)(2p-3) \ldots 5 \times 3 \times 1$ different terms.

2.3 Perturbed Gaussian measure. Connected contributions

Even in the favourable situations in which the central limit theorem applies, the Gaussian measure is only the limiting distribution. It is thus useful to study the effect of corrections to the Gaussian distribution.

2.3.1 Perturbed Gaussian measure

We consider a more general normalized distribution $e^{-A(\mathbf{x}, \lambda)}/\mathcal{Z}(\lambda)$ where the function $A(\mathbf{x}, \lambda)$ is the sum of the quadratic form (2.5) and a polynomial $\lambda V(\mathbf{x})$ in the variables x_i:

$$A(\mathbf{x}, \lambda) = A(\mathbf{x}) + \lambda V(\mathbf{x}), \qquad (2.21)$$

the parameter $\lambda > 0$ characterizing the amplitude of the deviation from the Gaussian distribution and the polynomial V being chosen such that the integral converges.

The normalization $\mathcal{Z}(\lambda)$ is given by the integral

$$\mathcal{Z}(\lambda) = \int \mathrm{d}^n x \, e^{-A(\mathbf{x}, \lambda)}. \qquad (2.22)$$

The integral can be calculated by expanding the integrand in a formal power series in λ and integrating term by term:

$$\mathcal{Z}(\lambda) = \sum_{k=0}^{\infty} \frac{(-\lambda)^k}{k!} \int d^n x\, V^k(\mathbf{x})\, e^{-A(\mathbf{x})} = \mathcal{Z}(0) \sum_{k=0}^{\infty} \frac{(-\lambda)^k}{k!} \langle V^k(\mathbf{x}) \rangle_0, \qquad (2.23)$$

where $\langle \bullet \rangle_0$ means expectation value with respect to the Gaussian measure defined by equation (2.15). Each term in the expansion, which is the Gaussian expectation value of a polynomial, can then be calculated with the help of Wick's theorem (2.19).

Using equation (2.18) with $F = e^{-\lambda V}$, one also obtains the formal representation of the normalization (2.22),

$$\mathcal{Z}(\lambda)/\mathcal{Z}(0) = \exp\left[-\lambda V\left(\frac{\partial}{\partial \mathbf{b}}\right)\right] e^{\Delta(\mathbf{b})}\bigg|_{\mathbf{b}=0}. \qquad (2.24)$$

Example. Consider the perturbation

$$V(\mathbf{x}) = \frac{1}{4!} \sum_{i=1}^{n} x_i^4. \qquad (2.25)$$

At order λ^2, one finds ($\boldsymbol{\Delta}\mathbf{A} = \mathbf{1}$)

$$\mathcal{Z}(\lambda)/\mathcal{Z}(0) = 1 - \frac{1}{4!}\lambda \sum_i \langle x_i^4 \rangle_0 + \frac{1}{2!(4!)^2}\lambda^2 \sum_i \sum_j \langle x_i^4 x_j^4 \rangle_0 + O(\lambda^3)$$

$$= 1 - \tfrac{1}{8}\lambda \sum_i \Delta_{ii}^2 + \tfrac{1}{128}\lambda^2 \sum_i \Delta_{ii}^2 \sum_j \Delta_{jj}^2$$

$$+ \lambda^2 \sum_{i,j} \left(\tfrac{1}{16}\Delta_{ii}\Delta_{jj}\Delta_{ij}^2 + \tfrac{1}{48}\Delta_{ij}^4\right) + O(\lambda^3). \qquad (2.26)$$

A simple verification of the factors is obtained by specializing to the case of only one variable. Then,

$$\mathcal{Z}(\lambda)/\mathcal{Z}(0) = 1 - \tfrac{1}{8}\lambda + \tfrac{35}{384}\lambda^2 + O(\lambda^3).$$

The alternating sign in all expansions corresponding to the example (2.25) is produced by the alternating sign in the expansion of the exponential function $e^{-\lambda x^4}$.

2.4 Feynman diagrams. Connected contributions

To each perturbative contribution generated by Wick's theorem, one can associate a graph called a Feynman diagram. All contributions can be derived from a subclass that contains only connected terms, represented by connected diagrams.

Fig. 2.1 Feynman diagram: x^4 vertex in example (2.25).

2.4.1 Feynman diagrams

Each monomial contributing to a perturbation $V(\mathbf{x})$ is represented by a point (a vertex) from which originates a number of lines equal to the degree of the monomial and each pairing is represented by a line joining the vertices to which belong the corresponding variables.

In example (2.25), $V(\mathbf{x})$ is a homogeneous polynomial of degree 4 and to vertices are thus attached four lines (see Figure 2.1).

The contribution of order λ to the normalization $\mathcal{Z}(\lambda)$ (2.22) then contains one vertex and the four lines originating from the vertex are connected pairwise as displayed in Figure 2.2.

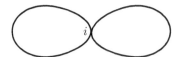

Fig. 2.2 Feynman diagram: Contribution to $\langle x^4 \rangle_0$ of order λ in example (2.25).

The order λ^2 involves the three diagrams displayed in Figures 2.3 and 2.4. They contain two vertices and are of two types, in the sense of graphs. In Figure 2.3 is displayed the contribution in which all pairings are internal either to the factor x_i^4 or to the factor x_j^4. It factorizes and is not connected in the sense of graphs.

Fig. 2.3 Feynman diagram: Non-connected contribution coming from $\langle x_i^4 x_j^4 \rangle_0$ at order λ^2 in example (2.25).

In Figure 2.4 are displayed the two other contributions, which are connected.

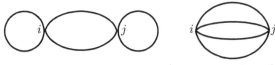

Fig. 2.4 Feynman diagrams: Connected contributions coming from $\langle x_i^4 x_j^4 \rangle_0$ at order λ^2 in example (2.25).

We have indicated explicitly the indices i and j corresponding to summations in expressions (2.26). In what follows, we will no longer indicate summation indices.

2.4.2 Connected contributions

We have observed that at second order, Feynman diagrams can be separated into connected and non-connected contributions. We now examine this question more systematically.

Considering expression (2.26), one notices that the two first terms can be exponentiated in such a way that $\ln \mathcal{Z}$ contains only *connected* contributions, that is, contributions that cannot be factorized into a product of sums and, thus, in the sense of graphs are connected:

$$\ln \mathcal{Z}(\lambda) - \ln \mathcal{Z}(0) = -\tfrac{1}{8}\lambda \sum_i \Delta_{ii}^2 + \lambda^2 \sum_{i,j} \left(\tfrac{1}{16} \Delta_{ii} \Delta_{jj} \Delta_{ij}^2 + \tfrac{1}{48} \Delta_{ij}^4 \right) + O(\lambda^3). \quad (2.27)$$

In the logarithm, the non-connected contribution of Figure 2.3 has cancelled.

The property that only connected terms contribute to $\ln \mathcal{Z}$ is a general property. To prove it, in the calculation of the Gaussian expectation value $\langle V^k(\mathbf{x}) \rangle$, we distinguish the contributions of Wick's theorem that contain pairings between all the k factors $V(\mathbf{x})$, from all others that can be factorized into products of expectation values of powers of $V(\mathbf{x})$ smaller than k. We use below the subscript (c) to indicate the connected part of an expectation value. With this notation, for example,

$$\langle V(\mathbf{x}) \rangle = \langle V(\mathbf{x}) \rangle_c, \quad \langle V^2(\mathbf{x}) \rangle = \langle V^2(\mathbf{x}) \rangle_c + \langle V(\mathbf{x}) \rangle_c^2$$

$$\langle V^3(\mathbf{x}) \rangle = \langle V^3(\mathbf{x}) \rangle_c + 3 \langle V^2(\mathbf{x}) \rangle_c \langle V(\mathbf{x}) \rangle_c + \langle V(\mathbf{x}) \rangle_c^3, \ldots$$

More generally, at order k, one finds

$$\tfrac{1}{k!} \langle V^k(\mathbf{x}) \rangle = \tfrac{1}{k!} \langle V^k(\mathbf{x}) \rangle_c + \text{non-connected terms}.$$

A non-connected term is a product of the form

$$\langle V^{k_1}(\mathbf{x}) \rangle_c \langle V^{k_2}(\mathbf{x}) \rangle_c \cdots \langle V^{k_p}(\mathbf{x}) \rangle_c, \quad k_1 + k_2 + \cdots k_p = k,$$

with a coefficient $1/k!$ coming from the expansion of the exponential, multiplied by a combinatorial factor corresponding to all possible ways to group k objects in subsets of $k_1 + k_2 + \cdots + k_p$ objects, when all k_i are distinct. One finds

$$\frac{1}{k!} \times \frac{k!}{k_1! k_2! \ldots k_p!} = \frac{1}{k_1! k_2! \ldots k_p!}.$$

When m powers k_i are equal, one must further divide by an additional combinatorial factor $m!$ because the same term has been counted $m!$ times.

One then notices that the perturbative expansion can be written as

$$\mathcal{W}(\lambda) = \ln \mathcal{Z}(\lambda) = \ln \mathcal{Z}(0) + \sum_k \frac{(-\lambda)^k}{k!} \langle V^k(\mathbf{x}) \rangle_c, \quad (2.28)$$

providing confirmation that the cancellation observed in expression (2.27) is general.

2.5 Expectation values. Generating function. Cumulants

The moments of the distribution $\mathrm{e}^{-A(\mathbf{x},\lambda)}/\mathcal{Z}(\lambda)$ where $A(\mathbf{x},\lambda)$ is the polynomial (2.21),
$$A(\mathbf{x}, \lambda) = A(\mathbf{x}) + \lambda V(\mathbf{x}),$$
that is, the expectation values $\langle x_{i_1} x_{i_2} \ldots x_{i_\ell}\rangle_\lambda$, are given by the ratios
$$\langle x_{i_1} x_{i_2} \ldots x_{i_\ell}\rangle_\lambda = \mathcal{Z}^{-1}(\lambda) \mathcal{Z}_{i_1 i_2 \ldots i_\ell}(\lambda), \tag{2.29a}$$
$$\mathcal{Z}_{i_1 i_2 \ldots i_\ell}(\lambda) = \int \mathrm{d}^n x\, x_{i_1} x_{i_2} \ldots x_{i_\ell} \exp\left[-A(\mathbf{x}, \lambda)\right]. \tag{2.29b}$$
In the language of statistical physics, they are called ℓ-point functions.

2.5.1 The two-point function

The two-point function $\langle x_{i_1} x_{i_2}\rangle_\lambda$ involves the calculation of the integral
$$\mathcal{Z}_{i_1 i_2}(\lambda) = \int \mathrm{d}^n x\, x_{i_1} x_{i_2} \exp\left[-A(\mathbf{x}, \lambda)\right].$$
In example (2.25) at order λ^2, one finds
$$\mathcal{Z}_{i_1 i_2}(\lambda)/\mathcal{Z}(0) = \Delta_{i_1 i_2} - \tfrac{1}{24}\lambda \Delta_{i_1 i_2} \sum_i \langle x_i^4\rangle_0 - \tfrac{1}{2}\lambda \sum_i \Delta_{i i_1}\Delta_{ii}\Delta_{i i_2}$$
$$+ \frac{\lambda^2}{2!(4!)^2} \sum_{i,j} \Delta_{i_1 i_2} \langle x_i^4 x_j^4\rangle_0 + \frac{\lambda^2}{2!4!} \sum_{i,j} \Delta_{i i_1}\Delta_{ii}\Delta_{i i_2} \langle x_j^4\rangle_0$$
$$+ \lambda^2 \sum_{i,j} \left(\tfrac{1}{4}\Delta_{i i_1}\Delta_{i i_2}\Delta_{ij}^2 \Delta_{jj} + \tfrac{1}{6}\Delta_{i_1 i}\Delta_{j i_2}\Delta_{ij}^3\right.$$
$$\left.+ \tfrac{1}{4}\Delta_{i_1 i}\Delta_{j i_2}\Delta_{ij}\Delta_{ii}\Delta_{jj}\right) + O(\lambda^3).$$

In the ratio of the two series
$$\langle x_{i_1} x_{i_2}\rangle_\lambda = \mathcal{Z}_{i_1 i_2}(\lambda)/\mathcal{Z}(\lambda),$$
the non-connected terms cancel and one obtains
$$\langle x_{i_1} x_{i_2}\rangle_\lambda = \Delta_{i_1 i_2} - \tfrac{1}{2}\lambda \sum_i \Delta_{i i_1}\Delta_{ii}\Delta_{i i_2} + \lambda^2 \sum_{i,j} \left(\tfrac{1}{4}\Delta_{i_1 i}\Delta_{j i_2}\Delta_{ij}\Delta_{ii}\Delta_{jj}\right.$$
$$\left.+ \tfrac{1}{4}\Delta_{i i_1}\Delta_{i i_2}\Delta_{ij}^2 \Delta_{jj} + \tfrac{1}{6}\Delta_{i_1 i}\Delta_{j i_2}\Delta_{ij}^3\right) + O(\lambda^3). \tag{2.30}$$
The Feynman diagrams corresponding to the contributions of order 1 and λ are displayed in Figure 2.5.

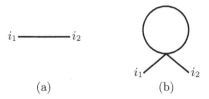

Fig. 2.5 The two-point function: Contributions of order 1 and λ in example (2.25).

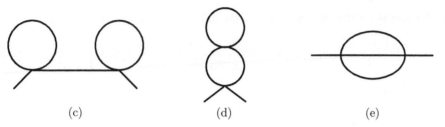

Fig. 2.6 The two-point function: Contributions of order λ^2 in example (2.25).

The diagrams of Figure 2.6 represent the three successive contributions of order λ^2.

As an illustration, let us justify, for example, the factor $1/6$ in front of diagram (e). At second order, the expansion of the exponential leads to the Gaussian expectation value

$$\frac{1}{2!\,(4!)^2} \sum_{i,j} \langle x_{i_1} x_{i_2} x_i^4 x_j^4 \rangle_0 \,,$$

which can be obtained from Wick's theorem.

First, x_{i_1} can be paired with any factor x in the product $x_i^4 x_j^4$; one has eight equivalent choices and one of the two vertices is distinguished. Then, x_{i_2} must be paired with a factor x of the remaining vertex: four choices. The remaining x^3 factors in the two vertices can be paired in all possible ways: $3!$ equivalent possibilities. Collecting all factors, one finds

$$\frac{1}{2}\frac{1}{(4!)^2} \times 8 \times 4 \times 3! = \frac{1}{6}\,.$$

It is worth mentioning that the factor $1/6$ that multiplies the diagram (e) also has an interpretation in terms of the isomorphisms of the graph. It is the inverse of the number $3!$ of permutations that exchange the three lines that join the two vertices. One can find general expressions that relate the combinatorial factors that multiply each diagram to the symmetries of the graph.

One could calculate by the same method the four-point function, the expectation value of a generic monomial of degree 4. One would find a large number of contributions. But the expressions of the cumulants of the distribution are much simpler. For this purpose, it is convenient to first define a generating function of moments or expectation values $\langle x_{i_1} x_{i_2} \ldots x_{i_p} \rangle_\lambda$.

2.5.2 Generating functions. Cumulants

We consider the function

$$\mathcal{Z}(\mathbf{b}, \lambda) = \int d^n x \, \exp\left[-A(\mathbf{x}, \lambda) + \mathbf{b}\cdot\mathbf{x}\right], \qquad (2.31)$$

which generalizes the function (2.9) of the Gaussian example. It is proportional to the generating function of the expectation values (2.29a) (see Section 2.1),

$$\langle e^{\mathbf{b}\cdot\mathbf{x}} \rangle_\lambda = \mathcal{Z}(\mathbf{b}, \lambda)/\mathcal{Z}(\lambda),$$

which generalizes the function (2.16). By differentiating, one finds

$$\langle x_{i_1} x_{i_2} \ldots x_{i_\ell} \rangle_\lambda = \mathcal{Z}^{-1}(\lambda) \left[\frac{\partial}{\partial b_{i_1}} \frac{\partial}{\partial b_{i_2}} \cdots \frac{\partial}{\partial b_{i_\ell}} \mathcal{Z}(\mathbf{b}, \lambda) \right]_{\mathbf{b}=0}. \tag{2.32}$$

We now introduce the function

$$\mathcal{W}(\mathbf{b}, \lambda) = \ln \mathcal{Z}(\mathbf{b}, \lambda). \tag{2.33}$$

In a probabilistic interpretation, $\mathcal{W}(\mathbf{b}, \lambda)$ is the generating function of the cumulants of the distribution. As a consequence of equation (2.28), the perturbative expansion of cumulants is much simpler since it contains only connected contributions. In particular, all contributions to the normalization (2.26) are contained in $\mathcal{W}(0, \lambda)$. Finally note that, in the Gaussian case, $\mathcal{W}(\mathbf{b})$ reduces to a form quadratic in $\mathbf{b}$.

Remark. In the framework of statistical physics, the cumulants

$$W^{(\ell)}_{i_1 i_2 \ldots i_\ell} = \left[\frac{\partial}{\partial b_{i_1}} \frac{\partial}{\partial b_{i_2}} \cdots \frac{\partial}{\partial b_{i_\ell}} \mathcal{W}(\mathbf{b}, \lambda) \right]_{\mathbf{b}=0},$$

are called *connected ℓ-point correlation functions*.

Examples. Expanding relation (2.33) in powers of $\mathbf{b}$, one finds that the one-point functions are identical:

$$W^{(1)}_i = \langle x_i \rangle_\lambda .$$

For the two-point function, one finds

$$W^{(2)}_{i_1 i_2} = \langle x_{i_1} x_{i_2} \rangle_\lambda - \langle x_{i_1} \rangle_\lambda \langle x_{i_2} \rangle_\lambda = \langle (x_{i_1} - \langle x_{i_1} \rangle_\lambda)(x_{i_2} - \langle x_{i_2} \rangle_\lambda) \rangle_\lambda . \tag{2.34}$$

Thus, it is the two-point function of the initial variables from which their expectation values have been subtracted.

In the case of an even perturbation, $V(\mathbf{x}) = V(-\mathbf{x})$, as in example (2.25),

$$W^{(2)}_{i_1 i_2} = \langle x_{i_1} x_{i_2} \rangle_\lambda ,$$
$$W^{(4)}_{i_1 i_2 i_3 i_4} = \langle x_{i_1} x_{i_2} x_{i_3} x_{i_4} \rangle_\lambda - \langle x_{i_1} x_{i_2} \rangle_\lambda \langle x_{i_3} x_{i_4} \rangle_\lambda - \langle x_{i_1} x_{i_3} \rangle_\lambda \langle x_{i_2} x_{i_4} \rangle_\lambda$$
$$- \langle x_{i_1} x_{i_4} \rangle_\lambda \langle x_{i_3} x_{i_2} \rangle_\lambda .$$

The connected four-point function, which vanishes exactly for a Gaussian measure, provides a first evaluation of the effect of a deviation from the Gaussian measure.

In example (2.25), at order λ^2 one then finds (see Figure 2.7)

$$W^{(4)}_{i_1 i_2 i_3 i_4} = -\lambda \sum_i \Delta_{i_1 i} \Delta_{i_2 i} \Delta_{i_3 i} \Delta_{i_4 i} + \tfrac{1}{2}\lambda^2 \sum_{i,j} \Delta_{i_1 i} \Delta_{i_2 i} \Delta_{i_3 j} \Delta_{i_4 j} \Delta^2_{ij}$$
$$+ \tfrac{1}{2}\lambda^2 \sum_{i,j} \Delta_{i_1 i} \Delta_{i_3 i} \Delta_{i_2 j} \Delta_{i_4 j} \Delta^2_{ij} + \tfrac{1}{2}\lambda^2 \sum_{i,j} \Delta_{i_1 i} \Delta_{i_4 i} \Delta_{i_3 j} \Delta_{i_2 j} \Delta^2_{ij}$$
$$+ \tfrac{1}{2}\lambda^2 \sum_{i,j} (\Delta_{ii} \Delta_{ij} \Delta_{i_1 i} \Delta_{i_2 j} \Delta_{i_3 j} \Delta_{i_4 j} + 3 \text{ terms}) + O(\lambda^3). \tag{2.35}$$

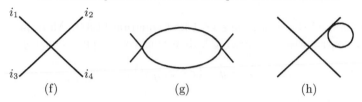

Fig. 2.7 Feynman diagrams, connected four-point function: Contributions of order λ and λ^2 in example (2.25).

2.6 Steepest descent method

To evaluate certain types of contour integrals in the complex domain, the steepest descent method, which reduces their evaluation to a succession of Gaussian expectation values, can sometimes be used.

We first describe the method in the case of a real simple integral, then generalize to the complex case. Finally, we generalize the method to an arbitrary number of variables.

2.6.1 Real integrals

We consider the integral

$$\mathcal{I}(\lambda) = \int_a^b dx \, e^{-S(x)/\lambda}, \tag{2.36}$$

where the function $S(x)$ is a real function, analytic in a neighbourhood of the segment (a, b), and λ a positive parameter. The goal is to evaluate the integral in the limit $\lambda \to 0_+$.

In this limit, the integral is dominated by the maxima of the integrand and thus the minima of $S(x)$. Two situations can arise:

(i) The minimum of $S(x)$ corresponds to a boundary of the integration domain. One then expands $S(x)$ near the minimum and one integrates. This is not the situation we are interested in here.

(ii) The function $S(x)$ has one or several minima in the interval (a, b). Since the function is regular, the minima correspond to points x_c characterized by

$$S'(x_c) = 0, \tag{2.37}$$

where generically $S''(x_c) > 0$ (the case $S''(x_c) = 0$ requires a separate analysis). For reasons that will become apparent later, these points are called saddle points (see the discussion of Section 2.6.2 and Figure 2.8). When several solutions are found, the leading contribution is given by the absolute minimum of $S(x)$.

An important remark is the following: if one neglects relative corrections of order $\exp[-\text{const.}/\lambda]$, one can restrict the integration to a vicinity $(x_c - \varepsilon, x_c + \varepsilon)$ of x_c, with ε arbitrarily small but finite. Indeed, the contributions outside the interval, divided by the leading contribution, are bounded by

$$(b-a)\,e^{-(S(x)-S(x_c))/\lambda} \sim (b-a)\,e^{-S''(x_c)\varepsilon^2/2\lambda},$$

where we have used the condition $\varepsilon \ll 1$ and equation (2.37). More precisely, the domain that contributes is of order $\sqrt{\lambda}$. It is thus convenient to change variables,

$$x \mapsto y = (x - x_c)/\sqrt{\lambda}.$$

The expansion of the function S then reads

$$S(x)/\lambda = S(x_c)/\lambda + \tfrac{1}{2}y^2 S''(x_c) + \tfrac{1}{6}\sqrt{\lambda}S'''(x_c)y^3 + \tfrac{1}{24}\lambda S^{(4)}(x_c)y^4 + O(\lambda^{3/2}).$$

At leading order, one can keep only the quadratic term. This reduces the calculation to a Gaussian integral over a finite interval:

$$\mathcal{I}(\lambda) \sim \sqrt{\lambda}\, e^{-S(x_c)/\lambda} \int_{-\varepsilon/\sqrt{\lambda}}^{\varepsilon/\sqrt{\lambda}} dy\, e^{-S''(x_c)y^2/2}.$$

In this integral, one can then extend the integration domain to $[-\infty, +\infty]$. Indeed, the contributions from $|y| > \varepsilon/\sqrt{\lambda}$ again are exponentially negligible. At leading order, the calculation thus reduces to a usual Gaussian integral and one finds

$$\mathcal{I}(\lambda) \sim \sqrt{2\pi\lambda/S''(x_c)}\, e^{-S(x_c)/\lambda}. \tag{2.38}$$

To calculate higher order corrections, one expands the integrand in powers of λ and one integrates term by term. Setting

$$\mathcal{I}(\lambda) = \sqrt{2\pi\lambda/S''(x_c)}\, e^{-S(x_c)/\lambda}\, \mathcal{J}(\lambda),$$

one finds, for example, at next order

$$\mathcal{J}(\lambda) = 1 - \frac{\lambda}{24} S^{(4)} \langle y^4 \rangle + \frac{\lambda}{2 \times 6^2} S'''^2 \langle y^6 \rangle + O(\lambda^2)$$

$$= 1 + \frac{\lambda}{24}\left(5\frac{S'''^2}{S''^3} - 3\frac{S^{(4)}}{S''^2}\right) + O(\lambda^2),$$

where $\langle \bullet \rangle$ means Gaussian expectation value.

Remarks.
(i) The steepest descent method generates a formal expansion in powers of λ:

$$\mathcal{J}(\lambda) = 1 + \sum_{k=1}^{\infty} J_k \lambda^k,$$

which, in general, diverges for all non-vanishing values of the expansion parameter. Nevertheless, the series is useful because, for λ small enough, one can prove that the partial sums satisfy

$$\exists \lambda_0 > 0, \{M_K\}: \quad \forall K \text{ and } 0_+ \leq \lambda \leq \lambda_0 \quad \left|\mathcal{J}(\lambda) - \sum_{k=0}^{K} J_k \lambda^k\right| \leq M_K \lambda^{K+1},$$

where the coefficients M_k generically increase like $k!$. Such a series is called an asymptotic series. At λ fixed, if K is chosen such that the bound is minimal, the function is determined up to an error of order $\exp[-\text{const.}/\lambda]$. Note that such a bound can be extended to a sector in the complex λ plane, $|\text{Arg}\,\lambda| < \theta$.

The divergence of the series can easily be understood: if one changes the sign of λ in the integral, the maximum of the integrand becomes a minimum and the selected saddle point can no longer give the leading contribution to the integral.

(ii) Often integrals have the more general form

$$\mathcal{I}(\lambda) = \int dx\, \rho(x)\, e^{-S(x)/\lambda}.$$

Then, provided $\ln \rho(x)$ is analytic at the saddle point, it is not necessary to take the factor $\rho(x)$ into account in the saddle point equation. Indeed, this factor would induce a shift of order $x - x_c$ to the saddle point position, a solution of

$$S''(x_c)(x - x_c) \sim \lambda \rho'(x_c)/\rho(x_c),$$

and, thus, of order λ while the contribution to the integral comes from a region of order $\sqrt{\lambda}$, which is much larger than the shift.

One can thus expand all expressions around the solution of $S'(x) = 0$. At leading order, one finds

$$\mathcal{I}(\lambda) \sim \sqrt{2\pi\lambda/S''(x_c)}\, \rho(x_c)\, e^{-S(x_c)/\lambda}.$$

To illustrate the method, we now apply it to two classical examples, a function related to Airy's functions and a representation of the Γ-function, which generalizes $n!$ to real and complex arguments.

Examples.

(i) Consider the integral, related to Airy functions,

$$\text{Hi}(x) = \frac{1}{\pi} \int_0^\infty e^{-t^3/3 + tx}\, dt,$$

in the limit $x \to +\infty$.

The integrand does not have immediately the canonical form, but the simple change of variables $t \mapsto t\sqrt{x}$ leads to

$$\text{Hi}(x) = \frac{\sqrt{x}}{\pi} \int_0^\infty e^{-x^{3/2}(t^3/3 - t)},$$

and $x^{3/2}$ then plays the role of $1/\lambda$.

In the notation of the general analysis $S(t) = t^3/3 - t$. The saddle point is given by

$$S'(t) = t^2 - 1 = 0 \Rightarrow t = t_c = 1.$$

Then,

$$S(t_c) = -2/3, \quad S''(t_c) = 2t_c = 2.$$

The leading order result is

$$\mathrm{Hi}(x) \sim \frac{1}{\sqrt{\pi}} x^{-1/4} e^{2x^{3/2}/3}.$$

(ii) The asymptotic evaluation of the function

$$\Gamma(s) = \int_0^\infty dx\, x^{s-1} e^{-x},$$

(generalization of $(s-1)!$ to real and complex values) for $s \to +\infty$ is a classical application of the steepest descent method. If s is not a positive integer, the integrand is singular at $x = 0$, but the contribution of the neighbourhood of the origin is negligible in this limit and thus the steepest descent method is applicable.

The integral has the canonical form (2.36) only after a linear change of variable, $x \mapsto x' = x/(s-1)$, and one also sets $s - 1 = 1/\lambda$. Then,

$$\Gamma(s) = (s-1)^{s-1} \int_0^\infty dx\, e^{-(x - \ln x)/\lambda}$$

and, thus, $S(x) = x - \ln x$. The saddle point position is given by

$$S'(x) = 1 - 1/x = 0 \Rightarrow x_c = 1.$$

The second derivative at the saddle point is $S''(x_c) = 1$. The result at leading order thus is

$$\Gamma(s) \underset{s\to\infty}{\sim} \sqrt{2\pi}(s-1)^{s-1/2} e^{1-s} \sim \sqrt{2\pi} s^{s-1/2} e^{-s}, \qquad (2.39)$$

an approximation also called Stirling's formula.

The complex steepest descent method, which we explain later, allows extending the result to s complex with $|\arg s| < \pi$.

2.6.2 Complex contour integrals

We consider the integral

$$\mathcal{I}(\lambda) = \oint_C dx\, e^{-S(x)/\lambda}, \qquad (2.40)$$

where $S(x)$ is an analytic function of the complex variable x and λ a real positive parameter. The contour C joins the point a to the point b in the complex plane, and is contained in the analyticity domain of S. As a limiting case, one can consider the situation where the points a and b go to infinity in the complex plane.

One wants to evaluate the integral for $\lambda \to 0_+$. One could think a priori that, again, the integral is dominated by the points where the modulus of the integrand is maximum and thus the real part of $S(x)$ is minimum. However, the contribution of the vicinity of such points can cancel because the phase varies very rapidly (an argument that leads to the stationary-phase method).

The steepest descent method consists in deforming the contour C in all possible ways within the domain of analyticity of the function (without crossing a singularity) in order to minimize the maximum modulus of the integrand along the contour, that is, maximize the minimum of $\operatorname{Re} S(x)$ along the contour.

If it is possible to deform the contour C into an equivalent contour on which $\operatorname{Re} S(x)$ is monotonic, then the integral is dominated by a boundary of the contour. Otherwise, the real part has a minimum. If the minimum on the optimal contour is not reached on the boundary of the domain of analyticity, it corresponds to a regular point. Then, these conditions imply that near the saddle point, $\operatorname{Re} S(x)$ has indeed a saddle point structure. As we show below, this can only happen at a point x_c where the derivative of S vanishes:

$$S'(x_c) = 0.$$

The structure of the integrand can be better understood if one remembers that the two set of curves $\operatorname{Re} S$ constant and $\operatorname{Im} S$ constant form two sets of bi-orthogonal curves. The only double points on these curves are singularities or saddle points. Indeed, let us expand the function at x_c,

$$S(x) - S(x_c) \sim \tfrac{1}{2} S''(x_c)(x - x_c)^2 \;\Rightarrow\; \operatorname{Re}[S(x) - S(x_c)] \sim \tfrac{1}{2}|S''(x_c)|(u^2 - v^2),$$

where the real coordinates u, v are defined by

$$u + iv = (x - x_c)\, e^{i \operatorname{Arg} S''(x_c)/2}.$$

Close to the saddle point, one can identify the contour with a curve $\operatorname{Im} S = \text{constant}$, locally $v = 0$ and, thus, the phase of the integrand remains constant: cancellations are no longer possible. The integral is dominated, up to relative corrections smaller than any power of λ, by the vicinity of the saddle point. The rest of the argument and the calculation are then the same as in the preceding real case, the real Gaussian integral being replaced by a complex Gaussian integral, which forces us to determine carefully the determination of the square root in expression (2.7).

The notion of saddle point. Let us describe more precisely the function $S(x)$ near the saddle point and justify the denominations *steepest descent method* and *saddle points*. For this purpose, we examine the modulus of the integrand of the function $\operatorname{Re}[S(x) - S(x_c)]$, in the complex plane near the saddle point $x = x_c$. After a rotation, in the coordinates u, v, the curves of constant modulus of the integrand are locally the curves $u^2 - v^2$ constant, that is, equilateral hyperbolae. The curve that passes through the saddle point degenerates into two straight lines. The saddle point, thus, is a double point (see Figure 2.8). Saddle points are the only regular double points. The modulus of the integrand has a saddle point structure in a topographical sense and the contour corresponds to leaving the saddle point by the steepest descent path.

Example. Hermite polynomials, which are related to eigenfunctions of the quantum harmonic oscillator, have an integral representation of the form

$$\mathcal{H}_n(x) = 2^{n/2}(n!/\sqrt{\pi})^{1/2} \frac{1}{2i\pi} \oint_C \frac{dp}{p^{n+1}}\, e^{px - p^2/4},$$

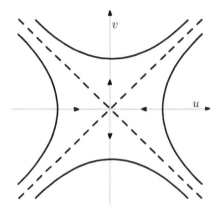

Fig. 2.8 Curves of constant modulus of the integrand near $u = v = 0$, which corresponds to the vicinity of the saddle point $x = x_c$.

where C is a closed simple contour enclosing the origin and oriented in the positive direction.

We want to evaluate the polynomial $P_n(z) = \mathcal{H}_n(z\sqrt{2n})$ for $n \to \infty$ and z real fixed by the steepest descent method. The polynomials $\mathcal{H}_n(x)$ being even or odd,

$$\mathcal{H}_n(-x) = (-1)^n \mathcal{H}_n(x),$$

we can restrict the study to $x \geq 0$.

The integrand is a meromorphic function with, as the only singularity, a multiple pole at $p = 0$. To reduce the integral to the standard form of the steepest descent method, it is convenient to set $p = s\sqrt{2n}$, which leads to

$$P_n(z) = \left(n!\, n^{-n}/\sqrt{\pi}\right)^{1/2} \frac{1}{2i\pi} \oint_C \frac{ds}{s^{ns+1}} e^{2nsz - ns^2/2}.$$

We also set

$$S(s) = \tfrac{1}{2}s^2 - 2sz + \ln s.$$

The saddle points are given by

$$S'(s) = s - 2z + 1/s = 0 \quad \Rightarrow \quad s_\pm = z \pm \sqrt{z^2 - 1}.$$

Also,

$$S''(s) = 1 - \frac{1}{s^2}.$$

It is then necessary to distinguish between the two cases $0 \leq z < 1$ and $z > 1$ ($z = 1$ requires a special analysis).

(i) $z > 1$. It is convenient to set $z = \cosh\theta$ with $\theta > 0$. Then,

$$s_\pm = e^{\pm\theta} \quad \Rightarrow \quad S(s_\pm) = -\tfrac{1}{2}e^{\pm 2\theta} \pm \theta - 1, \quad S''(s_\pm) = 1 - e^{\mp 2\theta}.$$

By moving the contour, it is simple to verify that the relevant saddle point is s_- because the contour near the saddle point is parallel to the imaginary axis and thus S'' must be negative. Taking into account the additional factor $1/s$ in the integrand, one concludes

$$P_n(z) \underset{n\to\infty}{\sim} \frac{(2n)^{-1/4}}{\sqrt{\pi(1-\mathrm{e}^{-2\theta})}} \exp\left[\tfrac{1}{2}n(\mathrm{e}^{-2\theta}+1) + n\theta\right],$$

where Stirling's formula (2.39) has been used.

(ii) $|z| < 1$. By contrast, in this case both saddle points, which are complex conjugate, contribute. Setting $z = \cos\theta$, one finds

$$s_\pm = \mathrm{e}^{\pm i\theta}, \quad S(s_\pm) = -\tfrac{1}{2}\mathrm{e}^{\pm 2i\theta} \pm i\theta - 1, \quad S''(s_\pm) = 1 - \mathrm{e}^{\mp 2i\theta}.$$

One concludes

$$P_n(z) \underset{n\to\infty}{\sim} \frac{(2n)^{-1/4}}{\sqrt{\pi(1-\mathrm{e}^{2i\theta})}} \exp\left[\tfrac{1}{2}n(\mathrm{e}^{2i\theta}+1) - ni\theta\right] + \text{complex conjugate}.$$

2.7 Steepest descent method: Several variables, generating functions

We now generalize the steepest descent method to the case of several variables, a generalization that in the case of complex integrals, technically, is not completely trivial.

2.7.1 Steepest descent method

We consider the general integral over n variables

$$\mathcal{I}(\lambda) = \int \mathrm{d}^n x \, \exp\left[-\frac{1}{\lambda}S(x_1,\ldots,x_n)\right], \tag{2.41}$$

where, to simplify, we first assume that S is a real entire function in all variables and that the initial integration domain is $\mathbb{R}^n$.

In the limit $\lambda \to 0_+$, the integral is dominated by the absolute minimum of the integrand, which is a saddle point solution of

$$\frac{\partial}{\partial x_i} S(x_1, x_2, \ldots, x_n) = 0 \quad \forall i. \tag{2.42}$$

To calculate the leading contribution of a given saddle point $\mathbf{x}^c$, we change variables, setting

$$\mathbf{x} = \mathbf{x}^c + \mathbf{y}\sqrt{\lambda}.$$

We then expand $S(\mathbf{x})$ in powers of λ (and thus of $\mathbf{y}$):

$$\frac{1}{\lambda}S(x_1,\ldots,x_n) = \frac{1}{\lambda}S(\mathbf{x}^c) + \frac{1}{2!}\sum_{i,j}\frac{\partial^2 S(\mathbf{x}^c)}{\partial x_i \partial x_j} y_i y_j + R(\mathbf{y}) \tag{2.43}$$

with

$$R(\mathbf{y}) = \sum_{k=3}^{\infty} \frac{\lambda^{k/2-1}}{k!} \sum_{i_1,i_2,\ldots,i_k} \frac{\partial^k S(\mathbf{x}^c)}{\partial x_{i_1} \cdots \partial x_{i_k}} y_{i_1} \cdots y_{i_k}. \quad (2.44)$$

The change of variables is such that the term quadratic in $\mathbf{y}$ is independent of λ. The integral becomes

$$\mathcal{I}(\lambda) = \lambda^{n/2} e^{-S\mathbf{x}^c)/\lambda} \int d^n y \exp\left[-\frac{1}{2!} \sum_{i,j} \frac{\partial^2 S(\mathbf{x}^c)}{\partial x_i \partial x_j} y_i y_j - R(\mathbf{y})\right]. \quad (2.45)$$

We then expand the integrand in powers of $\sqrt{\lambda}$: at each order, the calculation reduces to a Gaussian expectation value of polynomials. At leading order, one finds

$$\mathcal{I}(\lambda) \underset{\lambda \to 0}{\sim} (2\pi\lambda)^{n/2} \left[\det \mathbf{S}_c^{(2)}\right]^{-1/2} e^{-S(\mathbf{x}^c)/\lambda}, \quad (2.46)$$

where $\mathbf{S}_c^{(2)}$ is the matrix of partial second derivatives at the saddle point whose determinant is assumed not to vanish:

$$[\mathbf{S}_c^{(2)}]_{ij} \equiv \left.\frac{\partial^2 S}{\partial x_i \partial x_j}\right|_{\mathbf{x}=\mathbf{x}^c}.$$

The situation $\det \mathbf{S}_c^{(2)} = 0$ requires a special analysis.

Complex integrals. In the complex case, when several saddle points exist, one must classify the saddle points according to the values of $\operatorname{Re} S$. The leading saddle point is often the one that corresponds to $\operatorname{Re} S$ minimum. However, this is not always true because all saddle points do not necessarily contribute. To find out, one must proceed by deformation of the initial integration domain, an operation that is not always simple. In the case of several variables, it can be difficult to select the relevant saddle points.

2.7.2 Calculation of generating functions

We introduce the generating function

$$\mathcal{Z}(\mathbf{b},\lambda) = \int d^n x \exp\left[-\frac{1}{\lambda}(S(\mathbf{x}) - \mathbf{b} \cdot \mathbf{x})\right], \quad (2.47)$$

where $S(\mathbf{x})$ is a real regular function. We define

$$\mathcal{N} = 1/\mathcal{Z}(0,\lambda).$$

The function $\mathcal{Z}(\mathbf{b},\lambda)$ has the general form (2.31), and is proportional to the generating function of the moments of the distribution $\mathcal{N} e^{-S(\mathbf{x})/\lambda}$.

The expectation values of polynomials with the measure $\mathcal{N}\,e^{-S(\mathbf{x})/\lambda}$,

$$\langle x_{k_1} x_{k_2} \ldots x_{k_\ell} \rangle \equiv \mathcal{N} \int d^n x\, x_{k_1} x_{k_2} \ldots x_{k_\ell}\, e^{-S(\mathbf{x})/\lambda}, \qquad (2.48)$$

are related to derivatives of $\mathcal{Z}$ by (see equation (2.32)):

$$\langle x_{k_1} x_{k_2} \ldots x_{k_\ell} \rangle = \lambda^\ell \mathcal{N} \left[\frac{\partial}{\partial b_{k_1}} \frac{\partial}{\partial b_{k_2}} \cdots \frac{\partial}{\partial b_{k_\ell}} \mathcal{Z}(\mathbf{b},\lambda) \right]_{\mathbf{b}=0}.$$

Steepest descent calculation. We now apply the steepest descent method to the integral (2.47) and calculate the two first orders. The saddle point equation is

$$b_i = \frac{\partial S}{\partial x_i} \quad \forall\, i. \qquad (2.49)$$

We expand $S(\mathbf{x})$ at the saddle point $\mathbf{x}^c$ as explained in Section 2.6 and use the result (2.46):

$$\mathcal{Z}(\mathbf{b},\lambda) \underset{\lambda \to 0}{\sim} (2\pi\lambda)^{n/2} \left[\det \mathbf{S}_c^{(2)} \right]^{-1/2} \exp\left[-\frac{1}{\lambda} (S(\mathbf{x}^c) - \mathbf{b}\cdot\mathbf{x}^c) \right]$$

with

$$[\mathbf{S}_c^{(2)}]_{ij} \equiv \left. \frac{\partial^2 S}{\partial x_i \partial x_j} \right|_{\mathbf{x}=\mathbf{x}^c}. \qquad (2.50)$$

We now introduce $\mathcal{W}(\mathbf{b},\lambda)$, the generating function of the cumulants of the distribution, which are also the connected correlation functions (equation (2.33)), with a convenient normalization,

$$\mathcal{W}(\mathbf{b},\lambda) = \lambda \ln \mathcal{Z}(\mathbf{b},\lambda).$$

Using the result (2.46) and the identity (A8), $\ln \det \mathbf{M} = \operatorname{tr} \ln \mathbf{M}$, valid for any matrix $\mathbf{M}$, one finds

$$\mathcal{W}(\mathbf{b},\lambda) = -S(\mathbf{x}^c) + \mathbf{b}\cdot\mathbf{x}^c + \tfrac{1}{2} n\lambda \ln(2\pi\lambda) - \tfrac{1}{2}\lambda \operatorname{tr}\ln \mathbf{S}_c^{(2)} + O(\lambda^2). \qquad (2.51)$$

Since

$$\langle x_{k_1} x_{k_2} \ldots x_{k_\ell} \rangle_c = \lambda^{\ell-1} \left[\frac{\partial}{\partial b_{k_1}} \frac{\partial}{\partial b_{k_2}} \cdots \frac{\partial}{\partial b_{k_\ell}} \mathcal{W}(\mathbf{b},\lambda) \right]_{\mathbf{b}=0},$$

the successive derivatives with respect to $\mathbf{b}$ of expansion (2.51) (remembering that $\mathbf{x}_c$ is a function of $\mathbf{b}$ through equation (2.49)), calculated at $\mathbf{b}=0$ yield the expansions of the connected correlation functions. At leading order, for example,

$$\langle x_i \rangle = x_i^c + O(\lambda),$$
$$\langle x_i x_j \rangle_c = \lambda\, [\mathbf{S}_c^{(2)}]_{ij}^{-1} + O(\lambda^2),$$

where $\mathbf{S}_c^{(2)}$ is the matrix (2.50).

Exercises

Exercise 2.1

One considers two random correlated variables x, y with a Gaussian probability distribution. One finds the following five values

$$\langle x \rangle = \langle y \rangle = 0, \quad \langle x^2 \rangle = a, \quad \langle xy \rangle = b, \quad \langle y^2 \rangle = c.$$

Infer the expectation values $\langle x^4 \rangle$, $\langle x^3 y \rangle$, $\langle x^2 y^2 \rangle$, $\langle xy^5 \rangle$, $\langle x^3 y^3 \rangle$.

Which conditions must the coefficients a, b, c satisfy to be possible expectation values?

Determine the Gaussian distribution that leads to these values in the special case where the coefficients are related by $ac - b^2 = 1$.

Solution. The expectation values are, respectively,

$$3a^2, \quad 3ab, \quad ac + 2b^2, \quad 15bc^2, \quad 6c^3 + 9bac.$$

Since clearly $a, c > 0$, for the matrix

$$\mathbf{A} = \begin{pmatrix} a & b \\ b & c \end{pmatrix}$$

to be positive, it must in addition satisfy the condition $\det \mathbf{A} = ac - b^2 > 0$.

The corresponding distribution in the case $ac - b^2 = 1$ is

$$\frac{1}{2\pi} e^{-(cx^2 - 2bxy + ay^2)/2}.$$

Exercise 2.2

One considers three correlated random variables x, y, z with a Gaussian probability distribution. One finds the nine following expectation values

$$\langle x \rangle = \langle y \rangle = \langle z \rangle = 0, \quad \langle x^2 \rangle = \langle y^2 \rangle = \langle z^2 \rangle = a,$$
$$\langle xy \rangle = b, \quad \langle xz \rangle = \langle zy \rangle = c.$$

Infer, as functions of a, b, c, the expectation values $\langle x^4 \rangle$, $\langle x^6 \rangle$, $\langle x^3 y \rangle$, $\langle x^2 y^2 \rangle$, $\langle x^2 yz \rangle$.

Determine for $a = 2, b = 1, c = 0$ the Gaussian distribution that leads to these values.

Solution.

$$\langle x^4 \rangle = 3a^2, \quad \langle x^6 \rangle = 15a^3, \quad \langle x^3 y \rangle = 3ab, \quad \langle x^2 y^2 \rangle = a^2 + 2b^2,$$
$$\langle x^2 yz \rangle = ac + 2bc.$$

For $a = 2, b = 1, c = 0$ the Gaussian distribution that leads to these values is

$$\exp\left[-\tfrac{1}{12}\left(4x^2 + 4y^2 + 3z^2 - 4xy\right)\right].$$

Exercise 2.3

Recursive derivation of result (2.6). The determinant of a general $n \times n$ complex matrix $\mathbf{A}^{(n)}$, with elements $A_{ij}^{(n)}$, can be calculated recursively by subtracting to all rows a multiple of the last row in order to cancel the last column (assuming $A_{nn}^{(n)} \neq 0$, otherwise one must choose another row and column). The method leads to the relation between determinants

$$\det \mathbf{A}^{(n)} = A_{nn}^{(n)} \det \mathbf{A}^{(n-1)},$$

where $\mathbf{A}^{(n-1)}$ is an $(n-1) \times (n-1)$ matrix with elements

$$A_{ij}^{(n-1)} = A_{ij}^{(n)} - A_{in}^{(n)} A_{nj}^{(n)} / A_{nn}^{(n)}, \quad i, j = 1 \ldots n - 1. \tag{2.52}$$

Use the evaluation of the simple integral

$$\int_{-\infty}^{+\infty} \mathrm{d}x \, \mathrm{e}^{-ax^2/2 + bx} = \sqrt{2\pi/a} \, \mathrm{e}^{b^2/2a}$$

to recover expression (2.6).

Solution. We now consider the integral (2.4). We integrate over the variable x_n (assuming $\operatorname{Re} A_{nn} > 0$), using the result (2.7):

$$\int \mathrm{d}x_n \exp\left(-\tfrac{1}{2} A_{nn} x_n^2 - x_n \sum_{i=1}^{n-1} A_{ni} x_i\right) = \sqrt{\frac{2\pi}{A_{nn}}} \exp\left(\tfrac{1}{2} \sum_{i,j=1}^{n-1} \frac{A_{in} A_{nj}}{A_{nn}} x_i x_j\right).$$

The remaining Gaussian integral becomes the integral over $n-1$ variables

$$\mathcal{Z}(\mathbf{A}) = \sqrt{\frac{2\pi}{A_{nn}}} \int \left(\prod_{i=1}^{n-1} \mathrm{d}x_i\right) \exp\left(-\sum_{i,j=1}^{n-1} \tfrac{1}{2} x_i \left(A_{ij} - A_{in} A_{nn}^{-1} A_{nj}\right) x_j\right).$$

Comparing with the identity (2.52), one observes that the result is consistent with the calculation of $1/\sqrt{\det \mathbf{A}}$. One concludes

$$\mathcal{Z}(\mathbf{A}) = (2\pi)^{n/2} (\det \mathbf{A})^{-1/2}. \tag{2.53}$$

Exercise 2.4

Evaluate, using the steepest descent method, the modified Bessel function

$$I_\nu(x) = \frac{1}{2\pi} \int_{-\pi}^{\pi} \mathrm{d}\theta \, \mathrm{e}^{i\nu\theta} \mathrm{e}^{x \cos \theta},$$

$(= J_\nu(ix))$ with $\nu \geq 0$ integer, for $x \to +\infty$.

Solution. The integral has a canonical form for the application of the steepest descent method ($x = 1/\lambda$), and the integrand is an entire function. To determine the position of the saddle point, one can omit the factor $e^{i\nu\theta}$.

The saddle points then are given by
$$\sin\theta = 0 \Rightarrow \theta = 0 \pmod{\pi}.$$

For $x \to +\infty$, the leading saddle point is $\theta = 0$. One expands at the saddle point
$$x\cos\theta = x - \tfrac{1}{2}x\theta^2 + \tfrac{1}{24}x\theta^4 + O(\theta^6).$$

The region contributing to the integral is of order $\theta = O(1/\sqrt{x})$. Thus,
$$I_\nu(x) = \frac{1}{2\pi} e^x \int_{-\infty}^{\infty} d\theta\, e^{i\nu\theta}\, e^{-x\theta^2/2} \left(1 + \tfrac{1}{24}x\theta^4\right) + O(e^x/x^2)$$
$$= \frac{1}{\sqrt{2\pi x}} e^x \left(1 + \frac{1 - 4\nu^2}{8x} + O\left(\frac{1}{x^2}\right)\right).$$

Exercise 2.5

Evaluate by the steepest descent method the Bessel function
$$J_0(x) = \frac{1}{2\pi} \int_{-\pi}^{\pi} d\theta\, e^{ix\cos\theta},$$
x real, $x \to +\infty$.

Solution. The saddle points are the same as in the preceding exercise:
$$\sin\theta = 0 \Rightarrow \theta = 0 \pmod{\pi}.$$

All saddle points now contribute, $\theta = 0$ and the two halves of $\theta = \pm\pi$, which by periodicity is equivalent to a complete saddle point, for example $\theta = \pi$.
We first expand the integrand at the saddle point $\theta = 0$:
$$\frac{1}{2\pi} e^{ix} \int_{-\varepsilon}^{+\varepsilon} d\theta\, e^{-ix\theta^2/2}.$$

To cross the saddle point following a line of constant phase, one must set
$$\theta = e^{-i\pi/4} s.$$

The contribution of the saddle point becomes
$$\frac{1}{2\pi} e^{ix - i\pi/4} \int ds\, e^{-xs^2/2} = \frac{1}{\sqrt{2\pi x}} e^{ix - i\pi/4}.$$

The second saddle point yields the complex conjugate contribution. One thus finds
$$J_0(x) \underset{x \to +\infty}{\sim} \sqrt{\frac{2}{\pi x}} \cos(x - \pi/4).$$

Exercise 2.6

Evaluate the Bessel function of the third kind given by the integral
$$K_\nu(z) = \frac{1}{2}\int_0^\infty dt\, t^{\nu-1} e^{-z(t+1/t)/2}, \qquad (2.54)$$
($\operatorname{Re}\nu > 0$) using the steepest descent method, for $z \to +\infty$.

Solution. One sets
$$S(t) = \tfrac{1}{2}(t+1/t).$$
The saddle point t_c is given by
$$S'(t_c) = \tfrac{1}{2}(1-1/t_c^2) = 0 \ \Rightarrow\ t_c = 1.$$
Then,
$$S''(t_c) = 1/t_c^3 = 1.$$
One concludes
$$K_\nu(z) \underset{z\to+\infty}{\sim} (\pi/2z)^{1/2} e^{-z}.$$

Exercise 2.7

Evaluate by the steepest descent method the integral
$$I_n(s) = \frac{1}{2\pi}\int_{-\pi/2}^{+\pi/2} d\theta\, (\cos\theta)^n\, e^{in\theta\tanh s}$$
as a function of the real parameter s in the limit $n \to \infty$. One will verify that the function is real.

Solution. One introduces the function
$$S(\theta) = -i\theta\tanh s - \ln\cos\theta.$$
The function is analytic except at the points $\theta = \pi/2 \bmod (\pi)$. The locations of the saddle points are given by
$$S'(\theta) = -i\tanh s + \tan\theta = 0,$$
and at the saddle point
$$S''(\theta) = 1 + \tan^2\theta = 1 - \tan^2 s = \frac{1}{\cosh^2 s}.$$
The equation has the solution
$$\theta = is + m\pi,\quad m \in \mathbb{Z}.$$
One infers
$$S(\theta) = -i(is+m\pi)\tanh s - \ln\cosh s - im\pi \ \Rightarrow\ \operatorname{Re} S(\theta) = s\tanh s - \ln\cosh s.$$
For all saddle points, the integrand thus has the same modulus, but after contour deformation, one verifies that only the saddle point $m=0$ contributes. Thus,
$$I_n(s) \sim \frac{1}{\sqrt{2\pi n}}(\cosh s)^{n+1} e^{-ns\tanh s}.$$

Exercise 2.8

Adapt the steepest descent method to the calculation of the integral

$$I_\nu(x) = \frac{(x/2)^\nu}{\sqrt{\pi}\,\Gamma(\nu + 1/2)} \int_0^\pi d\theta (\sin\theta)^{2\nu}\, e^{x\cos\theta}$$

for $x \to +\infty$, where the expression generalizes the modified Bessel function defined in Exercise 2.4 to complex values $\operatorname{Re}\nu > -1/2$.

Solution. For $x \to +\infty$, the integral is dominated by the vicinity of $\theta = 0$. One can thus expand the integrand:

$$\int_0^\pi d\theta (\sin\theta)^{2\nu} e^{x\cos\theta}$$

$$= e^x \int_0^\infty d\theta\, \theta^{2\nu} \left(1 - \tfrac{1}{3}\nu\theta^2 + \tfrac{1}{24}x\theta^4\right) e^{-x\theta^2/2} + O(e^x/x^{\nu+5/2})$$

$$= e^x \int_0^\infty ds\, (2s)^{\nu-1/2} \left(1 - \tfrac{2}{3}\nu s + \tfrac{1}{6}xs^2\right) e^{-xs} + O(e^x/x^{\nu+5/2})$$

$$= 2^{\nu-1/2}\Gamma(\nu+1/2)x^{-\nu-1/2}e^x \left[1 + \frac{1 - 4\nu^2}{8x} + O\left(\frac{1}{x^2}\right)\right].$$

One verifies that this expression is a continuation of the expression obtained for the function I_ν in exercise 2.4.

Exercise 2.9

Evaluate by the steepest descent method the integral (see also definition (A5))

$$F_{\alpha,\beta}(s) = \int_0^1 t^{s\alpha - 1}(1-t)^{s\beta - 1} dt, \quad \alpha, \beta > 0,$$

$(\alpha, \beta > 0)$ for $s \to +\infty$.

Solution. The integrand is singular at $t = 0$ and $t = 1$ but, as we shall verify, the saddle point is at a finite distance from the boundary and the vicinity of these points does not contribute. Then, we rewrite the integrand as

$$t^{s\alpha - 1}(1-t)^{s\beta - 1} = \frac{1}{t(1-t)} e^{-sS(t)}$$

with

$$S(t) = -\alpha \ln t - \beta \ln(1-t).$$

The saddle point is given by

$$S'(t_c) = 0 \Rightarrow t_c = \frac{\alpha}{\alpha + \beta}, \quad S(t_c) = (\alpha + \beta)\ln(\alpha + \beta) - \alpha \ln\alpha - \beta \ln\beta$$

and, finally, from $S''(t_c) = (\alpha + \beta)^3/\alpha\beta$,

$$F_{\alpha,\beta}(s) \sim \sqrt{\frac{2\pi(\alpha+\beta)}{\alpha\beta s}}\, e^{-sS(t_c)}.$$

3 Universality and the continuum limit

We now begin the discussion of two of the main topics of this work, the related questions of *universality* and macroscopic *continuum limit* in random systems with a large number of degrees of freedom.

We first explain the notion of universality using the classical example of the central limit theorem in probability theory. Then, we discuss the properties of the random walk on a lattice, where universality is directly related to the continuum limit.

In both examples, we are interested in the collective properties of an infinite number of random variables in a situation where the probability of large deviations with respect to the mean value decreases fast enough.

They differ in the sense that a random walk is based on a spatial structure that does not necessarily exist in the case of the central limit theorem.

From the study of these first examples emerges the importance of Gaussian distributions, and this justifies the technical considerations of Chapter 2.

We then introduce some transformations, acting on distributions, which decrease the number of random variables. We show that Gaussian distributions are attractive fixed points for these transformations. This will provide us with the first, extremely simple, applications of the renormalization group (RG) ideas and allow us to establish the corresponding terminology.

Finally, in this context of the random walk, a path integral representation is associated to the existence of a continuum limit.

3.1 Central limit theorem of probabilities

Elementary probability theory provides a first example of *universality*: the central limit theorem, which we prove here in a strong version under conditions that are rather restrictive, but well adapted to the questions we want to investigate later.

We consider a real random variable q characterized by a probability distribution $\rho(q) \geq 0$.

We choose, for reasons of technical simplicity, $\rho(q)$ to belong to one of the two classes that will be useful later:

(i) $\rho(q)$ is a positive function (and not only a distribution), piecewise differentiable with a summable derivative.

(ii) $\rho(q)$ is a discrete distribution, q taking only integer values.

Moreover, in both cases, we assume $\rho(q)$ to be bounded with asymptotic exponential decay: we assume that there exists two positive constants M, A such that

$$\rho(q) \leq M \mathrm{e}^{-A|q|}, \quad A > 0. \tag{3.1}$$

With this assumption, large values of $|q|$ have a very small probability.

With these assumptions (rather restrictive from the sole viewpoint of the central limit theorem), one proves that, when $n \to \infty$, the asymptotic distribution of the mean value of n independent random variables, with the same distribution ρ, is a Gaussian distribution.

3.1.1 Fourier transformation

We first consider the situation where the distribution $\rho(q)$ is a positive function, piecewise differentiable with a summable derivative,

$$\int dq \, |\rho'(q)| < \infty, \qquad (3.2)$$

and with bounded variation as a consequence of positivity and the normalization condition $\int dq \rho(q) = 1$.

We denote by $\langle f \rangle$ the expectation value of a function $f(q)$:

$$\langle f \rangle \equiv \int_{-\infty}^{+\infty} dq \, \rho(q) f(q), \quad \langle 1 \rangle = \int_{-\infty}^{+\infty} dq \, \rho(q) = 1. \qquad (3.3)$$

To prove the theorem, it is convenient to introduce the Fourier transform $\tilde{\rho}(k)$ of the function $\rho(q)$, which is also a generating function of the moments of the distribution, as well as the function $w(k) = \ln \tilde{\rho}(k)$, which is the generating function of the cumulants.

Fourier transform. The Fourier transform

$$\tilde{\rho}(k) = \int dq \, e^{-ikq} \rho(q) \;\Rightarrow\; \rho(q) = \frac{1}{2\pi} \int dk \, e^{ikq} \tilde{\rho}(k) \qquad (3.4)$$

of the distribution $\rho(q)$ satisfies (* denotes here complex conjugation)

$$\tilde{\rho}^*(k) = \tilde{\rho}(-k), \quad \tilde{\rho}(k=0) = 1, \quad |\tilde{\rho}(k)| \leq 1$$

and, moreover, with our assumptions, $|\tilde{\rho}(k)| = 1$ is possible only for $k = 0$.

With the bound (3.1), the integral defining $\tilde{\rho}(k)$ also converges for k complex in the domain

$$|\operatorname{Im} k| < A.$$

The function $\tilde{\rho}(k)$ is, moreover, differentiable in this strip and, thus, is analytic.

Fourier transforms analytic in a strip of this form correspond to distributions with exponential decrease. This property illustrates the *duality between the asymptotic behaviour of a function and the regularity of its Fourier transform*. It plays an important role in the study of universality and continuum limit.

In particular, $\tilde{\rho}(k)$ has a Taylor series expansion convergent for $|k| < A$. It is also a generating function of the moments of the distribution $\rho(q)$ since

$$\tilde{\rho}(k) = 1 + \sum_{p=1}^{\infty} \frac{(-i)^p}{p!} \langle q^p \rangle k^p$$

with
$$\langle q^p \rangle = \int dq\, q^p \rho(q), \quad |\langle q^p \rangle| < \frac{2M}{A^{p+1}} p!\,.$$

Finally, the differentiability properties of $\rho(q)$ and the assumptions (3.2) imply that $\tilde{\rho}(k)$ decreases like $1/k$ for $|k| \to \infty$.

It is also convenient to introduce the function
$$w(k) = \ln \tilde{\rho}(k) = \sum_{p=1}^{\infty} \frac{(-i)^p}{p!} w_p k^p, \tag{3.5}$$

(note that $w^*(k) = w(-k)$) which is a generating function of the cumulants of the distribution:
$$w_1 = \langle q \rangle, \quad w_2 = \langle q^2 \rangle - (\langle q \rangle)^2, \quad w_3 = \langle q^3 \rangle - 3\langle q^2 \rangle \langle q \rangle + 2(\langle q \rangle)^3 \ldots$$

The function w satisfies
$$w(0) = 0, \quad \operatorname{Re} w(k) < 0 \text{ for } k \neq 0\,.$$

The coefficient w_2, which can also be written as
$$w_2 = \langle (q - \langle q \rangle)^2 \rangle \geq 0\,,$$

(the square of the mean quadratic deviation) is positive, except for a certain distribution, which we exclude. In the case of a Gaussian distribution, all cumulants w_p with $p > 2$ vanish since $\tilde{\rho}(k)$ is also a Gaussian function.

The singularities of the analytic function $w(k)$ are the singularities of $\tilde{\rho}(k)$ and its zeros. Since $\tilde{\rho}(0) = 1$, there exists a zero-free disk centred at the origin in which $w(k)$ is analytic.

3.1.2 Central limit theorem and consequences

We now consider n independent random variables q_i with the same distribution $\rho(q)$.

Distribution of the sum of n variables. We first recall the following result: if Q_1 and Q_2 are two independent random variables with distributions ρ_1 and ρ_2, respectively, the sum $Q = Q_1 + Q_2$ has the distribution
$$R(Q) = \int dQ'\, \rho_1(Q') \rho_2(Q - Q'). \tag{3.6}$$

The distribution of the sum of n variables is given by
$$P_n(Q) = \langle q_1 + q_2 + \cdots + q_n \rangle \quad \text{with} \quad \sum_{i=1}^{n} q_i = Q\,.$$

Using the result (3.6), one can infer the distribution P_{n+1} of $n+1$ variables from the distribution P_n of n variables and the distribution ρ of the last variable. The distribution P_n thus satisfies the recursion relation

$$P_{n+1}(Q) = \langle q_1 + q_2 + \cdots + q_n + q_{n+1}\rangle \quad \text{with} \quad \sum_{i=1}^{n+1} q_i = Q$$

$$= \int dQ' \rho(Q') P_n(Q - Q'). \tag{3.7}$$

After Fourier transformation, this convolution equation takes an algebraic form. Setting

$$\tilde{P}_n(k) = \int dQ \, e^{ikQ} P_n(Q),$$

one finds

$$\tilde{P}_{n+1}(k) = \tilde{\rho}(k) \tilde{P}_n(k)$$

and, thus,

$$\tilde{P}_n(k) = \tilde{\rho}^n(k) = e^{nw(k)}.$$

One infers

$$P_n(Q) = \frac{1}{2\pi} \int dk \, e^{ikQ} \tilde{P}_n(k) = \frac{1}{2\pi} \int dk \, e^{ikQ + nw(k)}. \tag{3.8}$$

The function $\tilde{P}_n(k)$ thus decreases like $1/|k|^n$ for $|k| \to \infty$, in such a way that $P_n(Q)$ is $n-1$ times differentiable.

Asymptotic distribution. Since $\operatorname{Re} w(k) < 0$ for $k \neq 0$, the contribution of the vicinity of $k = 0$ dominates the integral (3.8). In the limit $n \to \infty$, one can thus restrict the integration to a domain $|k| \leq \varepsilon$, $\varepsilon > 0$ arbitrarily small but fixed, neglecting corrections decreasing exponentially with n. As in the steepest descent method, one can thus replace the analytic function $w(k)$ by the first terms of its Taylor series expansion (3.5):

$$P_n(Q) = \frac{1}{2\pi} \int dk \, \exp\left(iQk - inw_1 k - \tfrac{1}{2} nw_2 k^2 + O(k^3)\right). \tag{3.9}$$

The Gaussian integration yields

$$P_n(Q) = \frac{1}{\sqrt{2\pi n w_2}} e^{-n(Q/n - w_1)^2 / 2w_2} (1 + O(1/n)), \tag{3.10}$$

where terms decreasing exponentially with n have been neglected.

At Q fixed, the probability converges exponentially to zero for all $w_1 \neq 0$.

By contrast, the random variable $Q = (q_1 + q_2 + \cdots + q_n)/n$, the mean value of the n variables, has the asymptotic distribution

$$R_n(Q) = n P_n(nQ) \underset{n \to \infty}{\sim} \sqrt{\frac{n}{2\pi w_2}} e^{-n(Q - w_1)^2 / 2w_2}. \tag{3.11}$$

The mean of the n variables is thus a random variable that converges toward a certain value, which is the expectation value

$$\langle Q \rangle = \frac{1}{n} \sum_{i=1}^{n} \langle q_i \rangle = \langle q \rangle = w_1. \tag{3.12}$$

Finally, the random variable

$$X = \left(\frac{1}{n} \sum_i q_i - w_1 \right) \sqrt{n}, \tag{3.13}$$

and thus $\langle X \rangle = 0$, has, as limiting distribution, the Gaussian distribution

$$L_n(X) = \sqrt{n} P_n(nw_1 + X\sqrt{n}) \sim \frac{1}{\sqrt{2\pi w_2}} e^{-X^2/2w_2}.$$

Universality. The asymptotic distribution of the mean of the n independent random variables of vanishing expectation value is a Gaussian distribution, independent of the initial distribution (in some class); in particular, it depends only on one parameter $w_2 = \langle (q - \langle q \rangle)^2 \rangle$. This independence from the initial distribution is a first example of *universality*. It shows a collective property of an infinite number of random uncorrelated variables.

It is easy to calculate corrections to the asymptotic result. If $w_3 \neq 0$, the main correction is of order $n^{-1/2}$ and one finds

$$L_n(X) = \frac{e^{-X^2/2w_2}}{\sqrt{2\pi w_2}} \left[1 + (X^3/6 - w_2 X/2) \frac{w_3}{w_2^3 \sqrt{n}} + O\left(\frac{1}{n}\right) \right].$$

The cumulants w_p, $p > 2$, then characterize the approach to the Gaussian distribution limit, the first non-vanishing cumulant giving the leading correction. They modify the Gaussian distribution by relative polynomial corrections.

Remarks.

(i) Since terms decaying exponentially in n have been neglected, numerically the asymptotic Gaussian form is a good approximation for n large but finite only for

$$|X| \ll \sqrt{n} \quad \text{for} \quad n \gg 1.$$

(ii) Only a special affine function, with coefficients depending explicitly on n, of the sum of the n initial random variables admits a non-trivial limiting distribution. The affine transformation (3.13) is a first example of a *renormalization*.

3.1.3 Various remarks

Steepest descent method. From the representation (3.8), one infers the representation of the distribution of the mean value $Q = \sum q_i/n$

$$R_n(Q) = \frac{n}{2\pi} \int dk\, e^{inQk+nw(k)}. \tag{3.14}$$

Since $w(k)$ is an analytic function, the integral can be calculated, for $n \to \infty$, by the steepest descent method. The saddle point equation is

$$w'(k) + iQ = 0.$$

The steepest descent method can only be justified if the saddle point is in the vicinity of the origin. In this case, one obtains

$$R_n(Q) \sim \sqrt{\frac{n}{-2\pi w''(k)}}\, e^{inQk+nw(k)}. \tag{3.15}$$

The combination $iQk + w(k)$, where k is a solution of the saddle point equation, is proportional to the Legendre transform of w (see Section 6.1.2).

If the saddle point is asymptotically close to the origin, one can expand w. The saddle point equation becomes

$$-iw_1 - w_2 k + iQ = O(k^2) \;\Rightarrow\; k = i(Q-w_1)/w_2 + O\!\left((Q-w_1)^2\right).$$

Substituting in expression (3.15), one recovers the evaluation (3.11).

Cumulants of a distribution. The Fourier transform $\tilde{R}_n(k)$ of the distribution $R_n(Q)$ is inferred from expression (3.14) by changing k into k/n. One obtains

$$\tilde{R}_n(k) = e^{nw(k/n)}. \tag{3.16}$$

The cumulants of the distribution $R_n(Q)$ are thus obtained simply from the cumulants (3.5) of the distribution $\rho(q)$,

$$\ln \tilde{R}_n(k) = \sum_{p=1}^{\infty} \frac{(-i)^p}{p!} \Omega_p k^p, \quad \Omega_p = n^{1-p} w_p.$$

This expression shows that, for $n \to \infty$, all cumulants of $R_n(Q)$ with $p > 1$ go to zero, which corresponds to the property that the mean value $Q = \langle Q \rangle$ becomes certain. Note that the behaviour of the moments of the distribution R_n is much less simple.

Similarly, the Fourier representation of the distribution $L_n(X)$ can be written as

$$L_n(X) = \frac{\sqrt{n}}{2\pi} \int dk\, e^{ik(X\sqrt{n}+nw_1)}\, e^{nw(k)}.$$

After the change of variables $\kappa = k\sqrt{n}$, one recognizes the Fourier transform $\tilde{L}_n(\kappa)$ of $L_n(X)$:

$$\ln \tilde{L}_n(\kappa) = nw(\kappa/\sqrt{n}) + iw_1\kappa\sqrt{n} = -\frac{1}{2}w_2\kappa^2 + \frac{(-i)^3}{3!}\frac{w_3}{n^{1/2}}\kappa^3$$
$$+ \frac{(-i)^4}{4!}\frac{w_4}{n}\kappa^4 + \cdots$$

and, thus, a generating function of the cumulants of the distribution of X. The cumulants converge toward the cumulants of a universal Gaussian distribution. Since the relations between moments and cumulants are algebraic, the same result applies to moments. Let us point out that the proof of this convergence requires only weaker assumptions.

Examples and counter-examples.
(i) The distribution uniform on the segment $[-1, +1]$ and vanishing outside,

$$\rho(q) = \tfrac{1}{2}(\mathrm{sgn}(q+1) - \mathrm{sgn}(q-1)), \quad \Rightarrow \langle q \rangle = 0, \langle q^2 \rangle = \tfrac{1}{3},$$

(sgn is the sign function) is centred around zero and satisfies the hypotheses of the central limit theorem. Its Fourier transform is

$$\tilde{\rho}(k) = \frac{1}{2}\int_{-1}^{1} dq\, e^{iqk} = \frac{\sin k}{k}.$$

The generating function of cumulants has the expansion

$$w(k) = \ln(\sin k/k) = -\tfrac{1}{6}k^2 - \tfrac{1}{180}k^4 + O(k^6).$$

The coefficient of k^4 characterizes the leading deviation from the asymptotic Gaussian distribution. This example shows clearly that the Gaussian form is not a good approximation for $|X| > \sqrt{n}$ where the exact distribution vanishes.

(ii) The distribution

$$\rho(q) = \tfrac{1}{2}m e^{-m|q|}, \quad q \in \mathbb{R}, m > 0,$$

also satisfies all hypotheses. Its Fourier transform is

$$\tilde{\rho}(k) = \frac{m^2}{k^2 + m^2} \quad \Rightarrow w(k) = -k^2/m^2 + k^4/2m^4 + O(k^6).$$

(iii) On the contrary, the Cauchy–Lorentz distribution

$$\rho(q) = \frac{1}{\pi}\frac{1}{1+q^2} \quad q \in \mathbb{R},$$

provides an example of a distribution that does not satisfy the conditions of the central limit theorem, even in its weakest form. It has no second moment since the integral $\int q^2 \rho(q) dq$ diverges. Its Fourier transform

$$\tilde{\rho}(k) = \tfrac{1}{2} e^{-|k|},$$

is not differentiable at $k = 0$. Notice that

$$[\tilde{\rho}(k/n)]^n = \tfrac{1}{2} e^{-|k|}$$

and, thus, $R(Q) = \rho(Q)$. This distribution is the asymptotic distribution of another universality class.

3.1.4 Random variables taking integer values

The central limit theorem, in the restricted form we have presented, can be extended to the situation where random variables take only integer values. The integrals are replaced by sums and Fourier transforms become Fourier series.

Let q be a random variable taking integer values whose probability measure $\rho(q) \geq 0$ again satisfies the exponential bound (3.1).

One then introduces the convergent Fourier series

$$\tilde{\rho}(k) = \sum_{q \in \mathbb{Z}} e^{-ikq} \rho(q), \quad \rho(q) = \frac{1}{2\pi} \int_{-\pi}^{\pi} dk\, e^{ikq} \tilde{\rho}(k),$$

and the function $w(k) = \ln \tilde{\rho}(k)$.

With the bound (3.1), the periodic function $\tilde{\rho}(k)$, with period 2π, is also analytic in the strip $|\operatorname{Im} k| < A$.

Moreover,

$$|\tilde{\rho}(k)| \leq 1.$$

In general, in the interval $-\pi \leq k \leq \pi$, the maximum $\tilde{\rho}(k) = 1$ is reached only for $k = 0$ except if $\rho(q)$ is non-vanishing only on a subset of the form

$$q = a + mb \quad \text{with} \quad a, b \text{ fixed and } b > 1, \quad m \in \mathbb{Z}.$$

A simple example of such a situation, with $a = 1$, $b = 2$, is

$$\rho(q) = \tfrac{1}{2} \quad \text{for} \quad q = \pm 1.$$

In a first analysis, we exclude this situation and comment on it later.

Then, the Fourier series has the properties required to prove the central limit theorem.

The distribution of the sum of n independent variables,

$$P_n(Q) = \sum_{q_1, q_2, \dots, q_n \in \mathbb{Z}} \prod_i \rho(q_i) \quad \text{with} \quad \sum_i q_i = Q,$$

satisfies the discrete form of the recursion relation (3.7):

$$P_{n+1}(Q) = \sum_{Q' \in \mathbb{Z}} \rho(Q') P_n(Q - Q'). \tag{3.17}$$

The corresponding Fourier series

$$\tilde{P}_n(k) = \sum_{q \in \mathbb{Z}} e^{-ikq} P_n(q), \quad P_n(q) = \frac{1}{2\pi} \int_{-\pi}^{\pi} dk\, e^{ikq} \tilde{P}_n(k),$$

satisfies the recursion relation

$$\tilde{P}_{n+1}(k) = \tilde{\rho}(k) \tilde{P}_n(k)$$

and, thus,
$$\tilde{P}_n(k) = \tilde{\rho}^n(k) = e^{nw(k)}.$$

In the Fourier representation, the situation thus is analogous to the one considered in Section 3.1.2. The random variable $(q_1 + q_2 + \cdots + q_n)/n$ converges toward a certain value, the expectation value $w_1 = \langle q \rangle$.

The distribution of the random variable

$$X = \left(\frac{1}{n}\sum_i q_i - w_1\right)\sqrt{n}, \qquad (3.18)$$

converge toward a universal Gaussian distribution characterized by the parameter w_2.

Let us point out, however, that the distribution $L_n(X)$ has no limit in the sense of functions, since it is non-vanishing only on a discrete set. By contrast, in the sense of measures or distributions, the initial discrete character is not relevant and, in the limit $n \to \infty$, one finds the same Gaussian limit.

From a physics viewpoint, one can interpret X as a macroscopic variable, the average of a large number of microscopic variables, which is measured only with a finite precision ΔX. The concept of convergence in the sense of measures is then appropriate. Moreover, as soon as $1/\sqrt{n} \gg \Delta X$ the discrete character of the q variables is no longer relevant. One can then consider X as a continuous variable. We thus introduce the notion of *continuum limit*, the continuous function toward which the points of $L_n(X)$ converge. The Gaussian distribution, considered as a continuous function, is the *continuum limit* of $L_n(X)$.

Example.
$$\rho(q) = \begin{cases} s/2 & \text{for } q = \pm 1 \\ 1 - s & \text{for } q = 0 \end{cases}$$

with $0 < s < 1$. Then,
$$\tilde{\rho}(k) = 1 - s + s\cos k$$

varies between 1 and $1 - 2s$ and has modulus 1 only for $k = 0$. One infers

$$w(k) = \ln(1 - s + s\cos k) = -\tfrac{1}{2}sk^2 + \tfrac{1}{24}s(s-3)k^4 + O(k^6).$$

The limiting Gaussian distribution then is

$$L_\infty(X) = \frac{1}{\sqrt{2\pi s}} e^{-X^2/2s}.$$

Other maxima. Let us consider the preceding example in the limit $s = 1$. In this case, the maximum of the modulus of $\tilde{\rho}(k) = \cos k$ is reached for the two values $k = 0$ and $k = \pi$. This reflects the property that, depending on the parity of n, only even or odd values of n have a non-vanishing probability. However, when calculating expectation values of continuous functions with the measure $L_\infty(X)$, one verifies that the maximum corresponding to $k \neq 0$, which generates oscillatory terms, does not contribute. More precisely,

$$L_n(X) = \frac{\sqrt{n}}{2\pi} \int_{-\pi}^{+\pi} dk \, (\cos k)^n \, e^{ikX\sqrt{n}}$$

$$\sim \frac{\sqrt{n}}{2\pi} \int_{-\infty}^{+\infty} dk \, e^{-nk^2/2+ikX\sqrt{n}}$$

$$\times \left[1 + (-1)^n \left(e^{-i\pi X\sqrt{n}} \theta(k) + e^{i\pi X\sqrt{n}} \theta(-k)\right)\right]$$

$$\sim \frac{1}{2\pi} \int_{-\infty}^{+\infty} dk \, e^{-k^2/2+ikX} \left[1 + (-1)^n \left(e^{-i\pi X\sqrt{n}} \theta(k) + e^{i\pi X\sqrt{n}} \theta(-k)\right)\right],$$

where $\theta(t)$ is the step function: $\theta(t) = 1$ for $t > 0$, $\theta(t) = 0$ for $t < 0$. One observes that the two additional contributions oscillate with X with a period $2/\sqrt{n}$ and, thus, have for $n \to \infty$ a vanishing expectation value on any finite interval.

This analysis generalizes to any other maximum.

3.2 Universality and fixed points of transformations

We now derive the universality property, that is, the existence of a limiting Gaussian distribution that is independent of the initial distribution, by a quite different method.

It is convenient to assume that initially the number of independent variables is of the form $n = 2^m$. One then averages recursively pairwise over these variables, decreasing the number of variables by a factor 2 at each iteration. This provides a first, very simple, application of RG ideas to the derivation of universality properties. Moreover, this example will allow us to introduce for the first time the relevant RG terminology.

The distribution of the sum of two independent random variables with the same distribution ρ is given by the transformation (equation (3.6))

$$[\mathcal{T}\rho](q) = \int dq' \, \rho(q')\rho(q - q').$$

One combines this transformation with a *renormalization* of the sum depending on a parameter $1/\lambda$, $\lambda > 0$, and this leads to the transformation

$$[\mathcal{T}_\lambda \rho](q) = \lambda \int dq' \, \rho(q')\rho(\lambda q - q').$$

For example, $\lambda = 1$ corresponds to the sum and $\lambda = 2$ to the mean of the two initial variables.

For the distribution $\rho(q)$, the form of the transformation is somewhat complicated but applied to the function $w(k)$ (defined by equation (3.5)), it becomes a linear transformation. Indeed, introducing the Fourier representation (3.4), one obtains

$$[\mathcal{T}_\lambda \rho](q) = \frac{\lambda}{(2\pi)^2} \int dq' \, dk_1 \, dk_2 \, e^{ik_1 q' + ik_2(\lambda q - q')} \tilde{\rho}(k_1)\tilde{\rho}(k_2).$$

The integral over q' yields $2\pi\delta(k_1 - k_2)$, where $\delta(k)$ is Dirac's distribution. Integrating over k_2, changing $k_1 = k/\lambda$, one finds

$$[\mathcal{T}_\lambda \rho](q) = \frac{1}{2\pi} \int \mathrm{d}k\, \mathrm{e}^{ikq}\, \tilde{\rho}^2(k/\lambda) \;\Rightarrow\; [\mathcal{T}_\lambda \tilde{\rho}](k) = \tilde{\rho}^2(k/\lambda).$$

In terms of $w(k) = \ln \tilde{\rho}(k)$, the transformation can be simply written as

$$[\mathcal{T}_\lambda w](k) \equiv 2w(k/\lambda).$$

This linear transformation has a property important for what follows: it is independent of m. In the language of dynamical systems, its repeated application generates a *stationary, or invariant under time translation, Markovian* dynamics. Our goal is to study the properties of the iterated transformation $\mathcal{T}_\lambda^m$ for $m \to \infty$ as a function of the parameter λ.

3.2.1 Generic situation

A limiting distribution necessarily is a fixed point of the transformation. Thus, it corresponds to a function $w_*(k)$ (where the notation $*$ is not related to complex conjugation) that satisfies

$$[\mathcal{T}_\lambda w_*](k) \equiv 2w_*(k/\lambda) = w_*(k).$$

For the class of distributions considered in this chapter, the functions $w(k)$ are regular at $k = 0$. Thus, one can expand $w_*(k)$ in powers of k, setting $(w(0) = 0)$

$$w_*(k) = -iw_1 k - \tfrac{1}{2} w_2 k^2 + \sum_{\ell=3} \frac{(-ik)^\ell}{\ell!} w_\ell .$$

At order k, one finds the equation

$$2w_1/\lambda = w_1 .$$

In the generic situation $w_1 \neq 0$, the equation implies $\lambda = 2$ and, thus, identifying the terms of higher degree,

$$2^{1-\ell} w_\ell = w_\ell \;\Rightarrow\; w_\ell = 0 \text{ for } \ell > 1 .$$

Therefore,
$$w_*(k) = -iw_1 k .$$

The fixed points form a family depending on one parameter w_1, which can be absorbed (if $w_1 \neq 0$) into a normalization of the random variable q.

The choice $\lambda = 2$ corresponds to calculating the asymptotic distribution of the mean. As expected, fixed points then correspond to the certain distribution $q = \langle q \rangle = w_1$. Indeed,

$$\rho_*(q) = \frac{1}{2\pi} \int \mathrm{d}k\, \mathrm{e}^{ikq - iw_1 k} = \delta(q - w_1).$$

Convergence and fixed point stability. One can now study the convergence of an arbitrary initial distribution toward the fixed point.

For a general non-linear transformation, it is in general impossible to perform a global analysis. One can only linearize the transformation near the fixed point and perform a local study.

Here, this is not necessary since the transformation is linear. Setting

$$w(k) = w_*(k) + \delta w(k),$$

then,

$$\mathcal{T}_\lambda w = \mathcal{T}_\lambda w_* + \mathcal{T}_\lambda \delta w = w_* + \mathcal{T}_\lambda \delta w.$$

Since δw is a regular function, one can expand it in a Taylor series:

$$\delta w(k) = \sum_{\ell=1} \frac{(-ik)^\ell}{\ell!} \delta w_\ell.$$

Then,

$$[\mathcal{T}_\lambda \delta w](k) = 2\delta w(k/\lambda) = 2 \sum_{\ell=1} \frac{(-ik)^\ell}{\ell!} \lambda^{-\ell} \delta w_\ell.$$

This expression shows that the functions k^ℓ, where ℓ is a positive integer, are the eigenvectors of the transformation $\mathcal{T}_\lambda$ and the corresponding eigenvalues are

$$\tau_\ell = 2\lambda^{-\ell} = 2^{1-\ell}. \tag{3.19}$$

Since at each iteration the number of variables is divided by two, one can relate the eigenvalues to the behaviour as a function of the initial number n of variables. One defines an associated *exponent*

$$l_\ell = \ln \tau_\ell / \ln 2 = 1 - \ell.$$

After m iterations, the component δw_ℓ is multiplied by $n^{1-\ell}$. Indeed,

$$\mathcal{T}_\lambda^m k^\ell = 2^{m(1-\ell)} k^\ell = n^{1-\ell} k^\ell.$$

The behaviour, for $n \to \infty$, of a component of δw on the eigenvectors thus depends on the sign of $1 - \ell$.

Since the transformation $\mathcal{T}_\lambda$ provides the simplest example of an *RG transformation*, a concept that we discuss in a more general framework in Chapter 9 in the framework of statistical physics, we now adopt the RG language to discuss eigenvalues and eigenvectors.

Let us examine the different values of ℓ:

(i) $\ell = 1 \Rightarrow \tau_1 = 1$, $l_\ell = 0$. If one adds a term δw proportional to the eigenvector k to $w_*(k)$, $\delta w(k) = -i\delta w_1 k$, one changes

$$w_1 \mapsto w_1 + \delta w_1$$

and, thus, the fixed point. Let us point out that this change has an interpretation as a simple linear transformation on k and, thus, on the random variable q.

An eigen-perturbation corresponding to the eigenvalue 1 and, thus, to a vanishing exponent, is called *marginal*.

More generally, the existence of a one-parameter family of fixed points implies the existence of an eigenvector associated to the eigenvalue $\tau = 1$ and, thus, the exponent $l = 0$. Indeed, let us assume the existence of a one-parameter family of fixed points $w(\alpha)$,

$$\mathcal{T}w(\alpha) = w(\alpha),$$

where $w(\alpha)$ is a differentiable function of the parameter α. Then,

$$\mathcal{T}\frac{\partial w}{\partial \alpha} = \frac{\partial w}{\partial \alpha}.$$

(ii) $\ell > 2 \Rightarrow \tau_\ell = 2^{1-\ell} < 1$, $l_\ell < 0$. The components of δw on such eigenvectors converge to zero for n or $m \to \infty$.

In the RG language, the eigen-pertubations that correspond to eigenvalues smaller in modulus than 1 and, thus, to negative exponents (more generally with a negative real part), are called *irrelevant*.

Universality, in this formalism, is a consequence of the property that all eigenvectors, but a finite number, are irrelevant.

Dimension of a random variable. To the random variable that has a limiting distribution, one can attach a dimension d_q defined by

$$d_q = \ln \lambda / \ln 2. \tag{3.20}$$

Here, one finds $d_q = 1$.

3.2.2 Centred distribution

For a centred distribution, $w_1 = 0$, the first equation is trivial, and one can expand to next order. At order k^2, one then finds the equation

$$w_2 = 2w_2/\lambda^2.$$

Since the variance w_2 is strictly positive, except for a certain distribution, a case that we now exclude, the equation implies $\lambda = \sqrt{2}$. Again, the coefficients w_ℓ vanish for $\ell > 2$ and the fixed points have the form

$$w_*(k) = -\tfrac{1}{2}w_2 k^2.$$

Therefore, one finds the Gaussian distribution

$$\mathcal{P}_*(q) = \frac{1}{2\pi}\int \mathrm{d}k\, e^{ikq - w_2 k^2/2} = \frac{1}{\sqrt{2\pi w_2}} e^{-q^2/2w_2}.$$

Since, in the transformation $\mathcal{T}$, the number n of variables is divided by two, this value of the renormalization factor λ corresponds to the factor $\sqrt{n}$ in the transformation (3.13). One can assign to the random variable the dimension (defined in (3.20))
$$d_q = \ln \lambda / \ln 2 = \tfrac{1}{2}.$$

Fixed point stability. The value of λ and, thus, the transformation $\mathcal{T}_\lambda$ being determined, it is easy to again study the fixed point stability. One sets
$$w(k) = w_*(k) + \delta w(k),$$
and looks for the eigenvectors and eigenvalues of the transformation
$$[\mathcal{T}_\lambda \delta w](k) \equiv 2\delta w(k/\sqrt{2}) = \tau \delta w(k).$$

Clearly, the eigenvectors still have the form
$$\delta w(k) = k^\ell \;\Rightarrow\; \tau_\ell = 2^{1-\ell/2}.$$

In terms of the number n of variables, this value corresponds to exponents
$$l_\ell = \ln \tau_\ell / \ln 2 = 1 - \ell/2.$$

Let us examine the different cases:

(i) $\ell = 1 \Rightarrow \tau_1 = \sqrt{2}$, $l_\ell = \tfrac{1}{2}$. This corresponds to an unstable direction; a component on such a vector diverges for $m \to \infty$.

In the RG language, a perturbation corresponding to a positive exponent l is called *relevant*. Each iteration increases its amplitude and, thus, such a perturbation moves away from the fixed point.

Here, this result has a simple interpretation: a perturbation linear in k violates the condition $w_1 = 0$. The relevant transformations and fixed points then are those studied in Section 3.2.1.

(ii) $\ell = 2 \Rightarrow \tau_2 = 1$, $l_2 = 0$. A vanishing eigenvalue characterizes a *marginal* perturbation. Here, this perturbation only modifies the value of w_2. Again, this has an interpretation as a linear transformation on the random variable.

(iii) $\ell > 2 \Rightarrow \tau_\ell = 2^{1-\ell/2} < 1$, $l_\ell = 1 - \ell/2 < 0$. Finally, all perturbations $\ell > 2$ correspond to stable directions in the sense that their amplitudes converge to zero for $m \to \infty$ and are *irrelevant*.

Remarks.

(i) In the examples examined here, the marginal perturbations correspond to simple changes in the normalization of the random variables. In many problems, this normalization plays no role. One can then consider that fixed points corresponding to different normalizations should not be distinguished. In such a situation, in each case one has found here really only one fixed point and the perturbation corresponding to the vanishing eigenvalue is no longer called *marginal* but *redundant*, in the sense that it changes only an arbitrary normalization.

(ii) By changing the value of λ, it is easy to find other fixed points of the form $|k|^\alpha$, $0 < \alpha < 2$ ($\alpha > 2$ is excluded because the coefficient of k^2 is strictly positive). But these fixed points are not regular functions of k and correspond to distributions that are less concentrated around the mean value. In particular, these distributions have no second moment $\langle q^2 \rangle$ and variance.

In the RG language, they correspond to other *universality classes*, that is, other sets of distributions, which converge toward these fixed points.

Integer random variables. In this case, the initial function $w(k)$ is periodic with period 2π. At each iteration, the period is multiplied by λ, and the possible values of the random variable q is divided by λ. The fixed point $w_*(k)$ corresponds to a function that is no longer periodic and thus, to a random variable q taking real values. This reflects the property that there is convergence only in the sense of measures.

3.3 Random walk and Brownian motion

We now consider a stochastic process, a random walk, in discrete times, first in $\mathbb{R}^d$, then on the lattice $\mathbb{Z}^d$ of points with integer coordinates.

Such a process is specified by a probability distribution P_0 at initial time $n = 0$ and a density of transition probability $\rho(\mathbf{q}, \mathbf{q}')$ from the point $\mathbf{q}'$ to the point $\mathbf{q}$, which we assume independent of the time n.

These conditions define a Markov chain, a Markovian process, in the sense that the displacement at time n depends only on the position at time n, but not on the positions at prior times, and homogeneous or stationary, that is, invariant under time translation.

Our goal is to show, first in the continuum, then on a lattice, that asymptotically in time, for the class of *local* random walks, a notion that we define below, one again recovers universal asymptotic Gaussian distributions.

This will allow us also introducing the notions of *continuum limit* and *path integrals*.

3.3.1 Walk in continuum space

Terminology. In what follows, because there will be no real ambiguity, we speak often of probabilities while obviously we mean probability densities.

Let $P_n(\mathbf{q})$ be the probability for a walker to be at point $\mathbf{q}$ at time n. For a stationary Markov chain, the probability distribution $P_n(\mathbf{q})$ satisfies an evolution equation or recursion relation of the form

$$P_{n+1}(\mathbf{q}) = \int d^d q' \, \rho(\mathbf{q}, \mathbf{q}') P_n(\mathbf{q}'), \quad \rho(\mathbf{q}, \mathbf{q}') \geq 0, \quad (3.21)$$

(in other fields also called the master equation) of Chapman–Kolmogorov type. Probability conservation implies the condition

$$\int d^d q \, \rho(\mathbf{q}, \mathbf{q}') = 1. \quad (3.22)$$

One then verifies, integrating equation (3.21) over **q**,

$$\int d^d q\, P_{n+1}(\mathbf{q}) = \int d^d q\, P_n(\mathbf{q}) = \int d^d q\, P_0(\mathbf{q}) = 1. \qquad (3.23)$$

More generally, an iteration of equation (3.21) yields

$$P_n(\mathbf{q}) = \int d^d q_0\, d^d q_1 \ldots d^d q_{n-1}\, \rho(\mathbf{q}, \mathbf{q}_{n-1}) \ldots \rho(\mathbf{q}_2, \mathbf{q}_1) \rho(\mathbf{q}_1, \mathbf{q}_0) P_0(\mathbf{q}_0). \qquad (3.24)$$

3.3.2 Translation invariance and locality

Translation symmetries. We have already assumed that the random walk transition probability is invariant under time translation by choosing ρ independent of n. We now assume, in addition, that the transition probability is invariant under space translation and, thus,

$$\rho(\mathbf{q}, \mathbf{q}') \equiv \rho(\mathbf{q} - \mathbf{q}'). \qquad (3.25)$$

As a consequence, the recursion relation now takes the the form of a convolution equation,

$$P_{n+1}(\mathbf{q}) = \int d^d q'\, \rho(\mathbf{q} - \mathbf{q}') P_n(\mathbf{q}'), \qquad (3.26)$$

a generalization of equation (3.7).

Locality. We consider only transition functions that generalize the distributions of Section 3.1, that is, piecewise differentiable and with bounded variation in each variable, and satisfying a property of exponential decay called *locality* in the random walk interpretation: qualitatively, large displacements have a very small probability. More precisely, we assume that the transition probabilities $\rho(\mathbf{q})$ satisfy a bound of type (3.1),

$$\rho(\mathbf{q}) \leq M\, e^{-A|\mathbf{q}|}, \quad M, A > 0. \qquad (3.27)$$

Moreover, we assume that the initial distribution $P_0(\mathbf{q})$ is also local, that is, satisfies the same exponential bound, at least for $|\mathbf{q}|$ large enough:

$$P_0(\mathbf{q}) \leq M\, e^{-A|\mathbf{q}|} \quad \text{for } |q| > R. \qquad (3.28)$$

As a limiting case, we admit certain initial distributions of the form

$$P_0(\mathbf{q}) = \delta^{(d)}(\mathbf{q} - \mathbf{q}_0),$$

where $\delta^{(d)}$ is Dirac's distribution in d dimensions.

Fourier representation. Equation (3.26) is a convolution equation that simplifies after Fourier transformation. We thus introduce

$$\tilde{P}_n(\mathbf{k}) = \int d^d q\, e^{-i\mathbf{k} \cdot \mathbf{q}}\, P_n(\mathbf{q}), \qquad (3.29)$$

which is also a generating function of the moments of the distribution $P_n(\mathbf{q})$.
From the reality of $P_n(\mathbf{q})$ and the condition (3.23), it follows that

$$\tilde{P}_n^*(\mathbf{k}) = \tilde{P}_n(-\mathbf{k}), \quad \tilde{P}_n(\mathbf{k}=0) = 1.$$

Similarly, we introduce

$$\tilde{\rho}(\mathbf{k}) = \int d^d q \, e^{-i\mathbf{k}\cdot\mathbf{q}} \rho(\mathbf{q}), \tag{3.30}$$

which is also a generating function of the moments of the distribution $\rho(\mathbf{q})$. With the notation

$$\langle f(\mathbf{q})\rangle \equiv \int d^d q \, f(\mathbf{q})\rho(\mathbf{q}), \tag{3.31}$$

the expansion in powers of $\mathbf{k}$ can be written as

$$\tilde{\rho}(\mathbf{k}) = \sum_{\ell=0}^{\infty} \frac{(-i)^\ell}{\ell!} \sum_{\mu_1,\mu_2,\ldots,\mu_\ell} k_{\mu_1} k_{\mu_2} \ldots k_{\mu_\ell} \langle q_{\mu_1} q_{\mu_2} \ldots q_{\mu_\ell}\rangle. \tag{3.32}$$

From the reality of $\rho(\mathbf{q})$ and the condition (3.22), it follows that

$$\tilde{\rho}^*(\mathbf{k}) = \tilde{\rho}(-\mathbf{k}), \quad \tilde{\rho}(0) = 1.$$

Taking the modulus of equation (3.30), one finds the bound

$$|\tilde{\rho}(\mathbf{k})| \le 1. \tag{3.33}$$

Moreover, the bound is reached only at the origin $\mathbf{k}=0$. Finally, the decay condition (3.27) implies that the function $\tilde{\rho}(\mathbf{k})$ is a function analytic in the strip $|\operatorname{Im}\mathbf{k}| < A$.

Eigenvectors and eigenvalues. Note that the Fourier representation (3.30) can be written as

$$\int d^d q' \, \rho(\mathbf{q}-\mathbf{q}') e^{-i\mathbf{k}\cdot\mathbf{q}'} = \tilde{\rho}(\mathbf{k}) e^{-i\mathbf{k}\cdot\mathbf{q}} \tag{3.34}$$

and, thus, the functions $e^{-i\mathbf{k}\cdot\mathbf{q}}$ and $\tilde{\rho}(\mathbf{k})$ can be considered as the eigenvectors and eigenvalues, respectively, of ρ considered as an integral operator. The operator has a continuous spectrum and the condition (3.33) is in agreement with the more general analysis of Section 3.4.1, but for somewhat different reasons: the moduli of the eigenvectors are not integrable. In particular, the eigenvector corresponding to the largest value $\tilde{\rho}(\mathbf{k}=0) = 1$ is $P(\mathbf{q}) = 1$ whose integral over $\mathbf{q}$ does not exist. With the hypothesis of translation invariance, the distribution $P_n(\mathbf{q})$ does not converge for $n \to \infty$ toward a limiting distribution. Like in the example of the central limit theorem, a renormalization of $\mathbf{q}$ is required.

3.3.3 Generating function of cumulants

It is also convenient to introduce the function

$$w(\mathbf{k}) = \ln \tilde{\rho}(\mathbf{k}) \Rightarrow w^*(\mathbf{k}) = w(-\mathbf{k}), \quad w(\mathbf{0}) = 0, \tag{3.35}$$

a generating function of the cumulants of $\rho(\mathbf{q})$. The regularity of $\tilde{\rho}$ and the condition $\tilde{\rho}(\mathbf{0}) = 1$ imply that $w(\mathbf{k})$ has a regular expansion at $\mathbf{k} = 0$. Setting

$$w(\mathbf{k}) = -i\mathbf{w}^{(1)} \cdot \mathbf{k} - \frac{1}{2} \sum_{\mu,\nu=1}^{d} w^{(2)}_{\mu\nu} k_\mu k_\nu + O(|k|^3)$$

and identifying the coefficients of the expansion of w with the moments of the distribution $\rho(\mathbf{q})$, one obtains the relation (2.34):

$$w^{(2)}_{\mu\nu} = \langle (q_\mu - \langle q_\mu \rangle)(q_\nu - \langle q_\nu \rangle) \rangle.$$

The matrix $\mathbf{w}^{(2)}$ is strictly positive, except for a certain distribution that we exclude. Indeed, for all vectors $|\mathbf{K}| = 1$,

$$\sum_{\mu,\nu} K_\mu w^{(2)}_{\mu\nu} K_\nu = \left\langle \left[\sum_\mu K_\mu (q_\mu - \langle q_\mu \rangle)\right]^2 \right\rangle > 0,$$

as expectation value of a strictly positive quantity.

The decrease of P_0 and $\tilde{P}_0(\mathbf{0}) = 1$ imply that $\ln \tilde{P}_0(\mathbf{k})$ is analytic at $\mathbf{k} = 0$. We also set

$$\ln \tilde{P}_0(\mathbf{k}) = -i\mathbf{q}_0 \cdot \mathbf{k} + O(|k|^2),$$

where $\mathbf{q}_0$ is the mean initial position.

3.3.4 Random walk: Asymptotic behaviour

In terms of Fourier components, equation (3.26) takes the form

$$\tilde{P}_{n+1}(\mathbf{k}) = \tilde{\rho}(\mathbf{k})\tilde{P}_n(\mathbf{k}) \Rightarrow \tilde{P}_n(\mathbf{k}) = \tilde{\rho}^n(\mathbf{k})\tilde{P}_0(\mathbf{k}).$$

It follows that

$$P_n(\mathbf{q}) = \frac{1}{(2\pi)^d} \int d^d k \, e^{i\mathbf{k} \cdot \mathbf{q}} \, \tilde{\rho}^n(\mathbf{k}) \tilde{P}_0(\mathbf{k}). \tag{3.36}$$

We now study the asymptotic behaviour of the distribution $P_n(\mathbf{q})$ for $n \to \infty$.

With the hypotheses satisfied by P_0 and ρ, the asymptotic behaviour can be derived from arguments very similar to those leading to the central limit theorem. For $n \to \infty$, the integral (3.36) is dominated by a vicinity of order $1/\sqrt{n}$ of $\mathbf{k} = 0$ and, thus,

$$P_n(\mathbf{q}) \sim \frac{1}{(2\pi)^d} \int d^d k \, \exp\left[i\mathbf{q} \cdot \mathbf{k} - i\mathbf{q}_0 \cdot \mathbf{k} - in\mathbf{w}^{(1)} \cdot \mathbf{k} - \frac{n}{2} \sum_{\mu,\nu=1}^{d} w^{(2)}_{\mu\nu} k_\mu k_\nu \right]$$

$$= \frac{1}{(2\pi n)^{d/2}\sqrt{\det \mathbf{w}^{(2)}}} \exp\left[-\frac{1}{2n} \sum_{\mu,\nu} Q_\mu [\mathbf{w}^{(2)}]^{-1}_{\mu\nu} Q_\nu \right]$$

with
$$Q = q - q_0 - nw^{(1)}.$$

The distribution has no limit since it goes to zero everywhere, but only an asymptotic behaviour.

Again, the neglected terms are of two types, multiplicative corrections of order $1/\sqrt{n}$ and additive corrections decreasing exponentially with n.

The probability of finding the walker is concentrated around the straight uniform trajectory
$$q_n = q_0 + nw^{(1)},$$

and $w^{(1)}$ thus has an interpretation as a mean velocity.

The random variable that characterizes the deviation with respect to the mean trajectory,
$$X = Q\sqrt{n},$$

has a universal limiting Gaussian distribution:

$$L_n(X) \underset{n\to\infty}{\sim} \frac{1}{(2\pi)^{d/2}} \frac{1}{\sqrt{\det w^{(2)}}} \exp\left[-\frac{1}{2}\sum_{\mu,\nu} X_\mu [w^{(2)}]^{-1}_{\mu\nu} X_\nu\right]. \qquad (3.37)$$

Only the interpretation differs here slightly from the central limit theorem, because n is a time variable. The result implies that the mean deviation from the mean trajectory increases as the square root of time. This is a characteristic property of *Brownian motion*.

3.3.5 Continuum time limit

To simplify the discussion, we now assume $\rho(q) = \rho(-q)$ and thus $w_1 = 0$, and we perform an affine change of coordinates such that

$$\sum_{\mu,\nu} Q_\mu [w^{(2)}]^{-1}_{\mu\nu} Q_\nu \mapsto \frac{1}{w_2} q^2,$$

where w_2 is a positive number.

The limiting Gaussian distribution becomes

$$P_n(q) \sim \frac{1}{(2\pi n w_2)^{d/2}} e^{-q^2/2nw_2}.$$

By changing the time-scale and by a continuous interpolation, one can define a diffusion process or Brownian motion in continuous time.

Let t and ε be two real positive numbers and n an integer such that

$$n = [t/\varepsilon], \qquad (3.38)$$

where $[\bullet]$ denotes an integer part. One then takes the limit $\varepsilon \to 0$ at t fixed and thus $n \to \infty$.

If the time t is measured with a finite precision Δt, as soon as $\Delta t \gg \varepsilon$, time can be considered as a continuous variable for what concerns all continuous functions of time.

One also performs the change of distance-scale

$$\mathbf{q} = \mathbf{x}/\sqrt{\varepsilon}. \tag{3.39}$$

Since the Gaussian function is continuous, the limiting distribution takes the form

$$\varepsilon^{-d/2} P_n(\mathbf{q}) \underset{\varepsilon \to 0}{\sim} \Pi(t, \mathbf{x}) = \frac{1}{(2\pi t w_2)^{d/2}} e^{-\mathbf{x}^2/2tw_2}. \tag{3.40}$$

(The change of variables $\mathbf{q} \mapsto \mathbf{x}$ implies a change of normalization of the distribution.) This distribution is a solution of the partial differential equation

$$\frac{\partial}{\partial t} \Pi(t, \mathbf{x}) = \tfrac{1}{2} w_2 \nabla_\mathbf{x}^2 \Pi(t, \mathbf{x}), \tag{3.41}$$

which has the form of a diffusion or heat equation. In the limit $n \to \infty$ and in adapted macroscopic variables, one thus obtains a random process that can entirely be described in continuous time.

The limiting distribution $\Pi(t, \mathbf{x})$ implies a *scaling property* characteristic of the Brownian motion. The moments of the distribution satisfy

$$\langle \mathbf{x}^{2m} \rangle = \int d^d x \, \mathbf{x}^{2m} \Pi(t, \mathbf{x}) \propto t^m. \tag{3.42}$$

The variable $\mathbf{x}/\sqrt{t}$ has time-independent moments. As the change (3.39) also indicates, one can thus assign to the vector $\mathbf{x}$ a *dimension* $1/2$ in unit of time (this also corresponds to assigning a Hausdorf dimension 2 to a Brownian trajectory).

3.3.6 Corrections to the continuum limit

It is simple to study how perturbations to the limiting Gaussian distribution decrease with ε.

We choose the boundary condition

$$P_0(\mathbf{q}) = \delta^{(d)}(\mathbf{q} - \mathbf{q}_0),$$

where $\delta^{(d)}(\mathbf{q})$ is Dirac's d-dimensional distribution. Then (equation (3.36))

$$P_n(\mathbf{q}) = \frac{1}{(2\pi)^d} \int d^d k \, e^{i\mathbf{k}\cdot\mathbf{q}} e^{nw(\mathbf{k})}.$$

We assume below that $\rho(\mathbf{q})$ is such that $\mathbf{w}_1 = 0$ and we have chosen a system of coordinates such that the expansion in powers of $\mathbf{k}$ of the regular function $w(\mathbf{k})$ defined in (3.35) can be written as

$$w(\mathbf{k}) = -\frac{1}{2} w_2 \mathbf{k}^2 + \sum_{r=3} \sum_{\mu_1,\dots,\mu_r} \frac{i^r}{r!} w^{(r)}_{\mu_1\dots\mu_r} k_{\mu_1} \dots k_{\mu_r}.$$

After the change of scales (3.38), (3.39), which for the Fourier variables corresponds to $\mathbf{k} = \kappa\sqrt{\varepsilon}$, one finds

$$nw(\mathbf{k}) = t\omega(\kappa),$$

$$\omega(\kappa) = -\frac{w_2}{2!}\kappa^2 + \sum_{r=3} \varepsilon^{r/2-1} \sum_{\mu_1,\ldots,\mu_r} \frac{i^r}{r!} w^{(r)}_{\mu_1\ldots\mu_r} \kappa_{\mu_1}\cdots\kappa_{\mu_r}.$$

One observes that, when $\varepsilon = t/n$ goes to zero, the contributions decrease faster when the powers of κ increase.

In the continuum limit, the distribution becomes

$$\Pi(t,\mathbf{x}) = \frac{1}{(2\pi)^d} \int d^d\kappa\, e^{-i\kappa\cdot\mathbf{x}} e^{t\omega(\kappa)}.$$

Differentiating with respect to time t, one verifies that $\Pi(t,\mathbf{x})$ satisfies the linear partial differential equation

$$\frac{\partial}{\partial t}\Pi(t,\mathbf{x}) = \left[\frac{w_2}{2!}\nabla_\mathbf{x}^2 - \sum_{r=3}\varepsilon^{r/2-1}\sum_{\mu_1,\ldots,\mu_r}\frac{i^r}{r!}w^{(r)}_{\mu_1\ldots\mu_r}\frac{\partial}{\partial x_{\mu_1}}\cdots\frac{\partial}{\partial x_{\mu_r}}\right]\Pi(t,\mathbf{x}),$$

where the property has been used that multiplication by κ of the Fourier transform corresponds to differentiation with respect to x of the function.

In the expansion, the contributions that contain more derivatives decrease faster to zero, since each additional derivative implies an additional factor $\sqrt{\varepsilon}$.

3.3.7 Random walk on a lattice

Very often a random walk is discussed on a lattice. We thus consider the d-dimensional lattice of points with integer coordinates $\mathbf{q} \equiv (q_1,\ldots,q_d)$.

We define $P_n(\mathbf{q})$ as the probability for a walker to be at point $\mathbf{q}$ at time n. The probability $P_n(\mathbf{q})$ satisfies a recursion relation of type (3.21) where integrals are replaced by sums:

$$P_{n+1}(\mathbf{q}) = \sum_{\mathbf{q}'\in\mathbb{Z}^d} \rho(\mathbf{q},\mathbf{q}')P_n(\mathbf{q}'). \tag{3.43}$$

Conservation of probabilities implies the conditions

$$\sum_{\mathbf{q}\in\mathbb{Z}^d} P_0(\mathbf{q}) = 1, \quad \sum_{\mathbf{q}\in\mathbb{Z}^d}\rho(\mathbf{q},\mathbf{q}') = 1 \Rightarrow \sum_{\mathbf{q}\in\mathbb{Z}^d} P_n(\mathbf{q}) = 1. \tag{3.44}$$

Moreover, we assume that the process is ergodic in the sense that there exists a non-zero probability to connect two arbitrary points on the lattice.

Translation invariance. Again, we restrict the analysis to transition probabilities invariant under space translations and, thus,

$$\rho(\mathbf{q},\mathbf{q}') \equiv \rho(\mathbf{q}-\mathbf{q}'). \tag{3.45}$$

Then, the recursion relation takes the form of a convolution equation,

$$P_{n+1}(\mathbf{q}) = \sum_{\mathbf{q}'\in\mathbb{Z}^d} \rho(\mathbf{q}-\mathbf{q}')P_n(\mathbf{q}'), \qquad (3.46)$$

a generalized form of equation (3.17).

Locality. We assume that the transition probability $\rho(\mathbf{q})$ has a locality property, that is, satisfies a bound of type (3.27):

$$\rho(\mathbf{q}) \le M\,\mathrm{e}^{-A|\mathbf{q}|}, \quad M, A > 0.$$

This includes the classical example where motion is restricted to nearest neighbours on the lattice.

Moreover, we assume that the initial distribution $P_0(\mathbf{q})$ is also local, that is, satisfies the same exponential bound:

$$P_0(\mathbf{q}) \le M\,\mathrm{e}^{-A|\mathbf{q}|}.$$

A special case is a non-random initial position $\mathbf{q} = \mathbf{q}_0$ and, thus,

$$P_0(\mathbf{q}) = \delta_{\mathbf{q}\mathbf{q}_0}.$$

Symmetry. For simplicity reasons, we assume that the transition probability $\rho(\mathbf{q})$ possesses the symmetries of the lattice corresponding to linear isometries, that is, is invariant under the cubic group, a finite group with $2^d d!$ elements generated by transpositions and one reflection

$$\begin{cases} \rho(q_1,\ldots,q_{\mu+1},q_\mu,\ldots,q_d) = \rho(q_1,\ldots,q_\mu,q_{\mu+1},\ldots,q_d) \quad \forall \mu, \\ \rho(-q_1,\ldots,q_d) = \rho(-q_1,\ldots,q_d). \end{cases} \qquad (3.47)$$

The cubic group is a subgroup of the orthogonal group $O(d)$ (rotations–reflections in $\mathbb{R}^d$). It admits as subgroups the symmetric or permutation group $q_\mu \mapsto q_{P(\mu)}$ generated by transpositions, as well as all reflections $q_\mu \mapsto -q_\mu$. For example, for the square lattice, its action on a point (q_1, q_2) generates the eight elements

$$(\epsilon_1 q_1, \epsilon_2 q_2) \quad \text{and} \quad (\epsilon_2 q_2, \epsilon_1 q_1),$$

with $\epsilon_1 = \pm 1$, $\epsilon_2 = \pm 1$.

All directions on the lattice thus are equivalent.

Theorem. With these assumptions, one proves that *in the asymptotic limit* $n \to \infty$, *the random variable* $\mathbf{X} = (\mathbf{q}-\mathbf{q}_0)/\sqrt{n}$, *where* $\mathbf{q}_0$ *depends on the initial distribution at time* $n = 0$, *again has a limiting Gaussian distribution given by*

$$L(\mathbf{X}) = \frac{1}{(2\pi w_2)^{d/2}}\,\mathrm{e}^{-\mathbf{X}^2/2w_2}, \qquad (3.48)$$

where w_2 depends on the transition probability. The proof of the result relies on a straightforward generalization of the method used in Section 3.3.4.

3.3.8 Fourier series

In the case of a transition probability invariant under space translations, the distribution $P_n(\mathbf{q})$ can be calculated by Fourier transformation. We set

$$\tilde{P}_n(\mathbf{k}) = \sum_{\mathbf{q}\in\mathbb{Z}^d} e^{-i\mathbf{k}\cdot\mathbf{q}} P_n(\mathbf{q}) \;\Rightarrow\; \tilde{P}_n^*(\mathbf{k}) = \tilde{P}_n(-\mathbf{k}),\quad \tilde{P}_n(\mathbf{k}=0) = 1, \qquad (3.49)$$

and

$$\tilde{\rho}(\mathbf{k}) = \sum_{\mathbf{q}\in\mathbb{Z}^d} e^{-i\mathbf{k}\cdot\mathbf{q}} \rho(\mathbf{q}) \;\Rightarrow\; \tilde{\rho}^*(\mathbf{k}) = \tilde{\rho}(-\mathbf{k}),\quad \tilde{\rho}(0) = 1, \qquad (3.50)$$

where the properties (3.44), (3.45) have been used. The functions $\tilde{P}_n$ and $\tilde{\rho}$ are periodic functions of the components k_μ of the vector $\mathbf{k}$ and these components can thus be restricted to $-\pi \leq k_\mu < \pi$ (a domain called in physics a *Brillouin zone*).

Taking the modulus of both sides of the defining equation, one finds the bound

$$|\tilde{\rho}(\mathbf{k})| \leq 1. \qquad (3.51)$$

If in the Brillouin zone the bound is reached for $\mathbf{k} \neq 0$, the asymptotic contributions from the other maxima yield oscillatory asymptotic contributions that vanish on average (see the discussion at the end of Section 3.1.4). We consider in what follows only the maximum at $\mathbf{k} = 0$.

The exponential bound (3.27) again implies that the function $\tilde{\rho}(\mathbf{k})$ is analytic in $|\operatorname{Im}\mathbf{k}| < A$. We set

$$\tilde{\rho}(\mathbf{k}) = e^{w(\mathbf{k})} \;\Rightarrow\; w^*(\mathbf{k}) = w(-\mathbf{k}),\quad w(0) = 0. \qquad (3.52)$$

The regularity of $\tilde{\rho}$ and the condition $\tilde{\rho}(0) = 1$ then imply that $w(\mathbf{k})$ has a convergent expansion at $\mathbf{k} = 0$.

One verifies that, because $\mathbf{k}\cdot\mathbf{q}$ has the form of a scalar product, the function $\tilde{\rho}(\mathbf{k})$ is invariant under the group of transformations (3.47) applied to the vector $\mathbf{k}$. The symmetry (3.47) implies that the function $\tilde{\rho}(\mathbf{k})$ is symmetric and even in all components of the vector $\mathbf{k}$ and, in particular, real. No term odd in $\mathbf{k}$ can be even in all vector components. Only one quadratic term is symmetric, the squared length of the vector. The function $w(\mathbf{k})$ admits a Taylor series expansion at $\mathbf{k} = 0$, which we parametrize as

$$w(\mathbf{k}) = -w_2 \mathbf{k}^2/2 + O(k^4),\quad w_2 > 0, \qquad (3.53)$$

the positivity of w_2 resulting from inequality (3.51).

3.3.9 Asymptotic behaviour. Continuum limit

After Fourier transformation, equation (3.46) leads to

$$\tilde{P}_{n+1}(\mathbf{k}) = \tilde{\rho}(\mathbf{k})\tilde{P}_n(\mathbf{k}) \Rightarrow \tilde{P}_n(\mathbf{k}) = \tilde{\rho}^n(\mathbf{k})\tilde{P}_0(\mathbf{k}).$$

Therefore,

$$P_n(\mathbf{q}) = \frac{1}{(2\pi)^d} \int d^d k \, e^{i\mathbf{k}\cdot\mathbf{q}} \, \tilde{\rho}^n(\mathbf{k})\tilde{P}_0(\mathbf{k}). \tag{3.54}$$

For $n \to \infty$, the integral (3.54) is dominated by the maximum values of $|\tilde{\rho}(\mathbf{k})|$ and, as explained in Section 3.1.4, only the vicinity of the maximum at $\mathbf{k} = \mathbf{0}$ contributes in the sense of measures. Because $w(\mathbf{k})$ has the form (3.53), for $\mathbf{k} \to \mathbf{0}$, only the values of $\mathbf{k}$ of order $1/\sqrt{n}$ contribute to the integral (3.54). We can thus neglect terms of order k^4 in $w(\mathbf{k})$.

Moreover, with the assumption (3.28) $\ln \tilde{P}_0(\mathbf{k})$, where $\tilde{P}_0(\mathbf{k})$ is the Fourier series associated to the initial distribution, is regular and we set

$$\tilde{P}_0(\mathbf{k}) = \exp\left[-i\mathbf{k}\cdot\mathbf{q}_0 + O(k^2)\right],$$

where $\mathbf{q}_0$ is the mean initial position. Terms of order k^2 or higher are negligible.

For the same reasons as in the steepest descent method, in the limit $n \to \infty$, in (3.54) one can integrate, in the Gaussian approximation, over all real values of $\mathbf{k}$ without restriction to a period (Brillouin zone). One obtains the asymptotic form

$$P_n(\mathbf{q}) \sim \frac{1}{(2\pi w_2 n)^{d/2}} \exp\left[-\frac{(\mathbf{q} - \mathbf{q}_0)^2}{2w_2 n}\right]. \tag{3.55}$$

The situation is now very similar to the one encountered in Section 3.3.2. For $n \to \infty$, the probability $P_n(\mathbf{q})$ takes a universal asymptotic Gaussian form that again depends only on general properties of the transition function $\rho(\mathbf{q} - \mathbf{q}')$.

Remarks.

(i) As in the case of the central limit theorem of probabilities discussed in Section 3.1.4, convergence toward the Gaussian distribution is only a convergence in the sense of measures.

(ii) The asymptotic distribution has a rotation symmetry (it depends only on the modulus of the vector $(\mathbf{q} - \mathbf{q}_0)$) and thus admits a symmetry group larger than the transition function $\rho(\mathbf{q})$, which has only a discrete lattice symmetry.

(iii) The asymptotic Gaussian form is valid at large times and for distances $|\mathbf{q}| \ll n$.

Continuum limit. We now rescale time and distance, setting

$$t = n\varepsilon, \quad \mathbf{x} = (\mathbf{q} - \mathbf{q}_0)\sqrt{\varepsilon}, \tag{3.56}$$

and take the limit $\varepsilon \to 0$ at t and $\mathbf{x}$ fixed and, thus, $n = O(1/\varepsilon)$ (more precisely n and the components of $\mathbf{q} - \mathbf{q}_0$ are the integer parts of the corresponding expressions).

Universality and the continuum limit

As we have already pointed out, if the *macroscopic* variables t and $\mathbf{x}$ are measured with a finite precision, from the viewpoint of any continuous function, they can be considered as continuous variables. The limiting distribution, defined for all real values by continuity, takes the form (3.40):

$$\Pi(t, \mathbf{x}) = \frac{1}{(2\pi w_2 t)^{d/2}} e^{-\mathbf{x}^2/2w_2 t}. \tag{3.57}$$

The time t and the coordinates $\mathbf{x}$ are the variables that describe, at the macroscopic scale, the Brownian motion generated by the microscopic dynamics (3.46) on the lattice. *Universality* has allowed defining a *continuum limit*, because the lattice structure and the details of the elementary process have disappeared.

Finally, the distribution (3.57) is a solution of equation (3.41), an equation of diffusion or heat type, isotropic in continuum space $\mathbb{R}^d$,

$$\frac{\partial}{\partial t}\Pi(t, \mathbf{x}) = \tfrac{1}{2} w_2 \nabla_{\mathbf{x}}^2 \Pi(t, \mathbf{x}).$$

Example. The properties derived in this section, can be verified in the example of the random walk with a motion limited to the nearest neighbours on the lattice. In this example, $\rho(\mathbf{q})$ vanishes except if $\mathbf{q} = \pm \mathbf{e}_\mu$, where $\mathbf{e}_\mu$ is the unit vector in the direction $\mu = 1, \ldots, d$. In the latter case $\rho(\mathbf{q})$ has the value $1/2d$. Then,

$$\tilde{\rho}(\mathbf{k}) = \frac{1}{2d}\sum_{\mu=1}^{d}\left(e^{-i\mathbf{k}\cdot\mathbf{e}_\mu} + e^{i\mathbf{k}\cdot\mathbf{e}_\mu}\right) = \frac{1}{d}\sum_{\mu=1}^{d} \cos k_\mu \;\Rightarrow\; w(\mathbf{k}) = -\mathbf{k}^2/2d + O(k^4),$$

where k_μ are the components of the vector $\mathbf{k}$. We assume a certain starting point $\mathbf{q} = 0$ and, thus,

$$P_0(\mathbf{q}) = \delta_{\mathbf{q},0} \;\Rightarrow\; \tilde{P}_0(\mathbf{k}) = 1.$$

Again, $\tilde{\rho}(\mathbf{k})$ has another maximum in the Brillouin zone, corresponding to $k_1 = k_2 = \cdots = k_d = \pi$. But this maximum does not contribute in the sense of the measure convergence. Therefore, the limiting distribution is

$$\Pi(t, \mathbf{x}) = \left(\frac{d}{2\pi t}\right)^{d/2} e^{-d\mathbf{x}^2/2t}.$$

3.3.10 Time-scale dilatation and fixed points

By generalizing the strategy used in Section 3.2 in the framework of the central limit theorem, one can directly justify the universal properties of the asymptotic behaviour of $P_n(\mathbf{q})$ for $n \to \infty$.

Here, the natural idea is to change the time-scale by a factor 2. More precisely, one starts from $n = 2^m$ and, at each iteration, one replaces ρ by $\mathcal{T}\rho$:

$$(\mathcal{T}\rho)(\mathbf{q} - \mathbf{q}') \equiv \sum_{\mathbf{q}'' \in \mathbb{Z}^d} \rho(\mathbf{q} - \mathbf{q}'')\rho(\mathbf{q}'' - \mathbf{q}').$$

This corresponds, in the Fourier representation, to the transformation

$$(\mathcal{T}w)(\mathbf{k}) \equiv 2w(\mathbf{k}).$$

Expanding in powers of $\mathbf{k}$, one verifies that such a transformation has, with our assumptions, only the trivial fixed point $w(\mathbf{k}) \equiv 0$. But the study of Section 3.2 has revealed that a larger class of fixed points becomes available if the transformation is combined with a renormalization of the distance-scale, $\mathbf{q} \mapsto \lambda \mathbf{q}$, with $\lambda > 0$. We thus consider the transformation

$$(\mathcal{T}_\lambda w)(\mathbf{k}) \equiv 2w(\mathbf{k}/\lambda).$$

Again, let us point out that at each iteration the periods of the functions $w(k)$ in all components k_μ are multiplied by λ, which from the viewpoint of the set of values of the coordinates $\mathbf{q}$ corresponds to a lattice whose spacing has been divided by λ.

The fixed points w_* of the transformation are solutions of

$$2w_*(\mathbf{k}/\lambda) = w_*(\mathbf{k}).$$

With the condition of cubic symmetry (equation (3.53))

$$w_*(\mathbf{k}) = -\tfrac{1}{2}w_2 \mathbf{k}^2 + O(|\mathbf{k}|^4).$$

For $|\mathbf{k}| \to 0$, the fixed point equation implies

$$2w_2/\lambda^2 = w_2,$$

which has a non-trivial solution only for $\lambda = \sqrt{2}$. Then, the coefficients of the terms of higher degree vanish. One finds the fixed points

$$w_*(\mathbf{k}) = -\tfrac{1}{2}w_2 \mathbf{k}^2,$$

which correspond to Gaussian transition probabilities.

Following the example of Section 3.2 in the case of distributions for variables with vanishing mean value, one can then study the stability of a fixed point and the decay of perturbations with time. The discussion is very similar, the only notable difference being that the eigenvectors are homogeneous even polynomials of d variables, invariant under the action of the cubic group. For example, beyond the marginal term associated with a translation of w_2, the most important irrelevant terms are the polynomials of degree four

$$\sum_\mu k_\mu^4, \quad (\mathbf{k}^2)^2.$$

Finally, since a scale transformation of time by a factor 2 corresponds to a scale transformation of distances by a factor $\sqrt{2}$, the *scaling property* (3.42) follows, $|x| \propto \sqrt{t}$, which has an interpretation as assigning the dimension $1/2$ in time units to position variables.

The two essential asymptotic properties of the random walk, convergence toward a Gaussian distribution and the scaling property, are thus reproduced by this RG-type analysis.

3.4 Random walk: Additional remarks

To complete this study, a few general remarks concerning stationary Markov chains, in the spirit of the Perron–Frobenius theorem, are useful. In this section, we assume that the stochastic process is ergodic and aperiodic, assumptions that are satisfied, for example, if the condition $\rho(\mathbf{q}, \mathbf{q}') > 0$ for all pairs $\mathbf{q}, \mathbf{q}'$ is fulfilled.

3.4.1 Asymptotic distribution

Asymptotically for $n \to \infty$, the behaviour of the $P_n(\mathbf{q})$ solution of equations (3.21) or (3.43) depends on the eigenvalues of largest modulus of $\rho(\mathbf{q}, \mathbf{q}')$ considered as an integral operator. We consider the lattice situation and, thus, equation (3.43), but the generalization to continuum space is simple.

The eigenvectors V and eigenvalues r of ρ satisfy the equation

$$\sum_{\mathbf{q}' \in \mathbb{R}^d} \rho(\mathbf{q}, \mathbf{q}') V(\mathbf{q}') = r V(\mathbf{q}). \tag{3.58}$$

Summing over $\mathbf{q}$ and using the condition (3.44), one infers

$$\sum_{\mathbf{q}' \in \mathbb{R}^d} V(\mathbf{q}') = r \sum_{\mathbf{q} \in \mathbb{R}^d} V(\mathbf{q}),$$

provided the sum converges. Moreover, if the sum does not vanish, the eigenvalue $r = 1$. Taking the absolute values, one obtains the inequality

$$\sum_{\mathbf{q}' \in \mathbb{R}^d} \rho(\mathbf{q}, \mathbf{q}') |V(\mathbf{q}')| \geq |r| |V(\mathbf{q})|$$

and, thus, summing over $\mathbf{q}$, if the sum converges,

$$\sum_{\mathbf{q} \in \mathbb{R}^d} |V(\mathbf{q})| \geq |r| \sum_{\mathbf{q} \in \mathbb{R}^d} |V(\mathbf{q})|,$$

which implies $|r| \leq 1$. If $|r| = 1$, $|V(\mathbf{q})|$ is a stationary distribution. Moreover, $V(\mathbf{q})$ must have a constant phase and can be chosen positive. Therefore, only the eigenvalue $r = 1$ has modulus 1 and the stationary solution of equation (3.43) is the asymptotic limit of P_n for $n \to \infty$, with an exponential convergence.

In the case of transition functions invariant under space translations, the condition of summability is not satisfied, which explains the difference in behaviour.

3.4.2 Detailed balance

To prepare for Chapter 4, we briefly present a class of random processes that satisfy a condition of detailed balance, although this topic is not directly useful here. This condition expressed in terms of the transition probability $\rho(\mathbf{q}, \mathbf{q}')$, takes the form of a relation between a process and its time reversal:

$$\rho(\mathbf{q}, \mathbf{q}') P_\infty(\mathbf{q}') = \rho(\mathbf{q}', \mathbf{q}) P_\infty(\mathbf{q}), \tag{3.59}$$

where P_∞ is a probability distribution:
$$P_\infty(\mathbf{q}) \geq 0, \quad \int d^d q\, P_\infty(\mathbf{q}) = 1.$$

Integrating equation (3.59) over $\mathbf{q}'$ and using the condition (3.22), one obtains
$$P_\infty(\mathbf{q}) = \int d^d q'\, \rho(\mathbf{q}, \mathbf{q}') P_\infty(\mathbf{q}').$$

Thus, the distribution $P_\infty(\mathbf{q})$ is the asymptotic distribution of the process (3.21) when $n \to \infty$.

If $P_\infty(\mathbf{q})$ is strictly positive, which, for example, is implied by the condition $\rho(\mathbf{q}, \mathbf{q}') > 0$, one can define the kernel
$$\mathcal{T}(\mathbf{q}, \mathbf{q}') = P_\infty^{-1/2}(\mathbf{q}) \rho(\mathbf{q}, \mathbf{q}') P_\infty^{1/2}(\mathbf{q}') = \mathcal{T}(\mathbf{q}', \mathbf{q}),$$

which corresponds to a real symmetric operator. With a few additional weak technical conditions, this operator has a discrete real spectrum and plays the role of the transfer matrix of Section 4.1.2, the time of the stochastic process becoming the space of the statistical lattice model.

In the translation-invariant example of Section 3.3.2, the condition of detailed balance is formally satisfied if $\rho(q) = \rho(-q)$, but then the function $P_\infty = 1$ is not normalizable. Nevertheless, the spectrum of ρ is still real.

3.5 Brownian motion and path integrals

If one is interested only in the asymptotic properties of the distribution, which have been shown to be independent of the initial transition probability, one can obtain them, in the continuum limit, starting directly from Gaussian transition probabilities of the form (assuming rotation symmetry)
$$\rho(\mathbf{q}) = \frac{1}{(2\pi w_2)^{d/2}} e^{-\mathbf{q}^2/2w_2}.$$

In the case of a certain initial position $\mathbf{q} = \mathbf{q}_0 = 0$, an iteration of the recursion relation (3.43) then leads to (equation (3.24))
$$P_n(\mathbf{q}) = \frac{1}{(2\pi w_2)^{nd/2}} \int d^d q_1\, d^d q_2 \ldots d^d q_{n-1}\, e^{-\mathcal{S}(\mathbf{q}_0, \mathbf{q}_2, \ldots, \mathbf{q}_n)} \tag{3.60}$$

with $\mathbf{q}_n = \mathbf{q}$ and
$$\mathcal{S}(\mathbf{q}_0, \mathbf{q}_2, \ldots, \mathbf{q}_n) = \sum_{\ell=1}^{n} \frac{(\mathbf{q}_\ell - \mathbf{q}_{\ell-1})^2}{2w_2}.$$

We then introduce macroscopic time variables,
$$\tau_\ell = \ell \varepsilon, \quad \tau_n = n\varepsilon = t,$$

and a continuous, piecewise linear path $\mathbf{x}(\tau)$ (Figure 3.1)

$$\mathbf{x}(\tau) = \sqrt{\varepsilon}\left[\mathbf{q}_{\ell-1} + \frac{\tau - \tau_{\ell-1}}{\tau_\ell - \tau_{\ell-1}}(\mathbf{q}_\ell - \mathbf{q}_{\ell-1})\right] \quad \text{for } \tau_{\ell-1} \leq \tau \leq \tau_\ell.$$

One then verifies that $\mathcal{S}$ can be written as (with the notation $\dot{\mathbf{x}}(\tau) \equiv d\mathbf{x}/d\tau$)

$$\mathcal{S}(\mathbf{x}(\tau)) = \frac{1}{2w_2}\int_0^t (\dot{\mathbf{x}}(\tau))^2 d\tau \tag{3.61}$$

with the boundary conditions

$$\mathbf{x}(0) = 0, \quad \mathbf{x}(t) = \sqrt{\varepsilon}\mathbf{q} = \mathbf{x}.$$

Moreover,

$$P_n(\mathbf{q}) = \frac{1}{(2\pi w_2)^{d/2}} \int \left(\prod_{\ell=1}^{n-1} \frac{d^d x(\tau_\ell)}{(2\pi w_2 \varepsilon)^{d/2}}\right) e^{-\mathcal{S}(\mathbf{x})}. \tag{3.62}$$

In the continuum limit $\varepsilon \to 0$, $n \to \infty$ with t fixed, expression (3.62) becomes a representation of the distribution of the continuum limit

$$\Pi(t, \mathbf{x}) \sim \varepsilon^{-d/2} P_n(\mathbf{q})$$

in the form of a *path integral*, which we denote symbolically

$$\Pi(t, \mathbf{x}) = \int [d\mathbf{x}(\tau)] e^{-\mathcal{S}(\mathbf{x}(\tau))}, \tag{3.63}$$

where $\int [d\mathbf{x}(\tau)]$ means sum over all continuous paths that start from the origin at time $\tau = 0$ and reach $\mathbf{x}$ at time t. The trajectories that contribute to the path integral correspond to a Brownian motion, a random walk in continuum time and space. This representation of the Brownian motion by path integrals, initially introduced by Wiener, is also called the Wiener integral.

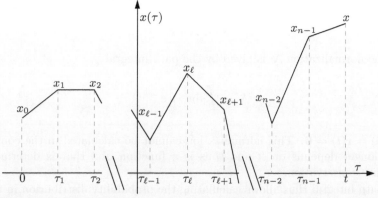

Fig. 3.1 Path contributing to the integral (3.60) ($d = 1$) with $x_\ell \equiv x(\tau_\ell)$.

Calculation of the path integral. We have introduced the formal concept of path integral. One could fear that its calculation requires returning always to its definition as a limit of integrals with discrete times. Fortunately, this is not the case. The path integral can often be calculated without referring to a limiting process, *except for a global normalization*. This is what we want now to show in this example.

To calculate the integral, we change variables,

$$\mathbf{x}(\tau) \mapsto \mathbf{r}(\tau) = \mathbf{x}(\tau) - \mathbf{f}(\tau),$$

where $\mathbf{f}(\tau)$ is the solution of the variational equation

$$\delta \mathcal{S} = 0 \quad \Leftrightarrow \quad \ddot{\mathbf{x}}(\tau) = 0$$

satisfying the boundary conditions $\mathbf{x}(0) = 0$, $\mathbf{x}(t) = \mathbf{x}$.

Here τ must be regarded as an index taking continuous values and $\mathbf{f}(\tau)$ as a set of constants parametrized by the index τ. For each time τ, the change of variables is a translation and, thus, the Jacobian is 1.

The path $\mathbf{f}(\tau)$ corresponds to a uniform rectilinear motion that joins the origin to the point $\mathbf{x}$, in the time t:

$$\mathbf{f}(\tau) = \mathbf{x}\tau/t.$$

The path $\mathbf{r}(\tau)$ then satisfies the boundary conditions

$$\mathbf{r}(0) = \mathbf{r}(t) = \mathbf{0}.$$

The translation leads to

$$\mathcal{S} = \frac{1}{2w_2}\left[\frac{\mathbf{x}^2}{t} + 2\int_0^t d\tau\, \dot{\mathbf{r}}(\tau)\cdot \mathbf{x}/t + \int_0^t \left(\dot{\mathbf{r}}(\tau)\right)^2 d\tau\right].$$

The term linear in $\mathbf{r}$ can be integrated explicitly and vanishes due to the boundary conditions. One infers

$$\Pi(t, \mathbf{x}) = \mathcal{N}\mathrm{e}^{-\mathbf{x}^2/2w_2 t},$$

where the normalization $\mathcal{N}$ is given by the path integral

$$\mathcal{N} = \int [\mathrm{d}\mathbf{r}(\tau)]\, \mathrm{e}^{-\mathcal{S}(\mathbf{r})}$$

with $\mathbf{r}(0) = \mathbf{r}(t) = \mathbf{0}$. This normalization cannot be calculated in the continuum, but no longer depends on $\mathbf{x}$ and thus is a function of t that is determined by probability conservation.

The path integral thus allows calculating the probability distribution in the continuum limit, by continuum methods.

Important remark.

(i) The notation $\dot{\mathbf{x}}$ seems to imply that paths contributing to the path integral are differentiable. This is not the case. In the continuum limit, we know that

$$\left\langle [\mathbf{x}(\tau) - \mathbf{x}(\tau')]^2 \right\rangle = \int d x'(x - x')^2 \Pi(|\tau - \tau'|, \mathbf{x} - \mathbf{x}') = w_2 |\tau - \tau'|.$$

Therefore, typical paths are continuous since the left-hand side vanishes for $\tau \to \tau'$.

The expectation value of the derivative squared is obtained by dividing by $(\tau - \tau')^2$ and taking the limit $|\tau - \tau'| \to 0$. The right-hand side then diverges. We conclude that typical paths of the Brownian motion are continuous but not differentiable; they only satisfy a Hölder condition of order $1/2$:

$$|\mathbf{x}(\tau) - \mathbf{x}(\tau')| = O(|\tau - \tau'|^{1/2}).$$

Nevertheless, the notation $\dot{\mathbf{x}}$ is useful because the paths that yield the leading contributions to the path integral are in the vicinity of differentiable paths.

(ii) As expression (3.62) shows, in the symbol $[d x(\tau)]$ is hidden a normalization that is independent of the trajectory, but that is difficult to handle in the continuum limit. Therefore, one calculates, in general, the ratio of the path integral and a reference path integral whose value is already known.

Exercises

Exercise 3.1

One considers a Markovian random walk on a two-dimensional square lattice. At each time step, the walker either remains motionless with probability $1 - s$, or moves by one lattice spacing in one of the four possible directions with the same probability $s/4$, where $0 < s < 1$. At initial time $n = 0$ the walker is at the point $\mathbf{q} = 0$.

Determine the asymptotic distribution of the walker position after n steps when $n \to \infty$.

Solution. The general study has shown that the asymptotic distribution is simply related to the Fourier series associated with the transition probability

$$\tilde{\rho}(k) = 1 - s + \tfrac{1}{2} s \cos k_1 + \tfrac{1}{2} s \cos k_2.$$

Then,

$$w(k) = \ln \tilde{\rho}(k) = -\tfrac{1}{4} s \left(k_1^2 + k_2^2 \right) + O(k^4).$$

One infers the asymptotic Gaussian distribution ($q \equiv |\mathbf{q}|$)

$$R_n(q) \sim \frac{1}{\pi n s} e^{-q^2/ns}.$$

Exercise 3.2

One considers a Markovian random walk on a cubic lattice, that is, in $\mathbb{Z}^3$. At each step the walker either remains motionless with probability $1 - s$, or moves by one lattice spacing in one of the six possible directions with the same probability $s/6$, where $0 < s < 1$. At initial time $n = 0$ the walker is at the point $\mathbf{q} = 0$.

Determine the asymptotic distribution of the position of the walker after n steps when $n \to \infty$.

Solution. The Fourier series associated with the transition probability is now
$$\tilde{\rho}(k) = 1 - s + \tfrac{1}{3}s \cos k_1 + \tfrac{1}{3}s \cos k_2 + \tfrac{1}{3}s \cos k_3.$$
Then,
$$w(k) = \ln \tilde{\rho}(k) = -\tfrac{1}{6}s\left(k_1^2 + k_2^2 + k_3^2\right) + O(k^4).$$
One infers the asymptotic Gaussian distribution
$$R_n(q) \sim \left(\frac{3}{2\pi ns}\right)^{3/2} e^{-3q^2/2ns}.$$

Exercise 3.3

Study the *local* stability of the Gaussian fixed point, corresponding to the probability distribution
$$\rho_G(q) = \frac{1}{\sqrt{2\pi}} e^{-q^2/2}$$
by the method of Section 3.2, but applied directly to the equation
$$[\mathcal{T}_\lambda \rho](q) = \lambda \int dq'\, \rho(q')\rho(\lambda q - q'). \tag{3.64}$$
First determine the value of the renormalization factor λ for which the Gaussian probability distribution ρ_G is a fixed point of $\mathcal{T}_\lambda$.

Setting $\rho = \rho_G + \delta\rho$, expand equation (3.64) to first order in $\delta\rho$. Show that the eigenvectors of the linear operator acting on $\delta\rho$ have the form
$$\delta\rho_p(q) = \left(\frac{d}{dq}\right)^p \rho_G(q), \quad p > 0.$$
Infer the associated eigenvalues.

Solution: A few indications. One recovers $\lambda = \sqrt{2}$. One linearizes the equation. One notes that probability conservation implies
$$1 = \int dq\, \rho(q) = \int dq\, [\rho_G + \delta\rho(q)] = 1 + \int dq\, \delta\rho(q) \Rightarrow \int dq\, \delta\rho(q) = 0.$$
Setting
$$[\mathcal{T}_\lambda(\rho_G + \delta\rho)] = \rho_G + \mathcal{L}\delta\rho + O\left(\|\delta\rho\|^2\right),$$
where the action of the linear operator $\mathcal{L}$ on a function $\delta\rho$ is given by
$$[\mathcal{L}\delta\rho](q) = 2\lambda \int dq'\, \rho_G(q')\delta\rho(\lambda q - q'),$$
one verifies that the eigenvectors of $\mathcal{L}$ have the proposed form by integrating several times by parts.

Exercise 3.4

Random walk on a circle. To exhibit the somewhat different asymptotic properties of a random walk on compact manifolds, it is proposed to study random walk on a circle. One still assumes translation invariance. The random walk is then specified by a transition function $\rho(q - q')$, where q and q' are two angles corresponding to positions on the circle. Moreover, the function $\rho(q)$ is assumed to be periodic and continuous. Determine the asymptotic distribution of the walker position. At initial time $n = 0$, the walker is at the point $q = 0$.

Solution. Due to translation invariance, the evolution equation is still a convolution, which simplifies in the Fourier representation. But, since the function $\rho(q)$ is periodic and continuous, it admits a Fourier series expansion of the form

$$\rho(q) = \sum_{\ell \in \mathbb{Z}} e^{iq\ell} \tilde{\rho}_\ell$$

with

$$\tilde{\rho}_\ell = \frac{1}{2\pi} \int_{-\pi}^{+\pi} dq \, e^{-iq\ell} \rho(q)$$

and, thus,

$$\tilde{\rho}_0 = 1/2\pi, \quad |\tilde{\rho}_\ell| < 1/2\pi \quad \text{for} \quad \ell \neq 0.$$

Thus, at time n the distribution of the walker position can be written as

$$P_n(q) = \sum_{\ell \in \mathbb{Z}} e^{iq\ell} \tilde{\rho}_\ell^n.$$

For $n \to \infty$, the sum converges exponentially toward the contribution $\ell = 0$ and, thus,

$$P_n(q) \underset{n \to \infty}{=} \frac{1}{2\pi},$$

which is a uniform distribution on the circle.

The maximum value of $|\tilde{\rho}_\ell|$ for $\ell \neq 0$, which yields the leading correction, defines a time

$$\tau = \max_{\ell \neq 0} -\frac{1}{\ln |\tilde{\rho}_\ell|},$$

which characterizes the exponential decay of corrections, called the *relaxation time*.

Exercise 3.5

One considers a Markovian process in continuum space with, as transition probability, the Gaussian function

$$\rho(q, q') = \frac{1}{\sqrt{2\pi}} \exp\left[-\tfrac{1}{2}(q - \lambda q')^2\right],$$

where λ is a real parameter with $|\lambda| < 1$.

Show that $\rho(q,q')$ satisfies the condition of detailed balance (3.59). Infer the asymptotic distribution at time n when $n \to \infty$. Associate to $\rho(q,q')$ a real symmetric operator as explained in Section 3.4.2. To determine the eigenvalues of the operator ρ of modulus smaller 1, one may then use the results of Section 4.6.1.

Solution. The ratio $\rho(q,q')/\rho(q',q)$ can be factorized in a ratio of two functions of q and q' that correspond to normalizable distributions. One infers (equation (3.59)) the asymptotic distribution

$$P_\infty(q) = \sqrt{(1-\lambda^2)/2\pi}\, e^{-(1-\lambda^2)q^2/2}.$$

One then introduces the real symmetric operator that has the same spectrum as ρ (see Section 3.4.2):

$$\mathcal{T}(q,q') = \frac{1}{\sqrt{2\pi}} \exp\left[-\tfrac{1}{4}(1+\lambda^2)(q^2+q'^2) + \lambda q q'\right].$$

For $\lambda > 0$, one can set $\lambda = e^{-\theta}$, $\theta > 0$ and, after a linear transformation on q, q' one recovers expression (4.33) and, thus, the eigenvalues are

$$\tau_k = e^{-k\theta}, \quad k \geq 0.$$

The case $\lambda < 0$ can be studied, for example, by setting $\lambda = -e^{-\theta}$ and the spectrum is

$$\tau_k = (-1)^k e^{-k\theta}, \quad k \geq 0.$$

More directly, one can study the action of ρ on the functions

$$\psi_k(q) = \frac{d^k}{(dq)^k} P_\infty(q).$$

The result is then obtained by successive integrations by parts.

4 Classical statistical physics: One dimension

In this chapter, within the framework of classical statistical mechanics, we discuss a family of models defined on one-dimensional lattices. Although from the physics viewpoint, some of these models are somewhat artificial, they have a practical utility as simplified versions of models in higher space dimensions and, thus, bring us a little closer to the physics that we want to investigate. In addition, their study will enable us to introduce a number of definitions and concepts needed in the following chapters.

These models are defined in the following way. To each site of the lattice, one associates one or several real random variables. The probability distribution of these random variables are then specified by a Boltzmann weight. The lattice structure is reflected in the form of the coupling between different lattice sites.

The quantities of physical interest are related to the partition function and, more generally, include correlation functions, which are the expectation values of products of random variables in different sites.

Again, the concept of *locality*, which we will define more precisely in the following chapters, plays an essential role: the direct couplings between different sites induced by the Boltzmann weight must decrease sufficiently fast with distance.

In this chapter, we study the simplest local examples: models that involve only interactions between nearest neighbours on the lattice. For such models, correlation functions can be calculated by a transfer matrix formalism.

Thus, we first describe some general properties of transfer matrices in one-dimensional models. We use this formalism to establish various properties of correlation functions, like the thermodynamic or infinite volume limit, the large-distance behaviour of the two-point correlation function, and introduce the very important concept of *correlation length*.

Connected correlation functions, cumulants of the distribution, play a particularly important role. Indeed, these functions satisfy the *cluster property*, which characterizes their decay at large distance.

We apply the transfer matrix formalism to the example of a Gaussian Boltzmann weight, which we study in detail. We calculate the partition function and correlation functions explicitly. We observe that, in *the low-temperature limit, the correlation length diverges, which makes it possible to define a continuum limit*.

We show that the results of the continuum limit can be reproduced directly by solving a partial differential equation in which all traces of the initial lattice structure have disappeared.

Finally, we exhibit a slightly more general class of models which share the same properties: divergent correlation length and continuum limit.

4.1 Nearest-neighbour interactions. Transfer matrix

We consider the one-dimensional lattice of points with integer coordinates on the real line. To each point $k \in \mathbb{Z}$ of the lattice, we associate one or more real variables q_k (e. g., the deviation of a particle from its equilibrium position). In what follows, we restrict the discussion to one variable per site, but the generalization to several variables is simple.

We first define systems on a lattice of finite size n, $k \in [0, n]$, imposing for convenience periodic boundary conditions: $q_n = q_0$, which give to the lattice a circle structure and makes it possible to construct a sequence of translation-invariant models. But our goal is to study the infinite size $n \to \infty$ limit, also called in this context the *thermodynamic limit*.

Statistical models are then specified by a statistical weight (a probability distribution), which we parametrize in the particular form

$$\mathcal{P}_n(\mathbf{q}) = \mathrm{e}^{-\mathcal{S}_n(\mathbf{q})}/\mathcal{Z}_n, \quad \mathbf{q} \equiv \{q_k\} \in \mathbb{R}^n, \tag{4.1}$$

where $\mathcal{Z}_n$ is a normalization, called the *partition function*, which is determined by the condition

$$\int \left(\prod_{k=1}^{n} \mathrm{d}q_k \right) \mathcal{P}_n(\mathbf{q}) = 1$$

and, thus,

$$\mathcal{Z}_n = \int \left(\prod_{k=1}^{n} \mathrm{d}q_k \right) \mathrm{e}^{-\mathcal{S}_n(\mathbf{q})}. \tag{4.2}$$

One may wonder why one singles out such a normalization, which is not a physical quantity. In fact, the partition function is useful for the following reason: its logarithmic derivatives with respect to different physical parameters (like the temperature) are expectation values and, thus, are physical quantities.

4.1.1 Nearest-neighbour interactions

The function $\mathcal{S}_n(\mathbf{q})$ can often be written as the sum of one-site terms $w(Q)$, which specify the distribution of the variables q_k in the absence of interactions, and of an interaction energy $\mathcal{E}_n(\mathbf{q})$ that couples the sites:

$$\mathcal{S}_n(\mathbf{q}) = \sum_{k=1}^{n} w(q_k) + \frac{\mathcal{E}_n(\mathbf{q})}{T}, \tag{4.3}$$

the parameter T having a physical interpretation as the temperature.

For a translation-invariant system on a lattice, $\mathcal{E}_n(\mathbf{q})$ is invariant under the substitution $q_i \mapsto q_{i+1}$. In addition, the interaction is *local* if the coupling between two sites decreases sufficiently fast with the distance. Interactions of the form

$$\mathcal{E}_n(\mathbf{q}) = \sum_{k=1}^{n} E(q_k, q_{k+1}, \ldots, q_{k+m}), \quad m > 0,$$

with the convention $q_k \equiv q_{k'}$ if $k = k' \pmod n$, are called finite-range interactions and provide an example of local interactions. Besides, such a particular form of the interaction gives a meaning to the concept of lattice and distinguishes the position k from a simple index describing one among n variables.

The particular example that we study in detail here is the *nearest-neighbour interaction*, which corresponds to $m = 1$, where each variable q_k is directly coupled only to the variables attached to the nearest-neighbour sites on the lattice. In this case, the function $\mathcal{S}_n(\mathbf{q})$ takes the particular form

$$\mathcal{S}_n(\mathbf{q}) = \sum_{k=1}^{n} S(q_{k-1}, q_k) \tag{4.4}$$

with $S(q'', q') = S(q', q'')$.

Moreover, we assume $S(q, q')$ piecewise continuous and, to ensure the convergence of the integrals, we impose

$$S(q, q') \geq \mu(|q| + |q'|), \quad \mu > 0. \tag{4.5}$$

This restriction is rather strong, but convenient for the systems that we want to study.

Remark. With the assumption (4.4), the partition function is given by an expression very similar to the probability distribution (3.24), apart from the periodic boundary conditions.

4.1.2 Transfer matrix and partition function

To discuss the physical properties of classical statistical models on a lattice in the case of nearest-neighbour interactions, a tool proves particularly useful: the *transfer matrix*.

Let us point out that the transfer matrix formalism can be generalized to classical statistical systems on lattices with arbitrary space dimensions and finite-range interactions (with a finite number of coupled sites).

We introduce the operator $\mathbf{T}$ associated with the real symmetric kernel

$$[\mathbf{T}](q'', q') = e^{-S(q'', q')}. \tag{4.6}$$

The operator $\mathbf{T}$ is called the *transfer matrix* of the statistical model. This denomination comes from examples where the variable q takes only a finite number of values and thus $\mathbf{T}$ is a matrix.

In the kernel representation, the product of two operators $\mathbf{O}_1$ and $\mathbf{O}_2$ represented by the kernels $\mathcal{O}_1(q', q)$ and $\mathcal{O}_2(q', q)$ is given by

$$[\mathbf{O}_2 \mathbf{O}_1](q', q) = \int dq'' \, \mathcal{O}_2(q', q'') \mathcal{O}_1(q'', q). \tag{4.7}$$

The trace is defined by

$$\operatorname{tr} \mathbf{O} \equiv \int dq \, \mathcal{O}(q, q), \tag{4.8}$$

a linear expression that, indeed, satisfies the cyclic property $\operatorname{tr} \mathbf{O}_2 \mathbf{O}_1 = \operatorname{tr} \mathbf{O}_1 \mathbf{O}_2$, as one can verify using the product rule (4.7).

The bound (4.5) implies, as one can verify, that the traces of all powers of $\mathbf{T}$ then exist.

Partition function. With these definitions, the partition function (4.2) can be expressed in terms of the transfer matrix as ($q_0 = q_n$):

$$\mathcal{Z}_n = \int \prod_{k=1}^{n} dq_k [\mathbf{T}](q_{k-1}, q_k) = \operatorname{tr} \mathbf{T}^n. \tag{4.9}$$

4.1.3 Hilbert space and transfer matrix

In this chapter, the properties of the transfer matrix are studied by methods directly inspired by quantum mechanics. In particular, for what follows, it is convenient to introduce an auxiliary Hilbert space $\mathcal{H}$: the real vector space of square integrable functions of a real variable q. The scalar product of two functions f, g is defined by

$$(f, g) = \int_{\mathbb{R}} dq\, f(q) g(q).$$

Hilbert spaces (but in general complex) play a central role in the construction of quantum mechanics.

The action of an operator $\mathbf{O}$ on functions is defined by

$$[\mathbf{O}\psi](q) = \int dq' [\mathbf{O}](q, q') \psi(q'),$$

a definition consistent with the product rule (4.7).

From the existence of $\operatorname{tr} \mathbf{T}^4$, one infers that, if f belongs to $\mathcal{H}$,

$$\|\mathbf{T} f\|^2 \leq \|f\|^2 (\operatorname{tr} \mathbf{T}^4)^{1/2}$$

and, thus, $\mathbf{T} f$ is also square integrable. Moreover, in $\mathcal{H}$ the operator $\mathbf{T}$ is bounded.

The operator $\mathbf{T}$ is real symmetric. It is diagonalizable in $\mathcal{H}$, its spectrum is real and its eigenvectors real and orthogonal. Since the trace of $\mathbf{T}^2$ is finite, the spectrum is discrete and the eigenvalues have zero as the only accumulation point.

Below, we denote by $\psi_\nu(q)$ the real normalized eigenvectors of $\mathbf{T}$, associated to the (all real) eigenvalues $\tau_0 \geq |\tau_1| \geq |\tau_2| \cdots$:

$$\int dq' [\mathbf{T}](q, q') \psi_\nu(q') = \tau_\nu \psi_\nu(q), \quad \int dq\, \psi_\nu^2(q) = 1.$$

Finally, $[\mathbf{T}](q, q')$ can then be expressed as a linear combination of projectors on eigenvectors in the form

$$[\mathbf{T}](q, q') = \sum_{\nu=0}^{\infty} \tau_\nu \psi_\nu(q) \psi_\nu(q'). \tag{4.10}$$

The partition function can also be expressed in terms of eigenvalues (equation (4.9)) as

$$\mathcal{Z}_n = \operatorname{tr} \mathbf{T}^n = \sum_{\nu=0}^{\infty} \tau_\nu^n. \qquad (4.11)$$

Position operator. To study correlation functions, it is useful to also introduce the (unbounded) operator $\hat{\mathbf{Q}}$, an analogue of the position operator in quantum mechanics, which acts multiplicatively as

$$[\hat{\mathbf{Q}}\psi](q) = q\psi(q) \quad \Rightarrow \quad [\hat{\mathbf{Q}}\mathbf{T}](q, q') = q\,[\mathbf{T}](q, q'). \qquad (4.12)$$

The kernel $\delta(q - q')$, where $\delta(q)$ is Dirac's distribution, satisfies

$$\int \mathrm{d}q\, \delta(q'' - q)\mathcal{O}(q, q') = \mathcal{O}(q'', q')$$

and, thus, is the kernel of the identity operator. The kernel corresponding to $\hat{\mathbf{Q}}$ is then

$$[\hat{\mathbf{Q}}](q, q') = q\,\delta(q - q').$$

4.2 Correlation functions

Correlation functions: Definition. We now define the correlation functions of the variables q_k. In what follows, the notation $\langle \bullet \rangle_n$ means expectation value of $\bullet$ with respect to the measure $\mathrm{e}^{-S_n}/\mathcal{Z}_n$. With this notation, we define the p-point correlation function as the moment of the distribution ($q_0 = q_n$)

$$\langle q_{\ell_1} q_{\ell_2} \cdots q_{\ell_p} \rangle_n \equiv \mathcal{Z}_n^{-1} \int \left(\prod_{s=1}^n \mathrm{d}q_s \right) q_{\ell_1} q_{\ell_2} \cdots q_{\ell_p}\, \mathrm{e}^{-S_n(q)}. \qquad (4.13)$$

4.2.1 One-point function and transfer matrix

The one-point function, which is the expectation value of q_ℓ, is given by

$$\langle q_\ell \rangle_n = \mathcal{Z}_n^{-1} \int \left(\prod_{s=1}^n \mathrm{d}q_s \right) q_\ell\, \mathrm{e}^{-S_n(q)}.$$

The integral can be written as

$$\int \left(\prod_{s=1}^n \mathrm{d}q_s \right) q_\ell\, \mathrm{e}^{-S_n(q)} = \int \mathrm{d}q_\ell\, q_\ell \int \mathrm{d}q_n \int \prod_{s=1}^{\ell-1} \mathrm{d}q_s \prod_{s=1}^{\ell} [\mathbf{T}](q_{s-1}, q_s)$$

$$\times \int \prod_{s=\ell+1}^{n-1} \mathrm{d}q_s \prod_{s=\ell+1}^{n} [\mathbf{T}](q_{s-1}, q_s).$$

Integrating over all variables except q_ℓ, one finds
$$\mathcal{Z}_n \langle q_\ell \rangle_n = \int dq_\ell \, q_\ell [\mathbf{T}^n](q_\ell, q_\ell).$$

Then, introducing the operator (4.12), one notices
$$[\hat{\mathbf{Q}}\,\mathbf{T}](q, q') = q[\mathbf{T}](q, q') \tag{4.14}$$

and, thus,
$$\mathcal{Z}_n \langle q_\ell \rangle_n = \int dq_\ell \, [\hat{\mathbf{Q}}\,\mathbf{T}^n](q_\ell, q_\ell).$$

The remaining integral yields the trace of the product of operators (equation (4.8)). The one-point function can thus be written as
$$\langle q_\ell \rangle_n = \operatorname{tr} \hat{\mathbf{Q}}\,\mathbf{T}^n / \operatorname{tr} \mathbf{T}^n. \tag{4.15}$$

Translation invariance. The expression (4.15) shows that the expectation value is independent of the point ℓ. This is a direct consequence of two properties: the transfer matrix is independent of the points on the lattice; we have chosen periodic boundary conditions that have given to the lattice the structure of a circle. This induces translation invariance on the lattice.

4.2.2 Multi-point correlation functions and the transfer matrix

The two-point correlation function, the expectation value of the product of the values of q in two points ℓ_1, ℓ_2 of the lattice, is defined by
$$\langle q_{\ell_1} q_{\ell_2} \rangle_n \equiv \mathcal{Z}_n^{-1} \int \left(\prod_{s=1}^{n} dq_s \right) q_{\ell_1} q_{\ell_2} \, e^{-\mathcal{S}_n(q)}.$$

The function $\mathcal{Z}_n^{(2)}(\ell_1, \ell_2)$ is symmetric in the exchange $\ell_1 \leftrightarrow \ell_2$. We assume, for example, $\ell_2 \geq \ell_1$. One can then write the expression as
$$\langle q_{\ell_1} q_{\ell_2} \rangle_n = \mathcal{Z}_n^{-1} \int q_{\ell_1} dq_{\ell_1} q_{\ell_2} dq_{\ell_2} \int dq_n \int \prod_{s=1}^{\ell_1 - 1} dq_s \prod_{s=1}^{\ell_1} [\mathbf{T}](q_{s-1}, q_s)$$
$$\times \int \prod_{s=\ell_1 + 1}^{\ell_2 - 1} dq_s \prod_{s=\ell_1 + 1}^{\ell_2} [\mathbf{T}](q_{s-1}, q_s) \int \prod_{s=\ell_2 + 1}^{n-1} dq_s \prod_{s=\ell_2 + 1}^{n} [\mathbf{T}](q_{s-1}, q_s).$$

Integrating over all variables except q_{ℓ_1} and q_{ℓ_2}, one finds
$$\langle q_{\ell_1} q_{\ell_2} \rangle_n = \mathcal{Z}_n^{-1} \int q_{\ell_1} dq_{\ell_1} q_{\ell_2} dq_{\ell_2} [\mathbf{T}]^{n - \ell_2 + \ell_1}(q_{\ell_2}, q_{\ell_1}) [\mathbf{T}]^{\ell_2 - \ell_1}(q_{\ell_1}, q_{\ell_2}).$$

Using then again the remark (4.14), one finds

$$\langle q_{\ell_1} q_{\ell_2}\rangle_n = \mathcal{Z}_n^{-1} \int dq_{\ell_1} dq_{\ell_2}\, [\hat{\mathbf{Q}}\,\mathbf{T}]^{n-\ell_2+\ell_1}(q_{\ell_2},q_{\ell_1})[\hat{\mathbf{Q}}\,\mathbf{T}]^{\ell_2-\ell_1}(q_{\ell_1},q_{\ell_2})$$
$$= \operatorname{tr} \hat{\mathbf{Q}}\,\mathbf{T}^{n-\ell_2+\ell_1} \hat{\mathbf{Q}}\,\mathbf{T}^{\ell_2-\ell_1} / \operatorname{tr} \mathbf{T}^n . \tag{4.16}$$

In particular, the two-point function depends only on the distance between the two points on the lattice, a property that, more directly, follows from translation invariance.

More generally, if one assumes the ordering $\ell_1 \le \ell_2 \le \dots \le \ell_p$,

$$\langle q_{\ell_1} q_{\ell_2} \dots q_{\ell_p}\rangle_n = \mathcal{Z}_n^{-1} \int \left(\prod_{s=1}^n dq_s\right) q_{\ell_1} q_{\ell_2}\dots q_{\ell_p}\, e^{-\mathcal{S}_n(q)}$$
$$= \operatorname{tr} \hat{\mathbf{Q}}\,\mathbf{T}^{n-\ell_p+\ell_1} \hat{\mathbf{Q}}\,\mathbf{T}^{\ell_2-\ell_1} \hat{\mathbf{Q}}\,\mathbf{T}^{\ell_3-\ell_2}\dots \hat{\mathbf{Q}}\,\mathbf{T}^{\ell_p-\ell_{p-1}} / \operatorname{tr} \mathbf{T}^n. \tag{4.17}$$

Generating function. As we have explained in Chapter 2, it is useful to introduce the generating function of correlation functions

$$\mathcal{Z}_n(\mathbf{b}) = \sum_{p=0}^{\infty} \frac{1}{p!} \sum_{\ell_1,\ell_2,\dots,\ell_p} b_{\ell_1} b_{\ell_2}\dots b_{\ell_p} \langle q_{\ell_1} q_{\ell_2}\dots q_{\ell_p}\rangle_n = \left\langle \exp\sum_\ell b_\ell q_\ell\right\rangle_n . \tag{4.18}$$

It is proportional to the partition function of a system where a term linear in q, different on each site, has been added to the configuration energy.

4.3 Thermodynamic limit

We now evaluate the partition function and correlation functions in the infinite volume limit, called in this context the thermodynamic limit. In one space dimension, the volume reduces to the size n of the lattice.

4.3.1 Partition function

The thermodynamic limit involves directly the spectrum and the eigenvectors of the operator $\mathbf{T}$ in the space $\mathcal{H}$ of square integrable functions.

Leading eigenvalue of the transfer matrix. For the one-dimensional systems that we study here, one shows that τ_0, the eigenvalue of $\mathbf{T}$ with largest modulus, is *positive and simple*, and the unique eigenvector $\psi_0(q)$ can be chosen positive (see Appendix A3.1).

Thermodynamic limit. For $n \to \infty$, the partition function, expressed in terms of eigenvalues of the transfer matrix (equation (4.11)), is dominated by the eigenvalue with largest modulus τ_0 and thus

$$\mathcal{Z}_n \underset{n\to\infty}{\sim} \tau_0^n .$$

The quantity
$$\mathcal{W}_n = T \ln \mathcal{Z}_n, \qquad (4.19)$$
where, in the context of statistical physics, T is the temperature and $\mathcal{W}_n$ the *free energy*, is thus asymptotically proportional to the length n and $\mathcal{W}_n/n$, the free energy density, has the finite limit
$$\lim_{n \to \infty} \frac{1}{n} \mathcal{W}_n = T \ln \tau_0. \qquad (4.20)$$

In several space dimensions, for the class of systems in statistical physics that satisfy the cluster property (see Section 4.4), this property has a generalization: in the thermodynamic limit, the free energy is proportional to the volume. More directly, this property is a consequence of the short range of interactions.

In what follows, we often omit the proportionality factor T, necessary for the physical interpretation but which, generally, plays no role for the questions we want to investigate.

4.3.2 One-point function

We now use expression (4.15) to evaluate the expectation value of q. In the thermodynamic limit $n \to \infty$, the operator $\mathbf{T}^n$ is dominated by its largest eigenvalues. Then, from expansion (4.10), one infers
$$[\mathbf{T}^n](q, q') \underset{n \to \infty}{=} \tau_0^n \psi_0(q) \psi_0(q') + O(\tau_1^n)$$
and, thus,
$$\operatorname{tr} \hat{\mathbf{Q}} \mathbf{T}^n \underset{n \to \infty}{\sim} \tau_0^n \int dq \, q \, \psi_0^2(q).$$
The expectation value of q_ℓ has the finite limit
$$\langle q_\ell \rangle = \lim_{n \to \infty} \operatorname{tr} \hat{\mathbf{Q}} \mathbf{T}^n / \operatorname{tr} \mathbf{T}^n = \int dq \, q \, \psi_0^2(q).$$

Spontaneous symmetry breaking. In general, the expectation value $\langle q_\ell \rangle$ plays no special role since it can be cancelled by the simple translation
$$q_\ell \mapsto q'_\ell = q_\ell - \langle q_\ell \rangle.$$
It is only when one value of q is privileged that the situation is different, for example, if a system is invariant under the reflection $q \mapsto -q$. This situation occurs when the interaction satisfies
$$S(q'', q') = S(-q'', -q') \Rightarrow [\mathbf{T}](q'', q') = [\mathbf{T}](-q'', -q').$$
A non-vanishing expectation value then indicates a *spontaneous symmetry breaking* (see Chapter 7). In one dimension with short-range interactions, such a spontaneous

symmetry breaking is impossible. The argument is the following: the expectation value
$$\langle q_\ell \rangle = \int dq \, q \, \psi_0^2(q)$$
can only be different from zero if the function $\psi_0^2(q)$ is not symmetric, that is, not even. The eigenvalue τ_0 then is degenerate (here double) because the symmetric state $\psi_0(-q)$ is an independent eigenvector:
$$\tau_0 = \int dq\, dq' \, \psi_0(q')[\mathbf{T}](q',q)\psi_0(q) = \int dq\, dq' \, \psi_0(-q')[\mathbf{T}](q',q)\psi_0(-q).$$

However, with the hypotheses of this chapter, one can prove that in one dimension the eigenvalue τ_0 cannot be degenerate (cf. A3.1).

Since reflection symmetry cannot be spontaneously broken, a phase transition with order is impossible.

4.3.3 Two-point function and correlation length

Thermodynamic limit. For the two-point function expressed by equation (4.16), the same argument implies
$$\langle q_{\ell_1} q_{\ell_2} \rangle = \frac{1}{\tau_0^{|\ell_2-\ell_1|}} \int dq\, dq' \, \psi_0(q)[\hat{\mathbf{Q}}\mathbf{T}^{|\ell_2-\ell_1|}\hat{\mathbf{Q}}](q,q')\psi_0(q'). \quad (4.21)$$

Large-distance behaviour. We now examine the behaviour of the two-point function at large distance $\ell = |\ell_1 - \ell_2| \to \infty$ (assuming for simplicity $|\tau_2| < |\tau_1|$). Using the expansion (4.10),
$$[\mathbf{T}^n](q,q') \underset{n\to\infty}{=} \tau_0^n \psi_0(q)\psi_0(q') + \tau_1^n \psi_1(q)\psi_1(q') + O(\tau_2^n),$$
one obtains
$$\langle q_\ell q_0 \rangle = \langle q \rangle^2 + A_1^2 (\tau_1/\tau_0)^\ell + O\left((\tau_2/\tau_0)^\ell\right)$$
with
$$A_1 = \int dq \, q \, \psi_1(q)\psi_0(q).$$

The leading term is a constant, the square of the expectation value of q. As noticed above, it can be eliminated by the shift of the variables
$$q_k \mapsto q'_k = q_k - \langle q \rangle$$
as well as
$$\hat{\mathbf{Q}}' = \hat{\mathbf{Q}} - \langle q \rangle .$$
The two-point function of q' then decreases exponentially for $\ell \to \infty$ (assuming $A_1 \neq 0$, otherwise one must take into account the next term):
$$\langle q'_\ell q'_0 \rangle = \langle (q_\ell - \langle q \rangle)(q_0 - \langle q \rangle) \rangle = A_1^2 (\tau_1/\tau_0)^\ell + O\left((\tau_2/\tau_0)^\ell\right).$$

In general, one characterizes this decay by the *correlation length*

$$\xi = \lim_{\ell \to \infty} -\frac{\ell}{\ln(|\langle q'_0 q'_\ell \rangle|)}. \qquad (4.22)$$

Here, one finds

$$\xi^{-1} = \ln(\tau_0/\tau_1). \qquad (4.23)$$

Continuum limit. Let us assume that the transfer matrix depends on a parameter in such a way that, for some value of the parameter, the correlation length diverges. Then, as we show explicitly in the Gaussian example, it is possible to define a continuum limit, and correlation functions in the continuum limit can be calculated by solving a partial differential equation or from a path integral (see Chapter 5).

4.4 Connected functions and cluster properties

In general, expression (4.17) shows that the p-point function has the thermodynamic limit (for $\ell_1 \leq \ell_2 \leq \cdots \leq \ell_p$)

$$\langle q_{\ell_1} q_{\ell_2} \cdots q_{\ell_p} \rangle = \frac{1}{\tau_0^{\ell_p - \ell_1}} \int dq \, dq' \, \psi_0(q) [\mathbf{O}](q, q') \psi_0(q') \qquad (4.24)$$

with

$$\mathbf{O} = \hat{\mathbf{Q}} \, \mathbf{T}^{\ell_2 - \ell_1} \hat{\mathbf{Q}} \, \mathbf{T}^{\ell_3 - \ell_2} \cdots \hat{\mathbf{Q}} \, \mathbf{T}^{\ell_p - \ell_{p-1}} \hat{\mathbf{Q}}.$$

The problem that we have solved in the case of the two-point function arises more generally: to identify the combination which decays at large distance and to characterize the decay of correlations when the correlation length is finite.

We have shown in Section 4.3.3 that it is the two-point function of the variable from which the expectation value has been subtracted,

$$\langle (q_{\ell_1} - \langle q_{\ell_1} \rangle)(q_{\ell_2} - \langle q_{\ell_2} \rangle) \rangle = \langle q_{\ell_1} q_{\ell_2} \rangle - \langle q_{\ell_1} \rangle \langle q_{\ell_2} \rangle,$$

that vanishes at large distance. This combination, which is the second cumulant of the distribution, is also called the *connected two-point function* and will be denoted by

$$\langle q_{\ell_1} q_{\ell_2} \rangle_\mathrm{c} \equiv \langle q_{\ell_1} q_{\ell_2} \rangle - \langle q_{\ell_1} \rangle \langle q_{\ell_2} \rangle.$$

The property of exponential decay generalizes to cumulants of higher degree. For example, the cumulant of degree three or three-point connected correlation function, is given by

$$\langle q_{\ell_1} q_{\ell_2} q_{\ell_3} \rangle = \langle q_{\ell_1} \rangle \langle q_{\ell_2} \rangle \langle q_{\ell_3} \rangle + \langle q_{\ell_1} q_{\ell_2} \rangle_\mathrm{c} \langle q_{\ell_3} \rangle + \langle q_{\ell_2} q_{\ell_3} \rangle_\mathrm{c} \langle q_{\ell_1} \rangle + \langle q_{\ell_3} q_{\ell_1} \rangle_\mathrm{c} \langle q_{\ell_2} \rangle$$
$$+ \langle q_{\ell_1} q_{\ell_2} q_{\ell_3} \rangle_\mathrm{c}.$$

Using the explicit expressions (4.24), one verifies that $\langle q_{\ell_1} q_{\ell_2} q_{\ell_3} \rangle_\mathrm{c}$ decreases exponentially when $|\ell_2 - \ell_1|$ or $|\ell_3 - \ell_2|$ go to infinity.

More generally, using the techniques of Section 2.5.2, one introduces the generating function of cumulants,

$$\mathcal{W}(\mathbf{b}) = \ln \mathcal{Z}(\mathbf{b}), \qquad (4.25)$$

where $\mathcal{Z}(\mathbf{b})$ is the thermodynamic limit of the generating function (4.18). By definition, it is the generating function of connected correlation functions:

$$\mathcal{W}(\mathbf{b}) = \sum_{p=0}^{\infty} \frac{1}{p!} \sum_{\ell_1, \ell_2, \ldots, \ell_p} b_{\ell_1} b_{\ell_2} \ldots b_{\ell_p} \langle q_{\ell_1} q_{\ell_2} \ldots q_{\ell_p} \rangle_{\mathrm{c}}.$$

Let us point out that it is also the free energy of a system where a term coupled linearly to q, different at each site, has been added to the configuration energy.

Cluster property. It is the connected correlation functions that vanish at large distance, a property called the *cluster property*.

More precisely, we denote by $\langle q_{\ell_1} q_{\ell_2} \ldots q_{\ell_p} \rangle_{\mathrm{c}}$ the connected p-point function. We then group the p points $\ell_1, \ell_2, \ldots, \ell_p$ into two non-empty, disjoint sets I and J. We call the distance between the two sets the quantity

$$D = \min_{\ell_i \in I, \ell_j \in J} |\ell_i - \ell_j|.$$

Then,

$$\lim_{D \to \infty} \langle q_{\ell_1} q_{\ell_2} \ldots q_{\ell_p} \rangle_{\mathrm{c}} = 0. \qquad (4.26)$$

More specifically, with the hypotheses of the chapter, one can show that connected correlation functions decay at least as $\mathrm{e}^{-D/\xi}$, where ξ is the correlation length (4.23).

This result is valid for statistical systems more general than those studied here. The proof starting from the explicit form (4.24) is analogous to the combinatorial argument in Section 2.4.2.

4.4.1 Expectation value and thermodynamic limit

It is instructive to study the behaviour of moments of the mean random variable

$$Q = \frac{1}{n} \sum_{\ell=1}^{n} q_\ell$$

in the thermodynamic limit $n \to \infty$.

First, from translation invariance,

$$\langle Q \rangle_n = \langle q \rangle_n.$$

The second moment is related to the two-point function and, thus, to the two-point connected function. Indeed,

$$\langle Q^2 \rangle_n = \frac{1}{n^2} \sum_{\ell_1, \ell_2} \langle q_{\ell_1} q_{\ell_2} \rangle_n = \frac{1}{n} \sum_{\ell=\ell_1-\ell_2} \langle q_0 q_\ell \rangle_n = \langle q \rangle_n^2 + \frac{1}{n} \sum_{\ell} (\langle q_0 q_\ell \rangle_n)_{\mathrm{c}}.$$

In the thermodynamic limit, the connected two-point function decays exponentially for $|\ell| \to \infty$ with a rate determined by the correlation length. If the correlation length is finite, the sum over ℓ converges. The second cumulant of the Q distribution then goes to zero:

$$\langle (Q - \langle Q \rangle_n)^2 \rangle_n \underset{n \to \infty}{\sim} \frac{1}{n} \sum_\ell \langle q_0 q_\ell \rangle_c \,.$$

More generally, we introduce the generating function

$$Z_n(b) = \langle e^{nbQ} \rangle_n$$

of the moments of the sum nQ. Comparing with expressions (4.1), (4.2) and (4.4), one verifies that $Z_n(b)$ can be expressed in terms of the partition function $\mathcal{Z}_n(b)$ associated to the function

$$S(q, q'; b) = S(q, q') - \tfrac{1}{2}b(q + q')$$

as

$$Z_n(b) = \mathcal{Z}_n(b)/\mathcal{Z}_n(0).$$

For $|b|$ small enough, the analysis of the thermodynamic limit still holds and, thus, asymptotically

$$\ln Z_n(b) \underset{n \to \infty}{\sim} nW(b),$$

where the function $W(b)$ is related to the generating function of connected correlation functions (4.25) by

$$W(b) = \mathcal{W}(b_\ell = b)/n \,.$$

Moreover, this equation relates the coefficients of the expansion in powers of b to sums of connected correlation functions. The cluster property implies, at least if the correlation length is finite, that all sums converge but one that corresponds to translation invariance and gives a factor n. The function $W(b)$ thus is expandable in powers of b at $b = 0$. One finds results quite analogous to the central limit theorem (see equation (3.8) with the correspondence $b = ik$). When the correlation length is finite, the moments of the mean of the variables q_ℓ, have the behaviour predicted by the central limit theorem, although the variables q_ℓ are not independent. This result indicates that a collective behaviour with strong correlations is only possible when the correlation length diverges.

4.5 Statistical models: Simple examples

Before discussing more thoroughly the Gaussian example, we illustrate the preceding analysis by a few other simple examples.

(i) *Infinite temperature or independent variables.* In expression (4.3), when the configuration energy $\mathcal{E}$ vanishes or the temperature T is infinite, variables at different sites become independent. The transfer matrix factorizes,

$$[\mathbf{T}](q, q') = e^{-\omega(q)/2} e^{-\omega(q')/2},$$

and becomes a projector on the vector $\mathrm{e}^{-w(q)/2}$. Thus, the eigenvalue corresponding to this vector is positive and all other eigenvalues vanish. Not surprisingly, the correlation length vanishes.

(ii) *Translation invariant systems.* The opposite limit corresponds to choosing

$$S(q,q') = E(q-q')/T,$$

where $E(q)$ is an even function.

In this example, the bound (4.5) is not satisfied and we replace it by

$$E(q) \geq \mu |q|.$$

We then recognize a version of the random walk studied in Section 3.3, in one dimension in continuum space, the transfer matrix playing the role of the transition probability. The transfer matrix is diagonalized by a Fourier transformation:

$$[\mathbf{T}](q,q') = \int \mathrm{d}k \, \mathrm{e}^{ik(q-q')} \tau(k)$$

with

$$\tau(k) = \frac{1}{2\pi} \int \mathrm{d}q \, \mathrm{e}^{-iqk - E(q)/T}.$$

Thus, the transfer matrix has a continuous spectrum with eigenvalues $\tau(k)$. The thermodynamic limit is dominated by the vicinity of $k=0$. The correlation length is infinite. One can then define a continuum limit that is directly related to the brownian motion of Section 3.3.1. In the statistical physics context, time is replaced by position on the lattice and the transition probability by the transfer matrix.

(iii) *Low-temperature limit.* Consider now the example

$$S(q,q') = \tfrac{1}{2}(q-q')^2/T + \tfrac{1}{2}\left(w(q) + w(q')\right).$$

where T, in the physical interpretation, is proportional to the temperature and the function $w(q)$ is a regular function (e.g. a polynomial) positive with a unique minimum at $q = q_0$:

$$w(q) = \tfrac{1}{2} w_2 (q-q_0)^2 + O\left((q-q_0)^3\right).$$

For $T \to 0$, the leading configuration to the partition function corresponds to all q_k equal. The p-point function reduces to

$$\langle q^p \rangle_n = \int \mathrm{d}q \, q^p \, \mathrm{e}^{-nw(q)} \bigg/ \int \mathrm{d}q \, \mathrm{e}^{-nw(q)}.$$

Then, for $n \to \infty$, only the vicinity of the configuration $q = q_0$ contributes and $\langle q^p \rangle_n \to q_0^p$. To calculate connected functions, one can introduce the generating function

$$\mathcal{Z}_n(b) = \langle \mathrm{e}^{bq} \rangle_n.$$

The limit $n \to \infty$ can be evaluated by the steepest descent method. One can thus approximate $w(q)$ by $\frac{1}{2}w_2(q-q_0)^2$. The generating function is given by

$$\mathcal{Z}_n(b) \propto \int dq \, e^{-nw_2(q-q_0)^2/2+bq} \propto e^{bq_0+b^2/2nw_2}$$

and, thus, the generating function of connected functions by

$$\mathcal{W}_n(b) = bq_0 + b^2/2nw_2,$$

a result analogous to the central limit theorem. In particular, since the two-point function is a constant, the correlation length is infinite, a result that can be easily understood because the configuration energy is dominated by the translation-invariant term.

Finally, to study the vicinity of $T = 0$, one can also replace $w(q)$ by a quadratic approximation. One is then led to a Gaussian distribution that we now discuss.

4.6 The Gaussian model

We now study more thoroughly the example of a Gaussian Boltzmann weight corresponding to a quadratic function (4.3).

To each point $k \in \mathbb{Z}$ of the lattice there is associated a real variable x_k (e.g., the deviation of a particle from its equilibrium position) and a statistical distribution

$$\rho_n(x) = e^{-\mathcal{S}_n(x)} / \mathcal{Z}_n,$$

where we choose the function $\mathcal{S}_n(x)$ of the special form

$$\mathcal{S}_n(x) = \sum_k \left[\frac{J}{2T}(x_k - x_{k-1})^2 + \frac{1}{2}\omega^2 x_k^2 \right], \quad \omega, J > 0. \tag{4.27}$$

In the physical interpretation, the positive parameter T is the temperature, the constant J characterizes the strength of the interaction between nearest neighbours and ω characterizes the distribution for $T = +\infty$, which can be generated by a harmonic potential $V(x) = \frac{1}{2}\omega^2 x^2$ corresponding to a force of return to equilibrium.

Following the analysis of Section 4.5, one expects that a collective behaviour can show up only at low temperature, where the coupling between neighbouring points is the strongest. Indeed, for $T \to 0$, the leading configuration is obtained by minimizing the energy and corresponds to all x_k equal.

It is convenient to change the normalization of q, setting

$$x_k = \lambda q_k \quad \text{with} \quad \lambda^2 = \frac{2}{\omega\sqrt{\omega^2 + 4T/J}},$$

and introducing a parameter $\theta > 0$ defined by

$$\cosh\theta = 1 + \omega^2 T/2J, \tag{4.28}$$

which is proportional to $\sqrt{T}$ for $T \to 0$:

$$\theta \underset{T\to 0}{\sim} \omega\sqrt{T/J}.$$

In these new variables, the function $\mathcal{S}_n$ (equation (4.3)) becomes

$$\mathcal{S}_n(\mathbf{q}) = \sum_{k=1}^{n} S(q_k, q_{k-1}) \tag{4.29}$$

with, in the notation (4.4),

$$S(q, q') = \frac{1}{2\sinh\theta}\left[(q^2 + q'^2)\cosh\theta - 2qq'\right]. \tag{4.30}$$

We first calculate quantities of interest for a finite lattice $k \in [0, n]$, imposing periodic boundary conditions: $q_n = q_0$, and then study the limit of the infinite system, $n \to \infty$.

To simplify expressions, it is also convenient to change the normalization of the partition function and write the probability distribution as

$$\rho_n(\mathbf{q}) = \frac{1}{\mathcal{Z}(n,\theta)} \frac{1}{(2\pi\sinh\theta)^{n/2}} e^{-\mathcal{S}_n(\mathbf{q})}. \tag{4.31}$$

The partition function is then given by

$$\mathcal{Z}(n,\theta) = \int \left(\prod_{k=1}^{n} \frac{dq_k}{\sqrt{2\pi\sinh\theta}}\right) \exp\left[-\mathcal{S}_n(\mathbf{q})\right]. \tag{4.32}$$

4.6.1 Gaussian transfer matrix: Algebraic properties

The transfer matrix $\mathbf{T}$ is now associated to the kernel (for more details see Section 4.1.2):

$$[\mathbf{T}](q, q') = \frac{1}{\sqrt{2\pi\sinh\theta}} \exp\left[-\frac{1}{2\sinh\theta}\left((q^2 + q'^2)\cosh\theta - 2qq'\right)\right]. \tag{4.33}$$

The operator $\mathbf{T}$ is real symmetric. The kernel $[\mathbf{T}](q, q')$ satisfies the bound (4.5) and, thus, the traces of all powers of $\mathbf{T}$ exist. Its spectrum is discrete, real and bounded and its eigenvectors are orthogonal.

Moreover, one verifies that $\mathbf{T}$ is positive (see also Section 4.8). Indeed, using the identity

$$\frac{1}{2\pi}\int dp\, e^{ip(q-q')} e^{-p^2\sinh\theta/2} = \frac{1}{\sqrt{2\pi\sinh\theta}} e^{-(q-q')^2/2\sinh\theta}, \tag{4.34}$$

one finds, for all functions f in Hilbert space $\mathcal{H}$,

$$(f, \mathbf{T}f) = \frac{1}{2\pi} \int dp \, e^{-p^2 \sinh\theta/2} \left| \int dq \, e^{ipq} e^{-\tanh(\theta/2)q^2/2} f(q) \right|^2 > 0.$$

The eigenvalues of $\mathbf{T}$ thus are positive.

In what follows, we use directly the notation and results of Section 4.3. In particular, we denote by ψ_ν the eigenvectors of $\mathbf{T}$ and by τ_ν the corresponding eigenvalues. We also introduce the position operator $\hat{\mathbf{Q}}$, defined by equation (4.12), and which acts multiplicatively on the functions in Hilbert space:

$$[\hat{\mathbf{Q}}\psi](q) = q\,\psi(q).$$

For reasons that become clearer later, eigenvectors and eigenvalues of $\mathbf{T}$ can be determined by an algebraic method inspired from the solution of the quantum harmonic oscillator.

Annihilation and creation operators. In the language of quantum mechanics, $\hat{\mathbf{Q}}$ is the *position operator*. It is useful to also introduce the operator $\hat{\mathbf{D}}$, the generator of q-space translations (and proportional to the *momentum operator* $\hat{\mathbf{P}}$ of quantum mechanics), which acts on the dense subset of differentiable vectors in Hilbert space by

$$[\hat{\mathbf{D}}\psi](q) = \frac{d\psi}{dq}. \tag{4.35}$$

One verifies the commutation relation (with the notation $[\mathbf{U}, \mathbf{V}] \equiv \mathbf{UV} - \mathbf{VU}$)

$$[\hat{\mathbf{D}}, \hat{\mathbf{Q}}] = \mathbf{1},$$

where $\mathbf{1}$ is the identity operator. Using the notation $\dagger$ to indicate Hermitian conjugation, with the definition for all operators $\hat{\mathbf{O}}$:

$$(f, \hat{\mathbf{O}}g) = (\hat{\mathbf{O}}^\dagger f, g),$$

one notes that

$$\hat{\mathbf{Q}} = \hat{\mathbf{Q}}^\dagger, \quad \hat{\mathbf{D}} = -\hat{\mathbf{D}}^\dagger, \quad \mathbf{T} = \mathbf{T}^\dagger.$$

For example,

$$(f, \hat{\mathbf{D}}g) = \int dq \, f(q)g'(q) = -\int dq \, f'(q)g(q) = -(\hat{\mathbf{D}}f, g).$$

One then defines two operators

$$\mathbf{A} = \left(\hat{\mathbf{D}} + \hat{\mathbf{Q}}\right)/\sqrt{2}, \quad \mathbf{A}^\dagger = \left(-\hat{\mathbf{D}} + \hat{\mathbf{Q}}\right)/\sqrt{2}, \tag{4.36}$$

called, in the quantum language, *annihilation* and *creation* operators, respectively. Conversely,

$$\hat{\mathbf{Q}} = \left(\mathbf{A} + \mathbf{A}^\dagger\right)\sqrt{2}. \tag{4.37}$$

The commutator of $\mathbf{A}$ and $\mathbf{A}^\dagger$ is

$$[\mathbf{A}, \mathbf{A}^\dagger] = 1. \tag{4.38}$$

Transfer matrix and operators $\mathbf{A}$, $\mathbf{A}^\dagger$. Using the explicit form (4.33), one first verifies the equation

$$\left[\frac{\partial}{\partial q} + q + e^{-\theta}\left(\frac{\partial}{\partial q'} - q'\right)\right][\mathbf{T}](q, q') = 0.$$

Acting then with the operators $\mathbf{AT}$ and $\mathbf{TA}$ on an arbitrary vector ψ, one obtains

$$\sqrt{2}\mathbf{AT}\psi](q) \equiv \int dq' \left(\frac{\partial}{\partial q} + q\right)[\mathbf{T}](q, q')\psi(q')$$

$$= e^{-\theta} \int dq' \, \psi(q')\left(q' - \frac{\partial}{\partial q'}\right)[\mathbf{T}](q, q')$$

$$= e^{-\theta} \int dq' \, [\mathbf{T}](q, q')\left(\frac{\partial}{\partial q'} + q'\right)\psi(q') \equiv \sqrt{2}\, e^{-\theta}[\mathbf{TA}\psi](q),$$

after an integration by parts. One infers the commutation relations

$$\mathbf{AT} = e^{-\theta}\,\mathbf{TA} \quad \Rightarrow \quad \mathbf{A}^\dagger\mathbf{T} = e^{\theta}\,\mathbf{TA}^\dagger. \tag{4.39}$$

4.6.2 Transfer matrix. Eigenvectors and eigenvalues

The commutation relations (4.39) allow a simple determination of the spectrum and the eigenvectors of the transfer matrix.

If ψ_ν is the eigenvector of $\mathbf{T}$ associated to the eigenvalue τ_ν, then,

$$\mathbf{A}^\dagger\mathbf{T}\psi_\nu = \tau_\nu \mathbf{A}^\dagger\psi_\nu = e^{\theta}\,\mathbf{TA}^\dagger\psi_\nu.$$

Thus, if $\mathbf{A}^\dagger\psi_\nu$ does not vanish and is square integrable, $\mathbf{A}^\dagger\psi_\nu$ is also an eigenvector, associated to the eigenvalue $e^{-\theta}\tau_\nu < \tau_\nu$.

Let us show that the first condition is always satisfied. The equation $\mathbf{A}^\dagger\psi = 0$ can be written explicitly as

$$-\psi'(q) + q\psi(q) = 0.$$

The solution is

$$\psi(q) \propto e^{q^2/2},$$

which is not square integrable. Thus, there exists no vector ψ in $\mathcal{H}$ such that $\mathbf{A}^\dagger\psi = 0$.

The same argument, applied to $\mathbf{A}$, yields

$$\mathbf{AT}\psi_\nu = \tau_\nu\mathbf{A}\psi_\nu = e^{-\theta}\,\mathbf{TA}\psi_\nu.$$

Thus, if $\mathbf{A}\psi_\nu$ is square integrable and does not vanish, $\mathbf{A}\psi_\nu$ is also an eigenvector, associated to the eigenvalue $e^\theta \tau_\nu > \tau_\nu$.

But since the operator $\mathbf{T}$ is bounded, there necessarily exists a maximum eigenvalue τ_0. The corresponding eigenvector ψ_0 is thus such that

$$\mathbf{A}\psi_0 = 0. \tag{4.40}$$

Writing equation (4.40) in a more explicit form, one finds the differential equation

$$\psi_0'(q) + q\psi_0(q) = 0,$$

whose solution

$$\psi_0(q) \propto e^{-q^2/2}$$

is square integrable.

One infers that the vectors

$$\psi_\nu \propto \left(\mathbf{A}^\dagger\right)^\nu \psi_0,$$

which, as one verifies, are normalizable, are the eigenvectors of $\mathbf{T}$ associated to the eigenvalues

$$\tau_\nu = e^{-\nu\theta} \tau_0.$$

Calculating the trace of $\mathbf{T}$ in two different ways, one then finds

$$\operatorname{tr} \mathbf{T} = \int dq\, [\mathbf{T}](q, q) = \frac{1}{2\sinh(\theta/2)} = \sum_{\nu=0}^\infty \tau_\nu = \frac{\tau_0}{1 - e^{-\theta}}.$$

One infers the eigenvalues

$$\tau_\nu = e^{-(\nu+1/2)\theta}, \quad \nu \geq 0. \tag{4.41}$$

Hamiltonian operator. The operator

$$\mathbf{H} = \mathbf{A}^\dagger \mathbf{A} + \tfrac{1}{2} \tag{4.42}$$

is analogous to the Hamiltonian operator of the quantum harmonic oscillator (in units where Planck's constant $\hbar = 1$ and with unit frequency). It commutes with the transfer matrix since (relations (4.39))

$$\mathbf{A}^\dagger \mathbf{A} \mathbf{T} = e^{-\theta} \mathbf{A}^\dagger \mathbf{T} \mathbf{A} = \mathbf{T} \mathbf{A}^\dagger \mathbf{A},$$

and thus has the same eigenvectors. To determine the corresponding eigenvalues, one acts with $\mathbf{A}^\dagger \mathbf{A}$ on an eigenvector:

$$\mathbf{A}^\dagger \mathbf{A} (\mathbf{A}^\dagger)^\nu \psi_0.$$

Commuting the operator $\mathbf{A}$ systematically with the operators $\mathbf{A}^\dagger$ placed to its right with the help of relation (4.38), and finally using equation (4.40), one infers

$$\mathbf{H}\psi_\nu = (\nu + \tfrac{1}{2})\psi_\nu. \tag{4.43}$$

From these eigenvalues, one then derives the relation between the transfer matrix and the quantum Hamiltonian

$$\mathbf{T} = \exp(-\theta \mathbf{H}). \tag{4.44}$$

Finally, one notes that $\mathbf{H}$, expressed in terms of the operators $\hat{Q}$ and $\hat{D}$, takes the form

$$\mathbf{H} = -\tfrac{1}{2}\hat{D}^2 + \tfrac{1}{2}\hat{Q}^2. \tag{4.45}$$

4.6.3 Partition function. Correlation functions

Partition function. The partition function can be written as (Section 4.1.2):

$$\mathcal{Z}(n,\theta) = \operatorname{tr} \mathbf{T}^n = \sum_{\nu=0}^{\infty} e^{-n\theta(\nu+1/2)} = \frac{e^{-n\theta/2}}{1 - e^{-n\theta}}. \tag{4.46}$$

Correlation functions. Since the measure is Gaussian, it suffices to calculate the one- and two-point functions, other functions following from Wick's theorem.

The explicit calculation is based on a few simple identities. First, using the commutation relations (4.39) and the cyclic property of the trace, one finds

$$\operatorname{tr} \mathbf{T}^n \mathbf{A}^m = \operatorname{tr} \mathbf{T}^n \mathbf{A}^{\dagger m} = 0 \quad \text{for} \quad m > 0.$$

With the same ingredients, one obtains

$$\operatorname{tr} \mathbf{A}\mathbf{A}^\dagger \mathbf{T}^n = e^{n\theta} \operatorname{tr} \mathbf{A}^\dagger \mathbf{A}\, \mathbf{T}^n.$$

Using the commutation relation (4.38), one finally obtains

$$\operatorname{tr} \mathbf{A}^\dagger \mathbf{A}\, \mathbf{T}^n = \frac{2}{e^{n\theta} - 1} \operatorname{tr} \mathbf{T}^n.$$

One-point function. The relation (4.37) implies that the one-point function is proportional to

$$\operatorname{tr} \hat{\mathbf{Q}} \mathbf{T}^n \propto \operatorname{tr}(\mathbf{A} + \mathbf{A}^\dagger)\mathbf{T}^n = 0.$$

Actually, this result directly follows from the symmetry $[\mathbf{T}](q,q') = [\mathbf{T}](-q,-q')$.

Two-point function. The two-point function is given by (equation (4.16))

$$\langle q_0 q_\ell \rangle_n = \operatorname{tr} \mathbf{T}^{n-|\ell|} \hat{\mathbf{Q}} \mathbf{T}^{|\ell|} \hat{\mathbf{Q}} / \operatorname{tr} \mathbf{T}^n.$$

Using equation (4.46), the relation (4.37) and the commutation relations (4.39), one obtains the explicit result

$$\langle q_0 q_\ell \rangle_n = \frac{1}{2} \frac{\cosh\bigl((n/2 - |\ell|)\theta\bigr)}{\sinh(n\theta/2)}. \tag{4.47}$$

Thermodynamic limit. In the thermodynamic limit $n \to \infty$ (see Section 4.3), the partition function (4.46) behaves like

$$\mathcal{Z}(n,\theta) \sim e^{-n\theta/2}$$

and, thus, the free energy density (defined by equation (4.19)) $\mathcal{W}/n$ has the finite limit

$$\lim_{n\to\infty} \frac{1}{n}\mathcal{W} \equiv \frac{T}{n}\ln \mathcal{Z} = -T\frac{\theta}{2}.$$

In the low-temperature limit $T \to 0$, the relation (4.28) leads to $\theta \sim \omega\sqrt{T/J}$ and thus
$$\lim_{n \to \infty} \frac{1}{n}\mathcal{W} \sim -\frac{1}{2}\omega T\sqrt{T/J}.$$
Similarly, the two-point function becomes
$$\langle q_0 q_\ell \rangle = \tfrac{1}{2} e^{-\theta|\ell|}. \tag{4.48}$$

Large-distance behaviour. Expression (4.48) shows that the two-point function decays exponentially at large distance. One characterizes this behaviour by the *correlation length* (defined by (4.22)), which here is given by
$$\xi = \lim_{\ell \to \infty} -\frac{\ell}{\ln(|\langle q_0 q_\ell \rangle|)} = \frac{1}{\ln(\tau_0/\tau_1)} = \frac{1}{\theta},$$
in agreement with the general result (4.23) and the spectrum (4.41).

For $T \to 0$, that is, at low temperature,
$$\xi = \frac{1}{\theta} \sim \frac{1}{\omega}\sqrt{\frac{J}{T}}$$
and thus the correlation length diverges.

Moreover,
$$\sum_\ell \langle q_0 q_\ell \rangle = \frac{1}{2\tanh(\theta/2)},$$
and thus the distribution $R_n(Q)$ of the mean variable $Q = \sum_\ell q_\ell/n$ has the asymptotic Gaussian form (Section 4.4.1)
$$R_n(Q) \sim \sqrt{n\tanh(\theta/2)/\pi}\, e^{-n\tanh(\theta/2)Q^2}.$$

4.7 Gaussian model: The continuum limit

Since for $\theta \to 0$, the correlation length diverges, we introduce the macroscopic scale $\beta = n\theta$ and take the limit $n \to \infty$, $\theta \to 0$ at $n\theta = \beta$ fixed. This corresponds to fixing the length of the system in units of the correlation length, since $n \sim \beta\xi$. This limit is called the continuum limit, in the sense that for quantities defined at the correlation length-scale, all traces of the initial lattice have disappeared.

Actually, in the Gaussian model all functions already have, on the lattice, their continuum form, provided one chooses the proper parametrization. For example, from equation (4.46) one infers
$$\mathcal{Z}(n,\theta) = \mathcal{Z}(\beta) \equiv \frac{e^{-\beta/2}}{1-e^{-\beta}} = \lim_{\theta \to 0,\, n\theta=\beta \text{ fixed}} \mathcal{Z}(n,\theta). \tag{4.49}$$

4.7.1 Continuum limit and quantum Hamiltonian

The relation (4.44) between the Hamiltonian operator of the quantum harmonic oscillator (4.45) and the transfer matrix shows, similarly, that

$$\mathbf{T}^n = e^{-\beta \mathbf{H}} = \lim_{\theta \to 0,\; n\theta = \beta \text{ fixed}} \mathbf{T}^n.$$

The partition function in the continuum limit can then be written as

$$\mathcal{Z}(\beta) = \operatorname{tr} e^{-\beta \mathbf{H}}. \tag{4.50}$$

Quite remarkably, the function $\operatorname{tr} e^{-\beta \mathbf{H}}$, which coincides with the continuum limit of a classical partition function, is also the quantum partition function associated with the Hamiltonian operator $\mathbf{H}$. In the quantum interpretation, β becomes the inverse of the temperature in a unit system where Boltzmann's constant $k_B = 1$ and Planck's constant $\hbar$ also equals 1.

Finally, in terms of the macroscopic position variable $t = \ell\theta$, $\ell \in \mathbb{Z}$, the two-point function (4.47) takes the form

$$\langle q(0)q(t) \rangle = \frac{1}{2} \frac{\cosh(\beta/2 - |t|)}{\sinh(\beta/2)}. \tag{4.51}$$

Therefore, it is equal to its limit when $\theta \to 0$, $\ell \to \infty$ with $t = \ell\theta$ fixed, a limit that again corresponds to fixing the distance between two points in units of the correlation length: $\ell \sim t\xi$.

We conclude that the partition function and correlation functions have a limit in which all traces of the initial lattice have disappeared, thus called the continuum limit. It has been possible to define such a continuum limit only because there exists a limit in which the correlation length diverges.

Finally, in the continuum limit, the transfer matrix, expressed in terms of the Hamiltonian quantum, $\mathbf{T}(\beta) = e^{-\beta \mathbf{H}}$, has a quantum interpretation: it is the density matrix at thermal equilibrium at the temperature $1/\beta$, which is used to calculate thermal expectation values. The equation

$$\frac{\partial}{\partial \beta} \mathbf{T}(\beta) = -\mathbf{H}\mathbf{T}(\beta),$$

written in terms of kernels in the representation where the operator $\hat{\mathbf{Q}}$ (position operator in quantum mechanics) acts multiplicatively, takes the form of the partial differential equation (see equation (4.45)):

$$\frac{\partial}{\partial \beta}[\mathbf{T}(\beta)](q,q') = -\frac{1}{2}\left(-\frac{\partial^2}{(\partial q)^2} + q^2\right)[\mathbf{T}(\beta)](q,q') \tag{4.52}$$

with the boundary condition

$$[\mathbf{T}(0)](q,q') = \delta(q - q').$$

This is a diffusion or heat type equation, or a Schrödinger equation in 'imaginary time'.

Divergence of the correlation length and continuum limit. We have just exhibited a first example of the *continuum limit* associated with the divergence of a correlation length. The existence of a continuum limit implies universality properties. In the limit $\theta \to 0$ (here low temperature), the macroscopic properties of the statistical model are independent of the detailed form of the transfer matrix. We could have somewhat complicated the nearest-neighbour interaction and the distribution at each site and have obtained the same continuum limit. Finally, the result is expressed in terms of the solution of a diffusion equation in continuum space in which all traces of the initial lattice have disappeared.

Thermodynamic limit. In the continuum limit, the limit $\beta \to \infty$ corresponds to the thermodynamic limit of the classical statistical model and thus

$$\lim_{\beta \to \infty} \frac{1}{\beta} \ln \mathcal{Z}(\beta) = -\frac{1}{2}.$$

This is also the zero-temperature limit of the quantum model. In this limit, the quantum partition function is dominated by the smallest eigenvalue, that is, the ground state energy, of the quantum Hamiltonian. The study of the thermodynamic limit of the classical system is thus related to the study of the ground state of the quantum system.

In this limit, the two-point function reduces to

$$\langle q(t)q(0) \rangle = \tfrac{1}{2} \mathrm{e}^{-|t|}. \tag{4.53}$$

Classical and quantum statistical physics. We have exhibited, in the continuum limit, a relation between quantum statistical mechanics in zero dimension (only one particle) and classical statistical mechanics in one dimension. Indeed, in the limit $\theta \to 0$, $n\theta = \beta$ fixed, the classical partition function tends, except for a trivial normalization, toward the partition function of a quantum particle. This type of relations is not special to the Gaussian example, extends to some correlation functions and generalizes to higher dimensions: quantum statistical mechanics in $(d-1)$ dimensions of space has relations, in the low-temperature limit, with classical statistical mechanics in d dimensions. Therefore, although in this work, we are interested mainly in classical statistical mechanics, its study in the continuum limit leads to results transposable to quantum theory.

4.7.2 Decimation and the continuum limit

One again considers the transfer matrix $[\mathbf{T}](q, q')$ (equation (4.33)). From the representation (4.44), one infers

$$\mathbf{T}(\theta)\mathbf{T}(\theta') = \mathbf{T}(\theta + \theta'),$$

where $\mathbf{T}(\theta)$ is the operator with kernel $[\mathbf{T}](q, q')$ and parameter θ, a result that one can also verify directly by explicit calculation.

To calculate the partition function, one can then use a method known as decimation which consists in recursively grouping the sites of the lattice pairwise. It is the analogue, for this problem, of the transformations discussed into Section 3.2. Decimation amounts for the transfer matrix to the transformation

$$\mathbf{T} \mapsto \mathbf{T}^2.$$

In the Gaussian example, since

$$\mathbf{T}^2(\theta) = \mathbf{T}(2\theta),$$

the parameter θ transforms into 2θ and the iteration of the transformation generates the sequence $\mathbf{T}_m(\theta) = \mathbf{T}(2^m\theta)$.

A fixed point must satisfy

$$\mathbf{T}(2\theta) = \mathbf{T}(\theta),$$

an equation that has only the solution $\theta = 0$ and thus $\mathbf{T} = 1$ (leaving aside the fixed point $\theta = \infty$, which corresponds to independent sites and vanishing correlation length). This is the limiting situation of vanishing temperature, where all random variables are equal and the correlation length is infinite, which we have discussed in example (iii) of Section 4.5.

One then notes that, for $\theta > 0$, the correlation length, which is finite, is divided by two at each iteration and thus converges to zero. Iterations move away from the fixed point that corresponds to an infinite correlation length and the parameter θ is thus associated to an unstable direction, that is, to a relevant perturbation.

However, universality reduced to one point $\theta = 0$ is of limited physical relevance. On the other hand, we have proved the existence of a continuum limit possessing universality properties. We now show how to recover, in this context, asymptotic universality in the vicinity of a fixed point, here for $0 < \theta \ll 1$, using an idea that will play an important role later in this work.

The idea is to iterate the transformation m times such that $2^m \gg 1$, but $2^m\theta \ll 1$, that is, such that the correlation length after m iterations is still large with respect to the lattice spacing. One way to realize this situation is to divide θ by two at each iteration and, thus, to multiply by two the initial correlation length. This corresponds to the combined transformation

$$\mathbf{T}_m(\theta) \mapsto \mathbf{T}_{m+1}(\theta) = \mathbf{T}_m^2(\theta/2) \tag{4.54}$$

and, thus, $\mathbf{T}_m(\theta) = \mathbf{T}(\theta)$.

This transformation is such that the correlation length of the transformed system remains constant, but diverges in the initial lattice scale.

In the Gaussian model, the transformation (4.54) is the identity. This reflects the property that even on the lattice, in this model, irrelevant perturbations are absent.

The strategy of adjusting the initial amplitude of a relevant perturbation in such a way that the iterated amplitude remains finite, has a direct analogue in quantum or statistical field theory, where it takes the form of a *mass renormalization*. The analogue of the asymptotic universality that we have just defined will then be the universality in the *critical domain* (see Section 9.4.4).

4.8 More general models: The continuum limit

The results obtained in the Gaussian example generalize to a particular class of functions (in the representation (4.4))

$$S(q',q) = (q-q')^2/2\varepsilon + \tfrac{1}{2}\varepsilon V(q) + \tfrac{1}{2}\varepsilon V(q'), \tag{4.55}$$

where $V(q) \geq 0$ is, for example, a polynomial and ε a parameter that plays the role of $\sqrt{T}$ in the Gaussian example. We are then interested in the $\varepsilon \to 0$ asymptotic behaviour.

We define the transfer matrix by

$$[\mathbf{T}](q',q) = \frac{1}{\sqrt{2\pi\varepsilon}} e^{-S(q',q)}. \tag{4.56}$$

The partition function (always with the periodic boundary condition $q_0 = q_n$) is given by

$$\mathcal{Z}_n(\varepsilon) = \operatorname{tr} \mathbf{T}^n.$$

By using, in particular, identity (4.34) one can express the operator $\mathbf{T}$ in terms of the operators (4.12) and (4.35). One finds

$$\mathbf{T} = e^{-\varepsilon V(\hat{\mathbf{Q}})/2} e^{\varepsilon \hat{\mathbf{D}}^2/2} e^{-\varepsilon V(\hat{\mathbf{Q}})/2},$$

where again

$$\mathbf{T} = \mathbf{U}\mathbf{U}^\dagger, \quad \mathbf{U} = e^{-\varepsilon V(\hat{\mathbf{Q}})/2} e^{\varepsilon \hat{\mathbf{D}}^2/4},$$

a form that also shows that the operator $\mathbf{T}$ is positive.

The trace of $\mathbf{T}$ is finite if $V(q)$ tends to infinity fast enough for $|q| \to \infty$. Indeed,

$$\operatorname{tr} \mathbf{T} = \int dq\, [\mathbf{T}](q,q) = \int dq\, e^{-\varepsilon V(q)}.$$

Thus, the sum of the eigenvalues τ_k of $\mathbf{T}$ (which are positive) is bounded,

$$\operatorname{tr} \mathbf{T} = \sum_k \tau_k < \infty \Rightarrow \tau_k \underset{k\to\infty}{\to} 0,$$

and this implies that eigenvalues accumulate to zero.

Continuum limit $\varepsilon \to 0$. Since $\mathbf{T}$ is positive, one can define the Hermitian operator $\ln \mathbf{T}$ (it has the same eigenvectors as $\mathbf{T}$ and has $\ln \tau_k$ as eigenvalues).

For $\varepsilon \to 0$, $\mathbf{T}$ tends toward the identity operator and, thus, $\ln \mathbf{T} \to 0$. To determine the leading behaviour of $\ln \mathbf{T}$, it suffices to estimate $\mathbf{T} - 1$. Expanding the product, one finds

$$\mathbf{T} - 1 = -\varepsilon \left[-\tfrac{1}{2}\hat{\mathbf{D}}^2 + V(\hat{\mathbf{Q}}) \right] + O(\varepsilon^2). \tag{4.57}$$

One infers
$$\mathbf{H} = -\lim_{\varepsilon \to 0} \frac{1}{\varepsilon} \ln \mathbf{T} = -\tfrac{1}{2}\hat{\mathbf{D}}^2 + V(\hat{\mathbf{Q}}) \;\Rightarrow\; \mathbf{T} = \mathrm{e}^{-\varepsilon \mathbf{H} + O(\varepsilon^2)}.$$

The operator $\mathbf{H}$ is the quantum Hamiltonian of a particle with unit mass, inside a potential $V(q)$ (in a unit system where $\hbar = 1$). In the limit $\varepsilon \to 0$ at $n\varepsilon = \beta$ fixed, one thus finds
$$\mathcal{Z}(\beta) = \mathrm{tr}\, \mathrm{e}^{-\beta \mathbf{H}}.$$

Again, the classical partition function, in the continuum limit, tends toward the quantum partition function of a one-particle system. In the thermodynamic limit
$$W = \lim_{\beta \to \infty} \frac{1}{\beta} \ln \mathcal{Z}(\beta) = -E_0$$
and, thus, the classical free energy density is proportional to the energy E_0 of the quantum ground state.

This continuum limit is associated with the divergence of the correlation length ξ. Indeed,
$$\xi^{-1} = \ln(\tau_0/\tau_1) \sim \varepsilon(E_1 - E_0),$$
where $E_0 < E_1$ are the two lowest eigenvalues of $\mathbf{H}$.

Again, the continuum limit can be reached by a generalization of the transformation (4.54), used in the Gaussian example. Here, the natural choice is
$$\mathbf{T}_m(\varepsilon) \mapsto \mathbf{T}_{m+1}(\varepsilon) = \mathbf{T}_m^2(\varepsilon/2), \quad \mathbf{T}_0(\varepsilon) = \mathbf{T}(\varepsilon),$$
which, asymptotically, corresponds to a constant correlation length in the transformed models. In a technical sense, the transformation also corresponds to Trotter's formula $\mathrm{e}^{\mathbf{A}} = \lim_{m \to \infty}(\mathrm{e}^{\mathbf{A}/m})^m$. The iteration of this transformation suppresses irrelevant corrections since in the expansion of $\ln \mathbf{T}$ in powers of ε, the coefficients of terms of degree larger than one all go to zero:
$$\lim_{m \to \infty} \mathbf{T}_m(\varepsilon) = \mathrm{e}^{-\varepsilon \mathbf{H}}.$$

Finally, we could have chosen a transfer matrix without irrelevant corrections, like in the Gaussian example, by solving the equation
$$\frac{\partial}{\partial \varepsilon} \mathbf{T}(\varepsilon) = -\mathbf{H}\mathbf{T}(\varepsilon),$$
which, in terms of the associated kernel, takes the form of the partial differential equation
$$\frac{\partial}{\partial \varepsilon}[\mathbf{T}(\varepsilon)](q,q') = -\left(-\frac{1}{2}\frac{\partial^2}{(\partial q)^2} + V(q)\right)[\mathbf{T}(\varepsilon)](q,q').$$

Let us point out that, since to each quantum Hamiltonian is associated a continuum limit, the fixed point $\mathbf{T} = \mathbf{1}$ has an infinite number of relevant perturbations.

Exercises

Exercise 4.1

The segment. Consider a function $S(q-q')$ (equation (4.4)) of the form

$$S(q,q') = |q-q'|/T.$$

where q belongs to the interval $[-1,1]$ and T is a positive parameter that, from the physics viewpoint, has the interpretation of a temperature.

The eigenvectors $\psi(q)$ of the transfer matrix corresponding to an eigenvalue τ then satisfy

$$\int_{-1}^{+1} dq'\, e^{-|q-q'|/T} \psi(q') = \tau \psi(q). \qquad (4.58)$$

(i) Determine the eigenvectors and eigenvalues of the transfer matrix. It will prove useful to differentiate the eigenvalue equation.
(ii) Infer the correlation length. What can be said of the limit $T \to 0$?

Solution. The eigenvectors are even or odd functions.
Differentiating once equation (4.58), one finds

$$\int_{-1}^{+1} dq'\, \text{sgn}(q'-q)\, e^{-|q-q'|/T} \psi(q') = T\tau \psi'(q),$$

where $\text{sgn}(q)$ is the function sign of q: $\text{sgn}(q>0) = 1$, $\text{sgn}(q<0) = -1$.
For $q = \pm 1$, one can relate the values of the function ψ and its derivative. One finds

$$q=1: \quad \psi(1) = -T\psi'(1), \qquad q=-1: \quad \psi(-1) = T\psi'(-1). \qquad (4.59)$$

Differentiating again and using equation (4.58), one obtains the differential equation

$$-2T\psi(q) + \tau\psi(q) = \tau T^2 \psi''(q).$$

It is convenient to set

$$k^2 = \frac{2}{T\tau} - \frac{1}{T^2} \quad \Rightarrow \quad \tau = \frac{2T}{1+k^2 T^2}.$$

The solutions satisfying the boundary conditions (4.59) are

$$\psi_+(q) = \cos kq \quad \text{with} \quad kT\tan k = 1,$$
$$\psi_-(q) = \sin kq \quad \text{with} \quad \tan k = -kT.$$

Combining the two equations, one finds the spectral condition

$$\tan(2k) = -\frac{2kT}{1-k^2 T^2}.$$

The interesting situation is $T \to 0$ where correlations are the largest. Then (m positive integer),

$$k = \tfrac{1}{2}m\pi(1-T) + O(T^2), \quad m > 0 \Rightarrow \tau_{m-1} = \frac{2T}{1 + T^2 m^2 \pi^2/4} + O(T^4).$$

One infers the correlation length

$$\xi \sim \frac{1}{\ln(\tau_0/\tau_1)} \sim \frac{4}{3\pi^2 T^2}.$$

The correlation length diverges for $T \to 0$. The complete spectrum of $-\ln \mathbf{T}$ in this limit takes the form

$$-\ln(\tau_m/2T) \sim (m-1)^2 \pi^2 T^2 / 4.$$

The divergence of the correlation length allows defining a continuum limit and a path integral of random walk type, but with reflection boundary conditions on two walls located at $q = \pm 1$.

Exercise 4.2

The circle. We now consider the situation where the random variable q belongs to a circle of unit radius and the transfer matrix is a periodic function defined by

$$S(q, q') = -\cos(q - q')/T.$$

This model possesses an $O(2)$ orthogonal symmetry, corresponding to the group of rotation–reflections in the plane:

$$S(q+a, q'+a) = S(q, q'), \quad S(-q, -q') = S(q, q').$$

(i) Determine the eigenvectors and eigenvalues of the transfer matrix.
(ii) Infer the correlation length. Study the limit $T \to 0$.
(iii) In this context, it is natural to consider correlation functions of periodic functions of the variable q, since q is defined only mod 2π. Therefore, calculate the expectation values of products of the functions $e^{\pm i q_\ell}$ for one and two points, first at finite size, then in the thermodynamic limit.

Solution. The eigenfunctions are $e^{i\nu q}/\sqrt{2\pi}$ with ν integer. Indeed, these functions diagonalize the transfer matrix:

$$\frac{1}{2\pi} \int_{-\pi}^{+\pi} dq\, dq'\, e^{i\nu q - i\nu' q'}\, e^{\cos(q-q')/T} = \delta_{\nu\nu'} \int_{-\pi}^{+\pi} dq\, e^{i\nu q}\, e^{\cos q/T}$$

and, thus, the eigenvalues are given by

$$\tau_\nu = \tau_{-\nu} = \int_{-\pi}^{+\pi} dq\, e^{i\nu q}\, e^{\cos q/T} = 2\pi I_\nu(1/T),$$

where the function I_ν is a modified Bessel functions (see Exercise 2.4). The transfer matrix can then be expanded as

$$[\mathbf{T}](q, q') = \frac{1}{2\pi} \sum_\nu \tau_\nu \, e^{i\nu(q-q')}.$$

In the limit $T \to 0$, the eigenvalues can be evaluated by the steepest descent method. One finds (see Exercise 2.4)

$$\tau_\nu = \sqrt{2\pi T} \, e^{1/T} \left(e^{-T(\nu^2 - 1/4)/2} + O(T^2) \right) = \tau_0 \left(e^{-T\nu^2/2} + O(T^2) \right).$$

(ii) In the low-temperature limit $T \to 0$, the correlation length (equation (4.23))

$$\xi \sim \frac{1}{\ln(\tau_0/\tau_1)} \sim \frac{2}{T}$$

diverges. One can define a continuum limit, which corresponds to the brownian motion on the circle.

Finally, let us point out that the asymptotic spectrum is the exact spectrum of another transfer matrix on the circle that can be written as

$$[\mathbf{T}](q, q') = \sum_{m \in \mathbb{Z}} e^{-(q - q' + 2m\pi)^2/2T}.$$

Then,

$$\int_{-\pi}^{+\pi} dq \, e^{i\nu q} [\mathbf{T}](q, 0) = \sum_{m \in \mathbb{Z}} \int_{(2m-1)\pi}^{(2m+1)\pi} dq \, e^{i\nu q - q^2/2T}$$

$$= \int_{-\infty}^{+\infty} dq \, e^{i\nu q - q^2/2T} = \sqrt{2\pi T} \, e^{-T\nu^2/2}.$$

(iii) Translation invariance on the circle implies

$$\langle e^{\pm i q_\ell} \rangle = 0$$

and, moreover, that the only non-vanishing two-point function is $\langle e^{i q_{\ell_1}} e^{-i q_{\ell_2}} \rangle$. Using expression (4.16), one finds ($\ell = |\ell_1 - \ell_2|$)

$$Z_n^{(2)}(\ell) = Z_n^{-1} \operatorname{tr} e^{i\hat{Q}} \mathbf{T}^\ell e^{-i\hat{Q}} \mathbf{T}^{n-\ell}$$

$$= Z_n^{-1} \frac{1}{4\pi^2} \int dq \, dq' \sum_{\nu, \nu'} \tau_{\nu'}^{n-\ell} e^{i\nu'(q-q')} e^{iq} \tau_\nu^\ell e^{i\nu(q'-q)} e^{-iq'}$$

$$= Z_n^{-1} \sum_\nu \tau_{\nu-1}^{n-\ell} \tau_\nu^\ell.$$

In the thermodynamic limit

$$Z^{(2)}(\ell) = (\tau_1/\tau_0)^\ell = e^{-\ell/\xi}.$$

Exercise 4.3

Ising model: One dimension. In the Ising model, the random variables S_i at each lattice site i, called spins, take only two values ± 1. The model also has a $\mathbb{Z}_2$ reflection symmetry corresponding to spin reversal $S_i \mapsto -S_i$. The partition function of the Ising model in one dimension, with nearest-neighbour interactions, and in the presence of a uniform magnetic field h breaking the $\mathbb{Z}_2$ symmetry, is given by

$$\mathcal{Z}_\ell = \sum_{\{S_i = \pm 1\}} \exp[-\beta \mathcal{E}(S)] \tag{4.60}$$

with the configuration energy

$$\mathcal{E}(S) = -\sum_{i=1}^{n} J S_i S_{i+1} - h S_i, \quad J > 0. \tag{4.61}$$

Solve the model by the transfer matrix method. Calculate the free energy and the correlation length.

Solution. The transfer matrix is a 2×2 matrix that can be written as

$$[\mathbf{T}](S', S) = \exp\left[\beta\left(JSS' + \tfrac{1}{2}h(S+S')\right)\right], \text{ or } \mathbf{T} = \begin{pmatrix} e^{\beta(J+h)} & e^{-\beta J} \\ e^{-\beta J} & e^{\beta(J-h)} \end{pmatrix}, \tag{4.62}$$

where the matrix elements correspond to the values ± 1 of S and S'.

The two eigenvalues are

$$\tau_\pm = e^{\beta J}\left[\cosh(\beta h) \pm \sqrt{\sinh^2(\beta h) + e^{-4\beta J}}\right].$$

The thermodynamic limit corresponds to the limit $\ell \to \infty$. The partition function is related to the largest eigenvalue of the transfer matrix. The free energy density then is

$$W = \beta J + \ln\left[\cosh(\beta h) + \sqrt{\sinh^2(\beta h) + e^{-4\beta J}}\right].$$

The correlation length is related to the ratio of the two eigenvalues (the two largest, but there are only two):

$$\xi^{-1} = \ln(\tau_+/\tau_-) = 2\tanh^{-1}\left[\frac{(\sinh^2 \beta h + e^{-4\beta J})^{1/2}}{\cosh \beta h}\right]. \tag{4.63}$$

The correlation length diverges only in zero field, at zero temperature: $\xi(\beta, h = 0) \propto e^{4\beta J}$.

Exercise 4.4

The $O(\nu)$ model: One space dimension. We now consider a (classical) spin model with $O(\nu)$ symmetry (the orthogonal group of rotations–reflections in ν space dimensions) and nearest-neighbour interactions in one space dimension, a generalization of the $O(2)$ example of Exercise 4.2.

More specifically, at each lattice site the random variables form a ν-component spin vector $\mathbf{S}$ with unit length ($\mathbf{S}$ belongs to the sphere $S_{\nu-1}$). One then considers the partition function

$$\mathcal{Z}_\ell = \int \prod_{i=1}^n d\mathbf{S}_i\, \delta(\mathbf{S}_i^2 - 1) \exp\left[-\beta \mathcal{E}(\mathbf{S})\right] \tag{4.64}$$

where the symbol $d\mathbf{S}\,\delta(\mathbf{S}^2-1)$ means uniform measure on the sphere and the configuration energy is

$$\mathcal{E}(\mathbf{S}) = -J \sum_{i=1}^\ell \mathbf{S}_i \cdot \mathbf{S}_{i+1}, \quad J > 0. \tag{4.65}$$

Determine the two leading eigenvalues of the transfer matrix at low temperature.

Some relevant geometric remarks on the $O(\nu)$ group can be found in Exercise 8.1. In particular, we consider the integral

$$z(\mathbf{h}) = \int d\mathbf{S}\, \delta(\mathbf{S}^2 - 1)\, e^{\mathbf{S} \cdot \mathbf{h}}, \tag{4.66}$$

where $\mathbf{h}$ is an arbitrary vector. Due to rotation invariance, one can always choose $\mathbf{h} = (\mathbf{0}, h)$ and, for $\alpha < \nu$, use the parametrization $S_\alpha = n_\alpha \sin\theta$, where $\mathbf{0}$ and $\mathbf{n}$ are $(\nu-1)$-component vectors with $\mathbf{n}^2 = 1$, and $S_\nu = \cos\theta$, $0 \le \theta < \pi$. Then, $\mathbf{S} \cdot \mathbf{h} = h\cos\theta$. For $\nu = 2$,

$$z(h) = 2 \int_0^\pi d\theta\, e^{h\cos\theta}.$$

More generally,
$$d\mathbf{S}\,\delta(\mathbf{S}^2 - 1) = d\mathbf{n}\,\delta(\mathbf{n}^2 - 1) d\theta\, (\sin\theta)^{\nu-2},$$

where $(\sin\theta)^\nu$ comes from the change of variables and $1/\sin^2\theta$ from the δ function.

After integration over $\mathbf{n}$, which yields the surface of the sphere $S_{\nu-2}$, one obtains the measure

$$\frac{2\pi^{\nu/2-1/2}}{\Gamma\left(\tfrac{1}{2}(\nu-1)\right)} d\theta (\sin\theta)^{\nu-2}$$

for the $O(\nu)$ group and, thus,

$$z(h) = \frac{2\pi^{\nu/2-1/2}}{\Gamma\left(\tfrac{1}{2}(\nu-1)\right)} \int_0^\pi d\theta (\sin\theta)^{\nu-2}\, e^{h\cos\theta}.$$

Solution. The transfer matrix now corresponds to the kernel (setting $v = \beta J$)

$$[\mathbf{T}](\mathbf{S}', \mathbf{S}) = e^{v\mathbf{S}\cdot\mathbf{S}'},$$

with, as integration measure, the uniform measure on the sphere $\mathbf{S}^2 = 1$.

For any transfer matrix function only of the scalar product $\mathbf{S}\cdot\mathbf{S}'$, the eigenvectors follow from purely geometric considerations. They are related to the irreducible representations of the $O(\nu)$ group, here, symmetric tensors with vanishing partial traces derived from tensor products of the vector $\mathbf{S}$. The eigenvector corresponding to the largest eigenvalue is invariant under the group $O(\nu)$ and thus is a constant. Identifying $v\mathbf{S}$ with $\mathbf{h}$ in (4.66), one finds

$$\tau_0 = \int d\mathbf{S}'\, \delta(\mathbf{S}'^2 - 1)\, e^{v\mathbf{S}\cdot\mathbf{S}'}$$

$$= \frac{2\pi^{\nu/2-1/2}}{\Gamma(\frac{1}{2}(\nu-1))} \int_0^\pi d\theta (\sin\theta)^{\nu-2}\, e^{v\cos\theta} = (2\pi)^{\nu/2} v^{1-\nu/2} I_{\nu/2-1}(v),$$

where $\cos\theta = \mathbf{S}\cdot\mathbf{S}'$ and $I_\nu(v)$ is a modified Bessel function (generalized to ν non--integer).

For $v \to \infty$, the integral can be evaluated by a slightly adapted version of the steepest descent method (Exercises 2.4 and 2.8). One finds

$$\tau_0(v) \sim \left(\frac{2\pi}{v}\right)^{(\nu-1)/2} e^v \left[1 - \tfrac{1}{8}(\nu-1)(\nu-3)/v + O(1/v^2)\right].$$

The second eigenvalue corresponds to an eigenvector proportional to the vector $\mathbf{S}$. One verifies

$$\tau_1(v)\mathbf{S} = \int d\mathbf{S}'\, \delta(\mathbf{S}'^2 - 1)\, e^{v\mathbf{S}\cdot\mathbf{S}'}\, \mathbf{S}'.$$

Taking the scalar product with $\mathbf{S}$, one infers the eigenvalue

$$\tau_1(v) = \frac{2\pi^{\nu/2-1/2}}{\Gamma(\frac{1}{2}(\nu-1))} \int_0^\pi d\theta (\sin\theta)^{\nu-2} \cos\theta\, e^{v\cos\theta} = \tau_0'(v)$$

$$= \tau_0 \left[1 - \tfrac{1}{2}(\nu-1)/v + O(1/v^2)\right].$$

The behaviour of the correlation length follows:

$$\xi \underset{v\to\infty}{\sim} \frac{v}{2(\nu-1)}.$$

Exercise 4.5

Gaussian model. In the case of the Gaussian model, the partition function is given by a Gaussian integral that can thus be calculated directly. For this calculation, it is suggested to first determine the eigenvalues of the symmetric matrix associated with the quadratic form (4.29).

Solution. The integral (4.32) is Gaussian. Denoting by Λ_{ij} the matrix such that expression (4.29) can be written as

$$S_n(\mathbf{q}) = \frac{1}{2\sinh\theta} \sum_{i,j=1}^n \Lambda_{ij} q_i q_j,$$

one finds

$$\mathcal{Z}(n,\theta) = (\det \Lambda)^{-1/2}.$$

One can calculate the determinant as a product of the eigenvalues λ of the matrix Λ. The eigenvalue equation reads

$$\sum_{k=1}^n \Lambda_{jk} q_k = \sinh\theta \frac{\partial S_n}{\partial q_j} \equiv 2\cosh\theta q_j - q_{j+1} - q_{j-1} = \lambda q_j,$$

with the condition $q_n = q_0$. Due to the translation symmetry of the lattice with periodic boundary conditions, the eigenvectors have the form $q_i = r^i$ where the constant r satisfies $r^n = 1$ and

$$\lambda = 2\cosh\theta - r - r^{-1}.$$

The parameter r is a root of 1, $r = e^{2i\pi\ell/n}$, and thus

$$\lambda = 2(\cosh\theta - \cos(2\pi\ell/n)), \quad 0 \leq \ell < n.$$

One infers the partition function

$$\mathcal{Z}(n,\theta) = \left[\prod_{\ell=0}^{n-1} 2(\cosh\theta - \cos(2\pi\ell/n))\right]^{-1/2}. \tag{4.67}$$

It is possible to calculate the product explicitly. Using the parametrization (4.28), one finds the identity

$$\prod_{\ell=0}^{n-1} 2(\cosh\theta - \cos(2\pi\ell/n)) = 2(\cosh n\theta - 1) = 4\sinh^2(n\theta/2).$$

(This can be proved by comparing the roots and normalization of the two polynomials in the variable $\cosh\theta$.) One then recovers the explicit form (4.46) of $\mathcal{Z}(n,\theta)$.

5 Continuum limit and path integrals

In Chapter 4, we have studied statistical models defined on one-dimensional lattices. In particular, in Section 4.7 within the framework of the Gaussian model, we have shown that in the limit of divergent correlation length, it is possible to define a continuum limit provided one considers only quantities defined at a distance-scale proportional to the correlation length. The quantities characteristic of the continuum limit can then be calculated using a formalism of quantum mechanics type, in which any trace of the initial lattice structure has disappeared.

For example, the classical partition function of the Gaussian model (equation (4.32)) converges, in the corresponding limit, toward the partition function of the quantum harmonic oscillator.

Finally, we have shown that the Gaussian example can be generalized and that for a whole class of transfer matrices a continuum limit can be defined. Again, the continuum limit has an interpretation in terms of a quantum partition function.

We show now that, as in the case of the random walk, one can associate to the continuum limit a path integral, which generalizes the path integral (3.63) of the Brownian motion. We first study the Gaussian example which is simpler, and then the general case. However, the purpose of this chapter is not to present a thorough introduction to path integrals, and the interested reader is referred to the specialized literature.

5.1 Gaussian path integrals

The partition function of the Gaussian model on a one-dimensional lattice, with length n and periodic boundary conditions, can be written as (equation (4.32))

$$\mathcal{Z}(n,\theta) = \int \prod_{k=1}^{n} \frac{\mathrm{d}q_k}{\sqrt{2\pi \sinh\theta}} \exp\left[-\mathcal{S}(n,\theta,\mathbf{q})\right]$$

$(q_n = q_0)$ with

$$\mathcal{S}(n,\theta,\mathbf{q}) = \sum_{k=1}^{n} S(q_k, q_{k-1}) \qquad (5.1)$$

and (equation (4.30))

$$S(q,q') = \frac{1}{2\sinh\theta}\left[(q^2 + q'^2)\cosh\theta - 2qq'\right]. \qquad (5.2)$$

In the same limit $\theta \to 0$, $\mathcal{S}(n,\theta,\mathbf{q})$ reduces to

$$\mathcal{S}(n,\theta,\mathbf{q}) \sim \sum_{k=1}^{n}\left[\frac{(q_k - q_{k-1})^2}{2\theta} + \frac{\theta}{2}q_k^2\right].$$

We now associate to the positions k on the lattice the values of a real coordinate t with
$$t_k = k\theta, \quad t_n = n\theta = \beta.$$
Notice that if $\theta \to 0$ at $t = k\theta$ fixed, then k is proportional to $\xi = 1/\theta$, that is, the positions t_k on the lattice are proportional to the correlation length.

We also introduce the function
$$q(t) = q_{k-1} + (q_k - q_{k-1})(t - t_{k-1})/\theta \quad \text{for} \quad t_{k-1} \le t \le t_k,$$
which provides a linear interpolation between the points q_k, as in the example of Figure 3.1.

The following two identities
$$\frac{1}{2} \sum_{k=1}^{n} \frac{(q_k - q_{k-1})^2}{\theta} = \frac{1}{2} \int_0^\beta dt \, \dot{q}^2(t),$$
as well as, for $\theta \to 0$,
$$\theta \sum_{k=1}^{n} q_k^2 \sim \int_0^\beta dt \, q^2(t),$$
allow taking the formal continuum limit, $\theta \to 0$ at $n\theta = \beta$ fixed. One finds
$$\mathcal{S}_0(q) = \lim_{\theta \to 0, \, n\theta = \beta \text{ fixed}} S(n, \theta, \mathbf{q}) = \int_0^\beta dt \left[\tfrac{1}{2}\dot{q}^2(t) + \tfrac{1}{2}q^2(t)\right]. \tag{5.3}$$
The partition function is then given by the Gaussian path integral
$$\mathcal{Z}_0(\beta) = \lim_{\theta \to 0, \, n\theta = \beta \text{ fixed}} \mathcal{Z}(n, \theta) = \int [dq(t)] \exp[-\mathcal{S}_0(q)] \tag{5.4}$$
with $q(0) = q(\beta)$ (periodic boundary conditions), where the notation $\int [dq(t)]$ means the sum over all paths.

The quantity $\mathcal{S}_0(q)$, which is proportional to the classical action of a harmonic oscillator in imaginary time, is also called, in the framework of quantum physics, the *Euclidean action*. The path integral also yields directly the partition function (4.49) of the quantum harmonic oscillator.

This path integral representation can be generalized to the kernel associated to the statistical operator $e^{-\beta \mathbf{H}}$ in the case of the more general quantum Hamiltonians considered in Section 4.8.

Remark. Here, as in the example of the path integral (3.63), the notation $\dot{q}$ seems to indicate that generic paths contributing to the path integral are differentiable. For the same reasons, this is not the case. For $\theta \to 0$, the typical paths that contribute satisfy
$$[q(t+\theta) - q(t)]^2/\theta \underset{\theta \to 0}{=} O(1),$$
like Brownian paths. In particular, $[dq(t)]$ does not represent a measure over paths. The factor $\exp[-\tfrac{1}{2} \int \dot{q}^2 dt]$ belongs to the measure and specifies the space of paths that contribute to the integral.

Nevertheless, the notation is useful because the paths that yields the leading contributions to the path integral are in the vicinity of classical paths, which themselves are differentiable.

5.1.1 Generating functional. Functional derivative. Correlation functions

As in the case of discrete variables, to discuss the algebraic properties of correlation functions, it is useful to introduce the concepts of generating functional, a generalization of the generating functions introduced in Section 2.1, and of functional derivatives.

Generating functional. We first introduce the notion of generating function (also called in this context the generating functional) of correlation functions.

Let $\{F^{(n)}(t_1,\ldots,t_n)\}$, $n = 0, 1, \ldots$, be a sequence of functions *symmetric* in all their arguments. We introduce an additional function of one variable $f(t)$ and consider the following formal series in f:

$$\mathcal{F}(f) = \sum_{n=0}^{\infty} \frac{1}{n!} \int dt_1 \ldots dt_n \, F^{(n)}(t_1,\ldots,t_n) f(t_1) \ldots f(t_n). \tag{5.5}$$

One calls $\mathcal{F}(f)$ the generating functional of the sequence of functions $F^{(n)}$.

More generally, we will admit distributions also for the $F^{(n)}$'s. In such a case, the function $f(t)$ must belong to the corresponding class of test functions and thus be considered implicitly as continuous or even sufficiently differentiable.

Functional derivative. To calculate a function $F^{(n)}$ starting from $\mathcal{F}(f)$, we then need the notion of *functional derivative*, denoted in this work by $\delta/\delta f(t)$.

The functional derivative is defined by the properties that it satisfies the usual algebraic rules of a differential operator:

$$\frac{\delta}{\delta f(t)} [\mathcal{F}_1(f) + \mathcal{F}_2(f)] = \frac{\delta}{\delta f(t)} \mathcal{F}_1(f) + \frac{\delta}{\delta f(t)} \mathcal{F}_2(f),$$

$$\frac{\delta}{\delta f(t)} [\mathcal{F}_1(f)\mathcal{F}_2(f)] = \mathcal{F}_1(f) \frac{\delta}{\delta f(t)} \mathcal{F}_2(f) + \mathcal{F}_2(f) \frac{\delta}{\delta f(t)} \mathcal{F}_1(f) \tag{5.6}$$

and, moreover, for any function $G(t)$,

$$\frac{\delta}{\delta f(u)} \int dt \, G(t) f(t) = G(u). \tag{5.7}$$

The derivative of $\mathcal{F}(f)$, for example, is

$$\frac{\delta}{\delta f(u)} \mathcal{F}(f) = \sum_{n=0}^{\infty} \frac{1}{n!} \int dt_1 \ldots dt_n \, F^{(n+1)}(u, t_1, \ldots, t_n) f(t_1) \ldots f(t_n). \tag{5.8}$$

Then, by differentiating p times and taking the limit $f \equiv 0$, one finds

$$F^{(p)}(t_1,\ldots,t_p) = \left\{ \left(\prod_{i=1}^{p} \frac{\delta}{\delta f(t_i)} \right) \mathcal{F} \right\} \bigg|_{f \equiv 0}. \tag{5.9}$$

Correlation functions. In what follows, we calculate correlation functions of the form
$$Z^{(n)}(t_1,\ldots,t_n) = \langle q(t_1)\ldots q(t_n)\rangle,$$
where $q(t)$ is the value of the random function q at point t, and $\langle \bullet \rangle$ indicates expectation value with respect to some functional measure for the paths $q(t)$.

It is then convenient to introduce the generating function of correlation functions (continuum generalization of the generating functions introduced in Section 2.1)

$$\begin{aligned}
\mathcal{Z}(f) &= \sum_{n=0}^{\infty} \frac{1}{n!} \int dt_1 \ldots dt_n \, Z^{(n)}(t_1,\ldots,t_n) f(t_1) \ldots f(t_n) \\
&= \sum_{n=0}^{\infty} \frac{1}{n!} \int dt_1 \ldots dt_n \, \langle q(t_1) \ldots q(t_n) \rangle \, f(t_1) \ldots f(t_n) \\
&= \left\langle \exp\left[\int dt \, q(t) f(t) \right] \right\rangle. \qquad (5.10)
\end{aligned}$$

One notes
$$\frac{\delta}{\delta f(t_1)} \exp\left[\int dt \, q(t) f(t) \right] = q(t_1) \exp\left[\int dt \, q(t) f(t) \right]$$
and, thus,
$$\left\{ \left(\prod_{i=1}^n \frac{\delta}{\delta f(t_i)} \right) \exp\left[\int dt \, q(t) f(t) \right] \right\} \bigg|_{f \equiv 0} = \prod_{i=1}^n q(t_i).$$

Applied to the functional (5.10) and combined with the identity (5.9), this yields
$$\left\{ \left(\prod_{i=1}^n \frac{\delta}{\delta f(t_i)} \right) \mathcal{Z}(f) \right\} \bigg|_{f \equiv 0} = Z^{(n)}(t_1,\ldots,t_n) = \langle q(t_1) \ldots q(t_n) \rangle. \qquad (5.11)$$

Remark. The formalism can also be extended to the situation where the $F^{(n)}$ are no longer functions but general distributions. For example,
$$\frac{\delta}{\delta f(u)} f(t) = \delta(u-t),$$
$$\frac{\delta}{\delta f(u)} \frac{df(t)}{dt} = \frac{\delta}{\delta f(u)} \frac{d}{dt} \int dz \, \delta(t-z) f(z)$$
$$= \frac{d}{dt} \delta(t-u),$$
where $\delta(t)$ is Dirac's function (more exactly distribution).

In this way, it is possible to derive the equations of the classical motion, in the Lagrangian formalism, by acting on the action with a functional derivative and the variational calculation then takes a purely algebraic form.

Consider, for example, the action, the integral of the Lagrangian,

$$\mathcal{S}(q) = \int dt \left[\tfrac{1}{2}(\dot q(t))^2 - V(q(t))\right].$$

Its functional derivative is

$$\frac{\delta \mathcal{S}}{\delta q(\tau)} = \int dt \left[\dot q(t)\frac{d}{dt}\delta(t-\tau) - V'(q(t))\delta(t-\tau)\right] = -\ddot q(\tau) - V'(q(\tau)).$$

The equation $\delta \mathcal{S}/\delta q(\tau) = 0$ is the equation of the classical motion.

5.1.2 Gaussian correlation functions

In Section 5.1, we have introduced the formal notion of path integral. In Section 2.2, we have shown that the Gaussian integral with a linear term is a generating function of expectation values of polynomials with a Gaussian measure. We now generalize the method to the path integral.

It is convenient to replace the interval $[0, \beta]$ by $[-\beta/2, \beta/2]$, which corresponds to a simple time translation, $t \mapsto t - \beta/2$, but leads to a more transparent discussion of the thermodynamic limit.

We thus consider the path integral

$$\mathcal{Z}_G(b, \beta) = \int_{q(\beta/2)=q(-\beta/2)} [dq(t)] \exp[-\mathcal{S}_G(q, b)] \tag{5.12}$$

with

$$\mathcal{S}_G(q, b) = \int_{-\beta/2}^{\beta/2} dt \left[\tfrac{1}{2}\dot q^2(t) + \tfrac{1}{2}q^2(t) - b(t)q(t)\right]. \tag{5.13}$$

Defining the expectation value of any functional of $q(t)$ by

$$\langle \mathcal{F}(b) \rangle \equiv \mathcal{Z}_0^{-1}(\beta) \int_{q(\beta/2)=q(-\beta/2)} [dq(t)] \mathcal{F}(b) \exp\left[-\mathcal{S}_0(q)\right],$$

we can write $\mathcal{Z}_G(b, \beta)$ as

$$\mathcal{Z}_G(b, \beta) = \mathcal{Z}_0(\beta) \left\langle \exp\left[\int_{-\beta/2}^{\beta/2} dt\, b(t) q(t)\right] \right\rangle.$$

Therefore, $\mathcal{Z}_G(b, \beta)$, as a functional of $b(t)$, is a generating function of correlation functions, generalizing the generating function (2.1) (for details see Section 5.1.1).

5.1.3 Gaussian integral: Explicit calculation

Generalizing the calculation of the Brownian motion, we now show that the Gaussian path integral (5.12) can be calculated without reference to the limit of a discrete process, up to normalization.

To eliminate the term linear in q in $\mathcal{S}_G(q,b)$, one changes variables (a translation for any time t):
$$q(t) \mapsto r(t) = q(t) - q_c(t),$$
where the function $q_c(t)$ is a solution of the classical equation of motion
$$\left.\frac{\delta \mathcal{S}_G}{\delta q(t)}\right|_{q=q_c} = -\ddot{q}_c(t) + q_c(t) - b(t) = 0, \tag{5.14}$$

with the periodic boundary conditions
$$q_c(\beta/2) = q_c(-\beta/2), \quad \dot{q}_c(\beta/2) = \dot{q}_c(-\beta/2). \tag{5.15}$$

The boundary conditions imply $r(\beta/2) = r(-\beta/2)$. Then,
$$\mathcal{S}_G(q,b) = \mathcal{S}_0(r) + \mathcal{S}_G(q_c, b) + \int_{-\beta/2}^{\beta/2} dt\, [\dot{r}(t)\dot{q}_c(t) + r(t)q_c(t) - b(t)r(t)].$$

One integrates by parts the term linear in $\dot{r}$:
$$\int_{-\beta/2}^{\beta/2} dt\, \dot{r}(t)\dot{q}_c(t) = r(\beta/2)\dot{q}_c(\beta/2) - r(-\beta/2)\dot{q}_c(-\beta/2) - \int_{-\beta/2}^{\beta/2} dt\, r(t)\ddot{q}_c(t).$$

The conditions (5.15) imply that the integrated term vanishes. The term linear in $r(t)$ then vanishes because the function $q_c(t)$ is the solution of the differential equation (5.14).

The solution of equation (5.14) can then be written as
$$q_c(t) = \int_{-\beta/2}^{\beta/2} \Delta(t-u) b(u) du, \tag{5.16}$$

where the function $\Delta(t)$ is also the solution of the differential equation
$$-\ddot{\Delta}(t) + \Delta(t) = \delta(t)$$

(δ is Dirac's function) with the periodic boundary conditions
$$\Delta(\beta/2) = \Delta(-\beta/2), \quad \dot{\Delta}(\beta/2) = \dot{\Delta}(-\beta/2).$$

The equation (5.16) is the continuum analogue of equation (2.10) and the kernel $\Delta(t-u)$ is the inverse of the differential operator $-d_t^2 + 1$ with periodic boundary conditions.

The solution of the differential equation is

$$\Delta(t) = \frac{1}{2\sinh(\beta/2)} \cosh(\beta/2 - |t|), \qquad (5.17)$$

where one recognizes the two-point function (4.51).

This can be verified either by inserting the expression into equation (5.16), or directly, in the sense of distributions (sgn(t) is the sign function),

$$\dot{\Delta}(t) = -\frac{\operatorname{sgn}(t)}{2\sinh(\beta/2)} \sinh(\beta/2 - |t|)$$

$$\ddot{\Delta}(t) = \Delta(t) - \frac{\delta(t)}{\sinh(\beta/2)} \sinh(\beta/2 - |t|) = -\delta(t).$$

One then infers

$$\mathcal{S}_{\mathrm{G}}(q_{\mathrm{c}}, b) = \int_{-\beta/2}^{\beta/2} dt \left[\tfrac{1}{2}\dot{q}_{\mathrm{c}}^2(t) + \tfrac{1}{2}q_{\mathrm{c}}^2(t) - b(t)q_{\mathrm{c}}(t) \right]$$

$$= \int_{-\beta/2}^{\beta/2} dt\, q_{\mathrm{c}}(t) \left[-\tfrac{1}{2}\ddot{q}_{\mathrm{c}}(t) + \tfrac{1}{2}q_{\mathrm{c}}(t) - b(t) \right]$$

$$= -\frac{1}{2} \int_{-\beta/2}^{\beta/2} dt\, q_{\mathrm{c}}(t)b(t) = -\frac{1}{2} \int_{-\beta/2}^{\beta/2} dt\, du\, b(t)\Delta(t-u)b(u).$$

The remaining integral

$$\mathcal{N} = \int_{r(\beta/2)=r(-\beta/2)} [dr(t)] \exp[-\mathcal{S}_0(r)],$$

where $\mathcal{S}_0$ is the function (5.3), cannot be evaluated completely in the continuum (see also the discussion of Section 3.5). However, it depends only on β and is equal to the partition function $\mathcal{Z}_0(\beta)$ of the harmonic oscillator (equations (4.49) and (4.50)). Thus,

$$\mathcal{Z}_{\mathrm{G}}(b, \beta) = \mathcal{Z}_0(\beta)\, \mathrm{e}^{-\mathcal{S}_{\mathrm{G}}(q_{\mathrm{c}}, b)}$$

$$= \mathcal{Z}_0(\beta) \exp\left[\frac{1}{2} \int_{-\beta/2}^{\beta/2} du\, dv\, \Delta(v-u)b(v)b(u) \right]. \qquad (5.18)$$

Thermodynamic limit. The function $\Delta(t)$ simplifies in the thermodynamic limit, that is, when $\beta \to \infty$. One then finds

$$\Delta(t) = \frac{1}{2}\mathrm{e}^{-|t|} = \frac{1}{2\pi} \int_{-\infty}^{+\infty} d\kappa\, \frac{\mathrm{e}^{i\kappa t}}{\kappa^2 + 1}. \qquad (5.19)$$

5.2 Gaussian correlations. Wick's theorem

Differentiating twice expression (5.18), one finds the two-point correlation function

$$\langle q(t)q(u)\rangle = \mathcal{Z}_0^{-1}(\beta) \frac{\delta^2}{\delta b(t) \delta b(u)} \mathcal{Z}_G(b,\beta)\bigg|_{b\equiv 0} = \Delta(t-u). \tag{5.20}$$

More generally,

$$\langle q(t_1)q(t_2)\ldots q(t_p)\rangle = \prod_{j=1}^{p} \frac{\delta}{\delta b(t_j)} \exp\left[\frac{1}{2}\int du\, dv\, \Delta(v-u)b(v)b(u)\right]\bigg|_{b\equiv 0}.$$

Each functional derivative, acting on the exponential of the quadratic form, generates a new factor b:

$$\frac{\delta}{\delta b(t_1)} \exp\left[\frac{1}{2}\int du\, dv\, \Delta(v-u)b(v)b(u)\right]$$
$$= \int du_1\, \Delta(t_1-u_1)b(u_1) \exp\left[\frac{1}{2}\int du\, dv\, \Delta(v-u)b(v)b(u)\right].$$

Wick's theorem. The arguments of Section 2.2 apply here again. The only terms that survive in the limit $b \equiv 0$ correspond to all possible pairings of functional derivatives. One recovers the general property of the centred Gaussian measure: all Gaussian correlation functions can be expressed in terms of the two-point function in a way specified by Wick's theorem:

$$\langle q(t_1)q(t_2)\ldots q(t_\ell)\rangle = \sum_{\substack{\text{all possible pairings}\\P\text{ of }\{1,2,\ldots\ell\}}} \Delta(t_{P_1}-t_{P_2})\ldots\Delta(t_{P_{\ell-1}}-t_{P_\ell})$$

$$= \sum_{\substack{\text{all possible pairings}\\P\text{ of }\{1,2,\ldots\ell\}}} \langle q(t_{P_1})q(t_{P_2})\rangle \ldots \langle q(t_{P_{\ell-1}})q(t_{P_\ell})\rangle.$$

(5.21)

5.3 Perturbed Gaussian measure

In various situations, we will be led to study the effect of small deviations from the Gaussian measure. Moreover, we have indicated in Section 4.8 how more general continuum limits can be generated.

For instance, the partition function with periodic boundary conditions, corresponding in the continuum limit to the quantum Hamiltonian

$$\mathbf{H} = \tfrac{1}{2}\hat{\mathbf{P}}^2 + \tfrac{1}{2}\hat{\mathbf{Q}}^2 + V_{\mathrm{I}}(\hat{\mathbf{Q}}), \tag{5.22}$$

is given by the path integral

$$\mathcal{Z}(\beta) = \int [\mathrm{d}q]\exp\left\{-\int_{-\beta/2}^{\beta/2}\left[\tfrac{1}{2}\dot{q}^2(t) + \tfrac{1}{2}q^2(t) + V_\mathrm{I}(q(t))\right]\mathrm{d}t\right\} \qquad (5.23)$$

with $q(-\beta/2) = q(\beta/2)$.

In what follows, we assume that the perturbation

$$V_\mathrm{I}(q) = \sum_{n=1} v_n q^n$$

is a polynomial in the variable q, even if some results generalize to all functions expandable in powers of q.

The expansion of the integrand (5.23) in powers of $V_\mathrm{I}(q)$, leads to

$$\frac{\mathcal{Z}(\beta)}{\mathcal{Z}_0(\beta)} = \sum_{k=0} \frac{(-1)^k}{k!}\left\langle\left[\int \mathrm{d}t\, V_\mathrm{I}(q(t))\right]^k\right\rangle_0$$

$$= \sum_{k=0} \frac{(-1)^k}{k!}\int \mathrm{d}t_1 \mathrm{d}t_2 \ldots \mathrm{d}t_k \langle V_\mathrm{I}(q(t_1))\ldots V_\mathrm{I}(q(t_k))\rangle_0,$$

where $\langle\bullet\rangle_0$ means expectation value with respect to the Gaussian measure $\mathrm{e}^{-S_0}/\mathcal{Z}_0$ (equation (5.3)) with periodic boundary conditions. The arguments given in Section 2.2 apply here also. If $V_\mathrm{I}(q)$ is a polynomial, the successive terms of the expansion can be calculated systematically with the help of Wick's theorem (2.20) in the form (5.21). This provides the basis of perturbation theory.

Remark. It is possible to give a perturbative justification of the path integral (5.23) by taking the continuum limit order by order in the discrete perturbative expansion. The Gaussian two-point function and Wick's theorem being identical in both cases, it is enough to show that sums converge toward integrals.

Perturbative expansion and minimum of the potential. The path integral really depends only on the sum

$$V(q) = \tfrac{1}{2}q^2 + V_\mathrm{I}(q), \qquad (5.24)$$

called the potential in the quantum context. To each decomposition of the function $V(q)$ into a sum of a quadratic term and a remainder $V_\mathrm{I}(q)$, is associated a perturbative expansion. However, the integrand is maximum in the vicinity of the paths that minimize the action. Clearly, the periodic functions that minimize the action are constant functions $q(t) \equiv q_0$, to minimize the kinetic term $\int \dot{q}^2$, whose value q_0 minimizes the potential $V(q)$ and thus

$$V'(q_0) = 0, \quad V''(q_0) \geq 0.$$

The optimal decomposition then consists in setting

$$V(q) = V(q_0) + \tfrac{1}{2}V''(q_0)(q - q_0)^2 + V_\mathrm{I}(q),$$

if $V''(q_0) > 0$. Special problems are associated with the case $V''(q_0) = 0$ as well as to situations where the potential $V(q)$ has several degenerate minima.

Correlation functions. Correlation functions can be expanded in the same way as the partition function. For example, the expansion of the two-point function can be written as

$$\langle q(t)q(u) \rangle = \frac{\mathcal{Z}_0(\beta)}{\mathcal{Z}(\beta)} \left[\langle q(t)q(u) \rangle_0 - \int d\tau \, \langle q(t)q(u) V_I(q(\tau)) \rangle_0 + \tfrac{1}{2} \cdots \right].$$

The expansion of correlation functions can also be derived from the expansion of the generating function corresponding to the perturbed Gaussian measure. This involves expanding directly the path integral

$$\mathcal{Z}(b, \beta) = \int_{q(\beta/2)=q(-\beta/2)} [dq(t)] \exp[-\mathcal{S}(q, b)] \tag{5.25}$$

with

$$\mathcal{S}(q, b) = \int_{-\beta/2}^{\beta/2} dt \left[\tfrac{1}{2}\dot{q}^2(t) + V(q(t)) - b(t)q(t) \right]. \tag{5.26}$$

Again, the function

$$\mathcal{W}(b, \beta) = \ln \mathcal{Z}(b, \beta)$$

generates the connected correlation functions, which have cluster properties (Section 4.4) in the thermodynamic limit $\beta \to \infty$.

5.4 Perturbative calculations: Examples

We consider the distribution $e^{-\mathcal{S}(q)}/\mathcal{Z}$, where

$$\mathcal{S}(q) = \int_{-\beta/2}^{\beta/2} dt \left[\tfrac{1}{2}\dot{q}^2(t) + \tfrac{1}{2}q^2(t) + \frac{1}{4!}\lambda q^4(t) \right], \tag{5.27}$$

and $\mathcal{Z}$ is the partition function:

$$\mathcal{Z}(\beta, \lambda) = \int [dq] \exp\left[-\mathcal{S}(q)\right].$$

Again, we choose periodic boundary conditions.

As an application of Wick's theorem, we calculate the first terms of the expansion of the partition function and of the two- and four-point functions in powers of the parameter λ.

5.4.1 Partition function

The algebra is the same as in Section 2.3. Applying Wick's theorem, at order λ one finds (see expression (2.26) and Figure 2.2)

$$\mathcal{Z}(\beta,\lambda) = \mathcal{Z}_0(\beta)\left[1 - (\lambda/24) \times 3 \times \beta\Delta^2(0)\right] + O(\lambda^2)$$

$$= \frac{1}{2\sinh(\beta/2)}\left[1 - \beta(\lambda/32)\left(\cotanh^2(\beta/2)\right)\right] + O\left(\lambda^2\right),$$

where the expression (5.17) of the Gaussian two-point function has been used.

In the thermodynamic limit, one can neglect terms decreasing exponentially for $\beta \to \infty$ and the expression simplifies drastically:

$$\mathcal{Z}(\beta,\lambda) = e^{-\beta/2}\left(1 - \tfrac{1}{32}\beta\lambda\right) + O(\lambda^2).$$

In particular,

$$\lim_{\beta \to \infty} \frac{1}{\beta}\mathcal{W}(\beta,\lambda) \equiv \frac{1}{\beta}\ln\mathcal{Z}(\beta,\lambda) = -\tfrac{1}{2} - \tfrac{1}{32}\lambda + O(\lambda^2).$$

At next order, one finds (see expression (2.26))

$$\mathcal{Z}(\beta,\lambda)/\mathcal{Z}_0(\beta) = 1 - \tfrac{1}{8}\lambda\beta\Delta^2(0) + \tfrac{1}{128}\lambda^2\beta^2\Delta^4(0) + \tfrac{1}{16}\beta\lambda^2\Delta^2(0)\int_{-\beta/2}^{\beta/2} dt\, \Delta^2(t)$$

$$+ \tfrac{1}{48}\lambda^2\beta\int_{-\beta/2}^{\beta/2} dt\, \Delta^4(t) + O\left(\lambda^3\right), \tag{5.28}$$

where the periodicity of $\Delta(t)$ has been used.

In the logarithm, the non-connected terms, which behave like β^2 for $\beta \to \infty$, cancel. One finds (see expression (2.27) and Figure 2.4)

$$\mathcal{W}(\beta,\lambda) = \ln\mathcal{Z}(\beta,\lambda)$$

$$= \mathcal{W}_0(\beta) - \tfrac{1}{8}\lambda\beta\Delta^2(0) + \tfrac{1}{16}\beta\lambda^2\Delta^2(0)\int_{-\beta/2}^{\beta/2} dt\, \Delta^2(t)$$

$$+ \tfrac{1}{48}\lambda^2\beta\int_{-\beta/2}^{\beta/2} dt\, \Delta^4(t) + O\left(\lambda^3\right),$$

where $\mathcal{W}_0(\beta) = \ln\mathcal{Z}_0(\beta)$.

For $\beta \to \infty$, one can replace the function Δ by its asymptotic form (5.19) because the corrections are exponentially small. Moreover, one can integrate over $t \in (-\infty, +\infty)$ with, again, exponential errors since $\Delta(t)$ decreases exponentially for $t \to \infty$. One finds

$$\lim_{\beta \to \infty} \frac{1}{\beta}\mathcal{W}(\beta,\lambda) = -\tfrac{1}{2} - \tfrac{1}{32}\lambda + \tfrac{7}{1536}\lambda^2 + O(\lambda^3).$$

5.4.2 Correlation functions

Two-point function. We now calculate the two-point function $\langle q(t)q(u)\rangle_\lambda$ corresponding to the measure specified by the function (5.27), at order λ^2, and in the limit $\beta \to \infty$. We perform the calculation with the same periodic boundary conditions.

The algebra is the same as in Section 2.5.1.

Order λ. The expansion at order λ can be written as

$$Z^{(2)}(t,u) \equiv \langle q(t)q(u)\rangle_\lambda$$
$$= \frac{\mathcal{Z}(\beta,0)}{\mathcal{Z}(\beta,\lambda)}\left[\Delta(t-u) - \frac{1}{24}\lambda\int_{-\beta/2}^{\beta/2} d\tau\, \langle q(t)q(u)q^4(\tau)\rangle_0\right] + O(\lambda^2),$$

where the partition function $\mathcal{Z}(\beta,\lambda)$ has been calculated at this order above and $\Delta(t)$ is given in (5.17).

An application of Wick's theorem and the cancellation of non-connected terms in the ratio lead to the expansion at order λ of the expression (2.30), the matrix Δ being replaced by the function $\Delta(t)$ and sums being replaced by integrals:

$$\Delta_{i_1 i_2} \longmapsto \Delta(t_1 - t_2), \quad \sum_i \longmapsto \int_{-\beta/2}^{\beta/2} d\tau.$$

One infers (see also Figure 2.5)

$$Z^{(2)}(t,u) = \Delta(t-u) - \frac{1}{2}\lambda\Delta(0)\int_{-\beta/2}^{\beta/2} d\tau\, \Delta(t-\tau)\Delta(\tau-u) + O(\lambda^2).$$

In the limit $\beta \to \infty$, using the form (5.19) of Δ, one finds

$$Z^{(2)}(t,u) = \tfrac{1}{2} e^{-|t-u|}\left[1 - \tfrac{1}{8}\lambda(1 + |t-u|)\right] + O(\lambda^2). \tag{5.29}$$

It has been shown quite generally in Section 4.3.3 that at large distance

$$Z^{(2)}(t,u) \underset{|t-u|\to\infty}{\sim} A e^{-|t-u|/\xi}.$$

One infers the correction to the correlation length:

$$\xi^{-1} = 1 + \tfrac{1}{8}\lambda + O(\lambda^2).$$

One concludes that the two-point function can, at this order, be written as

$$Z^{(2)}(t,u) = \tfrac{1}{2}(1 - \tfrac{1}{8}\lambda)\, e^{-|t-u|/\xi} + O(\lambda^2).$$

Fourier representation. At this order, the two-point function still has the form of the Gaussian model. Its Fourier transform can be written as

$$\tilde{Z}^{(2)}(\kappa) = \int dt \, e^{-i\kappa t} Z^{(2)}(t,0) = \frac{1}{\kappa^2 + 1/\xi^2} + O(\lambda^2). \tag{5.30}$$

Order λ^2. The order λ^2 can also be derived directly from expression (2.30). One finds (see Figure 2.6)

$$\int_{-\beta/2}^{\beta/2} d\tau_1 \, d\tau_2 \left[\tfrac{1}{4}\Delta^2(0)\Delta(t-\tau_1)\Delta(\tau_2-\tau_1)\Delta(\tau_2-u) + \tfrac{1}{4}\Delta(0)\Delta^2(\tau_2)\Delta(t-\tau_1) \right.$$
$$\left. \times \Delta(u-\tau_1) + \tfrac{1}{6}\Delta(t-\tau_1)\Delta^3(\tau_2-\tau_1)\Delta(\tau_2-u) \right].$$

In the limit $\beta \to \infty$, it is convenient to pass directly to the Fourier representation:

$$\tfrac{1}{4}\Delta^2(0)\left(\kappa^2+1\right)^{-3} + \tfrac{1}{4}\Delta(0) \int d\tau \, \Delta^2(\tau)\left(\kappa^2+1\right)^{-2}$$
$$+ \tfrac{1}{6}\left(\kappa^2+1\right)^{-2} \int d\tau \, e^{-i\kappa\tau} \Delta^3(\tau)$$
$$= \tfrac{1}{16}\left(\kappa^2+1\right)^{-3} + \tfrac{1}{32}\left(\kappa^2+1\right)^{-2} + \tfrac{1}{8}\left(\kappa^2+1\right)^{-2}\left(\kappa^2+9\right)^{-1}.$$

The expansion then takes the form

$$\tilde{Z}^{(2)}(\kappa) = \frac{1}{\kappa^2+1} - \frac{\lambda/4}{(\kappa^2+1)^2} + \frac{\lambda^2/16}{(\kappa^2+1)^3}\left(1 + \tfrac{1}{2}(\kappa^2+1) + 2\frac{\kappa^2+1}{\kappa^2+9}\right) + O(\lambda^3).$$

One shows, quite generally, that the two-point function can be written as a sum of simple poles in κ^2 with positive residues and that the sum of residues is 1. Here one verifies

$$\tilde{Z}^{(2)}(\kappa) = \frac{1-\lambda^2/512}{\kappa^2+1/\xi^2} + \frac{\lambda^2/512}{\kappa^2+9} + O(\lambda^3),$$

with

$$\xi^{-2} = 1 + \tfrac{1}{4}\lambda - \tfrac{3}{64}\lambda^2 + O(\lambda^3) \Rightarrow \xi^{-1} = 1 + \tfrac{1}{8}\lambda - \tfrac{1}{32}\lambda^2 + O(\lambda^3).$$

Connected four-point function. The form of the connected four-point function can, for example, be inferred from expression (2.35). One finds (see also Figure 2.7)

$$W^{(4)}(t_1,t_2,t_3,t_4)$$
$$= -\lambda \int d\tau \, \Delta(t_1-\tau)\Delta(t_2-\tau)\Delta(t_3-\tau)\Delta(t_4-\tau)$$
$$+ \tfrac{1}{2}\lambda^2 \int d\tau_1 \, d\tau_2 \, \{\Delta^2(\tau_1-\tau_2)\left[\Delta(t_1-\tau_1)\Delta(t_2-\tau_1)\Delta(t_3-\tau_2)\Delta(t_4-\tau_2)\right.$$
$$+ \Delta(t_1-\tau_1)\Delta(t_3-\tau_1)\Delta(t_2-\tau_2)\Delta(t_4-\tau_2)$$
$$\left.+\Delta(t_1-\tau_1)\Delta(t_4-\tau_1)\Delta(t_2-\tau_2)\Delta(t_3-\tau_2)\right]$$
$$+\Delta(0)\Delta(\tau_1-\tau_2)\left[\Delta(t_1-\tau_1)\Delta(t_2-\tau_2)\Delta(t_3-\tau_2)\Delta(t_4-\tau_2) + 3 \text{ terms}\right]\}$$
$$+ O(\lambda^3). \tag{5.31}$$

The integrals can be evaluated explicitly, but a more interesting expression is obtained after the Fourier transformation. Since the function $W^{(4)}(t_1, t_2, t_3, t_4)$ is invariant under translation, it is determined by its value at $t_4 = 0$. One then sets

$$\widetilde{W}^{(4)}(\kappa_1, \kappa_2, \kappa_3, \kappa_4) = \int dt_1\, dt_2\, dt_3\, e^{i(\kappa_1 t_1 + \kappa_2 t_2 + \kappa_3 t_3)}\, W^{(4)}(t_1, t_2, t_3, 0), \quad (5.32)$$

where the variable $\kappa_4 = -\kappa_1 - \kappa_2 - \kappa_3$ has been introduced to restore the permutation symmetry between the four points (see Section 6.2.2). One then finds

$$\widetilde{W}^{(4)}(\kappa_1, \kappa_2, \kappa_3, \kappa_4) = \frac{1}{(\kappa_1^2 + 1)(\kappa_2^2 + 1)(\kappa_3^2 + 1)(\kappa_4^2 + 1)}$$

$$\times \left\{ -\lambda + \tfrac{1}{2}\lambda^2 \left[\left(\frac{1}{(\kappa_1 + \kappa_2)^2 + 4} + \frac{1}{(\kappa_1 + \kappa_3)^2 + 4} + \frac{1}{(\kappa_1 + \kappa_4)^2 + 4} \right) \right. \right.$$

$$\left. \left. + \frac{1}{2}\left(\frac{1}{\kappa_1^2 + 1} + \frac{1}{\kappa_2^2 + 1} + \frac{1}{\kappa_3^2 + 1} + \frac{1}{\kappa_4^2 + 1} \right) \right] \right\} + O(\lambda^3). \quad (5.33)$$

The four last terms combined with the term of order λ have the effect of replacing $\kappa^2 + 1$ by $\kappa^2 + 1/\xi^2$ in the denominators.

Exercises

Exercise 5.1

One considers the measure associated with the function

$$S(q) = \int_{-\beta/2}^{\beta/2} dt\, \left[\tfrac{1}{2}\dot{q}^2(t) + \tfrac{1}{2}q^2(t) + \lambda\gamma q^3(t) + \tfrac{1}{2}\lambda^2 q^4(t) \right],$$

where γ is an arbitrary constant. Determine the free energy density in the thermodynamic limit (which is the opposite of the ground state energy of the corresponding quantum Hamiltonian) at order λ^2.

Solution. Expanding in powers of λ, one finds

$$\mathcal{Z}(\beta)/\mathcal{Z}_0(\beta) = 1 - \gamma\lambda \int_{-\beta/2}^{\beta/2} dt\, \langle q^3(t) \rangle - \tfrac{1}{2}\lambda^2 \int_{-\beta/2}^{\beta/2} dt\, \langle q^4(t) \rangle$$

$$+ \tfrac{1}{2}\gamma^2\lambda^2 \int_{-\beta/2}^{\beta/2} dt_1\, dt_2\, \langle q^3(t_1) q^3(t_2) \rangle + O(\lambda^3),$$

where $\mathcal{Z}_0(\beta)$ is given by equation (5.4). Since the Gaussian measure is even, $\langle q^3(t) \rangle = 0$. Introducing the two-point function (5.17) and applying Wick's theorem, one obtains

$$\mathcal{Z}(\beta)/\mathcal{Z}_0(\beta) = 1 - \tfrac{3}{2}\lambda^2 \int_{-\beta/2}^{\beta/2} dt\, \Delta^2(0)$$

$$+ \tfrac{1}{2}\gamma^2\lambda^2 \int_{-\beta/2}^{\beta/2} dt_1\, dt_2\, \left[6\Delta^3(t_1 - t_2) + 9\Delta(t_1 - t_2)\Delta^2(0) \right] + O(\lambda^4)$$

$$= 1 - \tfrac{3}{2}\lambda^2 \beta \Delta^2(0) + 3\gamma^2\lambda^2 \beta \int_{-\beta/2}^{\beta/2} dt\, \left[\Delta^3(t) + \tfrac{3}{2}\Delta(t)\Delta^2(0) \right] + O(\lambda^4),$$

where the periodicity of $\Delta(t)$ has been used. Then,

$$\lim_{\beta \to \infty} \frac{1}{\beta} \ln \mathcal{Z}(\beta) = -\tfrac{1}{2} - \tfrac{1}{8}(3 - 11\gamma^2)\lambda^2 + O(\lambda^4).$$

Exercise 5.2

Determine the free energy density in the thermodynamic limit at order λ for

$$S(q) = \int_{-\beta/2}^{\beta/2} dt \left[\tfrac{1}{2}\dot{q}^2(t) + \tfrac{1}{2}q^2(t) + \lambda q^6(t) \right].$$

Solution.

$$\lim_{\beta \to \infty} \frac{1}{\beta} \ln \mathcal{Z}(\beta) = -\tfrac{1}{2} - \tfrac{15}{8}\lambda + O(\lambda^2).$$

Exercise 5.3

One considers now the action

$$S(\mathbf{q}) = \int_{-\beta/2}^{\beta/2} dt \left[\tfrac{1}{2}\dot{\mathbf{q}}^2(t) + \tfrac{1}{2}\mathbf{q}^2(t) + \tfrac{1}{4}\lambda(\mathbf{q}^2)^2 \right],$$

where $\mathbf{q}$ is a two-component vector (q_1, q_2) and $\mathbf{q}^2$ its length squared. Determine the Gaussian two-point function (i.e., for $\lambda = 0$) $\langle q_i(t) q_j(u) \rangle$. Then, apply Wick's theorem to the calculation of the Gaussian four-point function $\langle q_i(t_1) q_j(t_2) q_k(t_3) q_l(t_4) \rangle$. Expand the path integral at first order in λ and calculate the free energy density in the thermodynamic limit.

Solution. The Gaussian two-point function is

$$\langle q_i(t) q_j(u) \rangle = \delta_{ij} \Delta(t - u),$$

where Δ is the function (5.17).
Then,

$$\langle q_i(t_1) q_j(t_2) q_k(t_3) q_l(t_4) \rangle = \delta_{ij}\delta_{kl}\Delta(t_1 - t_2)\Delta(t_3 - t_4) + \delta_{ik}\delta_{jl}\Delta(t_1 - t_3)\Delta(t_2 - t_4)$$
$$+ \delta_{il}\delta_{jk}\Delta(t_1 - t_4)\Delta(t_2 - t_3).$$

Finally,

$$\lim_{\beta \to \infty} \frac{1}{\beta} \ln \mathcal{Z}(\beta) = -1 - \lambda/2 + O(\lambda^2).$$

6 Ferromagnetic systems. Correlation functions

We now apply a number of concepts and tools that we have presented in the preceding chapters to more general models in classical (i.e., non-quantum) statistical physics, in particular, in arbitrary space dimensions.

The concept of generating functions (or functionals) of correlation functions will again be very useful. This chapter thus introduces various types of correlation functions together with the corresponding generating functions, and recalls some of their properties.

In Section 4.4, we have already mentioned that connected correlation functions, in one space dimension for systems with finite range interactions, have a cluster property. This property generalizes to arbitrary dimensions for systems with short--range interactions (a concept that we will define).

To study phase transitions, it is particularly useful to introduce the thermodynamic potential, Legendre transform of the free energy. Let us recall that the relation between the Lagrangian and Hamiltonian in classical mechanics also has the form of a Legendre transformation. The relation by Legendre transformation between free energy and potential thermodynamic generalizes to a relation between the generating function of connected correlation functions and the generating function of vertex functions (also called proper vertices in the literature). Vertex functions, which, from the viewpoint of Feynman diagrams, receive only one-line irreducible contributions, are also called in quantum field theory 1PI functions (for one-particle irreducible). In the Fourier representation, these functions are more regular than the connected correlation functions from which they are derived, which explains why they are privileged quantities to study.

Finally, we examine the relation between the Legendre transformation and the steepest descent method. We calculate the first terms of the expansions of the two- and four-point vertex functions in a simple example.

Starting with this chapter we adopt, for convenience, a *ferromagnetic language*, even though many physical systems to which these considerations also apply, are not magnetic.

6.1 Ferromagnetic systems: Definition

We consider statistical models defined on a lattice, restricting the discussion to the $\mathbb{Z}^d$ lattice of points with integer coordinates (the generalization of square and cubic lattices) in d-dimensional space. The lattice can represent a physical crystal, but can also be a technical tool in situations where it is useful to approximate continuum space by a lattice.

To each lattice site are associated one or several real random variables, now called *spins* (but 'classical' spins) and denoted by S_i where the subscript i characterizes a site on the lattice and is a symbolic notation for the d integer coordinates:

$$i \equiv (n_1, n_2, \ldots, n_d), \quad n_\mu \in \mathbb{Z}.$$

We denote by $\mathcal{E}(S)$ the configuration energy of the spins in zero external magnetic field, β the inverse of the temperature and $\rho(S)$ the (normalized) spin distribution which weights the spin configurations at each site.

The configuration energy contains interaction terms that couple spins from different sites. We assume that the *interactions are short range or local*, a notion that we define more precisely later, but for which nearest-neighbour interactions provide a simple example. Moreover, we assume that the configuration energy is translation-invariant on the lattice.

Finite volume. We first define systems on a finite subset $\mathcal{C}$ of the lattice $\mathbb{Z}^d$, which we will call a *cube* irrespective of the space dimension d, of the points of integer coordinates

$$0 \leq n_\mu < L, \quad \text{for } 1 \leq \mu \leq d, \quad L \in \mathbb{Z}.$$

The number of points of $\mathcal{C}$ (the volume) thus is $\Omega = L^d$. Moreover, except if stated explicitly otherwise, we assume *periodic boundary conditions* in all directions in such a way that translation invariance is preserved for L finite.

In what follows, all expressions are written for one variable per lattice site but the generalization to several variables is simple.

The partition function, in a uniform magnetic field H/β and for a cube $\mathcal{C}$ of Ω sites, can then be written as

$$\mathcal{Z}_\Omega(H) = \int \left(\prod_{i \in \mathcal{C}} \rho(S_i) \mathrm{d} S_i \right) \exp\left[-\beta \mathcal{E}(S) + H \sum_i S_i \right]. \tag{6.1}$$

We assume that the statistical distribution is such that the partition function exists, for a finite volume Ω and for all values of H. Then, the partition function $\mathcal{Z}_\Omega(H)$ is an entire function of H.

The expectation values of functions of the spins are given by

$$\langle \Phi(S) \rangle_\Omega = \frac{1}{\mathcal{Z}_\Omega} \int \left(\prod_{i \in \mathcal{C}} \rho(S_i) \mathrm{d} S_i \right) \Phi(S) \exp\left[-\beta \mathcal{E}(S) + H \sum_i S_i \right],$$

where $\langle \bullet \rangle_\Omega$ means expectation value in a volume Ω.

Remark. One may wonder why we single out the partition function, which is only the normalization of the statistical distribution? The reason is that, if $\mathcal{Z}_\Omega$ indeed has no direct physical meaning, its logarithmic derivatives with respect to all parameters are expectation values, as the coming section will illustrate.

6.1.1 Mean-spin distribution and free energy

On a finite lattice with Ω sites, the moments of the distribution of the mean spin on the lattice

$$\sigma = \frac{1}{\Omega} \sum_{i \in \mathcal{C}} S_i, \qquad (6.2)$$

corresponding to the partition function (6.1), are given by

$$\langle \sigma^n(H) \rangle_\Omega = \Omega^{-n} \mathcal{Z}_\Omega^{-1}(H) \frac{\partial^n \mathcal{Z}_\Omega(H)}{(\partial H)^n}.$$

The expectation value $\langle \sigma \rangle_\Omega$ is the magnetization.

We have assumed translation invariance on the infinite lattice. Moreover, in the case of a cube $\mathcal{C}$ with periodic boundary conditions, translation invariance is preserved on the finite lattice and, thus, the expectation value $\langle S_i \rangle$ is independent of the site i. One infers

$$\langle \sigma \rangle_\Omega = \frac{1}{\Omega} \sum_{j \in \mathcal{C}} \langle S_j \rangle_\Omega = \langle S_i \rangle_\Omega \quad \forall i.$$

The corresponding free energy density $W_\Omega(H)$ (our definition of the free energy differs from usual definitions, here and later, by a temperature factor because it is generally irrelevant for the questions we want to study) is defined by

$$W_\Omega(H) = \frac{1}{\Omega} \ln \mathcal{Z}_\Omega(H). \qquad (6.3)$$

The positivity of the partition function implies that the function $W_\Omega(H)$ is also infinitely differentiable on a finite lattice. It is a generating function of the cumulants $(\langle \sigma^n \rangle_\Omega)_{\text{conn.}}$ of the mean spin distribution. One verifies

$$(\langle \sigma^n \rangle_\Omega)_{\text{conn.}} = \Omega^{1-n} \frac{\partial^n W_\Omega(H)}{(\partial H)^n}, \qquad (6.4)$$

In particular,

$$\langle \sigma \rangle_\Omega = (\langle \sigma \rangle_\Omega)_{\text{conn.}} = \frac{\partial W_\Omega(H)}{\partial H}.$$

Positivity. As we have already shown, the second derivative $\partial^2 W_\Omega/(\partial H)^2$, proportional to the second cumulant of the distribution of the mean spin, is strictly positive. Indeed, the second derivative can be written as

$$\frac{\partial^2 W_\Omega}{(\partial H)^2} = \frac{\partial \langle \sigma \rangle_\Omega}{\partial H} = \Omega^{-1} \mathcal{Z}_\Omega^{-1}(H) \frac{\partial^2 \mathcal{Z}_\Omega}{(\partial H)^2} - \Omega^{-1} \left(\mathcal{Z}_\Omega^{-1}(H) \frac{\partial \mathcal{Z}_\Omega}{\partial H} \right)^2$$

$$= \Omega \left\langle (\sigma - \langle \sigma \rangle_\Omega)^2 \right\rangle_\Omega. \qquad (6.5)$$

The second derivative vanishes only for a certain value of σ, a trivial situation that we exclude.

6.1.2 Legendre transformation

As we show in Chapter 7, in the study of phase transitions, it is more convenient to work at fixed magnetization than at fixed magnetic field. It is then natural to pass from the free energy to the thermodynamic potential $\mathcal{G}(M)$. The two functions $W(H)$ and $\mathcal{G}(M)$ are related by a Legendre transformation, a transformation that is discussed in more general terms in Section 6.3.

The concept of Legendre transform appears also in other parts of statistical mechanics or, more generally of physics, as in classical mechanics where it relates the Hamiltonian and Lagrangian. The conjugate variables then are the velocity $\dot{q}(t)$ and the conjugate momentum $p(t)$:

$$\mathcal{L}(\dot{q}, q) + H(p, q) = p(t)\dot{q}(t), \quad p(t) = \frac{\partial \mathcal{L}}{\partial \dot{q}(t)}.$$

Legendre transformation: Global definition. Let $W(H)$ be a function of H, everywhere defined and having a continuous and strictly positive second derivative. One calls the Legendre transform of $W(H)$, the function $\mathcal{G}(M)$ defined by the following relations

$$W(H) + \mathcal{G}(M) = HM, \tag{6.6a}$$

$$M = \frac{\partial W(H)}{\partial H}. \tag{6.6b}$$

With the hypothesis about the second derivative,

$$\frac{\partial H}{\partial M} = \left(\frac{\partial M}{\partial H}\right)^{-1} = \left(\frac{\partial^2 W}{(\partial H)^2}\right)^{-1}$$

exists. The transformation then is an involution. Indeed, the first relation implies

$$\frac{\partial \mathcal{G}(M)}{\partial M} = H + \frac{\partial H}{\partial M}\frac{\partial}{\partial H}\bigg|_M (HM - W(H)).$$

Equation (6.6b) then implies that the second term vanishes. Thus,

$$H = \frac{\partial \mathcal{G}(M)}{\partial M}. \tag{6.7}$$

Moreover, differentiating equation (6.6b) with respect to M and using equation (6.7), one finds

$$\frac{\partial^2 W}{(\partial H)^2}\frac{\partial^2 \mathcal{G}}{(\partial M)^2} = 1. \tag{6.8}$$

The second derivative of $\mathcal{G}(M)$ is thus also continuous and positive.

Where it is defined, the Legendre transformation is bijective since the function $W'(H)$ is increasing and continuous. Let us point out, however, that in contrast with $W(H)$, the function $\mathcal{G}(M)$ is not necessarily defined for all values of M but since $W'(H)$ is continuous, the domain where it is defined is connected.

Stationarity. An important algebraic property of the Legendre transformation is the following: let us assume that $W(H)$ is a differentiable function of an additional parameter ε. Then,

$$\frac{\partial \mathcal{G}(M)}{\partial \varepsilon} = \frac{\partial}{\partial \varepsilon}\left[\sum HM - W(H)\right]$$
$$= -\frac{\partial W(H)}{\partial \varepsilon} + \frac{\partial H}{\partial \varepsilon}\frac{\partial}{\partial H}[HM - W(H)]$$

and, thus, from (6.6b)

$$\frac{\partial W(H)}{\partial \varepsilon} + \frac{\partial \mathcal{G}(M)}{\partial \varepsilon} = 0. \tag{6.9}$$

In particular, if $W(H)$ is expandable at $\varepsilon = 0$,

$$W(H) = W_0(H) + \varepsilon W_1(H) + O(\varepsilon^2),$$

and if $\mathcal{G}_0(M)$ is the Legendre transform of $W_0(H)$ and $H^{(0)}(M)$ its derivative, then

$$\mathcal{G}(M) = \mathcal{G}_0(M) - \varepsilon W_1(H^{(0)}(M)).$$

Application. In the physical systems we are studying, the variable M is the expectation value of the spin or magnetization:

$$M = \langle \sigma \rangle.$$

The function $\mathcal{G}(M)$ is the thermodynamic potential density. At finite volume, the free energy $W(H)$ is a regular function and equation (6.5) implies that its second derivative is positive. The Legendre transformation is then globally defined and invertible.

Thermodynamic limit. In the case of local interactions, in the limit of infinite volume Ω (or thermodynamic limit): $\mathcal{C} \to \mathbb{Z}^d$, the free energy density $W(H)$ has a finite limit, a property that generalizes the result obtained in one dimension. From the relation (6.4) between cumulants and derivatives of $W(H)$, one concludes, in particular, that the mean spin tends toward a certain value $\langle \sigma(H) \rangle$. Moreover, for the distributions and values of H for which the second derivative of $W(H)$ is finite, the cumulants of the σ distribution have the behaviour predicted by the central limit theorem, even though the spins are not independent variables.

We have seen (Section 4.4.1) that in one dimension this condition is satisfied if the correlation length is finite and we will show that this property is general.

This question has a direct relation with the invertibility of the Legendre transformation and is intimately connected to the problem of phase transitions. Indeed, if the second derivative of $W(H)$ diverges for some values of H (this implies, in

particular, that the correlation length, defined in (6.13) diverges), $\mathcal{G}''(M)$ vanishes and the relation (6.7)
$$H = \frac{\partial \mathcal{G}(M)}{\partial M},$$
no longer has necessarily a unique solution in M.

Another situation can be realized: the derivative of $W(H)$ has discontinuities, and the domain where $\mathcal{G}(M)$ is defined, is not connected. Then, the Legendre transformation is no longer bijective: this is the situation realized in systems having a phase transition, in the several-phase region.

6.1.3 Mean-spin distribution and thermodynamic potential

The distribution $R_\Omega(\sigma)$ of the mean spin (6.2) is the Fourier transform of the expectation value of $\exp[-ik \sum_i S_i]$ (see Section 3.1):
$$R_\Omega(\sigma) = \frac{\Omega}{2\pi} \int dk\, e^{ik\Omega\sigma} \left\langle e^{-ik \sum_i S_i} \right\rangle.$$

This expectation value is obtained directly from $W_\Omega(H)$ by analytic continuation:
$$\left\langle e^{-ik \sum_i S_i} \right\rangle = \mathcal{Z}_\Omega^{-1}(H)\, e^{\Omega W_\Omega(H-ik)}.$$

Thus,
$$R_\Omega(\sigma) = \frac{\Omega}{2\pi} \mathcal{Z}_\Omega^{-1}(H) \int dk\, e^{\Omega[i\sigma k + W_\Omega(H-ik)]}.$$

In the thermodynamic limit $\Omega \to \infty$, $W_\Omega(H)$ tends toward a finite limit $W(H)$ that remains analytic at least if the correlation length remains finite (see Section 6.2.1). The integral can thus be calculated by the steepest descent method (Section 2.6). The saddle point is given by the equation
$$\sigma = W'(H - ik).$$

The equation has a solution at least for σ and H small enough. At leading order, the distribution $R_\Omega(\sigma)$ then reads
$$R_\Omega(\sigma) \underset{\Omega \to \infty}{\sim} \mathcal{Z}_\Omega^{-1}(H) \left(\frac{\Omega}{2\pi W''(H-ik)} \right)^{1/2} e^{\Omega H \sigma}\, e^{\Omega[-(H-ik)\sigma + W(H-ik)]}.$$

Comparing these expressions with equations (7.3), (7.13) and (7.14), one infers
$$R_\Omega(\sigma) \underset{\Omega \to \infty}{\sim} \mathcal{Z}_\Omega^{-1}(H) \sqrt{\Omega \mathcal{G}''(\sigma)/2\pi}\, e^{\Omega[H\sigma - \mathcal{G}(\sigma)]}, \qquad (6.10)$$

which is an asymptotic relation between the thermodynamic potential and the mean-spin distribution. Let us point out, however, that this relation is valid only in the domain in which $\mathcal{G}(\sigma)$ is defined, a subtlety important in the case of phase transitions.

In the thermodynamic limit, the value of σ tends toward a certain value which is the magnetization M, the absolute minimum of $H\sigma - \mathcal{G}(\sigma)$ and thus a solution of

$$H = \mathcal{G}'(M).$$

If the second derivative $\mathcal{G}''(M)$, which is always non-negative, does not vanish, the distribution can be approximated by a Gaussian distribution

$$R_\Omega(\sigma) \underset{\Omega\to\infty}{\sim} \sqrt{\Omega \mathcal{G}''(M)/2\pi}\, e^{-\Omega \mathcal{G}''(M)(\sigma-M)^2/2}, \qquad (6.11)$$

and σ behaves like the mean of independent random variables. We will show that this condition is fulfilled if the correlation length is finite (see equation (6.13)). Exceptions to this behaviour, infinite correlation length and degenerate saddle points, are all related to the existence of phase transitions.

6.2 Correlation functions. Fourier representation

We consider the partition function $\mathcal{Z}(\mathbf{H})$ of a classical ferromagnetic system of the kind defined in Section 6.1 (equation (6.1)), but with a site-dependent magnetic field:

$$\mathcal{Z}(\mathbf{H}) = \int \left(\prod_{i\in\mathcal{C}} \rho(S_i)\mathrm{d}S_i\right) \exp\left[-\beta\mathcal{E}(S) + \sum_{i\in\mathcal{C}} H_i S_i\right]. \qquad (6.12)$$

We assume that the statistical distribution is such that the partition function exists for all values of $\mathbf{H}$.

The partition function $\mathcal{Z}(\mathbf{H})$ is a generating function of spin correlation functions in a space-dependent field. Indeed,

$$\langle S_{i_1} S_{i_2} \ldots S_{i_n}\rangle = \mathcal{Z}^{-1}(\mathbf{H}) \frac{\partial^n \mathcal{Z}(\mathbf{H})}{\partial H_{i_1} \partial H_{i_2} \ldots \partial H_{i_n}}.$$

In what follows, we consider only correlation functions in a uniform field, which correspond to the limit $H_i = H$ and, in particular, in zero field, $H_i = 0$.

6.2.1 Connected functions and cluster property
The associated free energy

$$\mathcal{W}(\mathbf{H}) = \ln \mathcal{Z}(\mathbf{H})$$

(we recall that our definition of free energy differs from the usual definitions by a temperature factor) is then the generating function of connected correlation functions (see Section 4.4):

$$W^{(n)}_{i_1 i_2 \ldots i_n}(H) = \langle S_{i_1} S_{i_2} \ldots S_{i_n}\rangle_{\text{connected}} = \left.\frac{\partial^n \mathcal{W}(\mathbf{H})}{\partial H_{i_1} \partial H_{i_2} \ldots \partial H_{i_n}}\right|_{H_i=H}.$$

In a constant field, the function $\mathcal{W}(\mathbf{H})$ differs from the function $W(H)$ defined in Section 6.1 by a volume factor.

Cluster property. We now consider the connected n-point function $W^{(n)}_{i_1 i_2 \ldots i_n}$, $n > 1$ in the thermodynamic limit, that is, in infinite volume: $\mathcal{C} \to \mathbb{Z}^d$. We separate the points $i_1, i_2, \ldots, i_n$ into two disjoint, non-empty, subsets E, E'. We define the distance D between these subsets by

$$D = \min_{i \in E, i' \in E'} |i - i'|,$$

where we have denoted by $|i - i'|$ the distance between the points i and i'. Then, the correlation function goes to zero when the separation D diverges:

$$\lim_{D \to \infty} W^{(n)}_{i_1 i_2 \ldots i_n} = 0,$$

a property called the cluster property.

In the absence of a phase transition, or in the disordered phase, the decay is exponential when the separation goes to infinity: one calls the *correlation length* the inverse ξ of the smallest decay rate. In terms of the connected two-point function, it can be defined by

$$\frac{1}{\xi} = \max \lim_{|i-j| \to \infty} -\frac{\ln |\langle S_i S_j \rangle_{\text{conn.}}|}{|i - j|}, \tag{6.13}$$

a definition that generalizes the definition (4.22). When a transfer matrix can be defined (see Sections 7.6 and 7.7) and its largest eigenvalue is not degenerate, it is still given by equation (4.23).

By contrast, in a statistical system where a phase transition occurs, at the transition temperature and in the several-phase region, the decay of correlations may be only algebraic (this corresponds to a divergent correlation length).

When the correlation length ξ is finite, it follows from the definition (6.2), equation (6.5) and translation invariance on the lattice, that

$$W''(H) = \frac{1}{\Omega} \sum_{i,j} \langle S_i S_j \rangle_{\text{conn.}} = \sum_i \langle S_i S_j \rangle_{\text{conn.}} < \infty.$$

When the correlation length is finite, the second derivative $W''(H)$ is thus finite.

The property of exponential decay generalizes to all connected correlation functions. In the case of finite-range interactions, for which a transfer matrix can be defined, the representation (4.17) generalizes and provides again a starting point for a proof.

6.2.2 Translation invariance and Fourier representation

In what follows, we assume that the thermodynamic limit has been taken. Moreover, we now denote by $x_1, x_2, \ldots \in \mathbb{Z}^d$ the points on the d-dimensional lattice. We explicitly use the translation invariance of the statistical model, which implies that for all $a \in \mathbb{Z}^d$, the connected n-point correlation function satisfies

$$W^{(n)}(x_1 + a, \ldots, x_n + a) = W^{(n)}(x_1, \ldots, x_n).$$

We introduce its Fourier transform

$$F^{(n)}(p_1, \ldots, p_n) = \sum_{x_1, \ldots, x_n \in \mathbb{Z}^d} W^{(n)}(x_1, \ldots, x_n) \exp\left(i \sum_{j=1}^n x_j \cdot p_j\right),$$

where the function $F^{(n)}$ is periodic, with period 2π, in all components of the vectors p_j.

Using translation invariance, we take as arguments x_n and $y_j = x_j - x_n$ for $j < n$:

$$W^{(n)}(x_1, x_2, \ldots, x_n) = W^{(n)}(x_1 - x_n, \ldots, x_{n-1} - x_n, 0) = W^{(n)}(y_1, \ldots, y_{n-1}, 0).$$

After this change of variables, the sum becomes

$$F^{(n)}(p_1, \ldots, p_n) = \sum_{x_n} \sum_{y_1, \ldots, y_{n-1}} W^{(n)}(y_1, \ldots, y_{n-1}, 0)$$

$$\times \exp\left(i \sum_{j=1}^{n-1} y_j \cdot p_j + i x_n \cdot \sum_{j=1}^{n} p_j\right).$$

In the sense of distributions

$$\sum_{n \in \mathbb{Z}} e^{in\theta} = 2\pi \delta(\theta),$$

where here $\delta(\theta)$ is Dirac's distribution on the circle, concentrated at $\theta = 0$ mod (2π).

In d dimensions,

$$\sum_{x_n \in \mathbb{Z}^d} e^{i x_n \cdot P} = (2\pi)^d \delta^{(d)}(P) \quad \text{with} \quad P = \sum_i p_i,$$

where the $\delta^{(d)}$ function has for each component P_μ of P the support $P_\mu = 0$ mod (2π).

We then factorize Dirac's distribution $\delta^{(d)}(p_1 + p_2 + \cdots + p_n)$, which is a direct consequence of translation invariance, and set

$$F^{(n)}(p_1, \ldots, p_n) = (2\pi)^d \delta^{(d)}\left(\sum_{i=1}^n p_i\right) \tilde{W}^{(n)}(p_1, \ldots, p_n),$$

where the function

$$\widetilde{W}^{(n)}(p_1,\ldots,p_n) = \sum_{x_1,\ldots,x_{n-1}} W^{(n)}(x_1,\ldots,x_{n-1},0)\exp\left(i\sum_{j=1}^{n-1} x_j\cdot p_j\right) \qquad (6.14)$$

is defined only on the surface $\sum_i p_i = 0$.

Since one point is fixed, the connected function $W^{(n)}(0,x_2\ldots,x_n)$ goes to zero when x_i goes to infinity and $\widetilde{W}^{(n)}(p_1,\ldots,p_n)$ has regularity properties in the variables p_i.

Finally, the relation between the n-point function and its Fourier transform can be written as

$$(2\pi)^d \delta^{(d)}\left(\sum_{i=1}^n p_i\right) \widetilde{W}^{(n)}(p_1,p_2,\ldots,p_n)$$

$$= \sum_{x_1,x_2,\ldots,x_n} W^{(n)}(x_1,x_2,\ldots,x_n)\exp\left(i\sum_{j=1}^n x_j\cdot p_j\right). \qquad (6.15)$$

Terminology. In what follows, as a reference to the quantum formalism, we often call *momenta* the arguments p_i of the Fourier transforms.

Fourier representation and free energy in a uniform field. We first consider a periodic cube $\mathcal{C}$ containing Ω points. In this notation, the generating function $\mathcal{W}_\Omega(H)$ can be expanded in the form

$$\mathcal{W}_\Omega(H) = \sum_{n=0}^\infty \frac{1}{n!}\sum_{x_1,\ldots,x_n\in\mathcal{C}} W_\Omega^{(n)}(x_1,\ldots,x_n) H(x_1)\ldots H(x_n).$$

In a uniform field H, due to translation invariance, the expression becomes

$$\mathcal{W}_\Omega(H) = \sum_{n=0}^\infty \frac{H^n}{n!}\sum_{x_1,\ldots,x_n\in\mathcal{C}} W_\Omega^{(n)}(x_1,\ldots,x_n)$$

$$= \sum_{n=0}^\infty \frac{H^n}{n!}\sum_{y_1,\ldots,y_{n-1},x_n\in\mathcal{C}} W_\Omega^{(n)}(y_1,\ldots,y_{n-1},0)$$

$$= \Omega\sum_{n=0}^\infty \frac{H^n}{n!}\sum_{y_1,\ldots,y_{n-1}\in\mathcal{C}} W_\Omega^{(n)}(y_1,\ldots,y_{n-1},0).$$

In the limit $\Omega\to\infty$, if the sums converge, using the definition (6.14) one obtains

$$W(H) = \lim_{\Omega\to\infty}\mathcal{W}_\Omega(H)/\Omega = \sum_{n=0}^\infty \frac{H^n}{n!}\widetilde{W}^{(n)}(0,\ldots,0). \qquad (6.16)$$

6.3 Legendre transformation and vertex functions

It is useful to generalize the Legendre transformation to the case of local (space-dependent) magnetic or magnetization fields and, thus, to define the Legendre transform for an arbitrary number of variables. This allows, in particular, defining generalized correlation functions called vertex functions (also proper vertices), which play an important technical role in statistical or quantum field theory as a consequence of their regularity properties.

6.3.1 Legendre transformation: Generalization

Let $\mathcal{W}(\mathbf{H})$ be a function of some variables H_i, everywhere defined and having continuous second partial derivatives, such that the matrix

$$W_{ij}^{(2)}(\mathbf{H}) = \frac{\partial^2 \mathcal{W}}{\partial H_i \partial H_j}$$

is strictly positive:

$$\sum_{i,j} X_i W_{ij}^{(2)}(\mathbf{H}) X_j > 0 \quad \forall \mathbf{X} \quad \text{with} \quad |\mathbf{X}| = 1.$$

One calls the Legendre transform of $\mathcal{W}(\mathbf{H})$, the function $\Gamma(\mathbf{M})$ defined by the following relations:

$$\mathcal{W}(\mathbf{H}) + \Gamma(\mathbf{M}) = \sum_i H_i M_i, \tag{6.17a}$$

$$M_i = \frac{\partial \mathcal{W}(\mathbf{H})}{\partial H_i}. \tag{6.17b}$$

One verifies immediately that this transformation is an involution. Indeed, the matrix $\partial H_j/\partial M_i$ exists and the first relation then implies

$$\frac{\partial \Gamma(\mathbf{M})}{\partial M_i} = H_i + \sum_j \frac{\partial H_j}{\partial M_i} \frac{\partial}{\partial H_j} \left[\sum_k H_k M_k - \mathcal{W}(\mathbf{H}) \right].$$

Using (6.17b), one finds that the second term vanishes. Thus,

$$H_i = \frac{\partial \Gamma(\mathbf{M})}{\partial M_i}. \tag{6.18}$$

Note, however, that unlike $\mathcal{W}(\mathbf{H})$, the function $\Gamma(\mathbf{M})$ is not necessarily defined for all values of $\mathbf{M}$. But, due to the continuity of first derivatives, the domain in which it is defined is simply connected.

138   *Ferromagnetic systems. Correlation functions*

From the Legendre transformation, one derives a relation between second derivatives. Differentiating equation (6.17b) with respect to M_j and using equation (6.18), one finds

$$\sum_k \frac{\partial^2 \mathcal{W}}{\partial H_i \partial H_k} \frac{\partial^2 \Gamma}{\partial M_k \partial M_j} = \delta_{ij}. \tag{6.19}$$

The matrix of second derivatives of $\Gamma(\mathbf{M})$ is thus also strictly positive.

Stationarity. The algebraic property (6.9) of the Legendre transformation generalizes immediately. If $\mathcal{W}(\mathbf{H})$ depends in a differentiable way on a parameter ε, then

$$\frac{\partial \mathcal{W}(\mathbf{H})}{\partial \varepsilon} + \frac{\partial \Gamma(\mathbf{M})}{\partial \varepsilon} = 0. \tag{6.20}$$

Invertibility. The Legendre transformation, where it is defined, is bijective. This property is true locally because the matrix of second derivatives is strictly positive.

To prove it globally, we assume that two values $\mathbf{H}$ and $\mathbf{H}'$ correspond to the same value of $\mathbf{M}$. We introduce the linear interpolation

$$\mathbf{H}(t) = \mathbf{H} + t\,(\mathbf{H}' - \mathbf{H}) \quad \Rightarrow \quad \mathbf{M}(\mathbf{H}(1)) = \mathbf{M}(\mathbf{H}(0)).$$

We then consider the function

$$\Phi(t) = \mathcal{W}(\mathbf{H}(t)).$$

Its derivative is given by

$$\Phi'(t) = \sum_i \frac{\partial \mathcal{W}(\mathbf{H}(t))}{\partial H_i} (\mathbf{H}' - \mathbf{H})_i = \sum_i M_i(\mathbf{H}(t))\,(\mathbf{H}' - \mathbf{H})_i.$$

It thus satisfies

$$\Phi'(1) = \Phi'(0).$$

The second derivative

$$\Phi''(t) = \sum_{i,j} (H'_i - H_i) \frac{\partial^2 \mathcal{W}(\mathbf{H}(t))}{\partial H_i \partial H_j} (H'_j - H_j) > 0,$$

is positive since the matrix is positive. The function $\Phi'(t)$ is thus strictly increasing and the condition $\Phi'(1) = \Phi'(0)$ cannot be satisfied.

Statistical physics. On a finite lattice, all the preceding hypotheses are satisfied. The function $\mathcal{W}(\mathbf{H})$ is indefinitely differentiable. Moreover,

$$\frac{\partial \mathcal{W}(\mathbf{H})}{\partial H_i} = \frac{1}{\mathcal{Z}(\mathbf{H})} \frac{\partial \mathcal{Z}}{\partial H_i}$$

and thus

$$W_{ij}^{(2)}(\mathbf{H}) = \frac{\partial^2 \mathcal{W}(\mathbf{H})}{\partial H_i \partial H_j} = \mathcal{Z}^{-1}(\mathbf{H})\frac{\partial^2 \mathcal{Z}}{\partial H_i \partial H_j} - \mathcal{Z}^{-2}(\mathbf{H})\frac{\partial \mathcal{Z}}{\partial H_i}\frac{\partial \mathcal{Z}}{\partial H_j}.$$

Using the definition (6.12), one finds that the connected two-point function can be written as (see also the relation (2.34) and Section 3.3.3)

$$W_{ij}^{(2)}(\mathbf{H}) = \langle (S_i - \langle S_i \rangle)(S_j - \langle S_j \rangle) \rangle,$$

where $\langle \bullet \rangle$ means expectation value in the presence of the field H_i. Taking the expectation value of this symmetric matrix in a non-vanishing vector X_i, one finds

$$\sum_{i,j} X_i X_j W_{ij}^{(2)}(\mathbf{H}) = \left\langle \left[\sum_i X_i (S_i - \langle S_i \rangle)\right]^2 \right\rangle \geq 0, \qquad (6.21)$$

equality being possible only for a certain value of S, a trivial situation that we exclude. The matrix $W_{ij}^{(2)}(\mathbf{H})$ is thus strictly positive, which implies that all its eigenvalues are strictly positive.

Thermodynamic limit. In the thermodynamic limit, the invertibility question becomes more subtle and is intimately related to the existence of phase transitions. Indeed, $W_{ij}^{(2)}(\mathbf{H})$ can have eigenvalues that accumulate at infinity for some values of $\mathbf{H}$. This corresponds to a phase transition point. Then, $\partial^2 \Gamma(\mathbf{M})/\partial M_i \partial M_j$ can have vanishing eigenvalues and the relation (6.18)

$$H_i = \frac{\partial \Gamma(\mathbf{M})}{\partial M_i},$$

is no longer locally invertible.

Another situation can be realized where the partial derivatives of $\mathcal{W}(\mathbf{H})$ have discontinuities. Then, the Legendre transformation is no longer bijective: this is the situation realized in systems having a phase transition in the several-phase region.

6.3.2 Vertex functions

The coefficients of the expansion of the thermodynamic potential in powers of the local magnetization $M(x)$,

$$\Gamma(\mathbf{M}) = \sum_{n=0} \frac{1}{n!} \sum_{x_1, x_2, \ldots, x_n} \Gamma^{(n)}(x_1, x_2, \ldots, x_n) M(x_1) M(x_2) \ldots M(x_n), \qquad (6.22)$$

are called in this work *vertex functions*. They are also called *1-irreducible* correlation functions (for reasons that will be explained in Chapter 12), or proper vertices.

In models with short-range interactions (i.e., the statistical models we consider in this work), these functions have better decay properties than connected correlation functions and thus, for translation-invariant systems, their Fourier transforms

$$(2\pi)^d \delta \left(\sum_{i=1}^n p_i \right) \tilde{\Gamma}^{(n)}(p_1, \ldots, p_n) = \sum_{x_1, \ldots, x_n} \Gamma^{(n)}(x_1, \ldots, x_n) \exp\left(i \sum_{j=1}^n x_j p_j \right) \tag{6.23}$$

have better regularity properties.

Two-point functions. In a translation-invariant system, an application of the relation (6.14) to the case $n = 2$, yields

$$\widetilde{W}^{(2)}(p) = \sum_x W^{(2)}(0, x) \, e^{ix \cdot p}, \tag{6.24}$$

where the second argument $-p$ is generally omitted.
The positivity condition simply becomes

$$\widetilde{W}^{(2)}(p) > 0.$$

Similarly,

$$\widetilde{\Gamma}^{(2)}(p) = \sum_x \Gamma^{(2)}(0, x) \, e^{ix \cdot p}. \tag{6.25}$$

The relation between Fourier transforms is then algebraic:

$$\widetilde{W}^{(2)}(p) \widetilde{\Gamma}^{(2)}(p) = 1. \tag{6.26}$$

Remark. This property generalizes to the n-point function. In a translation-invariant system, in the Fourier representation, the relations between connected and vertex functions involve no momentum integration.

Thermodynamic potential in a uniform field and Fourier representation. Using the arguments of Section 6.2.2, which have led to the representation (6.16), one relates the thermodynamic potential density and Fourier components:

$$\mathcal{G}(M) = \sum_{n=0}^\infty \frac{M^n}{n!} \tilde{\Gamma}^{(n)}(0, \ldots, 0). \tag{6.27}$$

6.3.3 Gaussian model

We consider the partition function given by the Gaussian integral

$$\mathcal{Z}(\mathbf{H}) = \int \left(\prod_i \mathrm{d}S_i \right) \exp\left[-\mathcal{H}(\mathbf{S}) + \sum_i H_i S_i \right]$$

with
$$\mathcal{H}(\mathbf{S}) = \frac{1}{2} \sum_{i,j} \mathfrak{S}_{ij} S_i S_j, \qquad (6.28)$$

where the matrix $\mathfrak{S}_{ij}$ is strictly positive. Then,

$$\mathcal{W}(\mathbf{H}) - \mathcal{W}(0) = \ln(\mathcal{Z}(\mathbf{H})/\mathcal{Z}(0)) = \frac{1}{2} \sum_{i,j} H_i \left[\mathfrak{S}^{-1}\right]_{ij} H_j.$$

One infers the local magnetization

$$M_i = \frac{\partial \mathcal{W}(\mathbf{H})}{\partial H_i} = \sum_j \left[\mathfrak{S}^{-1}\right]_{ij} H_j \;\Rightarrow\; H_i = \sum_j \mathfrak{S}_{ij} M_j.$$

Finally, the thermodynamic potential is given by

$$\Gamma(\mathbf{M}) = -\mathcal{W}(0) + \frac{1}{2} \sum_{i,j} M_i \mathfrak{S}_{ij} M_j.$$

The Legendre transform of a quadratic form is again a quadratic form. Moreover, $\Gamma(\mathbf{M})$ is directly related to the quadratic form (6.28) that appears in the initial integral, up to an additive constant. Indeed,

$$\Gamma(\mathbf{M}) = -\mathcal{W}(0) + \mathcal{H}(\mathbf{M}).$$

Translation invariance. In a translation-invariant system and in a uniform field $H_i = H$, the magnetization is uniform. Moreover, for a cube $\mathcal{C}$ in $\mathbb{Z}^d$ with volume Ω and periodic boundary conditions

$$\mathfrak{S}_{ij} \equiv \mathfrak{S}(x_i - x_j).$$

The thermodynamic potential density is then

$$\mathcal{G}(M) = \frac{1}{\Omega}\Gamma(M) = \mathcal{G}(0) + \frac{1}{2\Omega} \sum_{x,x' \in \mathcal{C}} \mathfrak{S}(x - x') M^2 = \mathcal{G}(0) + \frac{1}{2} \sum_{x \in \mathcal{C}} \mathfrak{S}(x) M^2.$$

If the sum over x converges, one can take the limit of infinite volume. Introducing the Fourier transform of the function $\mathfrak{S}(x)$,

$$\tilde{\mathfrak{S}}(k) = \sum_x e^{ik \cdot x} \mathfrak{S}(x),$$

one can rewrite the expression as

$$\mathcal{G}(M) = \mathcal{G}(0) + \tfrac{1}{2}\tilde{\mathfrak{S}}(0) M^2. \qquad (6.29)$$

6.4 Legendre transformation and steepest descent method

The relation between the Legendre transformation and the steepest descent method explains partially the important role played by the thermodynamic potential.

We consider the partition function in an external field H_i,

$$\mathcal{Z}(\mathbf{H}) = \int \left(\prod_i \mathrm{d}S_i\right) \mathrm{e}^{-\mathcal{H}(\mathbf{S}) + \sum_i H_i S_i},$$

where $\mathcal{H}$, as a function of the variables S_i, has analytic properties such that the steepest descent method is applicable to the calculation of the integral.

The saddle points are given by (see Section 2.7.2)

$$H_i = \frac{\partial \mathcal{H}}{\partial S_i}. \tag{6.30}$$

We assume that the equation has a unique solution. The leading order is obtained by replacing $\mathbf{S}$ in the integrand by its saddle point value. One finds

$$\mathcal{W}(\mathbf{H}) = \mathcal{W}_0(\mathbf{H}) = -\mathcal{H}(\mathbf{S}) + \sum_i H_i S_i,$$

where $\mathbf{S}$ is a function of $\mathbf{H}$ through the saddle point equation (6.30). One observes that the relation between $\mathcal{H}$ and $\mathcal{W}$ is a Legendre transformation. Moreover,

$$M_i = \frac{\partial \mathcal{W}}{\partial H_i} = S_i(\mathbf{H})$$

and, thus,

$$\Gamma(\mathbf{M}) = \Gamma_0(\mathbf{M}) = \mathcal{H}(\mathbf{M}).$$

The second order of the steepest descent method is then given by the Gaussian integration (equation (2.51)). Neglecting factors 2π, one obtains

$$\mathcal{W}_1(\mathbf{H}) = -\frac{1}{2} \operatorname{tr} \ln \frac{\partial^2 \mathcal{H}}{\partial S_i \partial S_j}.$$

Using the stationarity property (6.20), one infers

$$\Gamma_1(\mathbf{M}) = \frac{1}{2} \operatorname{tr} \ln \frac{\partial^2 \mathcal{H}(\mathbf{M})}{\partial M_i \partial M_j}. \tag{6.31}$$

Applied to the Gaussian example, the steepest descent method is clearly exact.

More generally, the functionals $\mathcal{W}(\mathbf{H})$ and $\Gamma(\mathbf{M})$ can be expanded to all orders of the steepest descent method:

$$\mathcal{W}(\mathbf{H}) = \sum_{\ell=0}^\infty \mathcal{W}_\ell(\mathbf{H}), \quad \Gamma(\mathbf{M}) = \sum_{\ell=0}^\infty \Gamma_\ell(\mathbf{M}).$$

We shall see in Section 12.4 that, from the viewpoint of Feynman diagrams, ℓ then counts the number of loops of diagrams.

6.5 Two- and four-point vertex functions

We have exhibited the relation between connected and vertex two-point functions. We now establish, as an exercise that will be useful later, the relation between four-point functions in the case of spin distributions invariant under the reflection $S_i \mapsto -S_i$.

One starts from the previously established identity

$$\sum_k \frac{\partial^2 \Gamma}{\partial M_i \partial M_k} \frac{\partial^2 \mathcal{W}}{\partial H_k \partial H_j} = \delta_{ij}.$$

One differentiates with respect to M_l,

$$\sum_k \frac{\partial^3 \Gamma}{\partial M_l \partial M_i \partial M_k} \frac{\partial^2 \mathcal{W}}{\partial H_k \partial H_j} + \sum_{k,m} \frac{\partial^2 \Gamma}{\partial M_i \partial M_k} \frac{\partial^3 \mathcal{W}}{\partial H_m \partial H_k \partial H_j} \frac{\partial^2 \Gamma}{\partial M_m M_l} = 0, \qquad (6.32)$$

where the relations

$$H_m = \frac{\partial \Gamma}{\partial M_m} \Rightarrow \frac{\partial H_m}{\partial M_l} = \frac{\partial^2 \Gamma}{\partial M_m \partial M_l},$$

have again been used. Identity (6.32) can then be rewritten as

$$\frac{\partial^3 \Gamma}{\partial M_{i_1} \partial M_{i_2} \partial M_{i_3}} = -\sum_{j_1,j_2,j_3} \frac{\partial^3 \mathcal{W}}{\partial H_{j_1} \partial H_{j_2} \partial H_{j_3}} \prod_{k=1}^{3} \frac{\partial^2 \Gamma}{\partial M_{j_k} \partial M_{i_k}}. \qquad (6.33)$$

We introduce the notation

$$W^{(4)}_{i_1 i_2 i_3 i_4} = \frac{\partial^4 \mathcal{W}}{\partial H_{i_1} \partial H_{i_2} \partial H_{i_3} \partial H_{i_4}}\bigg|_{H=0},$$

$$\Gamma^{(4)}_{i_1 i_2 i_3 i_4} = \frac{\partial^4 \Gamma}{\partial M_{i_1} \partial M_{i_2} \partial M_{i_3} \partial M_{i_4}}\bigg|_{M=0}.$$

One again differentiates (6.33) with respect to M and takes the limit $M = 0$. Due to reflection symmetry, correlation functions in zero field of an odd number of spins vanish in the symmetric phase. This applies here to the three-point function. One infers

$$\Gamma^{(4)}_{i_1 i_2 i_3 i_4} = -\sum_{j_1,j_2,j_3,j_4} W^{(4)}_{j_1 j_2 j_3 j_4} \prod_{k=1}^{4} \Gamma^{(2)}_{i_k j_k}, \qquad (6.34)$$

or equivalently

$$W^{(4)}_{j_1 j_2 j_3 j_4} = -\sum_{i_1,i_2,i_3,i_4} \Gamma^{(4)}_{i_1 i_2 i_3 i_4} \prod_{k=1}^{4} W^{(2)}_{i_k j_k}.$$

The general identity, including a non-trivial three-point function, has the graphic representation displayed in Figure 12.4 in Section 12.1.

Example. In Section 2.5, we have expanded the connected two- and four-point correlation functions with the distribution $e^{-\mathcal{H}(x)}/\mathcal{Z}$, where

$$\mathcal{H}(x) = \sum_{i,j=1}^{n} \tfrac{1}{2}x_i A_{ij} x_j + \frac{1}{4!}\lambda \sum_{i=1}^{n} x_i^4, \qquad (6.35)$$

up to order λ^2. We have expressed them in terms of the two-point function $\boldsymbol{\Delta}$ of the Gaussian limit, the inverse of the matrix $\mathbf{A}$ with elements A_{ij}:

$$\boldsymbol{\Delta} = \mathbf{A}^{-1}.$$

The application of identity (6.34) to the perturbative expansions (2.30) and (2.35) of the two- and four-point functions leads to

$$\Gamma^{(2)}_{i_1 i_2} = A_{i_1 i_2} + \tfrac{1}{2}\lambda \delta_{i_1 i_2}\Delta_{i_1 i_1} - \lambda^2 \left(\tfrac{1}{4}\delta_{i_1 i_2}\Delta^2_{i_1 j}\sum_{j}\Delta_{jj} + \tfrac{1}{6}\Delta^3_{i_1 i_2} \right) + O(\lambda^3).$$

$$\Gamma^{(4)}_{i_1 i_2 i_3 i_4} = \lambda \delta_{i_1 i_2}\delta_{i_1 i_3}\delta_{i_1 i_4} - \tfrac{1}{2}\lambda^2 \delta_{i_1 i_2}\delta_{i_3 i_4}\Delta^2_{i_1 i_3} - \tfrac{1}{2}\lambda^2 \delta_{i_1 i_3}\delta_{i_2 i_4}\Delta^2_{i_1 i_4}$$
$$- \tfrac{1}{2}\lambda^2 \delta_{i_1 i_4}\delta_{i_2 i_3}\Delta^2_{i_1 i_2} + O(\lambda^3). \qquad (6.36)$$

One notes that in the vertex functions all one-reducible Feynman diagrams, that is, diagrams that can be disconnected by cutting only one line, have cancelled (Figures 2.5, 2.6, 2.7).

If the subscripts are associated with the sites of a cubic lattice with coordinates x and if translation invariance is assumed,

$$A_{ij} \equiv A(x(i) - x(j)) \;\Rightarrow\; \Delta_{ij} \equiv \Delta(x(i) - x(j)),$$

the equations can be rewritten as

$$\Gamma^{(2)}(x) = A(x) + \tfrac{1}{2}\lambda \delta(x)\Delta(0) - \lambda^2 \left(\tfrac{1}{4}\delta(x)\Delta(0)\sum_{y}\Delta^2(y) + \tfrac{1}{6}\Delta^3(x) \right) + O(\lambda^3)$$

and

$$\Gamma^{(4)}(x_1, x_2, x_3, x_4) = \lambda \delta(x_1 - x_2)\delta(x_1 - x_3)\delta(x_1 - x_4)$$
$$- \tfrac{1}{2}\lambda^2 \delta(x_1 - x_2)\delta(x_3 - x_4)\Delta^2(x_1 - x_3)$$
$$- \tfrac{1}{2}\lambda^2 \delta(x_1 - x_3)\delta(x_2 - x_4)\Delta^2(x_1 - x_4)$$
$$- \tfrac{1}{2}\lambda^2 \delta(x_1 - x_4)\delta(x_2 - x_3)\Delta^2(x_1 - x_2) + O(\lambda^3). \qquad (6.37)$$

Introducing the Fourier representation of the Gaussian two-point function,

$$\Delta(x) = \frac{1}{(2\pi)^d}\int d^d p \; e^{-ip\cdot x}\, \tilde{\Delta}(p), \qquad A(x) = \frac{1}{(2\pi)^d}\int d^d p \; e^{-ip\cdot x}\, \tilde{A}(p),$$

and thus
$$\tilde{A}(p)\tilde{\Delta}(p) = 1,$$

(the components of the vector p vary in the interval of size 2π), one can rewrite the expansions, in the Fourier representation, as

$$\tilde{\Gamma}^{(2)}(p) = \tilde{A}(p) + \tfrac{1}{2}\lambda O - \tfrac{1}{4}\lambda^2 OB(0) - \tfrac{1}{6}\lambda^2 C(p) + O(\lambda^3), \quad (6.38a)$$

$$\tilde{\Gamma}^{(4)}(p_1, p_2, p_3, p_4) = \lambda - \tfrac{1}{2}\lambda^2 \left[B(p_1+p_2) + B(p_1+p_3) + B(p_1+p_4) \right]$$
$$+ O(\lambda^3) \quad (6.38b)$$

with

$$O = \frac{1}{(2\pi)^d} \int d^d k\, \tilde{\Delta}(k), \qquad (6.39a)$$

$$B(p) = \frac{1}{(2\pi)^d} \int d^d k\, \tilde{\Delta}(k)\tilde{\Delta}(p-k), \qquad (6.39b)$$

$$C(p) = \frac{1}{(2\pi)^{2d}} \int d^d k_1\, d^d k_2\, \tilde{\Delta}(k_1)\tilde{\Delta}(k_2)\tilde{\Delta}(p-k_1-k_2). \qquad (6.39c)$$

Exercises

Exercise 6.1

Steepest descent method and perturbative expansion. Recover the contributions to the two- and four-point vertex functions at order λ and λ^2, respectively, in the expansion (6.36), by expanding in powers of λ expression (6.31), derived by the steepest descent method.

Solution. In example (6.35), expression (6.31) reads

$$\Gamma_1(\mathbf{M}) = \frac{1}{2} \operatorname{tr} \ln \frac{\partial^2 \mathcal{H}(\mathbf{M})}{\partial M_i \partial M_j} = \frac{1}{2} \operatorname{tr} \ln \left(A_{ij} + \tfrac{1}{2}\lambda M_i^2 \delta_{ij} \right).$$

Since $\operatorname{tr} \ln = \ln \det$,

$$\operatorname{tr} \ln \mathbf{AB} = \operatorname{tr} \ln \mathbf{A}\, \operatorname{tr} \ln \mathbf{B} \quad \Rightarrow \quad \operatorname{tr} \ln(\mathbf{A}+\mathbf{B}) = \operatorname{tr} \ln \mathbf{A} + \operatorname{tr} \ln(1 + \mathbf{BA}^{-1}).$$

Thus,
$$\Gamma_1(\mathbf{M}) - \Gamma_1(0) = \tfrac{1}{2} \operatorname{tr} \ln \left(1 + \tfrac{1}{2}\lambda \mathbf{K}\right)$$

with $(\mathbf{\Delta A} = 1)$
$$K_{ij} = M_i^2 \Delta_{ij}.$$

Using
$$\operatorname{tr} \ln(1 + \tfrac{1}{2}\lambda \mathbf{K}) = \tfrac{1}{2}\lambda \operatorname{tr} \mathbf{K} - \tfrac{1}{8}\lambda^2 \operatorname{tr} \mathbf{K}^2 + O(\lambda^3),$$

one obtains

$$\Gamma_1(\mathbf{M}) - \Gamma_1(\mathbf{0}) = \tfrac{1}{4}\lambda \sum_i M_i^2 \Delta_{ii} - \tfrac{1}{16}\lambda^2 \sum_{i,j} M_i^2 \Delta_{ij}^2 M_j^2 + O(\lambda^3).$$

Differentiating first twice then four times with respect to M_i, one recovers, in the limit $M_i = 0$, the contributions of order λ to $\Gamma^{(2)}$ and λ^2 to $\Gamma^{(4)}$ in (6.36):

$$\left.\frac{\partial^2 \Gamma_1}{\partial M_{i_1} \partial M_{i_2}}\right|_{\mathbf{M}=0} = \tfrac{1}{2}\lambda \Delta_{i_1 i_1} \delta_{i_1 i_2},$$

$$\left.\frac{\partial^4 \Gamma_1}{\partial M_{i_1} \partial M_{i_2} \partial M_{i_3} \partial M_{i_4}}\right|_{\mathbf{M}=0} = -\tfrac{1}{2}\lambda^2 \delta_{i_1 i_2} \delta_{i_3 i_4} \Delta_{i_1 i_3}^2 - \tfrac{1}{2}\lambda^2 \delta_{i_1 i_3} \delta_{i_2 i_4} \Delta_{i_1 i_4}^2$$
$$- \tfrac{1}{2}\lambda^2 \delta_{i_1 i_4} \delta_{i_2 i_3} \Delta_{i_1 i_2}^2.$$

7 Phase transitions: Generalities and examples

Before beginning the study of continuous or second-order phase transitions within the framework of classical statistical mechanics, we want to point out a few subtleties of the general concept of phase transition. For what follows, it is useful, but not essential, to have some knowledge of the elementary phenomenology of phase transitions in simple systems like liquid–vapour or magnetic systems.

A first important remark is that phase transitions are possible only in infinite volume systems, that is, systems having an infinite number of degrees of freedom. This shows already that the concept of phase transitions is in itself not completely trivial.

The models that we will examine, and which exhibit phase transitions, have the following property: according to the value of a control parameter, in general the temperature, the system can be in a region with one or several phases. These phases are characterized by different sensitivities to boundary conditions. The region with only one phase does not keep track of the specific way the thermodynamic limit, that is, the infinite volume limit, is taken. The situation is different in the region with several phases where, for example, some correlation functions depend on the way the thermodynamic limit is taken. Each distinct limit corresponds to a phase.

For the simple models that we will study, it is possible to find local observables whose values discriminate between the various phases. Such observables are called *order parameters*. For example, the spin is an order parameter for ferromagnetic transitions.

Moreover, in these models, the phase transition is associated with a *spontaneous breaking of symmetry*. For example, the statistical distribution of the Ising model does not change when all spins are reversed. One would thus expect that the expectation value of a spin vanishes.

However, if one adds to the interaction energy a term which breaks the symmetry of the system explicitly (a magnetic field for a ferromagnetic system), one takes the infinite volume limit and then the limit of vanishing breaking term, two situations may occur: in the single-phase region, symmetry is restored in the sense that all correlation functions have the symmetry of the model; on the contrary, in a phase with spontaneous symmetry breaking, the thermodynamic limit and the limit of vanishing breaking term do not commute. In the case of spins, one finds a non-vanishing spin expectation value, that is, a *spontaneous magnetization*. The direction of the spontaneous magnetization depends on the direction of the magnetic field when it goes to zero.

Another simple and rather general characterization of a phase transition is dynamical. One calls here configuration space the whole set of possible configurations of a system. For example, in the Ising model, a lattice model with classical spins that can take only two values ±1, the configuration space for Ω spins has 2^Ω elements.

Then, we consider random or deterministic dynamics in the space of configurations that admits, as an asymptotic distribution (one also speaks of an *equilibrium distribution*), the Boltzmann weight of a model exhibiting a phase transition. An example of such dynamics is the evolution equation (3.21) of the random walk, which leads to a Gaussian distribution (but which does not exhibit a phase transition). The transitions which we will consider then have the following property: as long as the volume of the system is finite, any element in configuration space has a non-vanishing probability to be reached during time evolution and this for any temperature (if the system is not discrete as in the Ising model, any element must be replaced by any arbitrarily small volume in configuration space) and for any starting point. One calls the system *ergodic*. If the system then converges toward a thermodynamic state of equilibrium, that is, a probability distribution invariant under the dynamics, time averages tend toward averages calculated by summing over all configurations in configuration space with the Boltzmann weight.

By contrast, in the infinite volume limit (with fixed density for a system of particles), according to the values of the temperature, the system can either remain ergodic, or on the contrary undergo a *breaking of ergodicity*. In the latter case, configuration space breaks up into disjoint subsets. When the system is prepared initially in one of these subsets, it remains. For example, for an Ising type system below the critical temperature, the two subsets correspond to the two opposite values of the spontaneous magnetization.

Our goal is to analyse the behaviour of thermodynamic quantities in the vicinity of a transition, in particular their singularities as functions of the temperature. We have already pointed out the role of the *correlation length*. The transitions that we will study in this work are those for which the correlation length diverges at the transition.

However, before discussing phase transitions with more elaborate methods, we illustrate some elementary aspects of the general properties described above by studying the behaviour of some simple ferromagnetic systems on a lattice. We begin the study of phase transitions with a simple, but slightly pathological model, corresponding, depending on the interpretation, to a limit of a space of infinite dimension or to long-range forces. This model can be solved by elementary methods and presents a phase transition of *mean-field* field type, sharing some properties with the *quasi-Gaussian model* that will be discussed in Chapter 8.

We exhibit the universal behaviour at the transition (called critical behaviour) of several thermodynamic quantities.

We then examine the properties of a ferromagnetic system with *nearest-neighbour interactions* at fixed space dimension. Such a system does not admit an exact solution in general. However, low- and high-temperature arguments (which can be made rigorous) make it possible to convincingly demonstrate, in the limit of infinite volume, the existence of phase transitions in Ising type systems, that is, with a reflection symmetry.

In the case of nearest-neighbour interactions, a transfer matrix can be defined. The thermodynamic (or infinite volume) limit is then dominated by the largest eigenvalue of the matrix. The possibility of a phase transition is related to the divergence of the correlation length and, thus, to a degeneracy of the leading (multiple) eigenvalue.

The Ising model is a characteristic example of models with discrete symmetries, that is, corresponding to finite groups. We then extend the analysis to ferromagnetic systems with continuous symmetries, whose properties are different in low space dimensions. An important conclusion, which relies on intuitive arguments (but which can be made rigorous) is the following: phase transitions with short-range interactions and spontaneous symmetry breaking, of the kind we want to study, are not possible below dimension 2 for discrete symmetry groups and 3 for continuous groups.

Ferromagnetic models of Ising type. The examples that we study in this chapter belong again to the class of classical ferromagnetic models introduced in Chapter 6. To each site of the cubic lattice are associated (real) random variables, classical spins S_i, where the index i characterizes the site of the lattice (i is a symbolic notation for the whole d coordinates, that is, an element of $\mathbb{Z}^d$).

These classical spins are already spins averaged over a physical volume large at the scale of microscopic interactions (which justifies their classical character), but small on the scale of the phenomena we want to investigate. In particular, this justifies also admitting spins with a continuous distribution even when the initial distribution is discrete.

The spin statistical distribution, or Boltzmann weight, has the form

$$\mathcal{Z}^{-1}(H) \left(\prod_i \rho(S_i) \right) \exp\left[-\beta \mathcal{E}(\mathbf{S}) + H \sum_i S_i \right]$$

where β is the inverse temperature and H, the magnetic field, includes a temperature factor, $\beta H \mapsto H$. $\rho(S)$ is the (normalized) spin distribution in each site and $\mathcal{E}(\mathbf{S})$ is the interaction energy of the spins in zero field. Finally, $\mathcal{Z}(H)$ is the partition function (see expression (6.1))

$$\mathcal{Z}(H) = \int \left(\prod_i \rho(S_i) \mathrm{d}S_i \right) \exp\left[-\beta \mathcal{E}(\mathbf{S}) + H \sum_i S_i \right]. \qquad (7.1)$$

Moreover, we assume,

(i) that the model in zero field has a reflection $\mathbb{Z}_2$ symmetry and, thus, $\mathcal{E}(\mathbf{S}) = \mathcal{E}(-\mathbf{S})$ together with $\rho(S) = \rho(-S)$,

(ii) that the spin distribution $\rho(S)$ decreases, for $|S| \to \infty$, faster than a Gaussian function:

$$\int_S^\infty \rho(S') \mathrm{d}S' \leq K \mathrm{e}^{-\mu S^\alpha}, \quad \mu > 0, \quad \alpha > 2. \qquad (7.2)$$

Later in the chapter, in Section 7.5.2, we generalize the analysis to systems where the spin is a ν-component vector and the symmetry group $\mathbb{Z}_2$ is replaced by the orthogonal group $O(\nu)$ of rotations–reflections in ν-dimensional space.

7.1 Infinite temperature or independent spins

To implement the preceding theoretical concepts, to establish some notation and a few elementary properties, we first consider a model at infinite temperature ($\beta = 0$) or without interaction. We are then led to a set of independent spins and thus to a situation similar to the central limit theorem of Section 3.1, but within a more specific framework.

7.1.1 One-site model

The partition function of a one-site model is

$$z(h) = \int ds\, \rho(s)\, e^{sh}, \quad z(0) = 1. \tag{7.3}$$

It is obtained from the Fourier transform of the spin distribution by analytic continuation $h \mapsto ih$ (see also Section 2.1). It is a generating function of the moments of the distribution in a field h.

For example, for the Ising model where s takes only two values, $s = \pm 1$, with probability $1/2$, one finds

$$z(h) = \cosh h.$$

The bound (7.2) implies that the integral converges for all h real or complex and $z(h)$ is thus an entire analytic function. It is even, due to the symmetry $s \mapsto -s$ of the distribution, and positive for h real. Moreover,

$$z'(h) = \int ds\, \rho(s) s\, e^{sh} = \int ds\, \rho(s) s \sinh(sh) \Rightarrow hz'(h) \geq 0.$$

The function $z(h)$ increases for $h > 0$ and decreases for $h < 0$.

Finally, the bound (7.2) implies that the integral

$$K(\lambda) = \int ds\, \rho(s)\, e^{\lambda |s|^\alpha}$$

converges for $\lambda < \mu$. Thus,

$$z(h) = \int ds\, \rho(s)\, e^{\lambda|s|^\alpha}\, e^{-\lambda|s|^\alpha + hs} \leq K(\lambda) \max_{\{s\}} e^{-\lambda|s|^\alpha + |h||s|}$$

$$\leq K(\lambda) \exp\left(\text{const.}\, |h|^{\alpha/(\alpha-1)}\right). \tag{7.4}$$

Since $\alpha > 2$, one finds the inequality

$$\alpha/(\alpha - 1) < 2.$$

The generating function of cumulants

$$A(h) = \ln z(h), \tag{7.5}$$

is even and, since $z(h)$ is strictly positive, analytic in the vicinity of the real axis. Its derivative is odd and vanishes at $h = 0$. Moreover, the remark (6.5) implies $A''(h) > 0$ and the function $A(h)$ is thus convex. It increases for $h > 0$.

Finally, as a consequence of condition (7.2), for a large class of distributions $\rho(s)$, the random variable s tends toward a certain value $s(h)$ when $|h| \to \infty$ and $A''(h)$, the second cumulant in a field, thus, goes to zero:

$$\lim_{|h|\to\infty} A''(h) = 0. \tag{7.6}$$

In what follows, we restrict the discussion to this class.

We now parametrize the expansion of A at $h = 0$ as

$$A(h) = \sum_{p=1} \frac{a_{2p}}{2p!} h^{2p}, \quad a_2 > 0. \tag{7.7}$$

The value $m = A'(h)$ is the magnetization in a field h.

In addition to these functions, already introduced in Section 3.1, we also define the Legendre transform $B(m)$ of $A(h)$ by

$$B(m) + A(h) = mh, \quad m = A'(h), \tag{7.8}$$

In the example of the Ising model ($s = \pm 1$), one finds

$$B(m) = \tfrac{1}{2}(1+m)\ln(1+m) + \tfrac{1}{2}(1-m)\ln(1-m). \tag{7.9}$$

The function $B(m)$ is even. The magnetization m has the sign of h.

The relation (6.8) implies

$$B''(m) = 1/A''(h). \tag{7.10}$$

Because $A(h)$ is a convex function of h, $B(m)$ is also convex. It is analytic in a neighbourhood of the origin. We can thus parametrize it as

$$B(m) = \sum_{p=1} \frac{b_{2p}}{2p!} m^{2p}, \quad b_2 > 0. \tag{7.11}$$

The coefficients b_p are related to the coefficients (7.7) of the expansion of $A(h)$. For example,

$$b_2 = 1/a_2, \quad b_4 = -a_4/a_2^4.$$

In the Ising model, one finds

$$b_2 = 1, \quad b_4 = 2.$$

Finally, the condition (7.6) implies either that $|m|$ is bounded or that $B(m)$ increases faster than m^2 for $|m| \to \infty$.

7.1.2 Independent spins

Partition function. In the absence of interactions ($\mathcal{E} = 0$), or for $\beta = 0$ (that is, at infinite temperature), the partition function (7.1) can be calculated immediately and one finds

$$\mathcal{Z}_0(H) = \int \left(\prod_i \rho(S_i) \mathrm{d}S_i\right) \exp\left(H \sum_i S_i\right) = \int \prod_i \mathrm{d}S_i\, \rho(S_i)\, \mathrm{e}^{HS_i}$$
$$= \prod_i \int \mathrm{d}S_i\, \rho(S_i)\, \mathrm{e}^{HS_i} = (z(H))^\Omega, \tag{7.12}$$

where Ω is the number of sites and $z(h)$ the function (7.3).

Free energy. The density free energy is then

$$W_0(H) = \frac{1}{\Omega} \ln \mathcal{Z}_0(H) = A(H), \tag{7.13}$$

where $A(h)$ is defined in (7.5). The magnetization follows:

$$M = \frac{1}{\Omega} \sum_i \langle S_i \rangle = A'(H).$$

Thermodynamic potential. The thermodynamic potential density is the Legendre transform of $W_0(H)$:

$$\mathcal{G}_0(M) = MH - W_0(H) \quad \text{with} \quad M = W_0'(H), \tag{7.14}$$

and, thus, $\mathcal{G}_0(M) = B(M)$ (definition (7.8)).

Mean-spin distribution and thermodynamic potential. In Section 6.1.3, we have related the distribution R_Ω of the mean spin σ on the lattice to the thermodynamic potential, with a regularity hypothesis for the free energy and the thermodynamic potential. This hypothesis is always satisfied in a finite volume, and also for independent spins. The function $B''(M)$ is always positive. The asymptotic evaluation (6.11) yields here

$$R_\Omega(\sigma) \underset{\Omega \to \infty}{\sim} \sqrt{\Omega B''(M)/2\pi}\; \mathrm{e}^{-\Omega B''(M)(\sigma - M)^2/2},$$

where M is the magnetization. In zero field, $M = 0$ and the width of the Gaussian distribution is related to $B''(0) = b_2$.

7.2 Phase transitions in infinite dimension

We now discuss a simple, slightly pathological model, but which can be solved exactly. In this model, all spins interact pairwise. The model has two interpretations. If the dimension of space is finite, it is a model with infinite-range interactions, that is, quite different from those we want to study. But it has also the interpretation of a model with nearest-neighbour interactions, in the limit where the dimension of space becomes infinite. Indeed, in infinite dimension, a site on a cubic lattice has an infinite number of neighbours.

In the model, each spin is coupled to an infinite number of other spins. We show that the interaction can be replaced by the action of a mean magnetic field, and in such a model, the so-called *mean-field* approximation is exact.

The model. The interaction energy, in zero field, of the model is given by

$$\beta\mathcal{E}(\mathbf{S}) = -\frac{v}{\Omega} \sum_{i,j=1}^{\Omega} S_i S_j , \qquad (7.15)$$

where the parameter v is proportional to β, the inverse temperature. We choose $v > 0$, which favours aligned spins: one calls such an interaction *ferromagnetic*.

In the model, the spatial distribution of spins plays no role and, thus, we number them simply $i = 1, \ldots, \Omega$. Moreover, because the number of terms that couple the spins is of order Ω^2, the thermodynamic limit exists only if the interaction is divided by a factor Ω.

We assume a normalized, even spin distribution at each site, $\rho(S) = \rho(-S)$, which has all the properties described in Section 7.1.1, in particular, which decreases faster than a Gaussian function for $|S| \to \infty$.

Since the spatial spin distribution plays no role, the only physically relevant quantity is the partition function in an external magnetic field H, which we calculate exactly in the thermodynamic limit $\Omega \to \infty$.

7.2.1 Mean-spin distribution. Thermodynamic functions

The interaction energy (7.15) can also be written as

$$\beta\mathcal{E}(\mathbf{S}) = -\frac{v}{\Omega} \left(\sum_{i=1}^{\Omega} S_i \right)^2 .$$

We then introduce the mean spin on the lattice

$$\sigma = \frac{1}{\Omega} \sum_i S_i . \qquad (7.16)$$

The function

$$R_\Omega(\sigma) = \Omega\, e^{\Omega v \sigma^2} \int \left(\prod_i \rho(S_i) \mathrm{d}S_i \right) \delta\left(\Omega\sigma - \sum_i S_i \right) \qquad (7.17)$$

is, *up to normalization*, the distribution of the mean spin σ.

It is possible to integrate over the spins by introducing the Fourier representation as in the case of the central limit theorem (representation (3.8)). One uses

$$\delta\left(\Omega\sigma - \sum_i S_i\right) = \frac{1}{2\pi} \int dk \, \exp\left[ik\left(\Omega\sigma - \sum_i S_i\right)\right].$$

Introducing the transform (7.3) of the distribution and its logarithm (equation (7.5)),

$$z(\lambda) = \int ds\, \rho(s)\, e^{\lambda s}, \quad A(\lambda) = \ln z(\lambda),$$

one derives the representation

$$R_\Omega(\sigma) = \frac{\Omega}{2\pi} e^{\Omega v \sigma^2} \int dk \, e^{\Omega[ik\sigma + A(-ik)]}. \tag{7.18}$$

In the thermodynamic limit $\Omega \to \infty$, this integral can be evaluated by the steepest descent method. The saddle point is given by

$$\sigma = A'(-ik).$$

At leading order of the steepest descent method, and using the definition (7.8) with $h = -ik$, $m = \sigma$ and the relation (7.10), one finds

$$R_\Omega(\sigma) \sim \sqrt{\Omega B''(\sigma)/2\pi} \exp\left[\Omega(v\sigma^2 - B(\sigma))\right]. \tag{7.19}$$

Setting

$$G(\sigma) = B(\sigma) - v\sigma^2, \tag{7.20}$$

one can rewrite the expression as

$$R_\Omega(\sigma) \sim \sqrt{\Omega B''(\sigma)/2\pi} \exp\left[-\Omega G(\sigma)\right]. \tag{7.21}$$

Thermodynamic functions. The partition function in a field is defined by

$$\mathcal{Z}_\Omega(H, v) = \int \left(\prod_i \rho(S_i) dS_i\right) \exp\left[-\beta \mathcal{E}(\mathbf{S}) + H \sum_{i=1}^{\Omega} S_i\right]. \tag{7.22}$$

Using equation (7.17), one then verifies that it is given by

$$\mathcal{Z}_\Omega(H, v) = \int d\sigma\, R_\Omega(\sigma)\, e^{\Omega H \sigma} \tag{7.23}$$

and, thus, for $\Omega \to \infty$,

$$\mathcal{Z}_\Omega(H, v) \sim \int d\sigma\, \sqrt{\Omega B''(\sigma)/2\pi}\, e^{\Omega[H\sigma - G(\sigma)]}.$$

For $\Omega \to \infty$, the integral can also be evaluated by the steepest descent method. At leading order, in the case of a *unique leading saddle point*, the free energy density is given by

$$W(H, v) = \lim_{\Omega \to \infty} \frac{1}{\Omega} \ln \mathcal{Z}_\Omega(H, v) = H\sigma - G(\sigma) \qquad (7.24)$$

with

$$H = G'(\sigma). \qquad (7.25)$$

This expression naturally has the form of a Legendre transform. The spin expectation value M, or magnetization, is then (using equation (7.25))

$$M = \langle S_i \rangle = \frac{1}{\Omega} \sum_i \langle S_i \rangle = \frac{\partial W(H, v)}{\partial H} = \sigma(H). \qquad (7.26)$$

The thermodynamic potential density follows immediately:

$$\mathcal{G}(M, v) = HM - W(H, v) = -vM^2 + B(M) = G(M). \qquad (7.27)$$

The magnetization M is also a solution of

$$H = \frac{\partial \mathcal{G}}{\partial M} = -2vM + B'(M). \qquad (7.28)$$

This relation between magnetic field, magnetization and temperature is called the *equation of state*.

We now analyse the solutions of the saddle point equation (7.25) and discuss their physical interpretation, as functions of the two parameters v and H.

7.2.2 Low- and high-temperature limits

The nature of the saddle points is related to the sign of the second derivative

$$G''(\sigma) = B''(\sigma) - 2v.$$

The function $B''(\sigma)$ is positive, continuous and, with our assumptions, tends toward infinity when $|\sigma| \to \infty$.

High temperature. For $v \to 0$, that is, at high temperature from the physics viewpoint and, more precisely, for

$$v \leq \min_\sigma \tfrac{1}{2} B''(\sigma),$$

the second derivative $G''(\sigma)$ is always positive and the saddle point equation has a unique solution, which corresponds to a maximum of the integrand.

For $H = 0$, it reduces to the trivial solution $\sigma = 0$. For $H \to 0$, the position of the saddle point and, thus, the magnetization M, become (in the parametrization (7.11))

$$M = \sigma(H) = \frac{1}{b_2 - 2v} H + O(H^2),$$

which thus vanish in zero field. Note that the limits $H \to 0$ and $\Omega \to \infty$ here commute.

Low temperature. In zero field, $H = 0$, the solution $\sigma = 0$ corresponds to a local maximum of the integrand as long as

$$v < v_c, \quad v_c = b_2/2,$$

but, for $v > v_c$, the second derivative becomes negative at $\sigma = 0$ and this solution corresponds to a minimum. By contrast, for $|\sigma| \to \infty$, $B''(\sigma)$ goes to infinity and the second derivative is always positive. Thus, for $v > v_c$, $G''(\sigma)$ has at least one zero and since it is an even function, it has at least two. The two opposite zeros correspond to a maximum of the integrand. One finds two degenerate leading saddle points $\pm\sigma_c$ that yield the same contribution to $W(0, v)$.

For $H \neq 0$, the two solutions $\pm\sigma_c$ are displaced, but the essential point is that the two saddle points are no longer degenerate however small $|H|$ is. For $\Omega \to \infty$, only one saddle point contributes and one finds a non-vanishing induced magnetization. Then, taking the limit $H \to 0$, one finds

$$\lim_{H \to 0} M \equiv \langle S_i \rangle = \lim_{H \to 0} \lim_{\Omega \to \infty} \frac{\partial W(H, v)}{\partial H} = \operatorname{sign}(H)|\sigma_c|.$$

Here, the limits no longer commute since the limits taken in the reverse order lead to a vanishing value for symmetry reasons.

The physical conditions of the problem single out the non-trivial result: indeed, the symmetric point is the equivalent of an unstable equilibrium point in classical mechanics. Any symmetry breaking term, however small, leads to one of the two non-vanishing solutions.

The model thus displays a phase transition that takes place at a value of $v \leq v_c$ between a situation with a unique and symmetric phase, and a situation with two different phases related by the symmetry. In the latter situation, one speaks of *spontaneous symmetry breaking*.

Let us point out that, in this situation, the thermodynamic potential density $\mathcal{G}(M, v)$ given by equation (7.27), is apparently non-convex, contradicting the general results (6.5), (6.8). In fact, it is not defined for values of M such that $G''(M) < 0$. Indeed,

$$\frac{\partial M}{\partial H} = \frac{\partial^2 W}{(\partial H)^2} > 0.$$

As a consequence, one always has $|M(H)| \geq |\sigma_c|$ and the intermediate values $|M| < |M(0)|$ cannot be reached.

7.2.3 Mean-spin distribution and phase transition

We now examine, in more detail, the thermodynamic properties in zero magnetic field.

Phase transitions: Generalities and examples

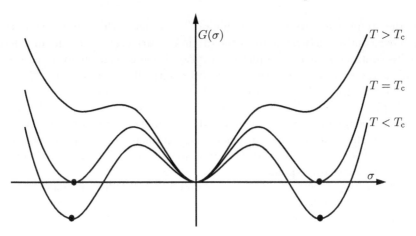

Fig. 7.1 Thermodynamic potential: First-order phase transition.

The analysis of saddle points then reduces to the study of the function R_Ω, which is proportional to the distribution of the mean spin σ, in the asymptotic form (7.21):

$$R_\Omega(\sigma) \sim \sqrt{\Omega B''(\sigma)/2\pi} \exp\left[-\Omega G(\sigma)\right]$$

and, thus, of the function $G(\sigma)$.

In the thermodynamic limit, the *spontaneous* magnetization M is given by the absolute minima of $G(\sigma)$ (equation (7.26)), whose location has to be determined when the parameter v, which is inversely proportional to the temperature T, varies.

First-order phase transition. The property (7.6) implies that for $|\sigma|$ large enough, $G(\sigma)$ is an increasing function. For v small (i.e., high temperature) $v\sigma^2$ is negligible and the right-hand side of equation (7.20) is convex. The minimum of $G(\sigma)$ is $\sigma = 0$; the magnetization vanishes. When v increases, in general, one meets a value of v for which other local minima appear, which then may become absolute minima of $G(\sigma)$ (Figure 7.1). When this happens, the magnetization M jumps discontinuously from zero to a finite value corresponding to one of these new absolute minima. The system undergoes a first-order phase transition. Fluctuations around the saddle point are governed by the value of the second derivative of the potential at the minimum. In this situation the second derivative is strictly positive.

Although first-order phase transitions are common, we will not discuss them in this work. One reason is that the mean-field approximation or the quasi-Gaussian model (Chapter 8 and, more specifically, Section 8.10), which share many properties with the model in infinite dimension, give in general a satisfactory qualitative description of the physical properties.

Continuous or second-order phase transition. By contrast, if no absolute minimum appears at a finite distance from the origin, eventually at a critical temperature T_c, corresponding to the value

$$v_c = \tfrac{1}{2}B''(0) = \tfrac{1}{2}b_2 \qquad (7.29)$$

of v, the origin ceases to be a minimum of G and, below this temperature, two minima move continuously away from the origin (Figure 7.2). Since the magnetization remains continuous at v_c, the phase transition is called continuous or second order.

This is the situation that we analyse systematically from now on.

At the transition, the mean-spin distribution behaves near $\sigma = 0$ like $e^{-\Omega b_4 \sigma^4/24}$. This implies $b_4 \geq 0$ otherwise $\sigma = 0$ would not be a minimum and a first-order transition would have been encountered at a higher temperature. We examine below only the generic situation where the parameter b_4 is strictly positive. The special situation $b_4 = 0$ but with $b_6 > 0$ requires a special analysis and corresponds to a so-called *tricritical* point.

The behaviour $e^{-\Omega b_4 \sigma^4/24}$ at the transition is clearly different from the behaviour of a sum of independent variables. The correlation between the infinite number of spins thus plays a crucial role.

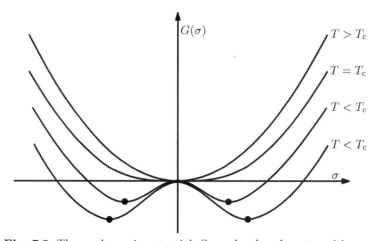

Fig. 7.2 Thermodynamic potential: Second-order phase transition.

Remark. In zero field, in a situation of continuous transition, exactly at the transition the simple steepest descent method does not apply since the function $G(\sigma)$ is of order σ^4. But this affects, for $\Omega \to \infty$, only corrections to the thermodynamic functions and not the leading order.

7.3 Universality in infinite space dimension

We now examine the behaviour of a few important thermodynamic quantities when the temperature T approaches T_c, that is, v approaches v_c in the case of a *continuous transition*. We verify in Chapter 8 that the universal results obtained in the framework of these models in infinite dimension are identical to those obtained in the framework of the quasi-Gaussian or mean-field approximations.

The magnetization goes to zero for $T \to T_c$ and $H \to 0$. In this limit we can expand $\mathcal{G}(M, v)$ (equation (7.27)), which like $B(M)$ is a regular even function, in a

Taylor series in M. In the parametrization (7.11),

$$\mathcal{G}(M, v) = -vM^2 + \frac{b_2}{2!}M^2 + \frac{b_4}{4!}M^4 + \cdots . \tag{7.30}$$

The convexity of $B(M)$ implies that b_2 is positive. We have shown above that the assumption of a generic continuous phase transition also implies that the parameter b_4 is strictly positive.

For v close to v_c and weak, uniform, applied magnetic field H, the first terms contributing to the equation of state (7.28) are

$$H = 2(v_c - v)M + \tfrac{1}{6}b_4 M^3 + O(M^5). \tag{7.31}$$

Spontaneous magnetization. In zero field, the equation of state reduces to

$$\frac{\partial \mathcal{G}(M)}{\partial M} = 0.$$

For $v > v_c$ but $|v - v_c| \ll 1$, this equation has two solutions

$$M \sim \pm [12(v - v_c)/b_4]^{1/2} \quad \text{for } |v - v_c| \to 0, \tag{7.32}$$

which are the two possible values of the *spontaneous magnetization*. At the critical temperature T_c, the magnetization thus has the *universal* power-law behaviour

$$M \propto (T_c - T)^\beta \quad \text{with} \quad \beta = 1/2, \tag{7.33}$$

where $\beta = \tfrac{1}{2}$ (not to be confused with the inverse temperature) is the quasi-Gaussian, or mean-field, or classical value of the magnetic exponent β.

Magnetic susceptibility. The inverse of the magnetic susceptibility χ (i.e., the response of the magnetization to a change of the magnetic field) is given by

$$\chi^{-1} = \left(\frac{\partial M}{\partial H}\right)^{-1} = \frac{\partial H}{\partial M} = 2(v_c - v) + \tfrac{1}{2}b_4 M^2 + O(M^4).$$

In zero field, one finds

$$\begin{aligned} \chi_+^{-1} &= 2(v_c - v) & T > T_c, \\ \chi_-^{-1} &= 4(v - v_c) & T < T_c, \end{aligned} \tag{7.34}$$

where the equation (7.32) has been used below T_c. The magnetic susceptibility thus diverges at T_c with the universal magnetic susceptibility exponents γ, γ'

$$\begin{aligned} \chi_+ &\sim C_+ (T - T_c)^{-\gamma}, & \gamma &= 1, \\ \chi_- &\sim C_- (T_c - T)^{-\gamma'}, & \gamma' &= 1, \end{aligned} \tag{7.35}$$

and the amplitude ratio of singularities has the universal value

$$C_+/C_- = 2. \tag{7.36}$$

Equation of state. We now return to the general equation (7.31). At T_c ($v = v_c$) for $H \to 0$, one finds

$$H \propto M^3. \tag{7.37}$$

Quite generally, one parametrizes the universal behaviour of H at T_c for $M \to 0$ as

$$H \propto M^\delta. \tag{7.38}$$

Here, the critical exponent δ has the quasi-Gaussian, or mean-field (or classical) value

$$\delta = 3. \tag{7.39}$$

More generally, for $H, T - T_c \to 0$ and thus $M \to 0$, after a change of the normalizations of magnetic field, temperature and magnetization, the equation of state takes the universal scaling form

$$H = M^\delta f\bigl((T - T_c)M^{-1/\beta}\bigr), \tag{7.40}$$

which shows that the ratio H/M^δ is not a function of the variables T and M independently, but only of the combination $(T - T_c)/M^{1/\beta}$.

The function $f(x)$ can be cast into the form

$$f(x) = 1 + x. \tag{7.41}$$

The value $x = -1$, where H vanishes with $M \ne 0$, corresponds to the coexistence curve in the two-phase region, and yields again the spontaneous magnetization.

Specific heat. In zero field, the derivative with respect to β (the inverse of the temperature) of the free energy density yields the average energy. Since v is proportional to β, one can differentiate with respect to v (a change of temperature unit). We have shown for any parameter (equation (6.9)), and this applies thus to v, the relation

$$\frac{\partial \mathcal{G}(M)}{\partial v} + \frac{\partial W(H)}{\partial v} = 0.$$

It follows that

$$\left.\frac{\partial W(H)}{\partial v}\right|_{H=0} = M^2(H = 0).$$

One finds that above T_c the average energy vanishes and that below T_c it is proportional to the square of the spontaneous magnetization. The derivative of the average energy with respect to the temperature is the specific heat C. Calculating the derivative with respect to v at v_c, one obtains a result proportional to the specific heat. Using expression (7.32), one finds

$$C(T \to T_{c+}) = 0, \qquad C(T \to T_{c-}) = 12/b_4. \tag{7.42}$$

In this model, like in the mean-field approximation, the specific heat has a discontinuity at T_c, which has a *non-universal* value.

Remarks.

(i) Other models can be solved in infinite dimension by the same method. For example, the spin can be a vector in $\mathbb{R}^N$ and the model invariant under the transformations of the $O(N)$ orthogonal group, rotations–reflections in N space dimensions. One verifies that most universal properties still hold. In infinite dimension, universal properties depend very little on the symmetry group.

(ii) More generally, one can define models with finite-range interactions in space dimension d and calculate thermodynamic quantities in the form of $1/d$ expansions. The corrections to the $d = \infty$ model do not affect universal quantities.

7.4 Transformations, fixed points and universality

As we have done in Chapters 3 and 4, we now associate the universality properties to the fixed points of a transformation. Since all thermodynamic properties can be derived from the thermodynamic potential density $\mathcal{G}(M)$, we express here the transformation directly on $\mathcal{G}$. We proceed in the following way: we group the spins pairwise, and for each pair we integrate over one spin, the mean of the two spins being fixed. We use the representation (7.18) that expresses the mean-spin distribution as an integral over independent spin models. We have already noted that in this case, the generating function of cumulants has a simple transformation (see Section 3.2). Here, one finds

$$\mathcal{T}: \quad \Omega \mapsto \Omega/2, \quad v \mapsto 2v, \quad A(h) \mapsto 2A(h/2).$$

The Legendre transform $B(M)$ of A is then multiplied by a factor 2. It follows that the transform of the thermodynamic potential is

$$[\mathcal{T}\mathcal{G}](M) = 2\mathcal{G}(M).$$

As in Section 3.2, we combine this decimation of the number of random variables with a renormalization of the magnetization: $M \mapsto \zeta M$ and the general transformation becomes

$$[\mathcal{T}\mathcal{G}](M) = 2\mathcal{G}(\zeta M).$$

The fixed point equation thus is

$$\mathcal{G}_*(M) = 2\mathcal{G}_*(\zeta M).$$

Gaussian fixed point. Since the function $\mathcal{G}(M)$ is analytic at $M = 0$, it can be expanded in a Taylor series. Identifying the coefficients of M^2, one obtains

$$v - b_2/2 = 2(v - b_2/2)\zeta^2.$$

For $v \neq v_c = b_2/2$ and the choice $\zeta = 1/\sqrt{2}$, one finds the fixed point

$$\mathcal{G}_*(M) = (b_2/2 - v)M^2.$$

Note that $\zeta < 1$, which explains why the fixed point corresponds to the first term of the expansion in powers of M.

The spin distribution corresponding to the fixed point is a Gaussian distribution, which is a singular limit in the class of distributions that we have assumed. Moreover, the fixed point makes sense only for $v < v_c$, a region in which the spontaneous magnetization vanishes.

The fixed point is stable in the sense that the coefficient b_{2p} of the term of order M^{2p} is multiplied by $2^{1-p} < 1$ for $p > 1$. All perturbations to the Gaussian model, even in the spins are thus irrelevant.

Fixed point at the critical temperature. For $v = v_c$, it becomes necessary to expand up to fourth order and for $\zeta = 2^{-1/4}$ one finds the fixed point

$$\mathcal{G}_*(M) = \frac{b_4}{4!} M^4.$$

Again, one notes that $\zeta < 1$.
The coefficient b_{2p} becomes

$$\mathcal{T} b_{2p} = 2^{1-p/2} b_{2p}.$$

From the stability analysis, one then concludes:

(i) An M^2 term is related to a relevant perturbation.
(ii) An M^4 term is related to a redundant perturbation.
(iii) All other terms correspond to irrelevant perturbations.

Universality in the critical domain. Following the strategy that has already been explained in Section 4.7.2, it is possible to establish asymptotic universality properties in an infinitesimal neighbourhood of the critical point, called the critical domain. One modifies the transformation $\mathcal{T}$ by renormalizing also the difference

$$v - v_c \mapsto (v - v_c)/\sqrt{2}.$$

This corresponds to decreasing the initial value of $v - v_c$ at each iteration. In this way, the coefficient of M^2 remains constant while the coefficients of all irrelevant terms go to zero. Asymptotically,

$$\mathcal{G}(M) = (v_c - v)M^2 + \frac{b_4}{4!} M^4.$$

Remark. In the interpretation of the system as a spin model in dimension d with infinite-range interaction, the number of spins is proportional to L^d where L characterizes the linear size of the lattice. Dividing the number of spins by 2 corresponds to dividing L by $2^{1/d}$. It is then convenient to express the behaviour of the various variables in terms of the inverse of a unit length (in order for the dimension of relevant variables to be positive). In this unit, considering the value of ζ and the renormalization of $v - v_c$, one infers that the magnetization has a dimension $d/4$, $v - v_c$ a dimension $d/2$ and $\mathcal{G}(M)$ is a linear combination of terms of dimension d. This concludes the analysis for $H = 0$.

Transformation in a field H. One adds to the free energy the term HM, which becomes $2\zeta HM$ and, then,

$$\mathcal{T}H = 2\zeta H = 2^{3/4}H.$$

The magnetic field is thus a relevant perturbation with dimension $3d/4$. Again, in order for the magnetic field term to have a finite limit, H must be renormalized into $H2^{-3/4}$. This corresponds to decreasing the initial magnetic field at each iteration.

Finally, taking into account the dimensions of $H, v - v_c, M$, one verifies that the scaling property (7.40) of the equation of state also follows from dimensional analysis.

7.5 Finite-range interactions in finite dimension

We now examine the problem of phase transitions in models with finite-range interactions in finite-dimensional space. Since such models can no longer, in general, be solved exactly, we first use low- and high-temperature arguments to establish the existence of transitions and to deduce some of their properties.

We first consider models on finite lattices $\mathcal{C}$ belonging to $\mathbb{Z}^d$, that is, containing the points of $\mathbf{R}^d$ with integer coordinates satisfying

$$\mathbf{r} \in \mathcal{C} \Leftrightarrow 0 \leq r_\mu \in \mathbb{Z} < L, \quad \mu = 1, \ldots, d.$$

It is convenient to choose periodic boundary conditions in the d dimensions, which, preserve on a finite lattice the translation invariance of the systems that we will study.

To illustrate the formalism by concrete examples, we consider spin models with nearest-neighbour interactions on the lattice, but the results obtained in this way are, more generally, valid for all systems with finite-range interactions. An important result that emerges from the heuristic analysis that follows, is the importance of the symmetry groups and, in particular, of their discrete or continuous nature.

7.5.1 Discrete symmetries: The Ising model

An example of a model with a discrete symmetry is the Ising model, where the spins take only the values ± 1, with a nearest-neighbour interaction on the lattice. The model has a $\mathbb{Z}_2$ reflection symmetry, corresponding to reversing all spins: $S \to -S$. The results that we derive below will be more generally valid for any model with a reflection symmetry and finite-range interactions, and representative of models with discrete symmetries.

The partition function of the Ising model in zero field in d dimensions with *nearest--neighbour* interactions (n.n.) can be written as

$$\mathcal{Z}_\Omega(\beta) = \sum_{\{S_\mathbf{r} = \pm 1\}} \exp\left(\beta \sum_{\mathbf{r},\mathbf{r}' \text{n.n.} \in \mathcal{C}} S_\mathbf{r} S_{\mathbf{r}'}\right), \qquad (7.43)$$

where β, the inverse of the temperature, also includes a factor characterizing the strength of the interaction, assumed ferromagnetic, and Ω is the number of sites in $\mathcal{C}$: $\Omega = L^d$. The number of spin configurations is then 2^Ω.

In the two limits of high and low temperatures, it is possible to evaluate the distribution of the mean spin $\sigma = \sum_i S_i/\Omega$ in the thermodynamic limit.

Mean-spin distribution: High temperature. In the high-temperature limit ($\beta \to 0$), spins become independent variables. The mean-spin distribution thus follows from the central limit theorem and has an asymptotic Gaussian form centred around $\sigma = 0$, a situation described in Section 7.1.2. For $\Omega = \infty$, $\sigma = 0$ is a certain value and the magnetization in zero field vanishes.

Low temperature. Since the total number of configurations is 2^Ω, at low temperature ($\beta \to \infty$), the mean-spin distribution is dominated by the interaction energy. The two most probable configurations correspond to all spins aligned: $S_i = S = \pm 1$. In the limit $\Omega \to \infty$, in order for the mean-spin distribution not to reduce to $\sigma = \pm 1$, configurations must exist with $+$ and $-$ spins occupying both a finite proportion of the volume $\Omega = L^d$, which have a non-vanishing probability. Let us assume that most spins have value $+1$. The relative probability of a configuration of the type represented in Figure 7.3 for the dimension 2, with respect to the probability of configurations with spins all aligned, is, at low temperature, of the order of $\mathcal{N} e^{-\beta \Delta \mathcal{E}}$, where $\Delta \mathcal{E}$ is the variation of interaction energy and $\mathcal{N}$ the corresponding number of configurations.

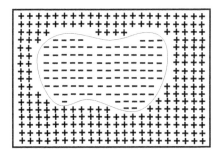

Fig. 7.3 Spin – domain in a spin + background, for $d = 2$.

With respect to the reference configuration with all spins equal to $+1$, the probability of a domain of spins -1 is proportional to $e^{-2\beta A}$, where A is the measure of the domain boundary. In d dimensions, the minimal boundary domain is a sphere. In terms of its radius ℓ, the measure of the boundary is of order $\Sigma_d \ell^{d-1}$ (Σ_d is the measure of the boundary of the sphere S_{d-1} with unit radius) and the probability is of order $\mathcal{N}(\ell)\,\mathrm{e}^{-2\Sigma_d \beta \ell^{d-1}}$. Moreover, one can verify that the number of configurations $\mathcal{N}(\ell)$ is bounded by $\mathrm{e}^{K\ell^{d-1}}$, where K is a constant, in such a way that for β large enough, the energy term always dominates.

In the thermodynamic limit, in order for the corresponding average spin to be smaller than 1, ℓ/L must remain finite. The conclusions thus are:

(i) For $d = 1$, the probability of $|\sigma| < 1$ remains finite and no phase transition is possible.

(ii) By contrast, for $d > 1$, the probability of $|\sigma| < 1$ goes to zero for $L \to \infty$ and thus, at low temperature, the distribution reduces to $\sigma = \pm 1$. This is a situation that we have already examined in the case of the model in infinite dimension. It corresponds to a two-phase region with non-vanishing spontaneous magnetization. Indeed, the addition of a magnetic field term changes the energy by $\pm HL^d$, favouring one of two configurations for arbitrarily small H. The limiting distribution is concentrated around only one value of σ, which is the spontaneous magnetization.

Clearly, in the infinite volume limit, there is no analytic continuation possible between a unique phase at high temperature and a region with two phases at low temperature and, thus, the thermodynamic quantities must have at least one singularity in β at a finite value β_c.

7.5.2 Continuous symmetries: The orthogonal group

We now briefly discuss a family of models with a continuous symmetry, to exhibit a few important differences with the case of discrete symmetries.

We again consider a classical spin system, but where the spins $\mathbf{S_r}$ are ν-component vectors with unit length, with two-spin nearest-neighbour ferromagnetic interactions. The partition function takes the form

$$\mathcal{Z}(\beta) = \int \left(\prod_{\mathbf{r}} \mathrm{d}\mathbf{S_r}\, \delta\left(\mathbf{S_r^2} - 1\right) \right) \exp\left[-\beta \mathcal{E}(\mathbf{S})\right] \qquad (7.44)$$

where $\mathrm{d}\mathbf{S}\delta(\mathbf{S}^2 - 1)$ is the uniform measure on the sphere $S_{\nu-1}$ and where we choose the configuration energy

$$\mathcal{E}(\mathbf{S}) = - \sum_{\mathbf{r},\mathbf{r}'\,\mathrm{n.n.}} \mathbf{S_r} \cdot \mathbf{S_{r'}}.$$

The model then has a continuous symmetry corresponding to the $O(\nu)$ orthogonal group of rotations–reflections in ν-dimensional space acting on the vectors $\mathbf{S_r}$. We consider below ν arbitrary and unrelated to the dimension d of space.

At high temperature ($\beta \to 0$), like in the case of the Ising model, the model tends toward a system of independent spins and the mean-spin distribution has a limit given by the central limit theorem. In particular, the spontaneous magnetization vanishes.

At low temperature, again the mean-spin distribution is dominated by minimal energy configurations, that is, where all spins are aligned in a given direction and thus $|\boldsymbol{\sigma}| = 1$. In order for the configurations with $|\boldsymbol{\sigma}| < 1$ to have a non-vanishing probability in the thermodynamic limit, spin configurations must exist that interpolate between two possible spin directions. The essential difference with the case of a discrete symmetry is that it is possible to interpolate between two asymptotic directions differing by an angle α by rotating the spins of an angle α/ℓ between adjacent sites. This must be contrasted with the case of a discrete symmetry, where the transition occurs between only two sites (or, more generally, a fixed number of sites). Thus, considering a sphere centred at $\mathbf{r} = 0$ and calling $\theta(r = |\mathbf{r}|)$ the angle of the spins in the plane formed by the initial and final spin directions, one chooses

$$\theta(r) = \alpha r/\ell, \quad 0 \le r \le \ell.$$

For ℓ large, the variation of the scalar product $\mathbf{S}_i \cdot \mathbf{S}_j$ between neighbour spins is proportional to $1 - \cos(\delta\theta(r)) \propto \alpha^2/\ell^2 (\delta r)^2$. The total variation of the configuration energy thus is proportional to

$$\alpha^2 \ell^{-2} \times \ell^d = \alpha^2 \ell^{d-2}.$$

Again, the spin expectation value is modified only if, in the infinite volume limit $L \to \infty$, the ratio ℓ/L remains finite. One concludes:

(i) For $d \le 2$, the variation of the energy remains finite and, thus, the mean-spin distribution has non-trivial support in the unit sphere. As a consequence, even at low temperature the spontaneous magnetization vanishes.

(ii) By contrast, in dimension $d > 2$, at low temperature the spin distribution reduces to the sphere $|\boldsymbol{\sigma}| = 1$. Again, an arbitrarily small magnetic field selects a spin direction and determines the direction of the spontaneous magnetization.

7.6 Ising model: Transfer matrix

In order to analyse more thoroughly the question of phase transition, we study again the Ising model with nearest-neighbour interactions, but within the transfer matrix formalism. We first consider a model on a finite lattice in which, in dimension $d > 1$, we distinguish one direction that we call the time direction, for convenience, when no confusion is possible. The lattice then corresponds to the points of integer coordinates $\mathbf{r} = (t, \boldsymbol{\rho})$ with

$$0 \le t \le \ell, \quad 0 \le \rho_\mu \le L \quad \text{for } 1 \le \mu \le d-1.$$

It is convenient to choose periodic boundary conditions in the $(d-1)$ transverse dimensions.

7.6.1 Transfer matrix

When interactions have a finite range, one can introduce the transfer matrix formalism that we have already discussed in Section 4.1.2, in the special case of one-dimensional statistical models.

Phase transitions: Generalities and examples

We define the transfer matrix in the time direction. In terms of the transfer matrix $\mathbf{T}$ that relates time t to time $t+1$, the partition function, in an arbitrary dimension d, with a periodic boundary condition in the time direction, then reads

$$\mathcal{Z}(\ell, L) = \operatorname{tr} \mathbf{T}^\ell. \tag{7.45}$$

We assume an isotropic interaction, which allows choosing a symmetric transfer matrix.

In the example of the Ising model in d dimensions with nearest-neighbour interactions, whose partition function is given by expression (7.43), the elements of the corresponding transfer matrix are

$$[\mathbf{T}](\mathbf{S}', \mathbf{S}) = \exp\left[\beta\left(\sum_\rho S_\rho S'_\rho + \frac{1}{2}\sum_{\rho,\rho'\text{ n.n.}} (S_\rho S_{\rho'} + S'_\rho S'_{\rho'})\right)\right], \tag{7.46}$$

where ρ is the position on the transverse $(d-1)$-dimensional lattice and $\mathbf{S}$ represents the set of all spins of the transverse lattice. A vector is associated to a spin distribution in the transverse $(d-1)$-dimensional space. For a lattice of transverse size L, the vector space has dimension $2^{L^{d-1}}$.

Symmetry $\mathbb{Z}_2$. The Ising model is characterized by a discrete $\mathbb{Z}_2$ symmetry, a reflection corresponding to changing the sign of all spins. Acting on the preceding vector space, the reflection is represented by a symmetric matrix $\mathbf{P}$, whose elements, in the notation (7.46), are

$$[\mathbf{P}](\mathbf{S}', \mathbf{S}) = \prod_\rho \delta_{S_\rho, -S'_\rho}. \tag{7.47}$$

It satisfies $\mathbf{P}^2 = 1$ and its eigenvalues are ± 1.

The matrix $\mathbf{P}$ commutes with the transfer matrix,

$$[\mathbf{T}, \mathbf{P}] = 0$$

and, thus, $\mathbf{P}$ and $\mathbf{T}$ can be diagonalized simultaneously: the eigenvectors of $\mathbf{T}$ can also be chosen eigenvectors of $\mathbf{P}$.

We thus denote by $\tau_{\pm,n}$ the eigenvalues of $\mathbf{T}$ corresponding to eigenvectors $\psi_{\pm,n}$ such that

$$\mathbf{P}\psi_{\pm,n} = \pm\psi_{\pm,n}, \quad \mathbf{T}\psi_{\pm,n} = \tau_{\pm,n}\psi_{\pm,n}.$$

The partition function can then be expressed in terms of eigenvalues as

$$\mathcal{Z}(\ell, L, \beta) = \sum_{n=0}^\infty \left(\tau_{+,n}^\ell + \tau_{-,n}^\ell\right). \tag{7.48}$$

It is useful for the discussion that follows to also introduce the partition function with anti-periodic boundary conditions in the time direction

$$\mathcal{Z}_a(\ell, L, \beta) = \operatorname{tr} \mathbf{P}\mathbf{T}^\ell = \sum_{n=0}^\infty \left(\tau_{+,n}^\ell - \tau_{-,n}^\ell\right). \tag{7.49}$$

Order parameter. The spin S_σ at site σ on the transverse lattice is an *order parameter*, in the sense that in the region where several phases coexist, its expectation value discriminates between the phases. Indeed, the corresponding matrix $\mathbf{S}_\sigma$, whose elements between two spin configurations are

$$[\mathbf{S}_\sigma](\mathbf{S}', \mathbf{S}) = S_\sigma \prod_\rho \delta_{S_\rho S'_\rho}, \qquad (7.50)$$

is odd under reflection:

$$\mathbf{P}^{-1}\mathbf{S}_\sigma\mathbf{P} = -\mathbf{S}_\sigma. \qquad (7.51)$$

Limit L finite, $\ell \to \infty$. If the size of the transverse lattice is finite, the general analysis of Section 4.1.2 remains valid. The unique eigenvector of the transfer matrix associated with the largest eigenvalue is symmetric

$$\mathbf{P}\psi_{+,0} = \psi_{+,0}.$$

Indeed, all components of the eigenvector $\psi_{+,0}$ are positive and equation (7.47) shows that $\mathbf{P}$ does not change the sign of the basis vectors.

As a consequence, the spin expectation value vanishes:

$$\langle S \rangle = (\psi_{+,0}, \mathbf{S}_\sigma \psi_{+,0}) = -(\psi_{+,0}, \mathbf{P}^{-1}\mathbf{S}_\sigma\mathbf{P}\psi_{+,0}) = -(\psi_{+,0}, \mathbf{S}_\sigma\psi_{+,0}) = 0. \qquad (7.52)$$

The symmetry $S \to -S$ remains unbroken at all temperatures.

Moreover, no level crossing is possible and the free energy $\mathcal{W}$, which in the limit $\ell \to \infty$ is given by

$$\mathcal{W} \sim \ell \ln \tau_{+,0}, \qquad (7.53)$$

is a regular function of the temperature $T = 1/\beta$ for $T > 0$. From this analysis, one concludes that in a spin model with finite-range interactions and reflection symmetry no phase transition is possible even without symmetry breaking; the free energy is a regular function of the temperature, and the correlation length ξ (defined by equation (6.13)) can diverge only at zero temperature. These results generalize to all short-range interactions.

Correlation length. The correlation length in the time direction, as defined by equation (6.13), is still given by equation (4.23):

$$\xi^{-1} = \ln(\tau_0/\tau_1),$$

where $\tau_0 > \tau_1$ are the two largest eigenvalues of the transfer matrix.

7.6.2 Infinite transverse dimension limit: Phase transitions

When the transverse dimension L diverges, new phenomena can occur, which an analysis of the high- and low-temperature limits reveals.

High temperature. At high temperature ($\beta \to 0$), spins on different sites decouple. At infinite temperature, spins become independent variables; all elements of the **T** matrix become equal and **T** becomes a projector on the eigenvector ψ_0 that has equal components on all spin configurations. All eigenvalues except one vanish. The correlation length also vanishes. These properties are independent of the volume and, thus, remain valid even for L infinite. At high temperature one finds, as expected, a disordered phase where $\langle S \rangle = 0$ and the reflection symmetry is unbroken.

Low temperature. At low temperature, that is, for $\beta \to \infty$, the leading contributions to the partition function correspond to two configurations where all spins are equal either to $+1$ or to -1:
$$S_\mathbf{r} = S = \pm 1 \quad \forall \mathbf{r}.$$

It follows that
$$\mathcal{Z}(\ell, L, \beta) = \operatorname{tr} \mathbf{T}^\ell \sim 2\, e^{\beta d\ell L^{d-1}}, \tag{7.54}$$

where the factor 2 corresponds to the two configurations.

At low temperature, the leading contributions to the partition function (7.49) with anti-periodic boundary conditions correspond to a region with spins $+1$ separated from a region with spins -1. With respect to the uniform contributions that dominate $\mathcal{Z}(\ell, L, \beta)$, the variation of the energy is proportional to the measure of the surface that separates the two regions. The minimal surface is a plane $t = t_0$ with $1 \le t_0 \le \ell$. The ratio $\mathcal{Z}_\mathrm{a}/\mathcal{Z}$ is then
$$r(\ell, L) = \frac{\operatorname{tr} \mathbf{P}\, \mathbf{T}^\ell}{\operatorname{tr} \mathbf{T}^\ell} = \frac{\mathcal{Z}_\mathrm{a}(\ell, L, \beta)}{\mathcal{Z}(\ell, L, \beta)} \sim \ell\, e^{-\beta L^{d-1}}, \tag{7.55}$$

where the factor ℓ corresponds to all possible positions of the interface.

For $\ell \to \infty$, the partition functions are dominated by the largest eigenvalues of the transfer matrix. Expression (7.54) shows that two eigenvalues dominate the sum (7.48). Since $\mathcal{Z}_\mathrm{a}(\ell, L, \beta)$ is asymptotically much smaller than $\mathcal{Z}(\ell, L, \beta)$, the two eigenvalues must be asymptotically equal and correspond to even and odd eigenvectors. Thus,
$$r(\ell, L) = \frac{\mathcal{Z}_\mathrm{a}(\ell, L, \beta)}{\mathcal{Z}(\ell, L, \beta)} \sim \frac{\tau_{+,0}^\ell - \tau_{-,0}^\ell}{\tau_{+,0}^\ell + \tau_{-,0}^\ell}. \tag{7.56}$$

In terms of the correlation length $\xi_L = 1/\ln(\tau_{+,0}/\tau_{-,0})$,
$$\frac{\tau_{+,0}}{\tau_{-,0}} - 1 \underset{\xi_L \to \infty}{\sim} \frac{1}{\xi_L},$$

and thus
$$r(\ell, L) \sim \frac{\ell}{2\xi_L} \sim \ell\, e^{-\beta L^{d-1}} \Rightarrow \xi_L \sim \tfrac{1}{2} e^{\beta L^{d-1}}. \tag{7.57}$$

Starting from the configurations where all spins are aligned, and taking into account configurations where a finite number of spins have been reversed, one can then determine eigenvalues and eigenvectors as low-temperature expansions. But the qualitative behaviour is not modified.

Phase transitions. For L finite, the correlation length ξ_L diverges only at zero temperature, like in one dimension. No phase transition is possible.

By contrast, for $d > 1$, at fixed sufficiently low temperature, the correlation length diverges in the limit $L \to \infty$, and the largest eigenvalue of the transfer matrix is double. The thermodynamic limit then depends on boundary conditions. In particular, an infinitesimal symmetry breaking selects not an even eigenvector but one of the eigenvectors corresponding to aligned spins. Then, the magnetization no longer vanishes. The two possible phases correspond to the two possible opposite values of the spontaneous magnetization.

In the infinite volume limit, the thermodynamic functions must have at least one singularity at a finite value β_c of β separating a high-temperature region with one phase from a low-temperature region with two phases.

Let us point out that the correlation length ξ_L has a meaning only in a finite volume. Indeed, it assumes a summation over all possible configurations. However, in a broken symmetry phase (a pure phase), the partition function must be calculated by summing only over a subset of configurations, the configurations for which the magnetization has a definite sign. The divergence of ξ_L is a precursor of the property that the two-spin correlation function in the infinite volume limit no longer vanishes at large distance, but converges toward the square of the spontaneous magnetization.

Remarks

(i) This analysis of the infinite volume limit is qualitatively correct in the whole low-temperature phase. However, at β_c the situation changes; an infinite number of eigenvalues accumulate at the same value $\tau_{+,0}$.

(ii) We have seen that the ratio (7.55) provides a criterion for spontaneous symmetry breaking. This analysis can be generalized to other symmetry groups. One then considers the ratio

$$r(\ell, L) = \frac{\operatorname{tr} \mathbf{R}\, \mathbf{T}^\ell}{\operatorname{tr} \mathbf{T}^\ell}, \qquad (7.58)$$

where $\mathbf{R}$ is an element of the symmetry group.

In the symmetric phase, the leading eigenvector of the transfer matrix is invariant under the transformations of the symmetry group and, thus, $r(\ell = \infty, L) = 1$.

On the contrary, if the symmetry is spontaneously broken, $\mathbf{R}$ shifts degenerate eigenvectors and, thus, $r(\ell = \infty, L)$ vanishes in the infinite volume limit $L \to \infty$.

Another limit of this ratio, $r(L, L)$ for $L \to \infty$, is interesting. It corresponds to a thermodynamic limit in which the sizes diverge in the same way in all dimensions. One is then led to calculate the ratio of two partition functions on a d-dimensional lattice of linear size L with different boundary conditions: the denominator corresponds to periodic boundary conditions in the time direction, the numerator corresponds to boundary conditions on the two sides that differ by transformations of the symmetry group. In Ising type systems, the only non-trivial group element is the reflection **P** and, thus, these boundary conditions are anti-periodic. This limit provides, at low temperature, a direct criterion of spontaneous symmetry breaking. The probability is then of order

$$r(L, L) = \frac{\operatorname{tr} \mathbf{P} \mathbf{T}^L}{\operatorname{tr} \mathbf{T}^L} \sim e^{-\beta L^{d-1}}. \tag{7.59}$$

7.7 Continuous symmetries and transfer matrix

We again discuss the family of models with orthogonal symmetry introduced in Section 7.5.2, to exhibit the differences between continuous and discrete symmetries in the framework of the transfer matrix formalism.

We consider classical spin systems whose partition function is given by expression (7.44). The spins $\mathbf{S_r}$ are ν-component vectors with unit length with two-spin, nearest-neighbour, ferromagnetic interactions.

These models have a continuous symmetry corresponding to the orthogonal group $O(\nu)$ of rotations–reflections in ν-dimensional space.

Transfer matrix. Here also, one can express the partition function (7.44) in terms of a transfer matrix **T**, whose elements are

$$[\mathbf{T}](\mathbf{S}', \mathbf{S}) = \exp\left[\beta\left(\sum_\rho \mathbf{S}_\rho \cdot \mathbf{S}'_\rho + \frac{1}{2} \sum_{\rho, \rho' \text{ n.n.}} (\mathbf{S}_\rho \cdot \mathbf{S}_{\rho'} + \mathbf{S}'_\rho \cdot \mathbf{S}'_{\rho'})\right)\right], \tag{7.60}$$

where ρ is the position in the $(d-1)$-dimensional transverse space.

At high temperature ($\beta \to 0$), like in the case of the Ising model, the eigenvector of the transfer matrix corresponding to the largest eigenvalue has uniform components on all configurations and thus is invariant under the transformations of the orthogonal $O(\nu)$ group.

To discuss the low-temperature phase, following the example of the Ising model, we introduce the generalization of the ratio (7.55). The symmetry operation is here a rotation $\mathbf{R}(\alpha)$ of angle α. We define

$$\mathcal{Z}(\alpha, \ell, L, \beta) = \operatorname{tr} \mathbf{R}(\alpha) \mathbf{T}^\ell \tag{7.61}$$

with periodic boundary conditions in all transverse directions. We study the behaviour of the ratio

$$r(\alpha, \ell, L) = \frac{\mathcal{Z}(\alpha, \ell, L, \beta)}{\mathcal{Z}(0, \ell, L, \beta)} \tag{7.62}$$

for $\beta \gg 1$ in the limit $L \to \infty$.

The function $\mathcal{Z}(\alpha, \ell, L, \beta)$ is the partition function on a d-dimensional lattice, with the modified boundary conditions in the time direction

$$\mathbf{S}_{t=\ell, \rho} \cdot \mathbf{S}_{t=0, \rho} = \cos \alpha. \tag{7.63}$$

At low temperature, minimal energy configurations correspond to taking the spins aligned in $(d-1)$ dimensions, and rotating by an angle α/ℓ between two adjacent sites along the time axis. This must be contrasted with the case of a discrete symmetry, where the transition between configurations forced by the boundary conditions occurs between only two sites (or, more generally, a finite number of sites).

The variation of the energy $\Delta\mathcal{E}$ resulting from the rotation is then

$$\Delta\mathcal{E} = \ell \times L^{d-1} \times [\cos(\alpha/\ell) - 1] \tag{7.64}$$

and, thus,

$$r(\alpha, \ell, L) \propto \exp\left[\beta\ell L^{d-1}(\cos(\alpha/\ell) - 1)\right].$$

For $\ell \gg 1$, one can expand

$$r(\alpha, \ell, L) \propto \exp\left[-\tfrac{1}{2}\beta L^{d-1}\alpha^2/\ell\right]. \tag{7.65}$$

The result is the same as for a $O(\nu)$ model in one dimension, at low temperature, with $\beta \mapsto \beta L^{d-1}$. Exercise 4.4 presents the calculation of the corresponding correlation length. Applying the result to the present example, one finds

$$\xi_L \sim \frac{\beta L^{d-1}}{2(\nu - 1)}. \tag{7.66}$$

An equivalent calculation consists in replacing, at low temperature, all spins $\mathbf{S}_\rho$ and all spins $\mathbf{S}'_\rho$ in the transfer matrix by two constant spins. Again, the result is the transfer matrix of a one-dimensional model, with an interaction $\beta \mapsto \beta L^{d-1}$.

The result seems to indicate, again, that the correlation length diverges for any dimension larger than or equal to 2, like in the discrete case. However, this property holds only in dimensions larger than 2.

In dimension 2, the estimate predicts that the correlation length increases linearly in L, that is, that ξ_L/L goes to a constant. This is a subtle situation where the fluctuations around the constant configurations affect the calculation even at low temperature and which must be studied with more refined arguments. Actually, the property holds only for $\nu = 2$ (the case where the rotation group $SO(2)$ is Abelian, i.e, commutative). For $\nu > 2$, the correlation length diverges only at zero temperature.

Moreover, even for $\nu = 2$, although the correlation length diverges, because ξ_L/L remains finite, the spontaneous magnetization in the low-temperature phase vanishes, $\langle \mathbf{S} \rangle = 0$, and the symmetry $O(2)$ is not broken. One finds a low-temperature phase characterized by an algebraic decay of connected correlation functions (the so-called Kosterlitz–Thouless phase). As an argument of ergodic type suggests, the spin expectation value can only be non-vanishing if $\xi_L/L \to \infty$ for $L \to \infty$.

Remark. As a symmetry breaking criterion, one can also use the ratio of partition functions (7.65) for $\ell = L$. One finds

$$r(\alpha, L, L) \propto \exp\left[-\tfrac{1}{2}\beta L^{d-2}\alpha^2\right]. \tag{7.67}$$

Therefore, in the case of a continuous symmetry, it is easier passer to pass from one energy minimum to another. This property has, as a direct consequence, that it is more difficult to break the symmetry and that Goldstone modes appear in the broken symmetry phase (see Section 7.8). For $d \le 2$, $r(\alpha, L, L)$ has a finite limit and the symmetry is never broken. This result, which we have justified by heuristic arguments, can be proved rigorously (the Mermin–Wagner–Coleman theorem).

For $d > 2$, by contrast, the symmetry is broken at low temperature. There must exist a finite temperature T_c, where a phase transition takes place.

7.8 Continuous symmetries and Goldstone modes

If the initial spin variable $\mathbf{S_r}$ is a ν-component vector and if both the interaction and the spin distribution have a continuous symmetry (associated with a compact Lie group), the appearance of several functions and correlation lengths when the magnetization is different from zero, induces a few new properties. In particular, in zero field, for all temperatures below T_c, some correlation lengths, associated with modes called *Goldstone modes* diverge.

Again, we illustrate these remarks with the example of $O(\nu)$ ($\nu > 1$) symmetric models.

Goldstone modes: $O(\nu)$ symmetry. The orthogonal symmetry $O(\nu)$ implies that the thermodynamic potential density $\mathcal{G}(\mathbf{M})$ in a uniform field is a function of the square of the magnetization vector and, thus, can be written as ($|\mathbf{M}| = M$)

$$\mathcal{G}(\mathbf{M}) = G(M^2/2).$$

As a consequence, the relation between components H_α and M_α of the magnetic field and the magnetization, respectively, takes the form

$$H_\alpha = M_\alpha G'(M^2/2) \tag{7.68}$$

and, thus, for the moduli,

$$|\mathbf{H}| \equiv H = MG'(M^2/2). \tag{7.69}$$

The connected two-point function is now the $\nu \times \nu$ matrix

$$W^{(2)}_{\alpha\beta}(\mathbf{r} - \mathbf{r}') = \langle S_\alpha(\mathbf{r}) S_\beta(\mathbf{r}') \rangle_{\text{conn.}}.$$

In the Fourier representation, the vertex function, the inverse of the connected two-point function, at vanishing argument, is given by (see equation (6.27))

$$\tilde{\Gamma}^{(2)}_{\alpha\beta}(\mathbf{k} = 0) = \frac{\partial^2 \mathcal{G}}{\partial M_\alpha \partial M_\beta} = M_\alpha M_\beta G''(M^2/2) + \delta_{\alpha\beta} G'(M^2/2). \tag{7.70}$$

The matrix $\tilde{\Gamma}^{(2)}_{\alpha\beta}(\mathbf{k}=0)$ has two eigenspaces corresponding to the vector M_α and to the $(\nu-1)$ vectors X_α orthogonal to M_α:

$$\sum_\beta \tilde{\Gamma}^{(2)}_{\alpha\beta}(0)M_\beta = M_\alpha \left[M^2 G''(M^2/2) + G'(M^2/2)\right],$$

$$\sum_\beta \tilde{\Gamma}^{(2)}_{\alpha\beta}(0)X_\beta = X_\alpha G'(M^2/2).$$

Denoting by $\tilde{\Gamma}^{(2)}_L$, $\tilde{\Gamma}^{(2)}_T$ the corresponding eigenvalues, respectively, one finds

$$\tilde{\Gamma}^{(2)}_L(0) = M^2 G''(M^2/2) + G'(M^2/2), \tag{7.71}$$
$$\tilde{\Gamma}^{(2)}_T(0) = G'(M^2/2) = H/M, \tag{7.72}$$

where in (7.72), equation (7.69) has been used.

The eigenvalues of the connected two-point function $\widetilde{W}^{(2)}(\mathbf{k}=0)$ are the inverses of those of $\tilde{\Gamma}^{(2)}$. The transverse eigenvalue of the connected two-point function at vanishing argument is thus

$$\widetilde{W}^{(2)}_T(k=0) = \left[\tilde{\Gamma}^{(2)}_T(k=0)\right]^{-1} = M/H. \tag{7.73}$$

Goldstone modes. Above T_c, the magnetization vanishes linearly with H and the two-point function has a finite limit for $H \to 0$. On the contrary, below T_c, when H goes to zero, M has the non-vanishing spontaneous magnetization as a limit and, thus, the transverse two-point function diverges. This implies the divergence of the transverse correlation length corresponding to $\nu - 1$ modes, called *Goldstone modes* (massless particles in the sense of the theory of fundamental microscopic interactions).

Another argument. This result can also be derived in the following way. In zero field, below T_c, the magnetization does not vanish although the potential is symmetric. If a vector $\mathbf{M}$ corresponds to a minimum of the potential, any vector deduced from $\mathbf{M}$ by a rotation (an orthogonal transformation), also corresponds to a minimum. This implies that the thermodynamic potential is minimum on a sphere $|\mathbf{M}| = M$. An infinitesimal rotation corresponds to adding to $\mathbf{M}$ a vector tangent to the sphere and thus orthogonal to $\mathbf{M}$. Denoting by $\mathbf{X}$ such a vector, $|\mathbf{X}| \ll 1$, and expanding the minimality condition to first order in $\mathbf{X}$, one obtains the condition

$$0 = \frac{\partial \mathcal{G}(\mathbf{M}+\mathbf{X})}{\partial M_\alpha} = \sum_\beta \frac{\partial^2 \mathcal{G}(\mathbf{M})}{\partial M_\alpha \partial M_\beta} X_\beta.$$

One recovers the $\nu - 1$ eigenvectors with zero eigenvalue, corresponding to the Goldstone modes.

Mathematical remark. More generally, one can consider models with a symmetry corresponding to a continuous group (Lie group) $\mathfrak{G}$ and such that this symmetry is spontaneously broken by a non-vanishing expectation value **M** of the order parameter. Let $\mathfrak{H}$ be the subgroup of $\mathfrak{G}$ that leaves the vector **M** invariant (the residual symmetry group). In the preceding example, $\mathfrak{G} \equiv O(\nu)$ and $\mathfrak{H} \equiv O(\nu-1)$. Finally, let g and h be the numbers of generators of the Lie algebra of $\mathfrak{G}$ and $\mathfrak{H}$, respectively. A simple generalization of the preceding arguments shows that there exists $g-h$ Goldstone modes, associated with the generators of $\mathfrak{G}$ that do not belong to $\mathfrak{H}$. In the orthogonal example

$$g = \tfrac{1}{2}\nu(\nu-1), \quad h = \tfrac{1}{2}(\nu-1)(\nu-2) \Rightarrow g - h = \nu - 1.$$

Exercises

Exercise 7.1

A model with $O(3)$ orthogonal symmetry in infinite dimension. One wants to generalize the study of Section 7.2 to a model where the spins **S** are three-component vectors belonging to the sphere S_2 with the same isotropic distribution as in Section 7.7, but with an interaction energy of the $O(3)$ invariant form

$$-\beta \mathcal{E}(\mathbf{S}) = \frac{v}{\Omega} \sum_{i,j=1}^{\Omega} \mathbf{S}_i \cdot \mathbf{S}_j, \quad v > 0.$$

(i) Calculate first the one-site partition function in a field, the corresponding free energy A and thermodynamic potential B. Expand B up to fourth order in the three-component magnetization vector.

(ii) It is then suggested to introduce the distribution of the mean-spin $\boldsymbol{\sigma}$, a three-component vector. Calculate the critical value v_c of v.

Solution. The one-site partition function (see Exercise 4.4) is given by

$$z(h) = \frac{1}{2} \int_0^\pi d\theta \, \sin\theta \, e^{h\cos\theta} = \frac{\sinh h}{h}.$$

Then, the one-site free energy is

$$A(h) = \ln(\sinh h / h) = \tfrac{1}{6} h^2 - \tfrac{1}{180} h^4 + O(h^6).$$

The Legendre transform $B(\mathbf{M})$ is a function only of the length $m = |\mathbf{M}|$ of the magnetization vector. One verifies

$$B(m) = mh - A'(h), \quad m = A'(h).$$

For $m \to 0$, one finds the expansion

$$B(m) = \tfrac{3}{2} m^2 + \tfrac{9}{20} m^4 + O(m^6).$$

The interaction energy can then be expressed in terms of the mean spin $\boldsymbol{\sigma}$ as
$$-\beta \mathcal{E}(\mathbf{S}) = \Omega v \sigma^2.$$

The mean-spin distribution is proportional to
$$R_\Omega(\boldsymbol{\sigma}) = \left(\frac{\Omega}{2\pi}\right)^3 e^{\Omega v \sigma^2} \int d^3k \; e^{\Omega[i\mathbf{k}\cdot\boldsymbol{\sigma}+A(-i\mathbf{k})]},$$

where $A(h = |\mathbf{h}|) = \ln z(h)$ is the free energy of the one-site model, derived from For $\Omega \to \infty$, this integral can be calculated by the steepest descent method. The saddle point is given by
$$\boldsymbol{\sigma} = \mathbf{k}A'(-i k)/k.$$

Then, one needs
$$\frac{\partial^2 B}{\partial \sigma_\alpha \partial \sigma_\beta} = \frac{B'(\sigma)}{\sigma}\left(\delta_{\alpha\beta} - \frac{\sigma_\alpha \sigma_\beta}{\sigma^2}\right) + B''(\sigma)\frac{\sigma_\alpha \sigma_\beta}{\sigma^2},$$

and thus
$$\det \frac{\partial^2 B}{\partial \sigma_\alpha \partial \sigma_\beta} = \left(\frac{B'(\sigma)}{\sigma}\right)^2 B''(\sigma).$$

It follows ($\sigma = |\boldsymbol{\sigma}|$) that
$$R_\Omega(\boldsymbol{\sigma}) \sim \left(\frac{\Omega}{2\pi}\right)^{3/2} [B''(\sigma)]^{1/2} \frac{B'(\sigma)}{\sigma} e^{\Omega v \sigma^2 - \Omega B(\sigma)}.$$

Again, the result can be expressed in terms of the Legendre transform B of the function A. The critical value of v thus is
$$v_c = \tfrac{3}{2}.$$

Exercise 7.2
A model with orthogonal symmetry in infinite dimension. Generalize the preceding exercise to a model with spins $\mathbf{S}$ belonging now to the sphere $S_{\nu-1}$ with the same isotropic distribution as in Section 7.7, but with an interaction energy sum of the $O(\nu)$ invariant term
$$-\beta\mathcal{E}(\mathbf{S}) = \frac{v}{\Omega}\sum_{i,j=1}^{\Omega} \mathbf{S}_i \cdot \mathbf{S}_j, \quad v > 0,$$

and the coupling to a magnetic field
$$\mathbf{H} \cdot \sum_i \mathbf{S}_i.$$

It is suggested to introduce the mean-spin distribution σ, a ν-component vector.
Solution. The one-site partition function is now given by (see Exercise 4.4)

$$z(h=|\mathbf{h}|) = \frac{2\pi^{\nu/2-1/2}}{\Gamma(\frac{1}{2}(\nu-1))} \int_0^\pi d\theta\, (\sin\theta)^{\nu-2} e^{h\cos\theta}.$$

The expansion of the free energy $A(h) = \ln z(h)$ of the one-site model is then

$$A(h) = \frac{h^2}{2\nu} - \frac{h^4}{4\nu(\nu+2)} + O(h^6).$$

The Legendre transform has the expansion

$$B(m) = \frac{\nu}{2}m^2 + \frac{\nu^2}{4(\nu+2)}m^4 + O(m^6).$$

The interaction energy can still be expressed in terms of the mean spin σ as $-\beta\mathcal{E}(\mathbf{S}) = \Omega v\sigma^2$. The mean-spin distribution is proportional to

$$R_\Omega(\boldsymbol{\sigma}) = \left(\frac{\Omega}{2\pi}\right)^\nu e^{\Omega v\sigma^2} \int d^\nu k\; e^{\Omega[i\mathbf{k}\cdot\boldsymbol{\sigma}+A(-i\mathbf{k})]}.$$

For $\Omega \to \infty$, this integral can be calculated by the steepest descent method. The saddle point is given by $\sigma = \mathbf{k}A'(-i\mathbf{k})/k$. Again, the result can be expressed in terms of the Legendre transform B of the function A.
   In particular the critical value of v is $v_c = \nu/2$.

8 Quasi-Gaussian approximation: Universality, critical dimension

In this chapter, we continue the study of phase transitions in statistical systems with short-range interactions begun in Chapter 7. We remain within the framework of ferromagnetic systems but, to a large extent, this is a restriction of terminology. Indeed, as a consequence of the *universality* property of critical phenomena, a property that we will describe and analyse, the results that will be derived apply to many other physical transitions that are not magnetic, like the liquid–vapour, binary mixtures, superfluid helium, and so on. To this list, one can add a problem that does not seem, at first sight, to be related to critical phenomena, the statistical properties of polymers, or from a more theoretical viewpoint, of self-avoiding random walk (SAW) on a lattice.

For reasons we have already explained, we are interested only in second-order phase transitions, in the vicinity of the transition temperature. For these transitions, the correlation length, which characterizes the decay at large distance of connected correlation functions, diverges at the transition (at the critical temperature). Thus, a distance scale, large with respect to the microscopic scales (range of forces, lattice spacing), is generated dynamically. Then, a non-trivial macroscopic or large-distance physics appears that has *universal* properties, that is, properties independent to a large extent of the details of the microscopic interactions.

We have already shown that, as long as the correlation length remains finite, macroscopic quantities, like the mean spin, have the behaviour predicted by the central limit theorem; in the infinite volume limit, they tend toward certain values with decreasing Gaussian fluctuations. This result can be understood in the following way: the initial microscopic degrees of freedom can be replaced by independent mean spins, attached to volumes having the correlation length as linear size. Therefore, it is natural to first study the properties of Gaussian models.

At the transition temperature T_c, and in the several-phase region, the arguments are no longer valid. Nevertheless, one may wonder whether the asymptotic Gaussian measure can then be simply replaced by a perturbed Gaussian measure, that is, whether the residual correlations between mean spins can be treated perturbatively. Such an approximation can be called *quasi-Gaussian*. The quasi-Gaussian approximation predicts remarkably universal large-distance properties, independent to a large extent of the symmetries, of the dimension of space, and so on. For homogeneous quantities, they coincide with those of the models in infinite dimension that we have studied in Section 7.2. But, in addition, the quasi-Gaussian approximation predicts that the singular behaviour of correlation functions for T close to T_c and in a weak magnetic field is also *universal*.

In the first part of the study, the discussion is restricted to models with a discrete symmetry. However, below T_c or in a magnetic field, models with continuous symmetries have special properties due, in particular, to the presence of Goldstone modes, which require a special analysis (see Section 7.8).

A systematic calculation of corrections to the quasi-Gaussian approximation allows the verification of its consistency and the determination of its domain of validity. The special role of dimension 4 emerges, which separates the higher dimensions where the approximation is justified, to lower dimensions where it cannot be valid.

The early and simplest description of phase transitions is the mean-field theory (MFT). It assumes that the infinite number of microscopic degrees of freedom can be replaced by a small number of macroscopic degrees of freedom and that the residual effects can be treated perturbatively. The MFT can also be qualified as quasi-Gaussian, in the sense that it predicts the same universal properties. The MFT can be introduced by several methods: partial summation of the high-temperature expansion, variational principle, leading order of the steepest descent method. The last method allows calculating corrections to the mean-field approximation and, thus, discussing its domain of validity, which indeed is the same as for the quasi-Gaussian approximation.

Ising type ferromagnetic systems. Again, we consider classical spins S_i on the d-dimensional lattice of points with integer coordinates, where i represents a lattice site. The partition function in a local magnetic field H has the form (7.1):

$$\mathcal{Z}(\mathbf{H}) = \int \left(\prod_i \rho(S_i) \mathrm{d}S_i \right) \exp\left[-\beta \mathcal{E}(S) + \sum_i H_i S_i \right], \qquad (8.1)$$

where, again, H includes a factor $\beta = 1/T$.

Moreover, we assume that the spin fluctuations around $S = 0$ are small and, thus, that the local spin distribution $\rho(S)$ satisfies the conditions of Section 7.1 and decays for $|S| \to \infty$ faster than a Gaussian function (condition (7.2)). Finally, we assume that the spin distribution in zero field has a reflection symmetry (of Ising type), that is, is invariant when $S_i \mapsto -S_i$. In particular, the spin distribution at each site is even: $\rho(S) = \rho(-S)$. We also choose a pair ferromagnetic interaction that generalizes the examples (7.22), (7.43), of the form

$$-\beta \mathcal{E}(S) = \sum_{i,j} V_{ij} S_i S_j, \qquad (8.2)$$

where the pair potential V_{ij} is a symmetric matrix with positive elements, invariant under space translations and short range (a notion that we define more precisely below).

The decay condition (7.2) implies that the partition function is defined, at least in a finite volume, for all pair interactions (8.2) and at any temperature.

8.1 Short-range two-spin interactions

In the model of Section 7.2, space plays no role, the infinite-range interaction connecting all spins. The systems we want to study, by contrast, have short-range interactions, a notion that we define more precisely here in the case of two-spin interactions.

We have assumed ferromagnetic pair interactions, which implies $V_{ij} = V_{ji} \geq 0$. Invariance under space translations implies that the pair potential V_{ij} has the form

$$V_{ij} \equiv V(\mathbf{r}_i - \mathbf{r}_j) = V(\mathbf{r}_j - \mathbf{r}_i) \geq 0 \,,$$

where $\mathbf{r}_i$ and $\mathbf{r}_j$ are the vectors joining the sites i and j to the origin.

We then define *short-range interactions* as interactions that decay exponentially with distance. In the case of a two-spin interaction, this implies

$$V(\mathbf{r}) \leq M \mathrm{e}^{-\kappa|\mathbf{r}|}, \quad \kappa > 0 \,. \tag{8.3}$$

It is convenient to introduce the parameter

$$v = \sum_{\mathbf{r}} V(\mathbf{r}) > 0 \,, \tag{8.4}$$

which is proportional to β, the inverse of the temperature (in what follows, we characterize the temperature by the value of $1/v$) and the function

$$U(\mathbf{r}) = V(\mathbf{r})/v \,. \tag{8.5}$$

Finally, we assume that $U(\mathbf{r})$ satisfies the ergodicity condition of the class of transition probabilities of the random walk of Section 3.3.7. With this condition, the minimum of the interaction energy is obtained for configurations where all spins have the same sign.

For simplicity, we consider only potentials that have the symmetry of the cubic lattice (as defined in (3.47)). The normalized potential, that is, the function $U(\mathbf{r})$, thus belongs to the class of transition probabilities in Section 3.3.7.

This assumption of cubic symmetry is not essential; the reflection symmetry $V(\mathbf{r}) = V(-\mathbf{r})$ suffices since, in the continuum limit, it remains always possible to perform linear coordinate transformations.

The interaction being translation invariant, it is natural to introduce the Fourier transforms

$$\tilde{V}(\mathbf{k}) = v\tilde{U}(\mathbf{k}) = \sum_{\mathbf{r} \in \mathbb{Z}^d} V(\mathbf{r}) \exp\left(i\mathbf{k} \cdot \mathbf{r}\right). \tag{8.6}$$

The function $\tilde{V}(\mathbf{k})$ is even and periodic in all components k_μ of the vector $\mathbf{k}$. One can thus restrict the components to what is called a Brillouin zone: $-\pi < k_\mu \leq \pi$. The vector $\mathbf{k}$ is sometimes called the *momentum* vector in analogy with quantum mechanics where position and momentum are dual in the Fourier transformation.

The short-range condition (8.3) implies that the Fourier transforms $\tilde V(\mathbf{k})$ and $\tilde U(\mathbf{k})$ are analytic for $|\operatorname{Im}\mathbf{k}| < \kappa$. Moreover, the positivity of the coefficients V_{ij} implies

$$|\tilde U(\mathbf{k})| \le \sum_{\mathbf{r}\in\mathbb{Z}^d} U(\mathbf{r}) = \tilde U(0) = 1. \tag{8.7}$$

For $\mathbf{k} \to 0$, the function $\tilde U(\mathbf{k})$ thus has an expansion of the form

$$\tilde U(\mathbf{k}) = 1 - a^2 k^2 + O(k^4), \tag{8.8}$$

where k is the length of the vector $\mathbf{k}$ and a is a positive constant.

Example: Nearest-neighbour interactions. A nearest-neighbour interaction on a lattice is an example of a short-range interaction. Denoting by $\mathbf{e}_\mu$ the d unit vectors corresponding to the links on the lattice, we can write the pair potential as

$$V(\mathbf{r}) = \frac{v}{2d}\sum_{\mu=1}^{d}\left(\delta_{\mathbf{r},\mathbf{e}_\mu} + \delta_{\mathbf{r},-\mathbf{e}_\mu}\right),$$

where δ is here the Kronecker symbol. Its Fourier transform is then

$$\tilde V(\mathbf{k}) = \sum_{\mathbf{r}\in\mathbb{Z}^d} V(\mathbf{r})\exp(i\mathbf{k}\cdot\mathbf{r}) = \frac{v}{2d}\sum_\mu\left[\exp(i\mathbf{k}\cdot\mathbf{e}_\mu) + \exp(-i\mathbf{k}\cdot\mathbf{e}_\mu)\right]$$

$$= \frac{v}{d}\sum_{\mu=1}^{d}\cos k_\mu,$$

where k_μ are the components of the vector $\mathbf{k}$. It is an analytic, entire function of all components k_μ.

For $|\mathbf{k}| \to 0$,

$$\tilde U(\mathbf{k}) = 1 - \frac{k^2}{2d} + \sum_{\mu=1}^{d}\frac{k_\mu^4}{24d} + O(k_\mu^6).$$

High-temperature expansion. It is possible to calculate the partition function and correlation functions by expanding expression (8.1) in powers of the potential V_{ij} and by evaluating the successive terms (high-temperature expansion). The expansion involves moments of the local distribution (Section 7.1)

$$\langle S_i^n\rangle = z^{-1}(H_i)\left(\frac{\partial}{\partial H_i}\right)^n z(H_i). \tag{8.9}$$

Like in the case of perturbations to the Gaussian measure, the free energy has a simpler expansion than the partition function because 'non-connected' contributions cancel:

$$\mathcal{W}(H) - \mathcal{W}_0(H) = \sum_{i,j} V_{ij}\langle S_i\rangle\langle S_j\rangle + \frac{1}{2!}\sum_{i,j,k,l} V_{ij}V_{kl}\langle S_iS_jS_kS_l\rangle_c + \cdots \tag{8.10}$$

where, for example,

$$\sum_{i,j,k,l} V_{ij}V_{kl}\langle S_iS_jS_kS_l\rangle_c = \sum_{i,j,k,l} V_{ij}V_{kl}\left[\langle S_iS_jS_kS_l\rangle - \langle S_iS_j\rangle\langle S_kS_l\rangle\right].$$

In the difference, all terms where the four indices i,j,k,l are different cancel. This property is related to the extensiveness of the free energy.

The high-temperature expansion diverges at the critical temperature where thermodynamic functions are singular. When one is able to calculate enough terms of the expansion, one can use it, combined with various mathematical techniques to analyse series (ratio method, Neville table extrapolation and generalizations, logarithmic Padé approximants, differential approximants, and so on), to obtain numerical information about the critical behaviour.

Remark. It is technically convenient, in the frameworks both of the high-temperature expansion and the mean-field approximation (see Section 8.10) to assume that V_{ii} vanishes (invariance under translation implies that V_{ii} is independent of the site i). This can be done without loss of generality since the corresponding one-site contribution can always be included in the measure $\rho(S)$.

8.2 The Gaussian model: Two-point function

We first discuss ferromagnetic systems in the disordered phase $T > T_c$. As we have already pointed out in Section 6.1.3, because the correlation length then is finite, one expects to be able to describe their macroscopic properties in terms of classical spins σ_i that are already averages of microscopic spins over small volumes. Moreover, one expects, in the spirit of the central limit theorem, that the fluctuations of the spins σ_i are small and that their distribution, at leading order, is Gaussian. These ideas lead to the model Gaussian that we now study.

A more formal derivation based on these ideas will be presented in Section 8.10.

Gaussian model. Since in the disordered phase spins fluctuate around an expectation value that vanishes as a consequence of the $\sigma \to -\sigma$ symmetry, a local one-site Gaussian distribution takes the form

$$\rho(\sigma) = e^{-b_2\sigma^2/2}, \quad b_2 > 0.$$

A two-spin interaction of the form (8.2) is directly quadratic. It can be considered as the first term in an expansion in powers of σ. With this interaction, the Gaussian partition function in a field can be written as

$$\mathcal{Z}(\mathbf{H}) = \int \left(\prod_i d\sigma_i\right) \exp\left[-\mathcal{H}(\sigma) + \sum_i H_i\sigma_i\right], \tag{8.11}$$

where

$$\mathcal{H}(\sigma) = \frac{1}{2}\sum_{i,j} \mathfrak{S}_{ij}\sigma_i\sigma_j, \quad \mathfrak{S}_{ij} = b_2\delta_{ij} - 2V_{ij} \tag{8.12}$$

and V_{ij} has the form discussed in Section 8.1. In particular, $\mathfrak{S}_{ij}$ depends only on $\mathbf{r}_i - \mathbf{r}_j$, where $\mathbf{r}_i, \mathbf{r}_j$ are the positions of the sites i and j:

$$\mathfrak{S}_{ij} \equiv \mathfrak{S}(\mathbf{r}_i - \mathbf{r}_j).$$

The model is defined only if the matrix $\mathfrak{S}_{ij}$ is strictly positive. In the thermodynamic limit, the eigenvalues of the matrix $\mathfrak{S}$ belong to a continuous spectrum given by the Fourier transform

$$\tilde{\mathfrak{S}}(\mathbf{k}) = \sum_{\mathbf{r}} e^{i\mathbf{k}\cdot\mathbf{r}}\,\mathfrak{S}(\mathbf{r}).$$

Indeed, this equation can be rewritten as the eigenvalue equation (see equation (3.34))

$$\tilde{\mathfrak{S}}(\mathbf{k})\, e^{i\mathbf{k}\cdot\mathbf{r}'} = \sum_{\mathbf{r}} e^{i\mathbf{k}\cdot\mathbf{r}}\,\mathfrak{S}(\mathbf{r} - \mathbf{r}').$$

With the definitions (8.6), (8.5), one finds

$$\tilde{\mathfrak{S}}(\mathbf{k}) = b_2 - 2v\tilde{U}(\mathbf{k}) \qquad (8.13)$$

and, as a consequence of the bound (8.7), the matrix positivity condition is thus

$$\tfrac{1}{2}b_2 \equiv v_c > v. \qquad (8.14)$$

As we shall verify, this condition implies that the correlation length is finite, and is consistent with the assumptions that have led to the Gaussian model.

We can then use the results of Section 6.3.3. The thermodynamic potential is simply

$$\Gamma(M) = \Gamma(0) + \frac{b_2}{2}\sum_i M_i^2 - \sum_{i,j} V_{ij} M_i M_j. \qquad (8.15)$$

8.2.1 Homogeneous quantities

In a uniform magnetic field $H_i = H$, the magnetization M is uniform. The function $\Gamma(M)$ is then proportional to the number of sites Ω, and we set

$$\Omega^{-1}\left[\Gamma(M) - \Gamma(0)\right] \equiv \mathcal{G}(M) = \tfrac{1}{2}(b_2 - 2v)M^2, \qquad (8.16)$$

where the definition (8.4) has been used.

The mean spin $\sigma = \sum_i S_i/\Omega$ distribution is Gaussian and given by

$$R_\Omega(\sigma) = \sqrt{(b_2 - 2v)/2\pi}\, e^{H\sigma - (b_2 - 2v)\sigma^2/2}. \qquad (8.17)$$

Again, the special value (8.14), $v_c = b_2/2$, appears. For $v < v_c$, $\mathcal{G}(M)$ is minimum at $M = 0$ and, thus, the magnetization vanishes in zero field:

$$M = \langle \sigma_i \rangle_{H=0} = 0.$$

The high-temperature phase is a disordered phase.

For $v = v_c$, the magnetization is undetermined and at lower temperature the model is meaningless since the distribution $R_\Omega(\sigma)$ is not summable.

The equation of state is linear:
$$H = \mathcal{G}'(M) = (v_c - v)M.$$

The magnetic susceptibility, which is the derivative of the magnetization with respect to the magnetic field, takes the form
$$\chi = \frac{\partial M}{\partial H} = \frac{1}{v_c - v}.$$

The magnetic susceptibility thus diverges for $v = v_c$, a value that can be interpreted as a transition point corresponding to a critical temperature T_c. The behaviour
$$\chi \propto (T - T_c)^{-1},$$
coincides with that obtained in Section 7.3 for the model in infinite dimension and zero field.

One does not expect, in general, such a divergence in non-zero field but, in the Gaussian model, the susceptibility is independent of the applied field.

8.2.2 Two-point function

The two-point function in zero field, for $v < v_c$,
$$\Delta_{ij} \equiv \langle \sigma_i \sigma_j \rangle = \langle \sigma_i \sigma_j \rangle_{\text{conn.}} = \left. \frac{\partial^2 \mathcal{W}(H)}{\partial H_i \partial H_j} \right|_{H=0},$$
is the inverse of the matrix $\mathfrak{S}$ (equation (8.12)):
$$\sum_k \Delta_{ik} \mathfrak{S}_{kj} = \delta_{ij}. \tag{8.18}$$

Translation invariance implies that both $\mathfrak{S}_{ij}$ and Δ_{ij} depend only on $\mathbf{r}_i - \mathbf{r}_j$, where $\mathbf{r}_i, \mathbf{r}_j$ are the positions of the sites i and j. Equation (8.18) thus is a convolution equation that simplifies after Fourier transformation. The Fourier transform
$$\tilde{\Delta}(\mathbf{k}) = \sum_{\mathbf{r}} e^{i\mathbf{k}\cdot\mathbf{r}} \Delta(\mathbf{r}) \quad \Leftrightarrow \quad \Delta(\mathbf{r}) = \int \frac{d^d k}{(2\pi)^d} e^{-i\mathbf{k}\cdot\mathbf{r}} \tilde{\Delta}(\mathbf{k}).$$

is the inverse of the Fourier transform (8.13):
$$\tilde{\Delta}(\mathbf{k}) = \tilde{\mathfrak{S}}^{-1}(\mathbf{k}) = \left[b_2 - 2\tilde{V}(\mathbf{k}) \right]^{-1}. \tag{8.19}$$

For $v < v_c$, the denominator does not vanish in a neighbourhood of the real axis. The function $\tilde{\Delta}(\mathbf{k})$ is thus analytic in a strip and the function $\Delta(\mathbf{r})$ decreases exponentially for $|\mathbf{r}| \to \infty$, confirming that the correlation length is finite.

By contrast, for $v = v_c$ (that is, at the transition temperature $T = T_c$), the denominator behaves for $|\mathbf{k}| \to 0$ like (using the parametrization (8.8))

$$2v_c\bigl(1 - \tilde{U}(\mathbf{k})\bigr) = 2v_c a^2 k^2 + O(k^4).$$

Moreover, the bound (8.7) implies that the denominator can vanish only at zero momentum. The correlation length then diverges, and this is consistent with the interpretation of v_c as corresponding to a transition temperature. The function $\tilde{\Delta}(\mathbf{k})$ is singular,

$$\tilde{\Delta}(\mathbf{k}) \underset{|\mathbf{k}| \to 0}{\sim} \frac{D}{k^2}, \quad D = 1/2v_c a^2, \tag{8.20}$$

and this leads to an algebraic decay of $\Delta(\mathbf{r})$ for $d > 2$.

For $d = 2$ and $v = v_c$, the integral

$$\Delta(\mathbf{r}) = \frac{1}{(2\pi)^2} \int d^2k \, e^{-i\mathbf{k}\cdot\mathbf{r}} \, \tilde{\Delta}(\mathbf{k})$$

diverges at $\mathbf{k} = 0$ and, thus, the Gaussian model cannot describe a transition point $T = T_c$ in dimension 2.

8.2.3 Critical behaviour

Although some properties of the Gaussian model at the critical point $v = v_c$ are clearly pathological, let us examine the behaviour of the two-point function

$$\Delta(\lambda\mathbf{r}) = \frac{1}{(2\pi)^d} \int d^d k \, e^{-i\lambda\mathbf{k}\cdot\mathbf{r}} \, \tilde{\Delta}(\mathbf{k}),$$

with a scaled momentum $\mathbf{r} \mapsto \lambda\mathbf{r}$, at the critical point, for $d > 2$, when $\lambda \to +\infty$. The explicit calculation of the integral is presented in Section 8.4.3, but the behaviour can be derived from simple arguments.

After the change $\mathbf{k} \mapsto \mathbf{k}/\lambda$, the integral becomes

$$\Delta(\lambda\mathbf{r}) = \frac{1}{(2\pi)^d} \lambda^{2-d} \int d^d k \, e^{-i\mathbf{k}\cdot\mathbf{r}} \lambda^{-2} \tilde{\Delta}(\mathbf{k}/\lambda),$$

where now the components k_μ of the vector $\mathbf{k}$ vary in the interval $-\lambda\pi \le k_\mu \le \lambda\pi$.

Because $\tilde{\Delta}(\mathbf{k})$ is regular for $\mathbf{k} \ne 0$, one can take the limit $\lambda \to \infty$ in the integrand and the integration bounds and, thus,

$$\Delta(\lambda\mathbf{r}) \underset{\lambda \to \infty}{\sim} \frac{1}{(2\pi)^d} \frac{\lambda^{2-d}}{2v_c a^2} \int \frac{d^d k}{k^2} e^{-i\mathbf{k}\cdot\mathbf{r}}.$$

Since the denominator is now rotation-invariant, in the limit $\Delta(\mathbf{r})$ is a function only of $r = |\mathbf{r}|$. One infers

$$\Delta(\mathbf{r}) \underset{r=|\mathbf{r}|\to\infty}{\propto} \frac{1}{r^{d-2}}. \tag{8.21}$$

The two-point function has an algebraic decay and, thus, the correlation length is infinite.

Moreover, asymptotically, $\Delta(\mathbf{r})$ has a $O(d)$ rotation symmetry (extending $\mathbf{r}$ to continuum space), which is larger than the discrete symmetries of the lattice (an analogous property has been obtained in the case of the random walk in Section 3.3.9).

It will shown later that, for a much more general class of models, the two-point function at the critical temperature $T = T_c$ behaves, for $r \to \infty$, like

$$\Delta(\mathbf{r}) \underset{|\mathbf{r}|\to\infty}{\propto} 1/r^{d-2+\eta}, \tag{8.22}$$

which corresponds, by the same arguments, for its Fourier transform to the behaviour

$$\tilde{\Delta}(\mathbf{k}) \underset{k\to 0}{\propto} 1/k^{2-\eta}. \tag{8.23}$$

Here, from the forms (8.20) or (8.21) of the Gaussian two-point function, one infers

$$\eta = 0, \tag{8.24}$$

which is the Gaussian, or classical, value of the exponent η.

8.2.4 Critical domain

For $v < v_c$, but $v_c - v \to 0$ and $\mathbf{k}$ of order $(v_c - c)^{1/2}$ (the so-called critical domain), the leading term of $1/\tilde{\Delta}(\mathbf{k})$ is

$$2v_c - 2v\tilde{U}(\mathbf{k}) = 2(v_c - v) + 2v_c a^2 k^2 + O(k^4, (v-v_c)k^2) \tag{8.25}$$

and, thus,

$$\tilde{\Delta}(\mathbf{k}) \sim \tfrac{1}{2}\left(v_c - v + v_c a^2 k^2\right)^{-1}.$$

Again, in this limit, the expansion (8.8) implies that $\tilde{\Delta}(\mathbf{k})$ has a $O(d)$ rotation symmetry, larger than the symmetries of the lattice.

Moreover, in this limit, the two-point function has the so-called Ornstein–Zernike or free-field form (see Section 8.4).

One infers that $\Delta(\mathbf{r})$ decays exponentially (equations (8.42) and (8.44)) with the asymptotic behaviour

$$\Delta(\mathbf{r}) \underset{r\to\infty}{\propto} \frac{1}{r^{(d-1)/2}} e^{-r/\xi},$$

where the length correlation diverges for $T \to T_c$ like

$$\xi \underset{v\to v_c}{\sim} a(1 - v/v_c)^{-1/2} \propto (T - T_c)^{-1/2}.$$

188 Quasi-Gaussian approximation: Universality, critical dimension

More generally, in a large class of models one finds

$$\xi(T) \propto (T - T_c)^{-\nu}, \qquad (8.26)$$

and $\nu = 1/2$, thus, is the Gaussian value of the correlation exponent.

In the restricted framework of the Gaussian theory on the lattice, we have derived the universal behaviour of the two-point function at large distance at T_c and near T_c when the correlation length is large with respect to the microscopic scale, as well as the universal singularity of the correlation length at T_c.

These various *universal behaviours* will be recovered in the quasi-Gaussian approximation (Section 8.5), from the more general assumptions of Landau's theory (Section 8.7), or from the mean-field approximation (Section 8.10).

8.3 Gaussian model and random walk

As an exercise, and because this property can be generalized, we show how the Gaussian two-point function can be related to a form of random walk, as defined in Section 3.3.7.

We expand expression (8.19),

$$\tilde{\Delta}(\mathbf{k}) = \left[2v_c - 2v\tilde{U}(\mathbf{k})\right]^{-1},$$

in powers of v:

$$\tilde{\Delta}(\mathbf{k}) = \frac{1}{2v_c} \sum_{n=0}^{\infty} \left(\frac{v}{v_c}\right)^n \tilde{U}^n(\mathbf{k}). \qquad (8.27)$$

Inverting the representation (8.6), we now express $\tilde{U}(\mathbf{k})$ in terms of its Fourier transform:

$$\tilde{U}(\mathbf{k}) = \sum_{\mathbf{r} \in \mathbb{Z}^d} e^{-i\mathbf{r}\cdot\mathbf{k}} U(\mathbf{r}).$$

Then,

$$\tilde{\Delta}(\mathbf{k}) = \frac{1}{2v_c} \sum_{n=0}^{\infty} \left(\frac{v}{v_c}\right)^n \sum_{\mathbf{r}_1,\ldots,\mathbf{r}_n} U(\mathbf{r}_1)\ldots U(\mathbf{r}_n) e^{-i\mathbf{k}\cdot(\mathbf{r}_1+\cdots+\mathbf{r}_n)}.$$

We now introduce an arbitrary initial point $\mathbf{q}_0$ on the lattice and change variables, setting

$$\mathbf{r}_\ell = \mathbf{q}_\ell - \mathbf{q}_{\ell-1}, \qquad 1 \leq \ell \leq n.$$

The expression becomes

$$\tilde{\Delta}(\mathbf{k}) = \frac{1}{2v_c} \sum_{n=0}^{\infty} \left(\frac{v}{v_c}\right)^n \sum_{\mathbf{q}_1,\ldots,\mathbf{q}_n} U(\mathbf{q}_1 - \mathbf{q}_0)\ldots U(\mathbf{q}_n - \mathbf{q}_{n-1}) e^{-i\mathbf{k}\cdot(\mathbf{q}_n - \mathbf{q}_0)}. \qquad (8.28)$$

We have already pointed out that the function $U(\mathbf{r})$ has all the properties of the transition functions of the translation invariant random walk. We then define the corresponding random process by the evolution equation (see equation (3.43))

$$P_n(\mathbf{q}) = \sum_{\mathbf{q}' \in \mathbb{Z}^d} U(\mathbf{q} - \mathbf{q}') P_{n-1}(\mathbf{q}'),$$

where $P_n(\mathbf{q})$ is the probability for a walker starting from point $\mathbf{q}_0$ at time 0 to be at point $\mathbf{q}$ at time n.

An iteration of the equation leads to

$$P_n(\mathbf{q}) = \sum_{\mathbf{q}_1,\ldots,\mathbf{q}_{n-2},\mathbf{q}_{n-1}} U(\mathbf{q} - \mathbf{q}_{n-1}) U(\mathbf{q}_{n-1} - \mathbf{q}_{n-2}) \ldots U(\mathbf{q}_1 - \mathbf{q}_0).$$

Expression (8.28) can thus be written as ($\mathbf{q}_n \mapsto \mathbf{q}$)

$$\tilde{\Delta}(\mathbf{k}) = \frac{1}{2v_c} \sum_{n=0}^{\infty} \left(\frac{v}{v_c}\right)^n \sum_{\mathbf{q}} P_n(\mathbf{q}) e^{-i\mathbf{k} \cdot (\mathbf{q} - \mathbf{q}_0)}. \tag{8.29}$$

The coefficient of v^n thus is a generating function of the moments of the distribution at time n. Also, the behaviour of the distribution $P_n(\mathbf{q})$ for $n \to \infty$ is related to the singularity of $\tilde{\Delta}(\mathbf{k})$, as a function of v, closest to the origin. Here, comparing with expression (8.27), one obtains directly

$$\tilde{U}^n(\mathbf{k}) = \sum_{\mathbf{q}} P_n(\mathbf{q}) e^{-i\mathbf{k} \cdot (\mathbf{q} - \mathbf{q}_0)}.$$

One can then expand in powers of $\mathbf{k}$. For example, the second moment is the coefficient of $\mathbf{k}^2$:

$$na^2 \mathbf{k}^2 = \frac{1}{2} \sum_{\mathbf{q}} P_n(\mathbf{q}) (\mathbf{k} \cdot (\mathbf{q} - \mathbf{q}_0))^2$$

and, thus,

$$\sum_{\mathbf{q}} P_n(\mathbf{q}) (\mathbf{q} - \mathbf{q}_0)_\alpha (\mathbf{q} - \mathbf{q}_0)_\beta = 2na^2 \delta_{\alpha\beta}.$$

One recovers, as was proved directly, that the asymptotic distribution of $\mathbf{q} - \mathbf{q}_0$ is isotropic and the expectation value of $(\mathbf{q} - \mathbf{q}_0)^2$ increases linearly with time.

8.4 Gaussian model and field integral

In Chapter 5, we have shown that the large-distance properties of one-dimensional lattice models can be described, when the correlation length diverges, by a path integral. In arbitrary dimensions, the universal properties of the Gaussian model on the lattice can be inferred from a statistical field theory (which replaces the distribution of spin configurations) in continuum isotropic space.

Let $\sigma(x)$ be a field in d-dimensional continuum space ($\mathbb{R}^d$) and $H(x)$ an arbitrary local magnetic field.

We consider the Gaussian *field integral*, or *functional integral*,

$$\mathcal{Z}(H) = \int [\mathrm{d}\sigma(x)] \exp\left[-\mathcal{H}(\sigma) + \int \mathrm{d}^d x\, \sigma(x) H(x)\right] \tag{8.30}$$

with

$$\mathcal{H}(\sigma) = \frac{1}{2} \int \mathrm{d}^d x \left[\sum_{\mu=1}^{d} (\partial_\mu \sigma(x))^2 + m^2 \sigma^2(x)\right] \tag{8.31}$$

and the notation $\partial_\mu \equiv \partial/\partial x_\mu$, where $m > 0$ is a parameter related to the deviation from the critical temperature, as will be verified below. The functional $\mathcal{H}(\sigma)$ is often called the *Hamiltonian* in the present context (a denomination borrowed from the statistical theory of classical gases).

Field integrals are the generalization to d dimensions of path integrals, and the symbol $[\mathrm{d}\sigma(x)]$ stands for integration over all fields $\sigma(x)$.

In the framework of quantum field theory that describes the fundamental interactions at microscopic scale, the Gaussian case corresponds to a free field theory. The form (8.31), quadratic in the fields, is then called the *Euclidean action* and the parameter m is the mass of the particle associated to the field σ.

As in the lattice case, when this seems necessary, we define the infinite volume or thermodynamic limit as the limit of a cube with periodic boundary conditions.

8.4.1 Maximum of the integrand and two-point function

The calculation of the Gaussian integral (8.30) is a simple generalization of the calculation of the path integral (5.13). One first looks for a maximum of the integrand and thus the minimum of

$$\mathcal{H}(\sigma, H) = \mathcal{H}(\sigma) - \int \mathrm{d}^d x\, \sigma(x) H(x). \tag{8.32}$$

One sets

$$\sigma(x) = \sigma_c(x) + \varepsilon(x) \tag{8.33}$$

and expands in ε. The field $\sigma_c(x)$ at the minimum is determined by the condition that the term linear in ε vanishes:

$$-\int \mathrm{d}^d x \left[\sum_{\mu=1}^{d} \partial_\mu \sigma_c(x) \partial_\mu \varepsilon(x) + m^2 \sigma_c(x) \varepsilon(x)\right] + \int \mathrm{d}^d x\, \varepsilon(x) H(x) = 0\,.$$

One integrates the term linear in $\partial_\mu \varepsilon(x)$ by parts (with respect to x_μ). Then,

$$\sum_{\mu=1}^{d} \int d^d x\, \partial_\mu \sigma_c(x) \partial_\mu \varepsilon(x) = -\sum_{\mu=1}^{d} \int d^d x\, \varepsilon(x) (\partial_\mu)^2 \sigma_c(x),$$

because the integrated terms cancel due to periodic boundary conditions. One finds the equation

$$(-\nabla_x^2 + m^2)\sigma_c(x) = H(x),$$

where ∇_x^2 is the Laplacian in d dimensions:

$$\nabla_x^2 = \sum_\mu \partial_\mu^2.$$

The solution can be written as

$$\sigma_c(x) = \int d^d x\, \Delta(x-y) H(y),$$

where $\Delta(x-y)$ is also the Gaussian two-point function in zero field (equations (2.10) and (2.19)):

$$\langle \sigma(x)\sigma(y)\rangle_{H=0} = \Delta(x-y).$$

It satisfies

$$(-\nabla_x^2 + m^2)\Delta(x) = \delta^{(d)}(x),$$

where $\delta^{(d)}$ is the Dirac distribution in d dimensions, as one verifies by acting with $-\nabla_x^2 + m^2$ on σ_c. The equation can be solved by Fourier transformation. In the infinite volume limit, one finds

$$\Delta(x) = \frac{1}{(2\pi)^d} \int d^d k\, e^{-ik\cdot x}\, \tilde{\Delta}(k) \tag{8.34}$$

with

$$\tilde{\Delta}(k) = \frac{1}{k^2 + m^2}, \tag{8.35}$$

as one verifies by acting with $-\nabla_x^2 + m^2$ on $\Delta(x)$ ($\int d^d k\, e^{-ik\cdot x} = (2\pi)^d \delta^{(d)}(x)$).

Finally, the functional (8.32) for $\sigma = \sigma_c$, after integration by parts, takes the form

$$\mathcal{H}(\sigma_c, H) = \int d^d x\, \sigma_c(x)\left[-\tfrac{1}{2}\nabla_x^2 + \tfrac{1}{2}m^2 - H(x)\right]\sigma_c(x)$$

$$= -\frac{1}{2}\int d^d x\, \sigma_c(x) H(x)$$

$$= -\frac{1}{2}\int d^d x\, d^d y\, H(x)\Delta(x-y)H(y). \tag{8.36}$$

8.4.2 Gaussian integration

One now performs the change of variable (8.33), $\sigma(x) \mapsto \varepsilon(x)$. The functional (8.32) becomes

$$\mathcal{H}(\sigma, H) = \mathcal{H}(\sigma_c, H) + \mathcal{H}(\varepsilon)$$

and, thus,

$$\mathcal{Z}(H) = e^{-\mathcal{H}(\sigma_c, H)} \int [d\varepsilon(x)] e^{-\mathcal{H}(\varepsilon)}.$$

The remaining Gaussian integration over $\varepsilon(x)$ yields a normalization,

$$\mathcal{Z}(0) = \int [d\varepsilon(x)] e^{-\mathcal{H}(\varepsilon)},$$

independent of H, and that can be explicitly evaluated only by replacing the continuum by a lattice (see also the discussion of Sections 3.5 and 5.1.3).

The generating functional of connected correlation functions is thus (equation (8.36))

$$\mathcal{W}(H) = \ln \mathcal{Z}(H) = \mathcal{W}(0) + \frac{1}{2} \int d^d x \, d^d y \, H(x) \Delta(x-y) H(y). \tag{8.37}$$

In a uniform field, the free energy density becomes

$$W(H) = (\mathcal{W}(H) - \mathcal{W}(0))/\text{volume}$$
$$= \tfrac{1}{2} H^2 \int d^d x \, \Delta(x) = \tfrac{1}{2} H^2 \tilde{\Delta}(0) = \tfrac{1}{2} H^2/m^2,$$

from which one infers the thermodynamic potential density

$$\mathcal{G}(M) = \tfrac{1}{2} m^2 M^2. \tag{8.38}$$

Comparing these expressions with the asymptotic universal parts of the corresponding expressions on the lattice, for example $\mathcal{G}(M)$ with expression (8.16), one identifies

$$m^2 \sim 2(v_c - v) \propto T - T_c. \tag{8.39}$$

8.4.3 Explicit calculation of the two-point function

We now calculate explicitly the two-point function in continuum space.
At T_c, $\tilde{\Delta}(\mathbf{k}) = 1/k^2$ and one must evaluate

$$\Delta(\mathbf{r}) = \frac{1}{(2\pi)^d} \int d^d k \, \tilde{\Delta}(\mathbf{k}) \, e^{-i\mathbf{k}\cdot\mathbf{r}} = \int \frac{d^d k}{(2\pi)^d} \frac{e^{-i\mathbf{k}\cdot\mathbf{r}}}{k^2}.$$

To calculate the integral, we use the identity

$$\frac{1}{k^2} = \int_0^\infty dt \, e^{-tk^2}.$$

The integral over k then becomes Gaussian. After integration,

$$\Delta(\mathbf{r}) = \frac{1}{(4\pi)^{d/2}} \int_0^\infty \frac{dt}{t^{d/2}} e^{-r^2/4t} .$$

After the change of variables $u = r^2/4t$, the integration over u yields

$$\Delta(\mathbf{r}) = \frac{2^{d-2}}{(4\pi)^{d/2}} \Gamma(d/2 - 1) \frac{1}{r^{d-2}} . \tag{8.40}$$

For the function $1/(k^2 + m^2)$ the strategy is the same. One then finds

$$\Delta(r) = \frac{1}{(4\pi)^{d/2}} \int_0^\infty \frac{dt}{t^{d/2}} e^{-r^2/4t - m^2 t} = \frac{2}{(4\pi)^{d/2}} \left(\frac{2m}{r}\right)^{d/2-1} K_{1-d/2}(mr), \tag{8.41}$$

where $K_\nu(r)$ is a Bessel function of the third kind (see definition (2.54)). For $z \to +\infty$, $K_\nu(z)$ can be evaluated by the steepest descent method (see Exercise 2.6). One infers

$$\Delta(\mathbf{r}) \underset{r \to \infty}{\sim} \frac{1}{2m} \left(\frac{m}{2\pi}\right)^{(d-1)/2} \frac{e^{-mr}}{r^{(d-1)/2}} . \tag{8.42}$$

The constant $\xi = 1/m$, which characterizes the exponential decay of the two-point function, is the correlation length.

Remark. To get an idea of the class of typical fields that contribute to the field integral, one can evaluate the two-point function in the limit of coinciding points:

$$\langle \sigma(\mathbf{x})\sigma(\mathbf{y}) \rangle \underset{|\mathbf{x}-\mathbf{y}| \to 0}{\sim} \Delta(\mathbf{x} - \mathbf{y}, m=0) = \frac{2^{d-2}}{(4\pi)^{d/2}} \Gamma(d/2 - 1) \frac{1}{|\mathbf{x} - \mathbf{y}|^{d-2}} .$$

One notices that this class of functions is so singular that the expectation value of $\sigma^2(\mathbf{x})$ diverges, and with a rate that increases with the dimension of space d. This singularity of the Gaussian measure corresponding to the Hamiltonian (8.31) will later be the source of new difficulties.

For $d = 2$, the short-distance behaviour takes the form

$$\langle \sigma(\mathbf{x})\sigma(\mathbf{y}) \rangle \underset{|\mathbf{x}-\mathbf{y}| \to 0}{\sim} -\frac{1}{2\pi} \ln(m|\mathbf{x} - \mathbf{y}|) .$$

8.4.4 Lattice and continuum limit

We now investigate how the two-point function on the lattice converges, at large distance, toward the continuum function when the correlation length diverges.

At T_c, the function on the lattice reads

$$\Delta(\mathbf{r}) = \frac{1}{2} \int_{|k_\mu| \leq \pi} \frac{d^d k}{(2\pi)^d} \frac{e^{-i\mathbf{k}\cdot\mathbf{r}}}{v_c - \tilde{V}(\mathbf{k})} \tag{8.43}$$

$$= \frac{1}{2} \int_{|k_\mu| \leq \pi} \frac{d^d k}{(2\pi)^d} e^{-i\mathbf{k}\cdot\mathbf{r}} \int_0^\infty dt\, e^{-t[v_c - \tilde{V}(\mathbf{k})]} .$$

If one restricts the integral over t to a finite interval $t < t_{\max}$, the function

$$\int_0^{t_{\max}} dt \, e^{-t[v_c - \tilde{V}(\mathbf{k})]}$$

is analytic in a strip and following the analysis of Section A2.2, its Fourier transform decays exponentially for $|\mathbf{r}| \to \infty$. Thus, the neighbourhood of $t = \infty$, which generates the singularity at $\mathbf{k} = 0$, yields the leading contribution at large distance. One can then use the steepest descent method. The saddle point is situated at $\mathbf{k} = 0$. One thus expands $v_c - \tilde{V}(\mathbf{k})$ for $\mathbf{k} \to 0$, and the leading term is proportional to k^2. Moreover, one can integrate freely over $\mathbf{k}$. One thus recovers the continuum result (8.40).

The analysis generalizes to the situation $|v - v_c| \ll 1$. After the change of variables $\mathbf{k} = (1 - v/v_c)^{1/2}\mathbf{k}'/a$ and thus (equation (8.25))

$$2v_c - 2v\tilde{U}(\mathbf{k}) = 2(v_c - v)(1 + k'^2) + O\left((v_c - v)^2\right),$$

the integral (8.43) becomes

$$\Delta(\mathbf{r}) \underset{v \to v_c}{\sim} (1 - v/v_c)^{d/2 - 1} \frac{1}{2v_c a^d} \int_{|k_\mu| \le \Lambda} \frac{d^d k}{(2\pi)^d} \frac{e^{-i\mathbf{k} \cdot \mathbf{r}/\xi}}{1 + k^2}, \qquad (8.44)$$

where we have set

$$\Lambda = \pi a (1 - v/v_c)^{-1/2}, \quad \xi = a(1 - v/v_c)^{-1/2}. \qquad (8.45)$$

When $v \to v_c$, the upper-bound $\Lambda \to \infty$ and one can thus integrate freely over the components of the vector $\mathbf{k}$. The calculation then becomes identical to the continuum calculation and ξ can be identified with the correlation length.

8.5 Quasi-Gaussian approximation

Below the transition point, the Gaussian model is clearly no longer valid since the quadratic form (8.12) is no longer positive and thus the Gaussian integral (8.11) is not defined. One also notices that the mean-spin distribution, in this quadratic limit, is not summable (e.g., see equation (7.21)).

However, even in the framework of the central limit theorem, the Gaussian distribution is only asymptotic. The analysis of the Gaussian model shows that below the transition point, corrections to the Gaussian distribution, that is, terms of higher degree in the effective spin distribution, even if their amplitude is small, can no longer be neglected.

Here, we show that the steepest descent method leads naturally to a quasi-Gaussian approximation that reproduces, at leading order, the results of the model in infinite dimension. However, unlike the model in infinite dimension, it allows also studying the behaviour of correlation functions at the transition.

Effective model. To go beyond the Gaussian model, we thus consider a more general one-site local distribution
$$\rho(\sigma) = e^{-B(\sigma)}$$
where, following the analysis of the end of Section 7.1.2 (see equation (6.10)), we choose a function $B(\sigma)$ of the form (7.11), that is, the thermodynamic potential of the one-site model. We have shown that such a function is analytic and we parametrize its expansion at $\sigma = 0$ in the form
$$B(\sigma) = \sum_{p=1} \frac{b_{2p}}{2p!} \sigma^{2p}, \quad b_2 > 0. \tag{8.46}$$

We also assume $b_4 > 0$ since we want to study continuous transitions.

The generating function of correlation functions can then be written as
$$\mathcal{Z}(H) = \int \left(\prod_i d\sigma_i\right) \exp\left[-\mathcal{H}(\sigma) + \sum_i H_i \sigma_i\right], \tag{8.47}$$
where the Hamiltonian $\mathcal{H}(\sigma)$ takes the form
$$\mathcal{H}(\sigma) = -\sum_{i,j} V_{ij} \sigma_i \sigma_j + \sum_i B(\sigma_i). \tag{8.48}$$

Quasi-Gaussian approximation. Since the integral (8.47) is not Gaussian, it can no longer be calculated exactly. But since $B(\sigma)$ is analytic, one can evaluate the integral over the spins by the steepest descent method. If one assumes that the fluctuations around the saddle point vary slowly, one can approximate the integral by the leading contribution, an approximation which one can call quasi-Gaussian. Such an assumption implies, in particular, that the spins σ_i are the sum of an average value M_i and a weakly correlated fluctuating part. This assumption goes beyond the idea of the central limit theorem in the sense that the average value M_i is no longer related only to the distribution in each site but also results from the interactions.

Steepest descent method. The maximum of the integrand in the integral (8.47) is given by a solution of the saddle point
$$H_i = \frac{\partial \mathcal{H}}{\partial \sigma_i} \tag{8.49}$$
and, at leading order,
$$\mathcal{W}(H) = -\mathcal{H}(\sigma) + \sum_i \sigma_i H_i,$$
where σ is a function of H through (8.49).

As we have already noticed, $\mathcal{W}(H)$ is the Legendre transform of $\mathcal{H}(\sigma)$. As a consequence, the thermodynamic potential $\Gamma(M)$, the Legendre transform of $\mathcal{W}(H)$, is simply

$$\Gamma(M) = \mathcal{H}(M) = -\sum_{i,j} V_{ij} M_i M_j + \sum_i B(M_i). \tag{8.50}$$

In the case of the models invariant under space translations that we study, the magnetization in a uniform field is uniform. The thermodynamic potential density is then

$$\mathcal{G}(M) = \Omega^{-1}\Gamma(M) = -vM^2 + B(M), \tag{8.51}$$

where v is the parameter (8.4).

The equation of state follows:

$$H = \frac{\partial \mathcal{G}}{\partial M} = -2vM + B'(M). \tag{8.52}$$

We recover exactly the expressions (7.27) and (7.28) of the model in infinite dimension. Homogeneous quantities thus have the universal properties already exhibited in Sections 7.2 and 7.3.

8.6 The two-point function: Universality

Unlike the model in infinite dimension, the quasi-Gaussian approximation generates a non-trivial two-point function whose properties we now study.

Divergence of the correlation length and continuous transition. A continuous transition is characterized by the property

$$\left.\frac{\partial^2 \mathcal{G}}{(\partial M)^2}\right|_{M=0} = 0 \tag{8.53}$$

and, thus, by the divergence of the magnetic susceptibility $\chi = \partial^2 \mathcal{W}/(\partial H)^2$ in zero field. Moreover,

$$\frac{\partial \mathcal{W}(H)}{\partial H} = \left.\frac{\partial \mathcal{W}}{\partial H_i}\right|_{H_i = H}$$

and, thus,

$$\frac{\partial^2 \mathcal{W}(H)}{(\partial H)^2} = \sum_j \left.\frac{\partial^2 \mathcal{W}}{\partial H_i \partial H_j}\right|_{H_i = H} = \sum_j W_{ij}^{(2)}$$

where $W_{ij}^{(2)}$ is the connected two-point function. Translation invariance in a uniform field implies

$$W_{ij}^{(2)} = W^{(2)}(\mathbf{r}_i - \mathbf{r}_j),$$

where $\mathbf{r}_i$ and $\mathbf{r}_j$ are the vectors joining the points i and j to the origin. Thus,

$$\frac{\partial^2 \mathcal{W}(H)}{(\partial H)^2} = \sum_{\mathbf{r}} W^{(2)}(\mathbf{r}).$$

We now introduce the Fourier transforms of the connected and vertex functions

$$\widetilde{W}^{(2)}(\mathbf{k}) = \sum_{\mathbf{r}} W^{(2)}(\mathbf{r}) \exp(i\mathbf{k} \cdot \mathbf{r}),$$

$$\tilde{\Gamma}^{(2)}(\mathbf{k}) = \sum_{\mathbf{r}} \Gamma^{(2)}(\mathbf{r}) \exp(i\mathbf{k} \cdot \mathbf{r}).$$

Then,

$$\frac{\partial^2 W(H)}{(\partial H)^2} = \sum_{\mathbf{r}} W^{(2)}(\mathbf{r}) = \widetilde{W}^{(2)}(\mathbf{k}=0) = 1/\tilde{\Gamma}^{(2)}(\mathbf{k}=0), \tag{8.54}$$

where the latter equation follows from equation (6.26).

The sum $\sum_{\mathbf{r}} W^{(2)}(\mathbf{r})$ diverges only if the correlation length diverges. The condition of continuous transition (8.53) thus implies the divergence of the correlation length for vanishing magnetization.

Two-point function. More generally, from equation (8.50) one infers the relation between local magnetic field and magnetization

$$H_i = \frac{\partial \Gamma}{\partial M_i} = -2 \sum_j V_{ij} M_j + B'(M_i). \tag{8.55}$$

By differentiating again, one obtains the two-point vertex function at fixed magnetization

$$\Gamma^{(2)}_{ij} \equiv \left. \frac{\partial^2 \Gamma}{\partial M_i \partial M_j} \right|_{M_i = M} = -2V_{ij} + B''(M)\delta_{ij}$$

or, in a more explicit notation,

$$\Gamma^{(2)}(\mathbf{r}) = -2V(\mathbf{r}) + B''(M)\delta^{(d)}(\mathbf{r}). \tag{8.56}$$

Its Fourier transform is given by

$$\tilde{\Gamma}^{(2)}(\mathbf{k}) = B''(M) - 2\tilde{V}(\mathbf{k}). \tag{8.57}$$

The Fourier transform of the connected two-point function follows (equation (6.26)):

$$\widetilde{W}^{(2)}(\mathbf{k}) = 1/\tilde{\Gamma}^{(2)}(\mathbf{k}) = \left[B''(M) - 2\tilde{V}(\mathbf{k})\right]^{-1}. \tag{8.58}$$

In zero field, above T_c, the magnetization vanishes and one recovers the form (8.19) of the *Gaussian model*

$$\widetilde{W}^{(2)}(\mathbf{k}) = \tfrac{1}{2}[v_c - \tilde{V}(\mathbf{k})]^{-1}, \tag{8.59}$$

where $v_c = b_2/2$. If the transition is second order, the expression remains valid up to $v = v_c$ ($T = T_c$) where the correlation length diverges because the second derivative of $\Gamma(M)$ vanishes, and this is the source of spatial universality properties. In particular, one recovers the Gaussian or classical values of the exponents (definitions (8.22) and (8.26)), $\eta = 0$, $\nu = 1/2$.

More generally, for $|v - v_c|$, $|\mathbf{k}|$, $M \ll 1$ (which also implies a weak magnetic field), in the parametrization (7.11), one finds

$$\widetilde{W}^{(2)}(\mathbf{k}) \sim \left[2v_c + \tfrac{1}{2}b_4 M^2 - 2v(1 - a^2 k^2)\right]^{-1}. \tag{8.60}$$

The correlation function keeps an Ornstein–Zernike or free-field form. Equations (8.44) and (8.45) generalize immediately and the correlation length for $M \neq 0$ follows:

$$\xi^{-2} = \frac{1}{2v_c a^2}\left(2v_c - 2v + \tfrac{1}{2}b_4 M^2\right). \tag{8.61}$$

In zero magnetic field, using below T_c the expression (7.32) of the spontaneous magnetization, one finds

$$\begin{aligned}\xi_+^{-2} &= a^{-2}(1 - v/v_c) && \text{for } T > T_c, \\ \xi_-^{-2} &= 2a^{-2}(v/v_c - 1) && \text{for } T < T_c.\end{aligned} \tag{8.62}$$

Introducing also quite generally a correlation length exponent ν' for $T \to T_{c-}$, and defining the critical amplitudes $f_\pm$ for $|T - T_c| \to 0$ by

$$\xi_+ \sim f_+ (T - T_c)^{-\nu}, \quad \xi_- \sim f_- (T_c - T)^{-\nu'}, \tag{8.63}$$

one infers from the relations (8.62) the quasi-Gaussian value of the exponent

$$\nu' = \tfrac{1}{2},$$

and the amplitude ratio

$$f_+/f_- = \sqrt{2}. \tag{8.64}$$

Notice that sometimes the correlation length is defined in terms of the second moment ξ_1^2 of $W_{ij}^{(2)}$ which is proportional to ξ^2, and thus has the same universal properties

$$\tilde{\Gamma}^{(2)}(\mathbf{k}) = \left[\widetilde{W}^{(2)}(\mathbf{k})\right]^{-1} \sim \tilde{\Gamma}^{(2)}(0)\left(1 + k^2 \xi_1^2 + O(k^4)\right). \tag{8.65}$$

One can find other universal amplitude ratios. For example, if for $v = v_c$, $H \to 0$, we set

$$\chi \sim C^c/H^{2/3} \Rightarrow 3C^c = (6/b_4)^{1/3}$$

and, in zero field,

$$M \sim M_0(v - v_c)^{1/2} \Rightarrow M_0^2 = 12/b_4.$$

Then, the combination

$$R_\chi = C^+ M_0^2 (3C^c)^{-3} = 1,$$

is universal.

8.7 Quasi-Gaussian approximation and Landau's theory

The universal results that we have obtained within the framework of the quasi-Gaussian approximation also follow from Landau's theory, which we recall here. Landau's theory is based on general assumptions concerning the properties of systems with short-range interactions, of which we have used some to justify the quasi-Gaussian approximation.

We suppose that in zero field the physical system is invariant under space translations. Landau's theory then takes the form of several regularity conditions of the thermodynamic potential as a function of temperature and local magnetization (more generally of a local order parameter):

(i) The thermodynamic potential $\Gamma(M)$, a function of the local magnetization $M(\mathbf{r})$ (generated by an inhomogeneous magnetic field), which is also the generating function of vertex functions, is expandable in powers of M at $M = 0$.

(ii) We introduce the Fourier representation of the magnetization field:

$$M(\mathbf{r}) = \int d^d k \, e^{i\mathbf{k}\cdot\mathbf{r}} \, \tilde{M}(\mathbf{k}).$$

The thermodynamic potential $\Gamma(M)$ is then expandable in powers of $\tilde{M}(\mathbf{k})$ (see the lattice definitions (6.22), (6.23)):

$$\Gamma(M) = \sum_n \frac{1}{n!} \int d^d k_1 \ldots d^d k_n \, \tilde{M}(\mathbf{k}_1) \ldots \tilde{M}(\mathbf{k}_n)$$
$$\times (2\pi)^d \delta^{(d)}\left(\sum_i \mathbf{k}_i\right) \tilde{\Gamma}^{(n)}(\mathbf{k}_1, \ldots, \mathbf{k}_n),$$

where the Dirac $\delta^{(d)}$ functions are the direct consequence of translation invariance which implies that the sum of Fourier variables must vanish.

Then, the vertex functions $\tilde{\Gamma}^{(n)}$, that appear in this expansion, are regular at $\mathbf{k}_i = 0$.

(iii) The coefficients of the expansion are regular functions of the temperature for T near T_c, the temperature at which the coefficient of $\tilde{\Gamma}^{(2)}(\mathbf{k} = 0)$ vanishes.

Finally, the positivity of $\tilde{\Gamma}^{(4)}(0, 0, 0, 0)$ is a necessary condition for the transition to be of second order.

These conditions are motivated by some general assumptions: the effective spins are microscopic averages of weakly coupled variables whose fluctuations can be treated perturbatively. This is also expressed as a decoupling of the various scales of physics, and leads to the conclusion that critical phenomena can be described, at leading order, in terms of a finite number of effective macroscopic variables, as in the mean-field approximation.

These remarks render even more puzzling the empirical observation that the universal results of the quasi-Gaussian or mean-field approximations are in quantitative disagreement (and sometimes even qualitative) with experimental results and with results, exact or numerical, coming from lattice models. An examination of the leading corrections to the Gaussian theory will indicate the origin of this difficulty.

8.8 Continuous symmetries and Goldstone modes

If the initial spin variables S_i are N-component vectors and if the interaction and the local spin distribution have a continuous symmetry (corresponding to a compact Lie group), most of the results obtained so far within the framework of the quasi-Gaussian approximation remain unchanged. However, the appearance of several types of correlation functions and correlation lengths when the magnetization is different from zero, induces some new properties. In particular in zero field, at any temperature below T_c some correlation lengths, associated with modes called *Goldstone modes*, diverge as we have shown quite generally in Section 7.8.

Let us verify these properties in the quasi-Gaussian approximation in the case of models having an orthogonal $O(N)$ symmetry ($N > 1$), that is, invariant under the group of space rotations–reflections in N dimensions acting on the N components of the vector $\mathbf{S}$.

Quasi-Gaussian approximation. The generalization of the quasi-Gaussian approximation to this more general situation is simple. The thermodynamic potential has a structure analogous to expression (8.50), except that the local magnetization $\mathbf{M}_i$ now is an N-component vector and the thermodynamic potential is invariant under orthogonal transformations acting on the vectors $\mathbf{M}_i$.

The $O(N)$ invariance implies that the thermodynamic potential can be expressed in terms of scalar products and, thus, can be written as

$$\Gamma(\mathbf{M}) = -\sum_{i,j} V_{ij} \mathbf{M}_i \cdot \mathbf{M}_j + \sum_i \mathcal{B}(\mathbf{M}_i^2), \tag{8.66}$$

where the function $\mathcal{B}(X)$ is expandable in powers of X with properties analogous to the function (7.11):

$$\mathcal{B}(X) = \frac{b_2}{2} X + \frac{b_4}{4!} X^2 + \cdots \quad \text{with} \quad b_2 > 0,$$

where the assumption of a continuous transition again implies $b_4 > 0$.

Equation of state. In a uniform field, the thermodynamic potential density then reads

$$\mathcal{G}(\mathbf{M}) = -v M^2 + \mathcal{B}(M^2).$$

The equation of state follows:

$$H_\alpha = \frac{\partial \mathcal{G}(\mathbf{M})}{\partial M_\alpha} = M_\alpha \bigl(-2v + \mathcal{B}'(M^2)\bigr).$$

Taking the modulus of both members, one obtains a form analogous to expression (7.31):

$$H = M\bigl(-2v + \mathcal{B}'(M^2)\bigr), \tag{8.67}$$

where H, M now are the lengths of the vectors $\mathbf{H}$, $\mathbf{M}$.

The existence of a continuous phase transition at $v_c = b_2/2$ and the universal properties of the equation of state follow then from arguments identical to those already presented in Section 7.3 in the case of the reflection symmetry $\mathbb{Z}_2$.

Two-point function. Differentiating expression (8.66), one finds

$$\frac{\partial \Gamma(\mathbf{M})}{\partial M_{\alpha,i}} = -2 \sum_j V_{ij} M_{\alpha,j} + 2 M_{\alpha,i} \mathcal{B}'(\mathbf{M}_i^2), \tag{8.68}$$

where $M_{\alpha,i}$ ($\alpha = 1, ..., N$) are the components of the local magnetization vector $\mathbf{M}_i$.

The derivative of equation (8.68) yields, for $\mathbf{M}$ uniform, the two-point vertex function, the inverse of the connected function,

$$\Gamma^{(2)}_{\alpha\beta,ij} \equiv \frac{\partial^2 \Gamma(\mathbf{M})}{\partial M_{\alpha,i} \partial M_{\beta,j}} \bigg|_{\mathbf{M}_i = \mathbf{M}}$$
$$= \left(-2 V_{ij} + 2\delta_{ij}\mathcal{B}'(M^2)\right) \delta_{\alpha\beta} + 4\delta_{ij} M_\alpha M_\beta \mathcal{B}''(M^2). \tag{8.69}$$

The Fourier components of the two-point vertex function then are

$$\tilde{\Gamma}^{(2)}_{\alpha\beta}(\mathbf{k}) = \left(-2\tilde{V}(\mathbf{k}) + 2\mathcal{B}'(M^2)\right)\delta_{\alpha\beta} + 4 M_\alpha M_\beta \mathcal{B}''(M^2). \tag{8.70}$$

The function $\tilde{\Gamma}^{(2)}_{\alpha\beta}(\mathbf{k})$ remains a matrix in the N-vector space, and its inverse in the sense of matrices is the connected correlation function $\widetilde{W}^{(2)}_{\alpha\beta}(\mathbf{k})$.

We introduce a unit vector along the direction of the magnetization:

$$\mathbf{M} = M\mathbf{u} \quad \text{with} \quad \mathbf{u}^2 = 1.$$

The matrix $\tilde{\Gamma}^{(2)}_{\alpha\beta}$ has two eigenspaces corresponding to the vector $\mathbf{u}$ and the vectors orthogonal to $\mathbf{u}$. The function (8.70) can then be decomposed into transverse and longitudinal parts:

$$\tilde{\Gamma}^{(2)}_{\alpha\beta} = u_\alpha u_\beta \tilde{\Gamma}^{(2)}_L + (\delta_{\alpha\beta} - u_\alpha u_\beta) \tilde{\Gamma}^{(2)}_T, \tag{8.71}$$

where $\tilde{\Gamma}^{(2)}_L$, $\tilde{\Gamma}^{(2)}_T$ are the two eigenvalues, respectively. They are given by

$$\tilde{\Gamma}^{(2)}_L(\mathbf{k}) = 2\mathcal{B}'(M^2) + 4M^2 \mathcal{B}''(M^2) - 2\tilde{V}(\mathbf{k}), \tag{8.72a}$$
$$\tilde{\Gamma}^{(2)}_T(\mathbf{k}) = 2\mathcal{B}'(M^2) - 2\tilde{V}(\mathbf{k}). \tag{8.72b}$$

The expressions (8.72a) and (8.60) are similar. Using equation (8.67), one can rewrite the second eigenvalue as

$$\tilde{\Gamma}^{(2)}_T(\mathbf{k}) = H/M + 2\left[\tilde{V}(0) - \tilde{V}(\mathbf{k})\right]. \tag{8.73}$$

This result clearly is consistent with the general result (7.72). Since $\tilde{\Gamma}_L^{(2)}$, $\tilde{\Gamma}_T^{(2)}$ are the eigenvalues of the matrix $\tilde{\Gamma}^{(2)}$, the matrix of connected functions has the inverse eigenvalues:

$$\widetilde{W}_L^{(2)}(\mathbf{k}) = \left[\tilde{\Gamma}_L^{(2)}(\mathbf{k})\right]^{-1}, \quad \widetilde{W}_T^{(2)}(\mathbf{k}) = \left[\tilde{\Gamma}_T^{(2)}(\mathbf{k})\right]^{-1}.$$

Goldstone modes. At any temperature $T < T_c$, the ratio H/M vanishes for $H \to 0$. This equation then shows that in zero field in the ordered phase, the transverse two-point correlation function diverges like $1/k^2$ for $k \to 0$, a result consistent with the general analysis of Section 7.8 that implies the existence of $N - 1$ Goldstone modes.

8.9 Corrections to the quasi-Gaussian approximation

To describe the low-temperature phase, it is necessary to go beyond the Gaussian model. But the quasi-Gaussian approximation is justified only if the steepest descent method is justified. Formally, this condition seems to be satisfied if all the coefficients b_{2p} of the expansion (7.11), except the coefficient b_2 of the quadratic term, are in some sense small.

However, it is also necessary that the unavoidable corrections to the leading order result change only the coefficients of the expansion of the thermodynamic potential, without affecting its regularity properties.

This is what we want to verify by calculating the first corrections to the second derivative $\mathcal{G}''(M)$ of the thermodynamic potential density in zero magnetization, that is, in the disordered phase above T_c ($v < v_c$), and in zero field.

We first determine the value of v for which $\mathcal{G}''(0)$ vanishes, in order to find the first correction to v_c and thus to the critical temperature. The value v_c not being universal, this correction does not play any role. We then calculate the first correction to the behaviour of $\mathcal{G}''(0)$, which is also the inverse of the magnetic susceptibility χ in zero field, for $v \to v_c$, which is the interesting quantity.

8.9.1 Calculation of the correction

In the disordered phase $v < v_c$, in zero field, the magnetization $M = \langle \sigma \rangle$ vanishes and the leading saddle point is simply $\sigma = 0$. The first correction to the steepest descent method then is also the first correction to the Gaussian model.

The corrections to the Gaussian result are obtained by expanding expression (8.47), separating in the Hamiltonian $\mathcal{H}(\sigma)$ the quadratic part from the remainder called the perturbation:

$$\mathcal{H}(\sigma) = -\sum_{i,j} V_{ij}\sigma_i\sigma_j + \sum_i B(\sigma_i) = \mathcal{H}_0(\sigma) + \sum_i \left(B(\sigma_i) - \tfrac{1}{2}b_2\sigma_i^2\right)$$

with

$$\mathcal{H}_0(\sigma) = \frac{1}{2}\sum_{i,j} \mathfrak{S}_{ij}\sigma_i\sigma_j, \quad \mathfrak{S}_{ij} = b_2\delta_{ij} - 2V_{ij}.$$

At leading order, the contributions come only from the quartic term in $B(\sigma)$. It thus suffices to consider the partition function

$$\mathcal{Z}(H) = \int \left(\prod_i d\sigma_i\right) \exp\left[-\mathcal{H}_0(\sigma) - \frac{b_4}{4!}\sum_i \sigma_i^4 + \sum_i H_i\sigma_i\right]$$

$$= \int \prod_i d\sigma_i \exp\left[-\mathcal{H}_0(\sigma) + \sum_i H_i\sigma_i\right] \sum_\ell \frac{(-1)^\ell}{\ell!}\left(\sum_i \frac{1}{4!}b_4\sigma_i^4\right)^\ell .$$

The second derivative of the thermodynamic potential is also the inverse of the Fourier transform of the connected two-point function, at vanishing argument. The first correction to the Gaussian form of the two-point function is given by the contribution of order b_4. Moreover, since the magnetization vanishes, the connected two-point function is equal to the complete two-point function. One can thus use the general result (2.30) with $\lambda \equiv b_4$ and $\Delta = \mathfrak{S}^{-1}$:

$$W_{ij}^{(2)} = \langle \sigma_i\sigma_j\rangle = \Delta_{ij} - \tfrac{1}{2}b_4 \sum_k \Delta_{ik}\Delta_{kk}\Delta_{kj} + O(b_4^2).$$

The inverse of the connected two-point function (in the sense of matrices) is the vertex function $\Gamma_{ij}^{(2)}$ (equation (6.19)). Here, one finds

$$\Gamma_{ij}^{(2)} = \mathfrak{S}_{ij} + \tfrac{1}{2}b_4\Delta_{ii}\delta_{ij} + O(b_4^2).$$

Due to translation invariance, Δ_{ii} is independent of the point i and thus $\Delta_{ii} \equiv \Delta(\mathbf{r} = 0)$.

In the Fourier representation (equations (8.13), (8.19)),

$$\Delta_{ii} \equiv \Delta(\mathbf{r} = 0) = \int \frac{d^d p}{(2\pi)^d}\tilde{\Delta}(\mathbf{p}) = \frac{1}{(2\pi)^d}\int \frac{d^d p}{b_2 - 2\tilde{V}(\mathbf{p})}.$$

The Fourier transform of $\Gamma_{ij}^{(2)}$ is then given by

$$\tilde{\Gamma}^{(2)}(\mathbf{k}) = b_2 - 2\tilde{V}(\mathbf{k}) + \tfrac{1}{2}b_4\frac{1}{(2\pi)^d}\int \frac{d^d p}{b_2 - 2\tilde{V}(\mathbf{p})} + O(b_4^2). \quad (8.74)$$

The coefficient of M^2 in the expansion of the thermodynamic potential density $\mathcal{G}(M)$, which is also the inverse of the magnetic susceptibility in zero field (see also equation (6.19)), is given by

$$\chi^{-1}(M=0) = \left.\frac{\partial^2 \mathcal{G}}{(\partial M)^2}\right|_{M=0} = \sum_j \Gamma_{ij}^{(2)} = \tilde{\Gamma}^{(2)}(\mathbf{k}=0)$$

$$= -2v + b_2 + \frac{b_4}{2(2\pi)^d}\int \frac{d^d p}{b_2 - 2v\tilde{U}(\mathbf{p})} + O(b_4^2), \quad (8.75)$$

where we have introduced the Fourier transform $\tilde{U} = \tilde{V}/v$ of the function (8.5) (equation (8.6)).

Steepest descent method: First correction. Alternatively, one can use the general result (6.31), which yields the first correction to the steepest descent method and which, here, takes the form

$$\Gamma(M) = -\sum_{i,j} V_{ij} M_i M_j + \sum_i B(M_i) + \tfrac{1}{2} \operatorname{tr} \ln\left[-2V_{ij} + B''(M_i)\delta_{ij}\right]. \tag{8.76}$$

If M is constant, translation invariance simplifies the calculation. The matrix

$$L_{ij} = \ln\left[-2V_{ij} + B''(M)\delta_{ij}\right],$$

in more explicit notation, has the form

$$L_{ij} = L(\mathbf{r}_i - \mathbf{r}_j).$$

Thus, in a volume Ω,

$$\operatorname{tr} \mathbf{L} = \sum_i L_{ii} = \Omega L(0).$$

The thermodynamic potential density follows:

$$\mathcal{G}(M) = -vM^2 + B(M) + \tfrac{1}{2} L(0).$$

In the infinite volume limit, due to translation invariance, the matrices $-2V_{ij} + B''(M)\delta_{ij}$ and L_{ij} are diagonalized by a Fourier transformation (equation (8.57)). The eigenvalues of L_{ij} are the logarithms of the eigenvalues of $-2V_{ij} + B''(M)\delta_{ij}$. One infers

$$L(0) = \frac{1}{(2\pi)^d} \int d^d p\, \tilde{L}(\mathbf{p}) = \int \frac{d^d p}{(2\pi)^d} \ln\left[-2\tilde{V}(\mathbf{p}) + B''(M)\right].$$

One then obtains

$$\mathcal{G}(M) = -vM^2 + B(M) + \frac{1}{2} \int \frac{d^d p}{(2\pi)^d} \ln\left[-2\tilde{V}(\mathbf{p}) + B''(M)\right]$$

and, for the second derivative at $M = 0$:

$$\chi^{-1}(M=0) = \mathcal{G}''(0) = -2v + b_2 + \frac{b_4}{2(2\pi)^d} \int \frac{d^d p}{b_2 - 2v\tilde{U}(\mathbf{p})} + O(b_4^2), \tag{8.77}$$

an expression that coincides with the result (8.75).

8.9.2 The critical behaviour

The criticality condition is now

$$\mathcal{G}''(0) = -2v + b_2 + \frac{b_4}{2(2\pi)^d} \int \frac{d^d p}{b_2 - 2v\tilde{U}(\mathbf{p})} + O(b_4^2) = 0. \quad (8.78)$$

The first effect of the correction is to modify the critical value v_c and, thus, the critical temperature. In the term of order b_4, one can replace v_c by $b_2/2$, its leading order value, and the equation becomes

$$2v_c = b_2 + \frac{b_4}{2b_2} \int \frac{d^d p}{(2\pi)^d} \frac{1}{1 - \tilde{U}(\mathbf{p})} + O(b_4^2),$$

For $\mathbf{p} \to 0$ (equation (8.8)),

$$1 - \tilde{U}(\mathbf{p}) \sim a^2 p^2.$$

One verifies again the pathological character of the model in dimension $d = 2$ where the integral diverges at $\mathbf{p} = 0$: continuous phase transitions in dimension 2 cannot be described by the Gaussian model and, thus, a perturbed Gaussian model.

One now differentiates $\mathcal{G}''(0)$ with respect to v:

$$\frac{\partial \mathcal{G}''(0)}{\partial v} = -2 + b_4 \int \frac{d^d p}{(2\pi)^d} \frac{\tilde{U}(\mathbf{p})}{\left(b_2 - 2v\tilde{U}(\mathbf{p})\right)^2} + O(b_4^2). \quad (8.79)$$

At this order, the value of the derivative for $v = v_c$ is obtained by replacing v_c by $b_2/2$ in the correction of order b_4. One finds

$$\left.\frac{\partial \mathcal{G}''(0)}{\partial v}\right|_{v=v_c} = -2 + \frac{b_4}{b_2^2} \int \frac{d^d p}{(2\pi)^d} \frac{\tilde{U}(\mathbf{p})}{\left(1 - \tilde{U}(\mathbf{p})\right)^2} + O(b_4^2). \quad (8.80)$$

If the integral has a finite limit when $v \to v_c$, the derivative exists at $v = v_c$ and the correction to $\mathcal{G}''(0)$, beyond the Gaussian contribution, remains proportional to $v - v_c$:

$$\mathcal{G}''(0) \underset{v \to v_c}{\sim} (v - v_c) \left.\frac{\partial \mathcal{G}''(0)}{\partial v}\right|_{v=v_c}.$$

Then, $\mathcal{G}''(0)$ still vanishes linearly at the critical point like $v - v_c$ or $T - T_c$, as in the quasi-Gaussian theory, and only the non-universal coefficient is weakly modified.

The special role of dimension 4. For $p \to 0$, the numerator of the integral in expression (8.80) tends toward 1 and the denominator behaves like p^4. The integral thus converges only for $d > 4$. This analysis thus exhibits the special role of the dimension 4:

For $d > 4$, the perturbation to the Gaussian theory is small, and modifies only non-universal quantities. The magnetic susceptibility still diverges like $1/(T - T_c)$ and the critical exponent γ (definition (7.35)) keeps its Gaussian value: $\gamma = 1$.

On the contrary, for $2 < d \le 4$, the integral diverges when $v \to v_c$. Therefore, however small the amplitude b_4 of the first correction to the Gaussian distribution is, for $d \le 4$ when the correlation length ξ diverges the contribution of order b_4 eventually becomes larger than the Gaussian term. Thus, the perturbative expansion cannot be valid close to T_c, and the universal predictions of the Gaussian model and the perturbed Gaussian model are not confirmed.

It is instructive to evaluate more precisely the behaviour of the integral (8.79) when $v \to v_c$. One notes that the leading contribution comes from the vicinity of $\mathbf{p} = 0$. One can thus approximate $1 - \tilde{U}(\mathbf{p})$ by $a^2 p^2$:

$$\frac{\partial \mathcal{G}''(0)}{\partial v} \sim -2 + \frac{b_4}{4(2\pi)^d} \int \frac{d^d p}{(v_c - v + v_c a^2 \mathbf{p}^2)^2} + O(b_4^2).$$

For $d < 4$, the integral in this approximation converges at infinity. An unrestricted integration modifies the result only by a negligible constant for $v \to v_c$. After the change of variables $\mathbf{p} = \mathbf{p}' \sqrt{v_c/v - 1}/a$, the integral becomes

$$\int \frac{d^d p}{(v_c - v + v_c a^2 \mathbf{p}^2)^2} = \frac{1}{v_c^2 a^d}(1 - v/v_c)^{d/2-2} \int \frac{d^d p}{(1+p^2)^2}.$$

Calculating the integral and introducing the Gaussian correlation length (8.45), $\xi = a/\sqrt{v_c/v - 1}$, one infers

$$\mathcal{G}''(0) = \chi^{-1} \underset{\xi \gg 1}{=} 2(v_c - v)\left[1 + \frac{b_4}{8 v_c^2 a^d} \frac{\Gamma(1-d/2)}{(4\pi)^{d/2}} \left(\frac{\xi}{a}\right)^{4-d}\right] + O(b_4^2). \qquad (8.81)$$

For $d = 4$, the correction has a logarithmic divergence:

$$\mathcal{G}''(0) = \chi^{-1} \underset{\xi \gg 1}{=} 2(v_c - v)\left[1 - \frac{b_4}{64\pi^2 v_c^2 a^4} \ln(\xi/a)\right] + O(b_4^2).$$

To summarize:

(i) For dimensions $d > 4$, the correction does not modify the universal predictions of the quasi-Gaussian approximation. One finds some singular corrections but they yield sub-leading contributions.

(ii) For dimensions $d \le 4$, singularities, also called 'infra-red' (IR), a denomination borrowed from quantum field theory, consequences of the large-distance behaviour of the Gaussian two-point function (also called propagator), or at vanishing argument of its Fourier transform, imply that the Gaussian predictions cannot be correct in general.

An inspection of the higher order corrections confirms these results. For $d \le 4$, the corrections are increasingly singular when the order increases, whereas for $d > 4$ they are less and less singular, which confirms the validity of the first-order analysis.

The perturbative terms responsible for this difficulty involve the ratio ξ/a between the correlation length and the microscopic scale. This gives some indication about the mechanism responsible for the failure of the quasi-Gaussian approximation: physics at the microscopic scale does not decouple from physics at large distance.

Indeed, for $d > 4$, the contribution from arguments $|\mathbf{p}| \leq \xi^{-1}$ is negligible when ξ diverges, which means that in direct space the degrees of freedom corresponding to distances of the order of the correlation length or larger play a negligible role. On the contrary, for $d \leq 4$, at T_c, all scales contribute. This is this property that invalidates ideas based on the central limit theorem, namely that a small number of degrees of freedom with a quasi-Gaussian distribution can replace the infinite number of initial microscopic degrees of freedom.

To solve this problem of coupling between all scales, a new tool has been introduced, the *renormalization group*.

Finally, note that the first singular contribution depends only on the coefficient of σ^4 in the expansion (8.46) and on the asymptotic form of Ornstein–Zernike type of the propagator (the Gaussian two-point function). The effect of the lattice has been limited to restrict the domain of integration in $\mathbf{p}$ to the Brillouin zone.

A systematic study then shows that the most singular terms in each order of the perturbative expansion can be reproduced, in the critical limit, by a statistical field theory with an interaction of σ^4 type, in continuum Euclidean space. As a consequence, if the sum of the most divergent terms suffices to determine the critical properties of the models under study, then the existence of a continuum limit and universal properties follow, since that it can be shown that the corresponding field theory depends only on a small number of parameters.

8.10 Mean-field approximation and corrections

We now present a more systematic formalism, based on the steepest descent method, which allows recovering the preceding results, starting from a rather general lattice microscopic model of type (8.1) with the interaction (8.2).

At leading order, it leads to the mean-field approximation which reproduces the results of the quasi-Gaussian approximation. The following terms of the expansion then allow studying the corrections to the mean-field approximation. Again, the leading corrections at the critical point and in its vicinity, the critical domain, have the form exhibited by the corrections to the quasi-Gaussian approximation.

We consider the partition function in a site-dependent field, the generating function of correlation functions,

$$\mathcal{Z}(\mathbf{H}) = \int \left(\prod_i \rho(S_i) \mathrm{d}S_i \right) \exp\left(-\beta \mathcal{E}(\mathbf{S}) + \sum_i H_i S_i \right), \qquad (8.82)$$

where we assume a pair interaction:

$$-\beta \mathcal{E}(\mathbf{S}) = \sum_{i,j} V_{ij} S_i S_j.$$

Note that here, as in the case of the high-temperature expansion, it is technically convenient to assume that V_{ii} vanishes. This implies no loss of generality since the corresponding one-site contribution can always be included in the local measure $\rho(S)$. Otherwise, it is necessary to slightly generalize the method that we explain below.

8.10.1 Mean-spin representation and steepest descent method

Since the partition function can easily be calculated when all sites are decoupled, a simple idea is to express the factor that in the sum over configurations couples the spins of different sites as an integral over a Boltzmann weight of decoupled spins. More explicitly, we introduce Dirac's δ-function, in its Fourier representation,

$$\delta(\sigma - S) = \frac{1}{2\pi} \int d\lambda \, e^{i\lambda(\sigma - S)},$$

in the representation (8.82) of the partition function. We use this identity for all points i, replacing S_i by σ_i in the interaction. The partition function then becomes

$$\mathcal{Z}(H) = \int \left(\prod_i \rho(S_i) dS_i \, d\sigma_i \, \frac{d\lambda_i}{2\pi} \right) \exp \left\{ -\beta \mathcal{E}(\sigma) + \sum_i [H_i \sigma_i + i\lambda_i(\sigma_i - S_i)] \right\}. \tag{8.83}$$

The integration over the variables S_i is then immediate. It leads to the replacement of the local spin distribution by its Fourier transform. Introducing the free energy $A(h)$ of the one-site model, defined in (7.5) (which is an analytic function), one finds

$$\mathcal{Z}(H) = \int \left(\prod_i d\sigma_i d\lambda_i \right) \exp \left[-\beta \mathcal{E}(\sigma) + \sum_i [(H_i + i\lambda_i)\sigma_i + A(-i\lambda_i)] \right], \tag{8.84}$$

Steepest descent method. We then evaluate the integral (8.84) by the steepest descent method, even though there is no associated parameter.

The derivatives with respect to λ_i and σ_i yield the two saddle point equations. Setting $h_i = -i\lambda_i$ because the saddle point value is imaginary, one finds

$$\sigma_i = A'(h_i), \tag{8.85a}$$

$$h_i = 2 \sum_j V_{ij} \sigma_j + H_i. \tag{8.85b}$$

The free energy, in the mean-field approximation, is then

$$\mathcal{W}(H) \equiv \ln \mathcal{Z}(H) = \sum_{i,j} V_{ij} \sigma_i \sigma_j + \sum_i ((H_i - h_i)\sigma_i + A(h_i)), \tag{8.86}$$

$$= -\sum_{i,j} V_{ij} \sigma_i \sigma_j + \sum_i A(h_i), \tag{8.87}$$

where σ_i, h_i are the solutions of the saddle point equations (8.85).

Since the expression (8.86) is stationary with respect to variations of σ_i and h_i, the local magnetization M_i in the mean-field approximation is given by the explicit derivative with respect to H:

$$M_i = \frac{\partial \mathcal{W}}{\partial H_i} = \sigma_i. \qquad (8.88)$$

Discussion. Comparing equations (7.26) and (8.85a), one notes that h_i has the meaning of an effective magnetic field. Equation (8.85b) determines the effective field as the sum of the applied field and a mean field generated by other spins. The meaning of this approximation thus is that the interaction between spins has been replaced by a mean magnetic field. A detailed analysis shows that the approximation becomes exact when the dimension d of space becomes large because a given site has an increasing number of neighbours whose action can indeed be replaced by a mean magnetic field. One then recovers a limiting model of type 7.2.

The advantage of this algebraic formulation of the mean-field approximation is that it allows a systematic discussion of corrections, unlike other classical methods like, for example, those based on variational principles.

Thermodynamic potential and phase transition. From equations (8.86) and (8.88) one infers the thermodynamic potential, the Legendre transform of $\mathcal{W}(H)$,

$$\Gamma(M) = \sum_i M_i H_i - \mathcal{W}(H) = -\sum_{i,j} M_i V_{ij} M_j + \sum_i B(M_i), \qquad (8.89)$$

where $B(M)$ is the Legendre transform (7.14) of $A(H)$. The expression is identical to expression (8.50). The mean-field approximation, thus, has the same properties as the quasi-Gaussian approximation, in particular, the same universal properties.

8.10.2 Steepest descent method: An expansion parameter

Before discussing the corrections to the mean-field approximation, it is instructive to introduce a parameter that characterizes the expansion around the mean field. We thus replace the coefficients V_{ij} in expression (8.2) by ℓV_{ij} and the spin S by a mean spin σ of ℓ independent spins with identical distribution $\rho(S)$:

$$\sigma = \frac{1}{\ell} \sum_{k=1}^{\ell} S^{(k)}.$$

The case $\ell = 1$ corresponds to the initial distribution. This modification realizes the idea that σ is an effective macroscopic spin, the local average of many weakly coupled microscopic spins.

By a method already used in the case of the central limit theorem, we express the distribution $R_\ell(\sigma)$ of the random variable σ in terms of the Fourier transform of the distribution ρ and thus the function (7.5):

$$R_\ell(\sigma) = \frac{\ell}{2\pi} \int d\lambda \, e^{\ell[i\lambda\sigma + A(-i\lambda)]}.$$

The partition function then becomes (up to a trivial change of normalization)

$$\mathcal{Z}(H) = \int \prod_i d\sigma_i d\lambda_i \, \exp\left\{-\ell\beta\mathcal{E}(\sigma) + \ell \sum_i [(H_i + i\lambda_i)\sigma_i + A(-i\lambda_i)]\right\}, \quad (8.90)$$

where we have assumed that the magnetic field is coupled to the sum of the ℓ spins.

This expression shows that the steepest descent method is formally justified *a priori* in the limit $\ell \to \infty$: the calculation of the partition function by the steepest descent method generates a formal expansion in powers of $1/\ell$.

It is convenient to define the free energy by

$$\mathcal{W}(H) \equiv \frac{1}{\ell} \ln \mathcal{Z}(H), \quad (8.91)$$

in such a way that the leading order is independent of ℓ.

Perturbative expansion. A systematic calculation of corrections to the mean-field result is now obtained by expanding around the saddle point. It is convenient to set

$$\int d\lambda \, e^{\ell[A(-i\lambda)+i\lambda\sigma]} = \mathcal{N} e^{-\ell\Sigma(\sigma,\ell)} \quad (8.92)$$

with

$$\mathcal{N} = \int d\lambda \, e^{\ell A(-i\lambda)}.$$

As we have already noted several times, for the integral (8.92) the saddle point is given by $\sigma = A'(-i\lambda)$, and the leading order of the steepest descent method thus involves the Legendre transform of A. One infers

$$\Sigma(\sigma, \ell) = B(\sigma) + O(1/\ell).$$

At order $1/\ell$, one thus finds two types of corrections:

First, the coefficients b_2 and b_4 that appear in $\Gamma(M)$ are modified. In particular, a correction to the critical temperature is generated. However, since the universal properties that we have exhibited are independent of these explicit values, they are not affected.

To calculate the other corrections, one can replace $\Sigma(\sigma, \ell)$ by $B(\sigma)$. In this approximation, and with a trivial change of normalization, the partition function reduces to

$$\mathcal{Z}(H) = \int \prod_i d\sigma_i \exp \ell \left[-\mathcal{H}(\sigma) + \sum_i H_i \sigma_i\right], \quad (8.93)$$

where the Hamiltonian takes the form

$$\mathcal{H}(\sigma) = -\sum_{i,j} V_{ij}\sigma_i\sigma_j + \sum_i B(\sigma_i). \quad (8.94)$$

In $\mathcal{H}(\sigma)$, we recognize the thermodynamic potential of the mean-field approximation, which is also a thermodynamic potential having the properties assumed in Landau's theory.

For $\ell = 1$, one recovers the model (8.48), which has been studied in the quasi-Gaussian approximation. For $\ell \neq 1$, after the change of variables $\sigma_i \mapsto \sigma_i/\sqrt{\ell}$, one also recognizes the model (8.48) where the function $B(\sigma)$ has modified coefficients, $b_{2p} \mapsto \ell^{1-p} b_{2p}$. All coefficients, except the Gaussian term b_2, tend toward zero for $\ell \to \infty$, and this formally justifies a perturbative treatment.

Above T_c, in zero field, the magnetization vanishes and the saddle point is $\sigma = 0$. The expansion of the steepest descent method corresponds to ordinary perturbation theory and the leading correction comes entirely from the quartic term of order σ^4 in $B(\sigma)$, expanded at first order. For example, expression (8.74) of the Fourier transform $\tilde{\Gamma}^{(2)}(\mathbf{k})$ of the vertex two-point function becomes

$$\tilde{\Gamma}^{(2)}(\mathbf{k}) = b_2 - 2\tilde{V}(\mathbf{k}) + \frac{b_4}{2\ell}\int \frac{d^d p}{(2\pi)^d}\tilde{\Delta}(\mathbf{p}) + O\left(\frac{1}{\ell^2}\right). \tag{8.95}$$

For v close to v_c and $2 < d < 4$, the leading contribution (8.81) to the magnetic susceptibility is replaced by

$$\chi^{-1} = 2(v_c - v)\left[1 + \frac{b_4}{8v_c^2 a^d \ell}\frac{\Gamma(1-d/2)}{(4\pi)^{d/2}}\left(\frac{\xi}{a}\right)^{4-d}\right] + O\left(\frac{1}{\ell^2}\right).$$

One observes that however large ℓ may be, for $d \leq 4$, the correction to the mean-field contribution diverges for $\xi \to \infty$.

8.11 Tricritical points

Up to now we have assumed that only one control parameter, in general the temperature, could be adjusted and, therefore, that the coefficient b_4 of the term M^4 in the expansion of $\Gamma(M)$ was generic, that is, a number of order unity. However, there are situations where other physical parameters can be varied and, for example, both the coefficients b_2 and b_4 of M^2 and M^4 can be cancelled. This happens, for example, in He^3–He^4 mixtures or in some metamagnetic systems. In the Ising type models that we have studied so far, this can be achieved by adjusting the spin distribution. If the coefficient b_6 of M^6 is positive, one finds a point called *tricritical* and a new analysis is required. More generally, one can study the vicinity of the tricritical point, where both the coefficients of M^2 and M^4 are small:

$$\Gamma(M) = -\sum_{i,j} V_{ij} M_i M_j + \sum_i \left(\frac{b_2}{2!}M_i^2 + \frac{b_4}{4!}M_i^4 + \frac{b_6}{6!}M_i^6 + \cdots\right). \tag{8.96}$$

In particular, for $v = v_c = b_2/2$, if starting from a positive value, b_4 decreases, one finds a line of ordinary critical points until b_4 vanishes at the tricritical point.

After the tricritical point, b_4 becomes negative and the phase transition becomes first order.

In the mean-field approximation, the exponents of the tricritical point have values different from those found for an ordinary critical point, for example $\beta = 1/4$, $\delta = 5$.

The corrections to the quasi-Gaussian or mean-field tricritical approximations can be studied by methods similar to that of Section 8.9. Analysing the perturbative expansion, one finds that above three dimensions the quasi-Gaussian approximation correctly predicts universal properties whereas it is definitively not valid in three dimensions and below: the upper-critical dimension is 3. Moreover, for $d \leq 3$, the most singular corrections are reproduced by a statistical field theory in continuum space with the ϕ^6 interaction.

Exercises

Exercise 8.1

Preliminary calculations. One considers the normalized Gaussian measure $(2\pi)^{-N/2} d\mathbf{S}\, e^{-\mathbf{S}^2/2}$, where $\mathbf{S} \in \mathbb{R}^N$, invariant under the orthogonal group $O(N)$ (rotations–reflections in N dimensions). Calculate, using Wick's theorem, the expectation values $\langle S_\alpha S_\beta \rangle_G$, $\langle S_\alpha S_\beta S_\gamma S_\delta \rangle_G$, where the S_α are the components of the vector $\mathbf{S}$.

Introducing radial and angular coordinates, infer the same expectation values, denoted below by $\langle \bullet \rangle_{\text{sph.}}$, with the uniform measure on the sphere S_{N-1}. The measure can be written as $\mathcal{N} d\mathbf{S}\, \delta(\mathbf{S}^2 - 1)$, where the normalization $\mathcal{N}$ is defined by the condition $\langle 1 \rangle = 1$. Verify the results by relating $\langle S_\alpha S_\beta S_\gamma S_\delta \rangle_{\text{sph.}}$ to $\langle S_\alpha S_\beta \rangle_{\text{sph.}}$ and $\langle S_\alpha S_\beta \rangle_{\text{sph.}}$ to $\langle 1 \rangle_{\text{sph.}} = 1$.

Remark. Using the identity

$$1 = \int_0^\infty d\rho\, \delta(\rho - \mathbf{S}^2) = \int_0^\infty 2R\, dR\, \delta(R^2 - \mathbf{S}^2),$$

where δ is the Dirac distribution, one derives

$$\int d\mathbf{S}\, f(\mathbf{S}) = \int_0^\infty 2R\, dR\, d\mathbf{S}\, \delta(R^2 - \mathbf{S}^2) f(\mathbf{S}).$$

One then changes variables, setting $\mathbf{S} = R\mathbf{S}'$. One notes

$$\delta(R^2 - \mathbf{S}^2) = R^{-2}\delta(1 - \mathbf{S}'^2).$$

The equation becomes (omitting primes)

$$\int d\mathbf{S}\, f(\mathbf{S}) = \int_0^\infty R^{N-1} dR \int d\mathbf{S}\, 2\delta(1 - \mathbf{S}^2) f(R\mathbf{S}),$$

which corresponds to a factorization of the initial measure $d\mathbf{S}$, in $\mathbb{R}^N$, into two radial and angular (or on the unit sphere) measures.

Solution. Directly,
$$\langle S_\alpha S_\beta \rangle_G = \delta_{\alpha\beta}$$
and, thus, from Wick's theorem,
$$\langle S_\alpha S_\beta S_\gamma S_\delta \rangle_G = \delta_{\alpha\beta}\delta_{\gamma\delta} + \delta_{\alpha\gamma}\delta_{\beta\delta} + \delta_{\alpha\delta}\delta_{\beta\gamma}.$$

One then converts to radial–angular coordinates, setting $\mathbf{S} = S\hat{\mathbf{S}}$ with $\hat{\mathbf{S}}^2 = 1$. One obtains
$$\langle S_\alpha S_\beta \rangle_G = \langle S_\alpha S_\beta \rangle_{\text{sph.}} I_1/I_0,$$
$$\langle S_\alpha S_\beta S_\gamma S_\delta \rangle_G = \langle S_\alpha S_\beta S_\gamma S_\delta \rangle_{\text{sph.}} I_2/I_0$$

with
$$I_p = \int_0^\infty dS\, S^{N-1+2p}\, e^{-S^2/2} = 2^{N/2+p-1}\Gamma(N/2+p),$$

where I_1/I_0 and I_2/I_0 are the radial expectation values.
In particular,
$$I_1/I_0 = N,\quad I_2/I_0 = N(N+2).$$

Thus, with the measure $\mathcal{N}\,d\mathbf{S}\,\delta(\mathbf{S}^2 - 1)$,

$$\langle S_\alpha S_\beta \rangle_{\text{sph.}} = \frac{1}{N}\delta_{\alpha\beta},\quad \langle S_\alpha S_\beta S_\gamma S_\delta \rangle_{\text{sph.}} = \frac{1}{N(N+2)}(\delta_{\alpha\beta}\delta_{\gamma\delta} + \delta_{\alpha\gamma}\delta_{\beta\delta} + \delta_{\alpha\delta}\delta_{\beta\gamma}).$$
(8.97)

Summing the two expectation values for $\alpha = \beta$ over α and using $\mathbf{S}^2 = 1$, one verifies the normalizations.

Exercise 8.2

The $O(2)$ model. Determine the expansion of the thermodynamic potential of the one-site model with an $O(2)$ symmetry, that is, for a two-component spin $\mathbf{S}$ belonging to the circle $\mathbf{S}^2 = 1$ with a uniform measure on the circle, up to order 4.

Repeat the calculations for a model with an $O(3)$ symmetry and a spin belonging to the sphere S_2.

Solution. In the case of the $O(N)$ group, the one-site partition function can be written as
$$z(\mathbf{h}) = \mathcal{N} \int d\mathbf{S}\,\delta(\mathbf{S}^2 - 1)\, e^{\mathbf{S}\cdot\mathbf{h}},$$

where the normalization $\mathcal{N}$ is defined by the condition $z(0) = 1$.

Due to rotation invariance, one can always choose $\mathbf{h} = (\mathbf{0}, h)$ and, for $\alpha < N$, use the parametrization $S_\alpha = n_\alpha \sin\theta$, where $\mathbf{0}$ and $\mathbf{n}$ are $(N-1)$-component vectors with $\mathbf{n}^2 = 1$, and $S_N = \cos\theta$, $0 \le \theta < \pi$. Then, $\mathbf{S}\cdot\mathbf{h} = h\cos\theta$. For $N = 2$,

$$z(h) = \frac{1}{\pi}\int_0^\pi d\theta\, e^{h\cos\theta} = 1 + \tfrac{1}{4}h^2 + \tfrac{1}{64}h^4 + O(h^6).$$

More generally,
$$d\mathbf{S}\,\delta(\mathbf{S}^2 - 1) = d\mathbf{n}\,\delta(\mathbf{n}^2 - 1)d\theta\,(\sin\theta)^{N-2},$$

where $(\sin\theta)^N$ comes from the change of variables and $1/\sin^2\theta$ from the δ function.

After integration over $\mathbf{n}$, one obtains the measure $d\theta(\sin\theta)^{N-2}$ for the $O(N)$ group. In particular, for $O(3)$,

$$z(h) = \frac{1}{2}\int_0^\pi d\theta\,\sin\theta\,e^{h\cos\theta} = \frac{1}{2}\int_{-1}^{+1} dt\,e^{th} = \frac{\sinh h}{h} = 1 + \tfrac{1}{6}h^2 + \tfrac{1}{120}h^4 + O(h^6).$$

An alternative method is based on a more geometric approach. Expanding,

$$z(\mathbf{h}) = 1 + \tfrac{1}{2}\sum_{\alpha,\beta} h_\alpha h_\beta\,\langle S_\alpha S_\beta\rangle_{\text{sph.}} + \tfrac{1}{24}\sum_{\alpha,\beta,\gamma,\delta} h_\alpha h_\beta h_\gamma h_\delta\,\langle S_\alpha S_\beta S_\gamma S_\delta\rangle_{\text{sph.}} + O(h^6),$$

where $\langle\bullet\rangle_{\text{sph.}}$ means averaging over the sphere S_{N-1} with a uniform, $O(N)$ invariant, measure. One can then use the results (8.97) that can also be inferred from rotation invariance. One concludes that the first terms of the expansions of $W(h) = \ln z(h)$ and of $B(m)$, its Legendre transform, are

$$W(h) = \frac{1}{2N}h^2 - \frac{1}{4N^2(N+2)}h^4 + O(h^6),$$

$$B(m) = \frac{N}{2}m^2 + \frac{N^2}{4(N+2)}m^4 + O(m^6).$$

Exercise 8.3

With the same distribution at each site as in the preceding exercise and in the case of a two-spin interaction,

$$-\beta\mathcal{E}(\mathbf{S}) = \sum_{i,j} V_{ij}\mathbf{S}_i\cdot\mathbf{S}_j,$$

calculate the thermodynamic potential in the mean-field approximation.

Solution. The result simply generalizes expression (8.89) and one finds

$$\Gamma(\mathbf{M}) = -\sum_{i,j} V_{ij}\mathbf{M}_i\cdot\mathbf{M}_j + \sum_i B(|\mathbf{M}_i|).$$

Exercise 8.4

Calculate, in the same model, the first correction to the mean-field approximation in the case of the two-point vertex function at vanishing argument, in the limit $T \to T_{c+}$.

Solution. The two-point function is diagonal:

$$\tilde{\Gamma}^{(2)}_{\alpha\beta}(0) = \delta_{\alpha\beta}\tilde{\Gamma}^{(2)}(0).$$

In the notation of Section 8.10, one then finds a result of the form

$$\tilde{\Gamma}^{(2)}(0) = \chi^{-1} = 2(v_c - v)\left[1 + \frac{b_4(N)}{8v_c^2 a^d \ell} \frac{\Gamma(1-d/2)}{(4\pi)^{d/2}} \left(\frac{\xi}{a}\right)^{4-d}\right],$$

where here $b_4(N) = N^2$.

Exercise 8.5

Again, one considers a two-component spin on a lattice. One assumes that a statistical model has the symmetry of the square, that is, is invariant under the three transformations

$$S_1 \leftrightarrow -S_1, \quad S_2 \leftrightarrow -S_2, \quad S_1 \leftrightarrow S_2.$$

$$S_1 \leftrightarrow S_2, \quad S_1 \mapsto -S_1.$$

Show that the densities $W(H_1, H_2)$ of the free energy and $\mathcal{G}(M_1, M_2)$ of the thermodynamic potential then have the same symmetry.

Write, in the framework of the quasi-Gaussian or mean-field approximation, the most general thermodynamic potential density $\mathcal{G}(M_1, M_2)$, in a uniform field, having the symmetry of the square, expanded up to fourth order in the magnetization $\{M_1, M_2\}$ for $|M| \to 0$.

One then considers a special example of such a thermodynamic potential

$$\mathcal{G}(M_1, M_2) = \tfrac{1}{2}\tau(M_1^2 + M_2^2) + \tfrac{1}{4}g\left(M_1^4 + M_2^4 + 2\cos\theta M_1^2 M_2^2\right),$$

with $\tau \propto T - T_c$, $g > 0$, $-\pi < \theta < \pi$. Calculate the spontaneous magnetization in zero magnetic field for $\tau < 0$. Infer the matrix of second derivatives

$$\mathcal{G}_{\alpha\beta} = \frac{\partial^2 \mathcal{G}}{\partial M_\alpha \partial M_\beta}$$

in this situation. One recalls that the matrix of magnetic susceptibilities

$$\chi_{\alpha\beta} = \frac{\partial^2 W}{\partial H_\alpha \partial H_\beta},$$

is the inverse of the matrix $\mathcal{G}_{\alpha\beta}$. Infer the matrix $\chi_{\alpha\beta}$ in zero magnetic field for $\tau < 0$. Show that a special value of θ can be found for which one of the eigenvalues of $\chi_{\alpha\beta}$ diverges for all $\tau < 0$. What is the interpretation of this special value?

Solution. One finds the spontaneous magnetization

$$M_1^2 = M_2^2 = -\frac{\tau}{g(1+\cos\theta)}.$$

The matrix of second derivatives of elements $\mathcal{G}_{\alpha\beta}$ is

$$\mathcal{G} = -\frac{2\tau}{1+\cos\theta}\begin{pmatrix} 1 & \pm\cos\theta \\ \pm\cos\theta & 1 \end{pmatrix}.$$

The eigenvalues of the matrix are

$$-2\tau, \quad -2\tau\frac{1-\cos\theta}{1+\cos\theta}.$$

For the value $\theta = 0$, the second eigenvalue vanishes and thus an eigenvalue of $\chi_{\alpha\beta}$ diverges. One then notes that the corresponding model then has a larger continuous symmetry corresponding to the group $O(2)$. The divergence of the transverse susceptibility is a manifestation of the Goldstone phenomenon.

9 Renormalization group: General formulation

In Chapter 8, we studied Ising type models (and, more generally, ferromagnetic models with $O(N)$ symmetry) with short-range interactions and determined the behaviour of the thermodynamic functions near a continuous phase transition, within the framework of the quasi-Gaussian or mean-field approximations. We have shown that these approximations predict a set of *universal properties*, that is, properties which are independent of the detailed structure of interactions or microscopic Hamiltonians.

However, many experimental observations as well as numerical and analytical results coming from model systems show that such results cannot be quantitatively correct, at least in dimensions 2 or 3. For example, the exact solution of the Ising model in two dimensions yields exponents like $\beta = 1/8$, $\eta = 1/4$ or $\nu = 1$, clearly different from the predictions of the quasi-Gaussian approximation.

By examining the leading corrections to the Gaussian approximation, we have identified the origin of the difficulty. We have found that above four dimensions these corrections do not affect universal quantities; on the contrary, below four dimensions, the corrections diverge at the critical temperature and, thus, the assumptions which are at the basis of the quasi-Gaussian approximation (or mean-field theory) can certainly not be correct.

Moreover, this analysis indicates that the coupling of degrees of freedom corresponding to very different scales of physics plays an essential role: it is impossible to consider only effective macroscopic degrees of freedom. One could then fear that in dimension $d \leq 4$ physics, even at large distance, is sensitive to the detailed microscopic structure of systems. Rather surprisingly, however, some universal properties survive, though different from those of the quasi-Gaussian approximation. But these properties are less universal in the following sense: the statistical systems which have the same properties in the quasi-Gaussian approximation, divide into *universality classes* characterized by the dimension of space, symmetries and some other qualitative features.

To explain this somewhat paradoxical situation, a completely new tool, initially suggested by Kadanoff (1966), has been developed by Wilson (1971), Wegner and then many other physicists: the *renormalization group* (RG). In Wilson's approach, the fundamental idea is to integrate successively over the degrees of freedom corresponding to the shortest scales. One thus obtains a succession of models which all describe the same large-distance physics but in which the short-distance structure is gradually eliminated. If this sequence has a limit, which implies that the RG transformations admit *fixed points*, then universality properties are explained: all statistical models which, after these transformations, converge toward the same fixed point, belong to the same universality class.

In this chapter, we first describe the fundamental ideas of the RG within a rather general framework, even if the formulation is not very precise and the arguments largely heuristic. The role of RG fixed points will be emphasized.

In practice, this RG is very difficult to construct since it acts on the infinite dimensional space of all possible statistical models. Only Gaussian models can be discussed systematically. We will thus identify the simplest fixed point, the *Gaussian fixed point*, which belongs to the class of Gaussian models discussed previously. Moreover, a complete local analysis in the vicinity of this fixed point is possible. It allows classifying perturbations to the Gaussian model as *relevant*, that is, which become increasingly important at large distance, *irrelevant* in the opposite case, and *marginal* in the limiting situation.

The general ideas of RG, though *a priori* somewhat vague, are extremely suggestive and, indeed, have been implemented in many approximate forms and have induced a large variety of practical calculations. Our goal here is not to review them. However, in the more specific context of quantum field theory, the assumptions at the basis of the RG have been clarified and verified in many examples of physical interest. The analysis has confirmed quite directly the major relation, initially recognized by Wilson, between the quantum field theory describing the physics of fundamental interactions at the microscopic scale and the theory of critical phenomena. (Actually, it is the quantum field theory in imaginary time which is also a classical statistical field theory.)

9.1 Statistical field theory. Landau's Hamiltonian

In what follows, we discuss statistical models directly formulated in continuum space, rather than on a lattice. Indeed, we have seen that even for models initially defined on the lattice, the universal properties of the quasi-Gaussian approximation are related to the existence of a continuum limit. The same applies to the most singular corrections which depend only on the properties of the Gaussian two-point function at large distance. This indicates that lattice models can be replaced by a *classical statistical field theory*. In fact, the role of the lattice has been limited to bound the arguments (momenta) of integrals in the Fourier representation. Within the context of the statistical field theory, this role is played by a cut-off, that is, a large scale at which the contribution of *momenta* is cut (as we have already explained, this denomination is borrowed from quantum field theory where the arguments of the Fourier transforms are momenta).

9.1.1 Effective statistical field theory

As in the Gaussian field theory of Section 8.4, the local order parameter thus is a field, which we denote by $\sigma(x)$ ($x \in \mathbb{R}^d$), and which plays the role of the lattice variables σ_i of Sections 8.2, 8.5 and 8.10. The field σ is a local mean spin and thus a continuum variable, even if the initial microscopic spin takes only discrete values, like in the Ising model.

The statistical model is then defined by a local functional $\mathcal{H}(\sigma)$ of the field σ, generalization of the quadratic form (8.31), and of the form (8.48) of the quasi-

Gaussian approximation on the lattice. The corresponding correlation functions are obtained by integrating over the fields $\sigma(x)$ with the weight $e^{-\mathcal{H}(\sigma)}/\mathcal{Z}$,

$$\langle \sigma(x_1)\sigma(x_2)\ldots\sigma(x_n)\rangle = \frac{1}{\mathcal{Z}}\int [d\sigma]\sigma(x_1)\sigma(x_2)\ldots\sigma(x_n)\,e^{-\mathcal{H}(\sigma)},$$

where $\mathcal{Z}$ is the partition function in zero field determined by the condition $\langle 1\rangle = 1$.

The partition function in an external variable field $H(x)$,

$$\mathcal{Z}(H) = \int [d\sigma]\exp\left[-\mathcal{H}(\sigma) + \int d^d x\, H(x)\sigma(x)\right],$$

(and thus $\mathcal{Z}\equiv \mathcal{Z}(0)$) generates correlation functions. The functional $\mathcal{H}(\sigma)$ is often called the *Hamiltonian* in the context of statistical physics. It is also a generalization of the *Euclidean action*, that is, the classical action in imaginary time, of the quantum theory of fundamental interactions.

It follows from the definition of the functional derivative that

$$\frac{\delta}{\delta H(y)}\exp\left[\int dx\,\sigma(x)H(x)\right] = \sigma(y)\exp\left[\int dx\,\sigma(x)H(x)\right]. \tag{9.1}$$

Correlation functions are thus obtained by acting with functional derivatives on $\mathcal{Z}(H)$:

$$\langle \sigma(x_1)\ldots\sigma(x_n)\rangle = \mathcal{Z}^{-1}(0)\left.\frac{\delta^n \mathcal{Z}(H)}{\delta H(x_1)\ldots\delta H(x_n)}\right|_{H(x)\equiv 0}.$$

9.1.2 Landau's Hamiltonian

The Hamiltonian $\mathcal{H}(\sigma)$ is a functional of the field $\sigma(x)$. We assume that it has the general properties of the thermodynamic potential of Landau's theory (Section 8.7). $\mathcal{H}(\sigma)$ thus generalizes the Hamiltonian of the quasi-Gaussian approximation or the effective Hamiltonian that leads to the mean-field approximation.

This hypothesis is consistent with the analysis of Section 8.10. Indeed, we have shown that the effective Hamiltonian (8.94) in expression (8.93) has the properties of the mean-field thermodynamic potential, and the latter can be inferred from the leading term in a calculation by the steepest descent method.

We thus assume that the Hamiltonian, which in this context is also called the Landau–Ginzburg–Wilson Hamiltonian, has the following properties:

(i) The Hamiltonian $\mathcal{H}$ is a regular function of all thermodynamic parameters like the temperature (except at zero temperature).

(ii) It is expandable in powers of the field σ:

$$\mathcal{H}(\sigma) = \sum_{n=0}^{\infty}\frac{1}{n!}\int d^d x_1 d^d x_2\ldots d^d x_n\,\mathcal{H}^{(n)}(x_1,x_2,\ldots,x_n)\sigma(x_1)\ldots\sigma(x_n). \tag{9.2}$$

(iii) We will discuss, in general, only translation-invariant systems and thus

$$\mathcal{H}^{(n)}(x_1,x_2,\ldots,x_n) = \mathcal{H}^{(n)}(x_1+a,x_2+a,\ldots,x_n+a)\quad \forall a\in\mathbb{R}^d.$$

Then, the Fourier transforms of the coefficients (functions or distributions) $\mathcal{H}^{(n)}$ take the form (see Section 6.2.2)

$$(2\pi)^d \delta^{(d)}\left(\sum_{i=1}^n p_i\right) \tilde{\mathcal{H}}^{(n)}(p_1,\ldots,p_n) = \int d^d x_1 \ldots d^d x_n \exp\left(i \sum_{j=1}^n x_j p_j\right)$$
$$\times \mathcal{H}^{(n)}(x_1,\ldots,x_n). \qquad (9.3)$$

(iv) The hypothesis of short-range interactions (with exponential decay), or of locality in the sense of the quantum field theory, implies that the coefficients $\tilde{\mathcal{H}}^{(n)}$ are analytic in strips of the form $|\operatorname{Im} p_i| < \kappa$.

Note that in terms of the Fourier components of the field σ:

$$\sigma(x) = \int d^d k \, e^{ikx} \, \tilde{\sigma}(k),$$

the expansion of $\mathcal{H}$ reads

$$\mathcal{H}(\sigma) = \sum_{n=0} \frac{1}{n!} \int d^d k_1 \ldots d^d k_n \, (2\pi)^d \delta^{(d)}\left(\sum_{i=1}^n k_i\right) \tilde{\mathcal{H}}^{(n)}(k_1,\ldots,k_n) \tilde{\sigma}(k_1) \ldots \tilde{\sigma}(k_n).$$
$$(9.4)$$

Remark. One may wonder why one immediately considers such a general class of Hamiltonians, whereas one seems unable to determine the critical behaviour of much simpler systems. Of course, in this way the analysis will apply to a larger class of systems. But the main reason is, as we shall see, that RG transformations generate such Hamiltonians, even when the initial Hamiltonian is much simpler.

9.2 Connected correlation functions. Vertex functions

In Chapter 6, we have already introduced, on the lattice, several generating functionals of correlation functions. Here, we only extend definitions and relations to generating functionals in continuum space. Most of the discussion of Chapter 6 applies also in the continuum.

In Section 6.2.1, we have introduced the generating functional (which is also the free energy in an inhomogeneous external field, up to a temperature factor)

$$\mathcal{W}(H) = \ln \mathcal{Z}(H)$$

of connected correlation functions

$$\mathcal{W}^{(n)}(x_1, x_2, \ldots, x_n) = \left.\frac{\delta^n \mathcal{W}(H)}{\delta H(x_1) \ldots \delta H(x_n)}\right|_{H(x)\equiv 0} \equiv \langle \sigma(x_1) \ldots \sigma(x_n) \rangle_{\text{connected}}.$$
$$(9.5)$$

In what follows, we consider only connected functions because they have large--distance decay properties, called cluster properties (see Section 6.2.1).

Fourier representation. We have assumed that the field theory is translation invariant. This hypothesis implies that for all $a \in \mathbb{R}^d$, the n-point connected correlation function satisfies

$$W^{(n)}(x_1 + a, \ldots, x_n + a) = W^{(n)}(x_1, \ldots, x_n).$$

The relation between an n-point function and its Fourier transform can be written as

$$(2\pi)^d \delta^{(d)}\left(\sum_{i=1}^n p_i\right) \widetilde{W}^{(n)}(p_1, \ldots, p_n)$$

$$= \int d^d x_1 \ldots d^d x_n \, W^{(n)}(x_1, \ldots, x_n) \exp\left(i \sum_{j=1}^n x_j p_j\right), \qquad (9.6)$$

which is the form in the continuum of the representation (6.15). The connected correlation function $W^{(n)}(x_1, x_2, \ldots, x_n)$ satisfies a cluster property (Section 6.2.1), which takes the form of decay properties when its arguments tends toward infinity. Regularity properties of $\widetilde{W}^{(n)}(p_1, \ldots, p_n)$ in the variables p_i follow.

Generating functional of vertex functions. In Section 6.3, we have also defined by Legendre transformation the generating functional $\Gamma(M)$ of vertex functions:

$$\Gamma(M) = \sum_{n=0} \frac{1}{n!} \int d^d x_1 \ldots d^d x_n \, \Gamma^{(n)}(x_1, \ldots, x_n) M(x_1) \ldots M(x_n).$$

In the continuum, the Legendre transformation takes the form

$$\mathcal{W}(H) + \Gamma(M) = \int d^d x \, H(x) M(x), \qquad (9.7a)$$

$$M(x) = \frac{\delta \mathcal{W}(H)}{\delta H(x)}. \qquad (9.7b)$$

In statistical physics, $\Gamma(M)$ is the thermodynamic potential, a function of the local magnetization.

The Fourier transform of the n-point vertex function is defined by (see definition (6.23))

$$(2\pi)^d \delta^{(d)}\left(\sum_{i=1}^n p_i\right) \tilde{\Gamma}^{(n)}(p_1, \ldots, p_n)$$

$$= \int d^d x_1 \ldots d^d x_n \, \Gamma^{(n)}(x_1, \ldots, x_n) \exp\left(i \sum_{j=1}^n x_j \cdot p_j\right). \qquad (9.8)$$

Vertex functions, due to the regularity properties of the $\tilde{\Gamma}^{(n)}$, play a central role in the general discussion of the perturbative expansion.

9.3 Renormalization group (RG): General idea

In Chapter 8, we have shown that universality properties emerge in the study of the asymptotic behaviour, at large distance, of the Gaussian connected two-point function in the vicinity of the critical point. Within the framework of a general model, we thus want to explore the properties of the large-distance behaviour of connected correlation functions.

9.3.1 RG equations

Technically, we want to determine the behaviour of the connected n-point function $W^{(n)}(\lambda x_1, \ldots, \lambda x_n)$, at the critical point, when the real positive dilatation parameter λ diverges. For reasons that will be understood later (but that are not unrelated to the remark at the end of Section 8.4.3), it is necessary to assume that *all points x_i are distinct*.

The RG idea then is to construct a Hamiltonian function $\mathcal{H}_\lambda(\sigma)$ of the dilatation parameter λ, such that

$$\mathcal{H}_{\lambda=1}(\sigma) \equiv \mathcal{H}(\sigma),$$

and for which the corresponding correlation functions $W_\lambda^{(n)}(x_i)$ satisfy

$$W_\lambda^{(n)}(x_1, \ldots, x_n) - Z^{-n/2}(\lambda) W^{(n)}(\lambda x_1, \ldots, \lambda x_n) = R_\lambda^{(n)}(x_1, \ldots, x_n) \qquad (9.9)$$

with

$$W_{\lambda=1}^{(n)}(x_1, \ldots, x_n) \equiv W^{(n)}(x_1, \ldots, x_n), \quad Z(1) = 1, \quad R_{\lambda=1}^{(n)}(x_1, \ldots, x_n) = 0.$$

The RG equation (RGE) (9.9) involves functions $R^{(n)}$ that satisfy only one condition: for $\lambda \to \infty$, they decay, for example, faster than any power of λ. In the absence of such a term, one speaks of a linear RG (see Section 16.2.4 for details).

The functions $Z^{-n/2}(\lambda) W^{(n)}(\lambda x_1, \ldots, \lambda x_n)$ are the connected correlation functions of the field $\sigma(\lambda x)/\sqrt{Z(\lambda)}$: the factor $\sqrt{Z(\lambda)}$ renormalizes the field $\sigma(x)$. One notes here a difference with the situations that we have encountered so far: the renormalization can *a priori* be an arbitrary function of λ and not only a power.

The Hamiltonian $\mathcal{H}_\lambda(\sigma)$ is also called the *effective Hamiltonian* at scale λ and the transformation $\mathcal{H}(\sigma) \mapsto \mathcal{H}_\lambda(\sigma)$ is called the RG transformation.

Various RG transformations may differ by the form of the functions $R^{(n)}$ and the function $Z(\lambda)$. In explicit constructions, the $R^{(n)}$'s are generated by the integration over the large momentum modes of $\sigma(x)$ (see Chapter 16). When $R^{(n)} \equiv 0$ and both space and fields are continuous variables, the RG transformation is realized by a rescaling of space and fields:

$$\mathcal{H}_\lambda\bigl(\sigma(x)\bigr) = \mathcal{H}_{\lambda=1}\bigl(Z^{1/2}(\lambda)\sigma(x/\lambda)\bigr). \qquad (9.10)$$

However, except in the case of Gaussian models, such transformations have, in general, no fixed points and, therefore, it is necessary to consider more general transformations. Below, we omit the terms $R^{(n)}$ and, thus, the relations between correlation functions must be understood modulo terms that decay faster than any power of λ. We thus discuss an *asymptotic linear RG*.

Fourier representation. Equation (9.9), after Fourier transformation (equation (9.6)) and neglecting R_n, can also be written as

$$(2\pi)^d \delta^{(d)}\left(\sum_{i=1}^n p_i\right) \widetilde{W}_\lambda^{(n)}(p_1, \ldots, p_n)$$

$$= Z^{-n/2}(\lambda) \int d^d x_1 \ldots d^d x_n \, W^{(n)}(\lambda x_1, \ldots, \lambda x_n) \exp\left(i \sum_{j=1}^n x_j p_j\right)$$

$$= Z^{-n/2}(\lambda)\lambda^{-nd} \int d^d x_1 \ldots d^d x_n \, W^{(n)}(x_1, \ldots, x_n) \exp\left(i \sum_{j=1}^n x_j p_j/\lambda\right)$$

$$= Z^{-n/2}(\lambda)\lambda^{-nd}(2\pi)^d \delta^{(d)}\left(\sum_{i=1}^n p_i/\lambda\right) \widetilde{W}^{(n)}(p_1/\lambda, \ldots, p_n/\lambda).$$

Using the property

$$\delta^{(d)}\left(\sum_{i=1}^n p_i/\lambda\right) = \lambda^d \delta^{(d)}\left(\sum_{i=1}^n p_i\right),$$

one infers

$$\widetilde{W}_\lambda^{(n)}(p_1, \ldots, p_n) = Z^{-n/2}(\lambda)\lambda^{(1-n)d}\widetilde{W}^{(n)}(p_1/\lambda, \ldots, p_n/\lambda). \quad (9.11)$$

In particular, for $n = 2$,

$$\widetilde{W}_\lambda^{(2)}(p) = Z^{-1}(\lambda)\lambda^{-d}\widetilde{W}^{(2)}(p/\lambda). \quad (9.12)$$

9.3.2 Generating functions and vertex functions

The asymptotic relation (9.9),

$$W_\lambda^{(n)}(x_1, \ldots, x_n) = Z^{-n/2}(\lambda)W^{(n)}(\lambda x_1, \ldots, \lambda x_n),$$

translates also into the relation between generating functions

$$\mathcal{W}_\lambda(H(x))$$

$$= \sum_{n=0}^\infty \frac{1}{n!} \int d^d x_1 \ldots d^d x_n \, Z^{-n/2}(\lambda) W^{(n)}(\lambda x_1, \ldots, \lambda x_n) H(x_1) \ldots H(x_n)$$

$$= \sum_{n=0}^\infty \frac{1}{n!} \int d^d x_1 \ldots d^d x_n \, Z^{-n/2}(\lambda)\lambda^{-nd} W^{(n)}(x_1, \ldots, x_n) H(x_1/\lambda) \ldots H(x_n/\lambda)$$

$$= \mathcal{W}(\lambda^{-d} Z^{-1/2}(\lambda) H(x/\lambda)).$$

Then, for the local magnetization,

$$M(y) = \frac{\delta \mathcal{W}_\lambda}{\delta H(y)}$$

$$= \sum_{n=0}^\infty \frac{1}{n!} \int d^d x_1 \ldots d^d x_n \, Z^{-(n+1)/2}(\lambda) W^{(n+1)}(\lambda y, \lambda x_1, \ldots, \lambda x_n) H(x_1)$$

$$\times \ldots H(x_n)$$

and, thus,

$$M(y/\lambda) = Z^{-1/2}(\lambda) \frac{\delta}{\delta H(y)} \mathcal{W}(\lambda^{-d} Z^{-1/2}(\lambda) H(x/\lambda)). \tag{9.13}$$

The inversion of this relation, using a property of the Legendre transformation, yields

$$\lambda^{-d} Z^{-1/2}(\lambda) H(x/\lambda) = \frac{\delta}{\delta M(x)} \Gamma(Z^{1/2}(\lambda) M(y/\lambda))$$

and thus

$$H(x/\lambda) = \lambda^d Z^{1/2}(\lambda) \frac{\delta}{\delta M(x)} \Gamma(Z^{1/2}(\lambda) M(y/\lambda)),$$

to be compared with

$$H(x) = \frac{\delta}{\delta M(x)} \Gamma_\lambda(M(y)).$$

Then comparing with equation (9.13), one observes that the passage from $\mathcal{W}, H$ to Γ, M corresponds also to the change $\lambda^{-d} Z^{-1/2} \mapsto Z^{1/2}$. One concludes

$$\Gamma_\lambda(M(y)) = \Gamma(Z^{1/2}(\lambda) M(y/\lambda)),$$

which is consistent with relation (9.10) and, thus, also

$$\Gamma_\lambda^{(n)}(x_1, \ldots, x_n) = Z^{n/2}(\lambda) \lambda^{nd} \Gamma^{(n)}(\lambda x_1, \ldots, \lambda x_n).$$

Replacing in equation (9.11) W by Γ and simultaneously $\lambda^{-d} Z^{-1/2}$ by $Z^{1/2}$, one obtains the form of the RGE for the Fourier transform of the n-point vertex function (see the definition (9.8)):

$$\tilde{\Gamma}_\lambda^{(n)}(p_1, \ldots, p_n) = Z^{n/2}(\lambda) \lambda^d \tilde{\Gamma}^{(n)}(p_1/\lambda, \ldots, p_n/\lambda). \tag{9.14}$$

In particular, for the two-point vertex function

$$\tilde{\Gamma}_\lambda^{(2)}(p) = Z(\lambda) \lambda^d \tilde{\Gamma}^{(2)}(p/\lambda), \tag{9.15}$$

a form consistent with the relation (9.12) since $\tilde{\Gamma}^{(2)}(p)$ is the inverse of $\widetilde{W}^{(2)}(p)$.

9.3.3 Fixed-point Hamiltonian

A general local Hamiltonian depends on an infinite number of parameters. The parameters that appear in $\mathcal{H}_\lambda$ are functions of λ. Let us assume that there exists an RG transformation such that, when $\lambda \to \infty$, the Hamiltonian $\mathcal{H}_\lambda(\sigma)$ has a limit $\mathcal{H}^*(\sigma)$, which is a fixed-point Hamiltonian. We denote by $W_*^{(n)}$ the corresponding correlation functions. Then, equation (9.9) implies

$$W^{(n)}(\lambda x_1, \ldots, \lambda x_n) \underset{\lambda \to \infty}{\sim} Z^{n/2}(\lambda) W_*^{(n)}(x_1, \ldots, x_n). \tag{9.16}$$

Introducing a second dilatation $\mu > 0$, one can calculate $W^{(n)}(\lambda \mu x_i)$ from equation (9.16) in two different ways:

$$W^{(n)}(\lambda \mu x_1, \ldots, \lambda \mu x_n) \underset{\lambda \to \infty}{\sim} Z^{n/2}(\lambda) W_*^{(n)}(\mu x_1, \ldots, \mu x_n)$$

$$\underset{\lambda \to \infty}{\sim} Z^{n/2}(\lambda \mu) W_*^{(n)}(x_1, \ldots, x_n).$$

Eliminating $W^{(n)}(\lambda \mu x_i)$, one obtains the relation

$$W_*^{(n)}(\mu x_1, \ldots, \mu x_n) = Z_*^{n/2}(\mu) W_*^{(n)}(x_1, \ldots, x_n) \tag{9.17}$$

with

$$Z_*(\mu) = \lim_{\lambda \to \infty} Z(\lambda \mu)/Z(\lambda). \tag{9.18}$$

Equation (9.17) being valid for arbitrary μ, this immediately implies that $Z_*(\lambda)$ forms a representation of the semi-group of dilatations:

$$Z_*(\lambda_1) Z_*(\lambda_2) = Z_*(\lambda_1 \lambda_2). \tag{9.19}$$

Thus, with reasonable assumptions, $Z_*(\lambda)$ has a power-law behaviour:

$$Z_*(\lambda) = \lambda^{-2d_\sigma}. \tag{9.20}$$

This implies the scaling behaviour

$$W_*^{(n)}(\lambda x_1, \ldots, \lambda x_n) = \lambda^{-nd_\sigma} W_*^{(n)}(x_1, \ldots, x_n). \tag{9.21}$$

The positive number d_σ that characterizes the power-law behaviour of correlation functions at a fixed point, is called the dimension of the field, or order parameter, $\sigma(x)$, expressed in inverse length units.

From equation (9.18), one then concludes that $Z(\lambda)$ also has asymptotically a power-law behaviour. Finally, equation (9.16) implies that connected correlation functions have a scaling behaviour at large distance of the form

$$W^{(n)}(\lambda x_1, \ldots, \lambda x_n) \underset{\lambda \to \infty}{\propto} \lambda^{-nd_\sigma} W_*^{(n)}(x_1, \ldots, x_n), \tag{9.22}$$

determined by the fixed point.

Since the right-hand side of equation (9.22), which determines the critical behaviour of correlation functions, depends only on the fixed-point Hamiltonian, the correlation functions corresponding to all Hamiltonians that converge after RG transformations toward the same fixed point, have the same critical behaviour. Such a *universality* property thus divides the space of Hamiltonians into *universality classes*. Universality, beyond the quasi-Gaussian approximation, relies on the existence of large-distance (IR) fixed points of the RG in the space of Hamiltonians.

Applied to the two-point function, this result shows, in particular, that if $2d_\sigma < d$, the correlation length ξ diverges and thus the corresponding Hamiltonians are necessarily critical. Critical Hamiltonians define in the space of Hamiltonians the *critical surface*, which is invariant under the RG. In the generic situation where ξ is finite, the correlation length ξ/λ corresponding to $\mathcal{H}_\lambda$ goes to zero. The Fourier components of correlation functions become momentum independent and thus correlation functions become δ-functions in direct space. This trivial fixed point of independent variables corresponds to $2d_\sigma = d$.

9.4 Hamiltonian flow: Fixed points, stability

We now assume that there exists a transformation on the space of Hamiltonians consistent with equation (9.9).

Moreover, we assume that this transformation is of Markovian type, that is, that the transformation depends on $\mathcal{H}_\lambda$ but not on the trajectory that has led from $\mathcal{H}$ to $\mathcal{H}_\lambda$. In continuum space, λ, the dilatation parameter, can be varied continuously (whereas on the lattice λ can only take a set of discrete values). Therefore, we can write the transformation in a differential form in terms of a transformation $\mathcal{T}$ in the space of Hamiltonians and a real function D_σ defined in the space of Hamiltonians. The RGE satisfied by the Hamiltonian then takes the general form

$$\lambda \frac{\mathrm{d}}{\mathrm{d}\lambda} \mathcal{H}_\lambda = \mathcal{T}[\mathcal{H}_\lambda], \qquad (9.23)$$

$$\lambda \frac{\mathrm{d}}{\mathrm{d}\lambda} \ln Z(\lambda) = -2D_\sigma[\mathcal{H}_\lambda]. \qquad (9.24)$$

We assume also that the Markovian process is stationary, in such a way that the right-hand side depends on λ only through $\mathcal{H}_\lambda$ (and thus not on λ explicitly). Finally, we assume, and this is also an important hypothesis, that the mapping $\mathcal{T}$ is sufficiently differentiable (for example, infinitely differentiable). These assumptions summarize the characteristic properties of these equations and are at the origin of many results that will be derived later.

The appearance of the derivative $\lambda \mathrm{d}/\mathrm{d}\lambda = \mathrm{d}/\mathrm{d}\ln\lambda$ reflects the multiplicative character of dilatations or scale changes. The RGE thus defines a dynamical process in the time $\ln\lambda$.

An explicit example of a transformation (9.23) is provided by equation (16.51).

9.4.1 Fixed points and linearized flow

A fixed-point Hamiltonian $\mathcal{H}^*$ is necessarily a solution of the fixed-point equation

$$\mathcal{T}[\mathcal{H}^*] = 0. \tag{9.25}$$

The dimension d_σ of the field σ is then

$$d_\sigma = D_\sigma[\mathcal{H}^*]. \tag{9.26}$$

Linearized RGE. The RG fixed point being determined, we apply the RGE (9.23) to a Hamiltonian $\mathcal{H}$ close to the fixed point $\mathcal{H}^*$. Setting $\mathcal{H}_\lambda = \mathcal{H}^* + \Delta\mathcal{H}_\lambda$, we linearize the RGE that then takes the form

$$\lambda \frac{d}{d\lambda} \Delta\mathcal{H}_\lambda = L^*(\Delta\mathcal{H}_\lambda), \tag{9.27}$$

where L^* is a linear operator independent of λ, acting in the space of Hamiltonians. It is then possible to study the RG flow in a neighbourhood of the fixed point and infer the local stability properties of the fixed point.

The formal solution of equation (9.27) is

$$\Delta\mathcal{H}_\lambda = \lambda^{L^*} \Delta\mathcal{H}_1. \tag{9.28}$$

Let us assume that L^* has a discrete spectrum with eigenvalues ℓ_i, corresponding to a set of eigenvectors $\mathcal{O}_i$. For $\lambda \to \infty$, the flow is dominated by the largest eigenvalues (of largest real parts if they are complex). We then expand $\Delta\mathcal{H}_\lambda$ on the eigenvectors $\mathcal{O}_i$ of L^*:

$$\Delta\mathcal{H}_\lambda = \sum_i h_i(\lambda) \mathcal{O}_i. \tag{9.29}$$

The flow equation (9.27) is equivalent to

$$\lambda \frac{d}{d\lambda} h_i(\lambda) = \ell_i h_i(\lambda), \tag{9.30}$$

which integrated yields

$$h_i(\lambda) = \lambda^{\ell_i} h_i(1). \tag{9.31}$$

9.4.2 Classification of eigenvectors

No general result implies that all eigenvalues ℓ_i must be real. Nevertheless, for the class of models that we study in this work, it is found by explicit calculation that this is indeed the case. In what follows, we thus discuss only this situation.

The eigenvectors, or eigen-perturbations, $\mathcal{O}_i$ can then be classified into four families according to the sign of the corresponding eigenvalues (see Section 3.2):

(i) *Positive eigenvalues.* The corresponding eigenvectors are called *relevant*. If $\Delta\mathcal{H}_\lambda$ has a component on such a vector, this component will grow with λ and $\mathcal{H}_\lambda$

will move away from $\mathcal{H}^*$. These vectors correspond to unstable directions in the vicinity of the fixed point. For example, the vectors associated with deviations from criticality are clearly relevant since a dilatation decreases the effective correlation length.

(ii) *Vanishing eigenvalues.* The eigenvectors corresponding to a vanishing eigenvalue are called *marginal*. In Section 9.5, we show that $\int dx\, \sigma^4(x)$ is marginal in four dimensions. To solve the RGE (9.23) and determine the behaviour of the corresponding component, it is necessary to expand beyond the linear approximation. Generically, one finds

$$\lambda \frac{d}{d\lambda} h_i(\lambda) \sim B h_i^2(\lambda). \tag{9.32}$$

Depending on the signs of the constant B and h_i, the fixed point is then marginally unstable or stable. In the latter case, for $\lambda \to \infty$, the solution behaves like

$$h_i(\lambda) \sim -1/(B \ln \lambda). \tag{9.33}$$

The presence of a marginal eigenvector thus leads, in general, to a logarithmic convergence toward the fixed point.

An exceptional situation is provided by the XY model in two dimensions (a model with a two-component spin and an $O(2)$ symmetry) that, instead of an isolated fixed point, has a line of fixed points. The eigenvector that corresponds to motion along the fixed line is clearly marginal (see Section 15.10.2).

(iii) *Negative eigenvalues.* The corresponding eigenvectors are called *irrelevant*. The components of $\Delta\mathcal{H}_\lambda$ on these eigenvectors decay for large dilatations.

(iv) Finally some eigenvectors do not affect physical properties. An example is provided by the eigenvector generating a constant multiplicative renormalization of the dynamical variables $\sigma(x)$. These eigenvectors are called *redundant*.

Fixed-point classification. Fixed points can be classified according to their local stability properties, that is, the number of linearly independent relevant eigenvectors. This number is also the number of parameters that it is necessary to fix to impose on a general Hamiltonian to belong to the surface that forms the basin of attraction of the fixed point. For a non-trivial fixed point corresponding to general critical Hamiltonians, it is the co-dimension of the critical surface.

9.4.3 RGE: Another form

Equation (9.9) can be written equivalently as (neglecting the remainder R_n)

$$Z^{n/2}(\lambda) W_\lambda^{(n)}(x_1/\lambda, \ldots, x_n/\lambda) = W^{(n)}(x_1, \ldots, x_n). \tag{9.34}$$

We now assume that the Hamiltonian $\mathcal{H}_\lambda$ is parametrized in terms of constants $h_i(\lambda)$ (in general, in infinite number). Then, one can change the notation and write

$$W_\lambda^{(n)}(x_1, \ldots, x_n) \equiv W^{(n)}\big(\{h(\lambda)\}; x_1, \ldots, x_n\big)$$

where, here, $\{h(\lambda)\}$ represents the set of all $h_i(\lambda)$. We then differentiate equation (9.34) with respect to λ. The right-hand side does not depend on λ and thus

$$\lambda \frac{\mathrm{d}}{\mathrm{d}\lambda}\left[Z^{n/2}(\lambda)W^{(n)}\left(\{h(\lambda)\};x_1/\lambda,\ldots,x_n/\lambda\right)\right]=0. \tag{9.35}$$

We introduce the differential operator

$$\mathrm{D}_{\mathrm{RG}} \equiv \sum_{k=1}^{n} x_k \cdot \nabla_{x_k} + \sum_i \beta_i(\{h\})\frac{\partial}{\partial h_i} + nD_\sigma(\{h\}),$$

with the definitions

$$\beta_i(\{h\}) = -\lambda\frac{\mathrm{d}}{\mathrm{d}\lambda}h_i(\lambda), \tag{9.36}$$

$$2D_\sigma(\{h\}) = -\lambda\frac{\mathrm{d}}{\mathrm{d}\lambda}\ln Z(\lambda). \tag{9.37}$$

These two equations are equivalent to equations (9.23) and (9.24). The *stationary Markovian* property implies that β_i and D_σ do not depend on λ explicitly, but only through the parameters $h_i(\lambda)$. Equation (9.35) can then be written as (using the chain rule in the differentiation)

$$-Z^{-n/2}(\lambda)\lambda\frac{\mathrm{d}}{\mathrm{d}\lambda}\left[Z^{n/2}(\lambda)W^{(n)}\left(\{h(\lambda)\};x_1/\lambda,\ldots,x_n/\lambda\right)\right]$$
$$=\mathrm{D}_{\mathrm{RG}}W^{(n)}\left(\{h(\lambda)\};x_1/\lambda,\ldots,x_n/\lambda\right)=0.$$

Finally, since in the new equation λ plays no explicit role anymore, we specialize to $\lambda = 1$. The equation becomes the partial differential equation for correlation functions

$$\left[\sum_{k=1}^{n} x_k \cdot \nabla_{x_k} + \sum_i \beta_i(\{h\})\frac{\partial}{\partial h_i} + nD_\sigma(\{h\})\right]W^{(n)}(\{h\};x_1,\ldots,x_n)=0, \tag{9.38}$$

another standard form of RGE. In this formalism, fixed points correspond to simultaneous solutions of all equations

$$\beta_i(\{h^*\}) = 0,$$

and equation (9.38) then implies the scaling behaviour already obtained directly since

$$\left(\sum_{k=1}^{n} x_k \cdot \nabla_{x_k} + nd_\sigma\right)W^{(n)}(\{h^*\};x_1,\ldots,x_n)=0. \tag{9.39}$$

Similarly, from equation (9.14) one infers

$$\left[\sum_{k=1}^{n} p_k \cdot \nabla_{p_k} - \sum_i \beta_i(\{h\})\frac{\partial}{\partial h_i} + nD_\sigma(\{h\}) - d\right]\tilde{\Gamma}^{(n)}(\{h\};p_1,\ldots,p_n)=0. $$
$$\tag{9.40}$$

9.4.4 The critical domain: Scaling properties

We have shown that the quasi-Gaussian approximation implies universality properties not only at the critical temperature where the correlation length is infinite, but also in a neighbourhood of the critical point, called the critical domain. The critical domain is characterized by the property that the correlation length is large with respect to the microscopic scale and, thus, $|T - T_c| \ll 1$ and magnetic field $|H| \ll 1$.

One expects that such universality generalizes also to other RG fixed points. Indeed, if the correlation length ξ is finite but large, one can perform scale transformations up to λ of order ξ. After these transformations, the correlation length has a size of order 1. The transformed Hamiltonian $\mathcal{H}_\lambda$ is no longer critical and the model can be solved by perturbative methods. But, simultaneously, all components of $\mathcal{H}_\lambda$ on irrelevant eigenvectors vanish asymptotically for $\xi \to \infty$. Thus, $\mathcal{H}_\lambda$ depends only on a number of parameters equal to the number of linearly independent relevant eigenvectors (see also Sections 4.7.2 and 7.4).

Universality is a consequence of the decay for $\lambda = O(\xi) \gg 1$ of all components on irrelevant eigenvectors. Here, we have implicitly assumed the absence of marginal perturbations that require a special analysis.

Technically, as we have already shown in examples, this universal asymptotic limit is obtained by renormalizing the amplitudes of relevant perturbations to cancel the effect of dilatations.

For this purpose, we rewrite equation (9.9) in the form

$$W^{(n)}(\{h(\lambda)\}; x_1, \ldots, x_n) = Z^{-n/2}(\lambda) W^{(n)}(\{h(1)\}; \lambda x_1, \ldots, \lambda x_n). \tag{9.41}$$

When $\lambda \to \infty$, the components on irrelevant perturbations go to zero. Also, we vary the initial components on relevant perturbations, as functions of λ, to cancel the dilatation factors:

$$h_i(1) \mapsto (h_i)_r \lambda^{-\ell_i} \quad \forall i \in E,$$

where E is the set of relevant components.

Then, the functions $W^{(n)}(\{h(\lambda)\}; x_1, \ldots, x_n)$ tend asymptotically toward functions $W_r^{(n)}(\{h(\lambda)\}; x_1, \ldots, x_n)$ that depend only on the amplitudes $(h_i)_r$ of relevant terms. The asymptotic relation

$$\lambda^{d_\sigma} W^{(n)}\left(h_j(1), j \notin E; (h_i)_r \lambda^{-\ell_i}, i \in E; \lambda x_1, \ldots, \lambda x_n\right)$$
$$\underset{\lambda \to \infty}{\propto} W_r^{(n)}\left((h_i)_r; x_1, \ldots, x_n\right)$$

follows. This equation applies directly to the asymptotic form and, thus,

$$W_r^{(n)}\left((h_i)_r; x_1, \ldots, x_n\right) = \lambda^{d_\sigma} W_r^{(n)}\left((h_i)_r \lambda^{-\ell_i}; \lambda x_1, \ldots, \lambda x_n\right). \tag{9.42}$$

This property of asymptotic universal correlation functions $W_r^{(n)}$ in the critical domain, is also called the *scaling property*.

9.5 The Gaussian fixed point

Within a rather general framework, we have shown that the existence of RG fixed points, in the space of Hamiltonians, implies universality properties of the critical behaviour. It now remains to construct these transformations explicitly and to determine their fixed points. In a limited class of models (but which includes important universality classes from the physics viewpoint), it has been possible to exhibit fixed points and to study their local stability. However, even in these examples, a complete analysis of the RG flow is lacking.

There exists, however, a subspace of the space of Hamiltonians which can be explored completely and for which RG transformations can be constructed explicitly: the subspace of quadratic Hamiltonians which corresponds to Gaussian models. In this case, one can take $R^{(n)} \equiv 0$ in the RG transformations.

The fixed points can then be identified: the critical Hamiltonians converge toward the *Gaussian fixed point* whose predictions above T_c coincide with the mean-field approximation. As we will show, the discussion has much in common with that of the random walk in Section 3.3.10. Note, finally, that in Gaussian models only one connected correlation function is different from zero, the two-point function, which notably simplifies the analysis.

9.5.1 The Gaussian fixed point

In Section 8.4, we have shown that the universal predictions of the Gaussian model can be inferred from the Hamiltonian

$$\mathcal{H}_G(\sigma) = \frac{1}{2} \int d^d x \left[\sum_{\mu=1}^{d} (\partial_\mu \sigma(x))^2 + m^2 \sigma^2(x) \right], \quad m^2 \propto T - T_c,$$

($\partial_\mu \equiv \partial/\partial_\mu$) defined in continuum space in d dimensions.

At T_c, the Hamiltonian reduces to

$$\mathcal{H}_G(\sigma)|_{m=0} = \frac{1}{2} \int d^d x \sum_{\mu=1}^{d} (\partial_\mu \sigma(x))^2,$$

and the two-point function is given by equation (8.40):

$$W^{(2)}(x) = \frac{2^{d-2}}{(4\pi)^{d/2}} \Gamma(d/2 - 1) \frac{1}{|x|^{d-2}}.$$

It has the scaling behaviour (9.21) characteristic of a fixed-point Hamiltonian. Indeed, the transformation of the function in a dilatation $x \mapsto \lambda x$ takes the form

$$W^{(2)}(\lambda x) = \lambda^{2-d} W^{(2)}(x).$$

From equations (9.9) and (9.20), one then infers the renormalization of the field

$$Z(\lambda) = \lambda^{-(d-2)} \tag{9.43}$$

and the dimension of the Gaussian critical field
$$d_\sigma = \tfrac{1}{2}(d-2). \tag{9.44}$$

The Hamiltonian
$$\mathcal{H}_G^*(\sigma) = \mathcal{H}_G(\sigma)|_{m=0}, \tag{9.45}$$
whose only non-vanishing connected correlation function has a scaling behaviour, is thus a fixed-point Hamiltonian: the *Gaussian fixed point*. It is invariant under the, here linear, RG transformations:
$$\sigma(x) \mapsto Z^{1/2}(\lambda)\sigma(x/\lambda) = \lambda^{(2-d)/2}\sigma(x/\lambda). \tag{9.46}$$
Indeed, the substitution leads to $\partial_\mu \mapsto \partial_\mu/\lambda$, and thus
$$\lambda^{2-d}\int d^d x \sum_{\mu=1}^{d}\lambda^{-2}(\partial_\mu \sigma(x/\lambda))^2 = \int d^d x \sum_{\mu=1}^{d}(\partial_\mu \sigma(x))^2,$$
after the change of variables $x_\mu \mapsto \lambda x_\mu$.

We have thus exhibited a fixed point, the Gaussian fixed point, whose predictions coincide with the universal predictions of the mean-field approximation.

9.5.2 General quadratic, isotropic Hamiltonian

We now consider a general Hamiltonian quadratic in σ, of the type envisaged in Section 8.2. In the case of short-range interactions, the Fourier transform of the coefficient of the term quadratic in σ is regular and can be expanded in a Taylor series. In direct space, this means that the term quadratic in σ can be expanded in powers of the differential operator ∂_μ. In continuum space, a Hamiltonian invariant under space rotations–reflections can depend only on the Laplacian operator in d dimensions, the square of the gradient operator ∇_x, and thus takes the general form
$$\mathcal{H}_G(\sigma) = \frac{1}{2}\int d^d x \sum_{r=0} u_r^{(2)}\sigma(x)(-\nabla_x^2)^r\sigma(x), \tag{9.47}$$
where the coefficients $u_r^{(2)}$ are constants and
$$\nabla_x^2 \equiv \sum_{\mu=1}^{d}\partial_\mu^2.$$

A lattice Hamiltonian can also involve terms having the lattice symmetry but not rotation invariant like $\sum_\mu \partial_\mu^4$, for example.

The RG transformations (9.46), which have led to the Gaussian fixed point, after the change of variables $x_\mu \mapsto \lambda x_\mu$ now lead to
$$\mathcal{H}_\lambda(\sigma) = \frac{1}{2}Z(\lambda)\int \lambda^d d^d x \sum_{r=0} u_r^{(2)}\sigma(x)(-\nabla_x^2/\lambda^2)^r\sigma(x)$$
$$= \frac{1}{2}\int d^d x \sum_{r=0} u_r^{(2)}\lambda^{2-2r}\sigma(x)(-\nabla_x^2)^r\sigma(x).$$

This expression thus is directly an expansion on eigenvectors. The coefficient of the term with $2r$ derivatives transforms like

$$u_r^{(2)} \mapsto u_r^{(2)}(\lambda) = \lambda^{2-2r} u_r^{(2)}. \tag{9.48}$$

Finally, differentiating $u_r^{(2)}(\lambda)$ (equation (9.48)) with respect to λ, one finds the equivalent of equation (9.23), which is here identical to equation (9.27) because the transformation $\mathcal{T}$ is linear, in the form (9.30):

$$\lambda \frac{\mathrm{d} u_r^{(2)}(\lambda)}{\mathrm{d}\lambda} = \ell_r u_r^{(2)}(\lambda) \text{ with } \ell_r = 2 - 2r.$$

The complete RG flow follows. In particular, for $\lambda \to \infty$, the leading eigenvectors have the smallest number of derivatives.

Relevant perturbation. For $r = 0$, $\ell_0 = 2$ and the coefficient u_0 increases, which is not surprising. The eigenvector is relevant for the Gaussian fixed point since it induces a finite correlation length. The addition of such a perturbation has the effect that the Hamiltonian then moves away from the fixed point.

Redundant perturbation. For $r = 1$, $\ell_1 = 0$ and the coefficient u_1 is invariant. The eigenvector is proportional to the fixed-point Hamiltonian and corresponds to a simple renormalization of σ, without any physical effect and thus is redundant.

Irrelevant perturbations. For $r > 1$, $\ell_r < 0$ and u_r decays for $\lambda \to \infty$. The eigenvectors are thus irrelevant.

The conclusion is the following: the Gaussian fixed point is stable in the class of Gaussian critical models. In the general class of Gaussian models, it has only one unstable direction, which corresponds to a deviation from the critical temperature.

Reflection symmetry breaking. Finally, if one extends the discussion to terms linear in σ which break the Ising $\mathbb{Z}_2$ symmetry, one can add a term $H \int \mathrm{d}^d x\, \sigma(x)$. In a dilatation, this magnetic term becomes

$$H \int \mathrm{d}^d x\, \sigma(x) \mapsto H\lambda^{(2-d)/2} \lambda^d \int \mathrm{d}^d x\, \sigma(x) = H\lambda^{d/2+1} \int \mathrm{d}^d x\, \sigma(x), \tag{9.49}$$

and thus is also relevant with eigenvalue $d/2 + 1 = d - d_\sigma$ (see equation (9.64)).

Operators. In the framework of quantum field theory, to the polynomials in the field $\sigma(x)$ are associated quantum *operators*. For this reason, one often calls these polynomials operators although here all fields are classical. For example, one speaks of relevant, marginal, irrelevant, operators, a traditional language that we also occasionally use.

Critical domain. The theory in the critical domain is obtained by renormalizing the coefficients of all relevant perturbations, in such a way that the RG transformed coefficients remain asymptotically constant. For example,

$$u_0^{(2)} \mapsto \lambda^{-2} u_0^{(2)}, \quad H \mapsto H\lambda^{-d/2-1}.$$

In this viewpoint, the coefficients at scale λ are fixed and one defines a flow of the initial parameters of the Hamiltonian. In the *renormalization* procedure of quantum field theory, this corresponds to fixing the parameters of the renormalized theory.

9.6 Eigen-perturbations: General analysis

We now study the transformation properties of general local functions, even in the field (i.e., having the Ising model $\mathbb{Z}_2$ symmetry). In this more general framework, we cannot discuss the global RG flow, but only its linearized version in the vicinity of the Gaussian fixed point.

9.6.1 Eigen-perturbations

The analysis of the preceding section, based on the assumption of short-range interactions, again allows performing a derivative expansion. One thus considers a perturbation of the general form

$$\Delta \mathcal{H}(\sigma) = \sum_{n=2}^{\infty} \sum_{r=0}^{\infty} \sum_{\alpha} \mathcal{O}_\alpha^{n,r}(\sigma),$$

where the $\mathcal{O}_\alpha^{n,r}(\sigma)$ are integrals of monomials $V_\alpha^{n,r}(\sigma)$ in $\sigma(x)$ and its derivatives, of degree n (even) in σ and with exactly r derivatives (r being also even):

$$\mathcal{O}_\alpha^{n,r}(\sigma) = \int d^d x \, V_\alpha^{n,r}\bigl(\sigma(x), \partial_\mu \sigma(x), \ldots\bigr).$$

The subscript α is there to remind us that, in general, one can find several linearly independent homogeneous polynomials at n, r fixed.

Beyond the quadratic terms already discussed, the terms $V_\alpha^{n,r}(\sigma)$ of lowest degree are, for example,

$$\sigma^4(x), \quad \sigma^6(x), \quad \sigma^2(x)\bigl(\partial_\mu \sigma(x)\bigr)^2, \quad \sigma^5(x)\nabla_x^2 \sigma(x) \ldots$$

Applying the RG transformation (9.46) and after the change $x \mapsto \lambda x$, one finds

$$\mathcal{O}_\alpha^{n,r}(\lambda, \sigma) = Z^{n/2}(\lambda) \lambda^{d-r} \mathcal{O}_\alpha^{n,r}(\sigma) = \lambda^{d-n(d-2)/2 - r} \mathcal{O}_\alpha^{n,r}(\sigma). \tag{9.50}$$

One sees the $\mathcal{O}_\alpha^{n,r}(\sigma)$ are eigenvectors, which can be classified according to decreasing eigenvalues $\ell_{n,r} = d - n(d-2)/2 - r$:

(i) For $n = 2$ and $r = 0$, $\ell_{2,0} = 2$, and thus the perturbation $\int \sigma^2(x) d^d x$, is relevant and corresponds to a deviation from the critical temperature.

(ii) The quadratic perturbation $n = 2$, $r = 2$, and thus $\ell_{2,2} = 0$, is redundant; it corresponds to a simple renormalization of the field σ.

(iii) Above dimension 4, for all other eigenvectors $\ell_{n,r} < 0$ and the perturbations are irrelevant: on the critical surface ($T = T_c$) the Gaussian fixed point is thus stable.

In dimension 4, the perturbation $n = 4$, $r = 0$, and thus $\ell_{4,0} = 4 - d = 0$, which is proportional to $\int \sigma^4(x) d^d x$, becomes marginal and all other perturbations remain irrelevant. A linear analysis is no longer sufficient to determine the stability properties of the Gaussian fixed point. It becomes necessary to construct an RG that takes into account the deviations from the Gaussian theory in order to be able to go beyond the linear approximation. If the Gaussian fixed point is marginally stable, one expects a logarithmic convergence.

Below dimension 4, $\ell_{4,0} = 4-d > 0$ and thus $\int \sigma^4(x)\mathrm{d}^d x$ is relevant: the Gaussian fixed point is unstable. When the dimension decreases, new perturbations corresponding to $\ell_{n,0} = d - n(d-2)/2$ also become relevant. For example, in dimension 3, $\int \sigma^6(x)\mathrm{d}^d x$ is marginal and for $d < 3$ relevant. However, let us point out that this classification is specific to the Gaussian fixed point and has no reason to remain valid for another fixed point.

Finally, note that this analysis leads to conclusions consistent with the behaviour of the leading corrections to the Gaussian approximation evaluated in Section 8.9.2.

Flow in differential form. Another form of equation (9.50) is obtained by expanding $\Delta \mathcal{H}_\lambda$ on a fixed basis, for example, the eigenvectors $\mathcal{O}_\alpha^{n,r}(\sigma)$, with scale-dependent coefficients:

$$\Delta \mathcal{H}_\lambda(\sigma) = \sum_{n=2}^{\infty} \sum_{r=0}^{\infty} \sum_\alpha h_\alpha^{(n,r)}(\lambda) \mathcal{O}_\alpha^{n,r}(\sigma),$$

where the coefficients $h_\alpha^{(n,r)}(\lambda)$, as a consequence of equation (9.50), satisfy

$$h_\alpha^{(n,r)}(\lambda) = \lambda^{d-n(d-2)/2-r} h_\alpha^{(n,r)}(1).$$

The differential form

$$\lambda \frac{\mathrm{d}}{\mathrm{d}\lambda} h_\alpha^{(n,r)} = L^* h_\alpha^{(n,r)} = \ell_{n,r} h_\alpha^{(n,r)} \tag{9.51}$$

with

$$\ell_{n,r} = d - \tfrac{1}{2} n(d-2) - r \tag{9.52}$$

provides an example of equation (9.30).

9.6.2 Fourier representation

In the Fourier representation, one obtains another differential form of RGE. We introduce the Fourier representation of the field

$$\sigma(x) = \int \mathrm{d}^d k \, \mathrm{e}^{ikx} \, \tilde{\sigma}(k).$$

We then expand $\Delta \mathcal{H}$ in powers of $\tilde{\sigma}$ (see equation (9.4)):

$$\Delta \mathcal{H}(\sigma) = \sum_{2}^{\infty} \frac{1}{n!} \int \mathrm{d}^d k_1 \ldots \mathrm{d}^d k_n \, \tilde{\sigma}(k_1) \ldots \tilde{\sigma}(k_n) (2\pi)^d \delta^{(d)}\left(\sum k_i\right) u^{(n)}(k_i). \tag{9.53}$$

Following the same line of arguments as before, one verifies that equation (9.9) leads to the transformation $k_i \mapsto k_i/\lambda$ and thus

$$u^{(n)}(k_i) \delta^d \left(\sum k_i\right) \mapsto u^{(n)}(k_i/\lambda) \lambda^d \delta^{(d)}\left(\sum k_i\right),$$

as well as
$$\tilde{\sigma}(k) \mapsto Z^{1/2}(\lambda)\tilde{\sigma}(k) = \lambda^{-(d-2)/2}\tilde{\sigma}(k).$$

One infers
$$u_\lambda^{(n)}(k_i) = \lambda^d Z^{n/2}(\lambda) u^{(n)}(k_i/\lambda) = \lambda^{d-n(d-2)/2} u^{(n)}(k_i/\lambda). \tag{9.54}$$

Differentiating with respect to λ, one finds the equivalent of equation (9.27), expanded in powers of $\tilde{\sigma}(k)$. Using
$$\lambda \frac{d}{d\lambda} u^{(n)}(k/\lambda) = -\sum_{i=1}^n k_i \frac{d}{dk_i} u^{(n)}(k/\lambda),$$

one obtains
$$\lambda \frac{d}{d\lambda} u_\lambda^{(n)}(k) = \left(-\sum_{i=1}^n k_i \frac{\partial}{\partial k_i} + d - \tfrac{1}{2} n(d-2) \right) u_\lambda^{(n)}(k). \tag{9.55}$$

This equation is written for the RG transformations that admit the fixed point (9.43), (9.45). The eigenvalue equation is then
$$L^* u_\lambda^{(n)}(k) = \left(-\sum_{i=1}^n k_i \frac{\partial}{\partial k_i} + d - \tfrac{1}{2} n(d-2) \right) u_\lambda^{(n)}(k) = \ell u_\lambda^{(n)}(k). \tag{9.56}$$

Equation (9.56) shows that the eigenvectors are homogeneous functions of the dynamical variables σ (the $u_\lambda^{(n)}(k)$'s for different values of n are not coupled) and the momenta k_i (Fourier representations of the space derivatives).

We have assumed that the functions $u_\lambda^{(n)}(k)$ are regular (hypothesis of short-range interactions). They can thus be expanded on a basis of homogeneous polynomials $P_\alpha^{(n,r)}$ of degree r in k:
$$\sum_{i=1}^n k_i \frac{\partial}{\partial k_i} P_\alpha^{(n,r)}(k) = r P_\alpha^{(n,r)}(k),$$

the subscript α recalling that there exists, in general, several homogeneous polynomials of degree r. Setting
$$u_\lambda^{(n)}(k) = \sum_{r,\alpha} h_\alpha^{(n,r)}(\lambda) P_\alpha^{(n,r)}(k),$$

one then finds that the coefficients $h_\alpha^{(n,r)}$ satisfy equation (9.51) with the eigenvalues $\ell_{n,r}$ given by equation (9.52).

9.7 A non-Gaussian fixed point: The ε-expansion

We have shown that the Gaussian fixed point is unstable for $d < 4$. We now assume that, at least within the framework of a perturbative expansion around the Gaussian fixed point, the *dimension of space d can be considered as a continuous parameter* (for technical details see Section 12.5). We then investigate the vicinity of dimension 4, considering the deviation $\varepsilon = 4 - d$ from dimension 4 as an expansion parameter. Moreover, we adjust one parameter (like the temperature) as a function of the dimension, in such a way that the system under study remains critical ($T = T_c$). This has the effect of suppressing the relevant perturbation that generates a finite correlation length ($n = 2, r = 0$ in the case of the Gaussian fixed point).

An additional assumption is also required, whose consistency, at least, will be later verified in explicit calculations: we have already assumed that the RGE (9.23) are expandable in the perturbations around fixed points and thus, here, in the deviations from the Gaussian theory. We assume, in addition, that the RGE are expandable in the parameter $\varepsilon = 4 - d$.

9.7.1 Fixed points

We assume that the initial Hamiltonian is critical ($T = T_c$) and is close enough to the Gaussian fixed point in such a way that, for not too large dilatations λ, equation (9.23) can still be approximated by the linearized form (9.27), where $\Delta \mathcal{H}$ is the deviation from the Gaussian fixed point.

As a consequence, the amplitudes $h_\alpha(\lambda)$ of all irrelevant eigenvectors decrease. Only the amplitude of $\int d^d x \, \sigma^4(x)$, that we now denote by $g(\lambda)$, increases. However, for $\varepsilon \ll 1$, $g(\lambda) \sim g(1)\lambda^\varepsilon$ increases slowly. One can then find values of λ large enough such that the amplitudes $h_\alpha(\lambda)$ become negligible with respect to $g(\lambda)$, but also small enough such that $g(\lambda)$ remains small and the expansion of equation (9.23) can still be used.

Note that for such values of λ, the condition that $\mathcal{H}(\sigma)$ is positive for large σ (required for the σ-field measure to be defined), implies that $g(\lambda)$ must be positive.

We can thus expand equation (9.23), projected on the corresponding eigenvector, in powers of $g(\lambda)$:

$$\lambda \frac{dg(\lambda)}{d\lambda} = (4 - d)g(\lambda) - \beta_2(d)g^2(\lambda) + O(gh_\alpha, g^3), \qquad (9.57)$$

and we redefine the dilatation parameter λ in such a way that for $\lambda = 1$ this asymptotic equation can already be used. The leading neglected terms are terms of order g^3 and quadratic terms proportional to $h_\alpha g$. With our assumptions, the terms proportional to $h_\alpha h_\beta$ are even smaller.

In particular, if the coefficients h_α initially vanish, their derivatives are dominated by terms of order at least g^2. The neglected terms gh_α are thus at least of order g^3.

The coefficient β_2 depends on the dimension d. We assume that it is regular at $d = 4$ and, thus, at leading order for $\varepsilon \to 0$, one can replace it by its value at $d = 4$.

One immediately notes that the sign of $\beta_2(4)$ plays a crucial role:

(i) $\beta_2(4) < 0$. In this case, for $d \leq 4$, the two terms have the same sign, the derivative of $g(\lambda)$ is positive and the Gaussian fixed point is unstable. If initially $g(1)$ does not vanish, $g(\lambda)$ grows with λ until the expansion in $g(\lambda)$ is no longer justified, and nothing more can be said. However, in dimension 4, the flow initially is slower than in lower dimensions. Indeed, for $g(1)$ small, $g(\lambda)$ increases first only logarithmically:

$$g(\lambda) = g(1) + |\beta_2(4)|\, g^2(1) \ln \lambda + O\left(g^3(1)\right).$$

For $d > 4$, and thus $\varepsilon < 0$, one finds a non-Gaussian fixed point

$$g^* = \varepsilon/\beta_2(4) + O(\varepsilon^2),$$

which is an unstable fixed point. If initially $g(1) < g^*$, $g(\lambda)$ converges toward the Gaussian fixed point. If $g(1) = g^*$, $g(\lambda) = g^*$ for all λ. If $g(1) > g^*$, $g(\lambda)$ increases and, again, nothing more can be concluded in this approach.

An interesting situation is the following: if $g(1) < g^*$ with $g^* - g(1) \ll 1$, the effective amplitude $g(\lambda)$ first moves very slowly away from the unstable fixed-point value. One then finds a slow transition (also called a *crossover*) between a behaviour of correlation functions first dominated by the unstable fixed point, and a behaviour dominated by the stable fixed point.

The case $\beta_2 < 0$ is not realized in the statistical models under study here, but is important for the physics of fundamental microscopic interactions ($d = 4$ in this case). Moreover, an example of such a situation, but where the dimension 4 is replaced by the dimension 2, will be met in Section 15.7.1 in the study of the non-linear σ model.

(ii) $\beta_2(4) = 0$. In this exceptional case, it becomes necessary to expand to next order and to take into account the terms of order g^3.

(iii) $\beta_2(4) > 0$. In this case, for $d > 4$, one finds only the Gaussian fixed point, which is stable.

In dimension 4, $\int \mathrm{d}^4 x\, \sigma^4(x)$ is marginal. The Gaussian fixed point is now stable since $g(1)$ is positive and the approach to the fixed point is then logarithmic,

$$g(\lambda) \sim \frac{1}{\beta_2(4) \ln \lambda},$$

in agreement with the general RG discussion.

On the contrary, for $d < 4$ (the relevant physical case), the Gaussian fixed point is unstable, but, for ε small, another fixed point is found,

$$g^* = \varepsilon/\beta_2(4) + O(\varepsilon^2),$$

which is stable. Indeed, if initially $g < g^*$, then $g(\lambda)$ increases, and if $g > g^*$ then $g(\lambda)$ decreases. This situation is specially interesting since one can then envisage calculating universal quantities in the form of an ε-expansion. Actually, for the models that we have studied so far, Ising type models and models with $O(N)$

symmetries, this is the situation that is realized (Wilson and Fisher 1972). This can be verified by calculating the first corrections to the quasi-Gaussian approximation contributing to the two- and four-point functions (see Section 10).

Linearizing equation (9.57) in the neighbourhood of the fixed point, one finds

$$\lambda \frac{\mathrm{d}}{\mathrm{d}\lambda}(g(\lambda) - g^*) \sim -\omega(g(\lambda) - g^*), \tag{9.58}$$

with (property of the parabola)

$$\omega = \varepsilon + O(\varepsilon^2). \tag{9.59}$$

Thus, at least for ε small, ω is positive and

$$g(\lambda) - g^* \underset{\lambda \to \infty}{\propto} \lambda^{-\omega}$$

goes to zero for $\lambda \to \infty$, which confirms that the new fixed point is stable with respect to a perturbation of $\int \mathrm{d}^d x \, \sigma^4(x)$ type, unlike the Gaussian fixed point.

The exponent ω characterizes the decay of the leading corrections to the universal critical behaviour predicted by the fixed-point Hamiltonian. It is also a universal quantity.

Finally, in the quadratic approximation, the flow equation (9.57) can be integrated exactly:

$$\frac{\mathrm{d}g}{\varepsilon g - \beta_2 g^2} = \frac{\mathrm{d}\lambda}{\lambda} \Rightarrow \ln \lambda = \int_g^{g(\lambda)} \frac{\mathrm{d}g'}{\varepsilon g' - \beta_2 g'^2},$$

and thus, for $g^* \propto \varepsilon \ll 1$,

$$g(\lambda) = \frac{g^* g}{g + \lambda^{-\varepsilon}(g^* - g)}.$$

Field or order parameter dimension. Once the fixed-point value g^* is determined, one can calculate the dimension of the field or order parameter $d_\sigma = D_\sigma(g^*)$. The calculation of the two-point function at second order in the parameter g shows that the deviation

$$\eta(g) = 2D_\sigma(g) - d + 2 \tag{9.60}$$

from the Gaussian value is of order g^2 and, thus, the deviation $\eta \equiv \eta(g^*)$ from the Gaussian (or mean-field) dimension of the field is of order ε^2.

9.7.2 Other eigenvectors

We have shown that, for $\beta_2(4) > 0$, a stable fixed point exists with $g^* = O(\varepsilon)$. But in the analysis we have neglected the components on the eigenvectors that are irrelevant relative to the Gaussian fixed point. The most important near dimension 4 are such that the total number of derivatives and factors σ is 6, like $\int \mathrm{d}^d x \, \sigma^6(x)$. To show the consistency of the analysis, we now examine the effect of these components on the flow.

The equations corresponding to the amplitudes of irrelevant eigenvectors $h_\alpha^{(n,r)}$ can depend on g starting from second order. We consider a parameter h_α for which this is the case. The flow equation then takes the form

$$\lambda \frac{\mathrm{d}}{\mathrm{d}\lambda} h_\alpha(\lambda) \equiv B_\alpha(g(\lambda), \{h(\lambda)\}) = \ell_\alpha h_\alpha(\lambda) - \beta_{h_\alpha g^2} g^2(\lambda) + \cdots, \qquad (9.61)$$

($\ell_\alpha < 0$) where the leading neglected terms are proportional to $h_{\alpha'} g$. At leading order in ε, one finds the fixed-point solution

$$h_\alpha^* = \frac{\beta_{h_\alpha g^2} \varepsilon^2}{\beta_2^2 \ell_\alpha} + O(\varepsilon^3),$$

where $\beta_{h_\alpha g^2}$, β_2 and the Gaussian eigenvalue ℓ_α can be replaced by their values in dimension 4. Therefore, the amplitudes of irrelevant terms at the fixed point are at least of order ε^2. Returning to equation (9.57), which takes the more general form

$$\lambda \frac{\mathrm{d}}{\mathrm{d}\lambda} g(\lambda) = -\beta(g(\lambda), \{h(\lambda)\}),$$

one notes that the neglected terms like g^3 and gh_α are all at least of order ε^3 and thus the corrections to g^* and to the slope ω are of order ε^2: at least the ε-expansion is a consistent procedure and the fixed point can be determined in the form of an ε-expansion.

Let us point out, however, that the fixed-point Hamiltonian reduces to a linear combination of the Gaussian term and of the $\int \mathrm{d}^d x\, \sigma^4(x)$ term only at leading order in ε since the other coefficients h_α do not vanish at the fixed point.

One can then calculate the eigenvalues corresponding to the flow in the vicinity of the new fixed point. One verifies that the eigenvalues differ from the eigenvalues associated with the Gaussian fixed point at order ε. Thus, for $0 \le \varepsilon \ll 1$, the classification of eigen-perturbations remains the same as in the Gaussian limit at $d = 4$ for all perturbations except one. Only one eigen-perturbation is very sensitive to the variation of the dimension: the perturbation that, at leading order, is $\int \mathrm{d}^d x\, \sigma^4(x)$, because it is marginal at $d = 4$. Relevant, for $d < 4$, from the viewpoint of the Gaussian fixed point, it becomes irrelevant for the new stable fixed point.

Perturbative RG. This analysis suggests that it should be possible to determine the RG flow equations from a perturbative expansion around the Gaussian model, whereas we have verified that this is impossible for correlation functions.

Flow between fixed points. In particular, it is then possible to determine perturbatively the RG trajectory that relates the Gaussian fixed point to the stable non--Gaussian fixed point. Assuming that the function $g(\lambda)$ is invertible, $g \mapsto \lambda = \lambda(g)$, a condition that is fulfilled at least for $\varepsilon \ll 1$, one can eliminate the dilatation parameter λ in favour of the amplitude g in the flow equations for the parameters h_α.

In equation (9.61), on a RG trajectory, h_α can be considered as a function of g and, thus, using the chain rule,

$$\lambda \frac{d}{d\lambda} h_\alpha = B_\alpha(g, \{h\}) = \lambda \frac{d}{d\lambda} g \frac{\partial}{\partial g} h_\alpha = -\beta(g, \{h\}) \frac{\partial}{\partial g} h_\alpha.$$

The left-hand side of equation (9.61) is of order ε^2 since the function β is of order ε^2, $\partial/\partial g$ is of order $1/\varepsilon$ and h_α of order ε^2 at least. At leading order, the derivative can thus be neglected, and one finds simply

$$h_\alpha(\lambda) = \beta_{h_\alpha g^2} g^2(\lambda)/\ell_\alpha + O(\varepsilon^3).$$

The derivative contributes only at next order. In this way, it is possible to determine the amplitudes of irrelevant eigen-perturbations as expansions in powers of g. Substituting these expansions in the function β, one obtains a unique equation for $g(\lambda)$:

$$\lambda \frac{d}{d\lambda} g(\lambda) = -\beta(g(\lambda)), \tag{9.62}$$

where $\beta(g)$ now is a formal series in g only. The solution of the latter equation determines completely the flow of the Hamiltonian. It depends only on one parameter, the initial value $g(\lambda = 1)$. The boundary conditions at $\lambda = 1$ for the amplitudes of the irrelevant terms are then all related to $g(1)$.

The zeros g^* of the function $\beta(g)$ determine the fixed points and the derivatives $\omega = \beta'(g^*)$ at fixed points are universal quantities. For a stable fixed point, ω characterizes the leading corrections to the scaling behaviour.

We have outlined here the strategy that will also be used in practical calculations.

9.8 Eigenvalues and dimensions of local polynomials

Once the asymptotic behaviour of correlation functions is determined, one could believe that one knows also the behaviour of expectation values of local polynomials of the fields like, for example, $\sigma^k(x)$ or $[\nabla \sigma(x)]^2$, because they can be obtained as limit of products of fields, for example,

$$\langle \sigma^2(y)\sigma(x_1) \ldots \sigma(x_n) \rangle = \lim_{y' \to y} \langle \sigma(y)\sigma(y')\sigma(x_1) \ldots \sigma(x_n) \rangle.$$

This is not the case because correlation functions corresponding to a fixed-point Hamiltonian (equation (9.21)) are in general singular when some arguments coincide as examples will show (see also the remark at the end of Section 8.4.3).

However, there exists one exceptional fixed point for which this property is true, with some qualification: the Gaussian fixed point that we have studied in Section 9.6.

Even then, since the expectation values of, for example, $\sigma^2(y)$ is not defined at the fixed point, it is necessary to first subtract from the product $\sigma(y)\sigma(y')$ a singular function of $y - y'$ to obtain a finite limit (a procedure called renormalization). This procedure generalizes to higher powers of the field σ and to local functions involving derivatives of the field.

In Section 9.4, we have discussed the RG eigen-perturbations at a fixed point. The study of the Gaussian example indicates that these eigen-perturbations take the form of integrals of local functions of the fields. Actually, due to the locality of the theory, there exists also local eigen-perturbations $\mathcal{O}_\alpha(\sigma, x)$ of the form of series in $\sigma(x)$ and its derivatives. They correspond to perturbations that break translation invariance. Their space integrals then yield the eigen-perturbations that we have discussed. These local functions $\mathcal{O}(\sigma, x)$ of fields have the following property: the correlation functions

$$W_\mathcal{O}^{(n)}(y; x_1, \ldots, x_n) = \langle \mathcal{O}(\sigma, y) \sigma(x_1) \ldots \sigma(x_n) \rangle_{\text{conn.}},$$

have an asymptotic power-law behaviour of the form

$$W_\mathcal{O}^{(n)}(\lambda y; \lambda x_1, \ldots, \lambda x_n) \underset{\lambda \to \infty}{\propto} \lambda^{-nd_\sigma - d_\mathcal{O}} W_\mathcal{O}^{*(n)}(y; x_1, \ldots, x_n), \tag{9.63}$$

where by definition $d_\mathcal{O}$ is the *dimension* of the local operator $\mathcal{O}(\sigma, x)$.

We then consider the space integral

$$\mathcal{V}(\sigma) = \int d^d x\, \mathcal{O}(\sigma, x).$$

Its correlation functions are obtained by integration:

$$\int d^d y\, W_\mathcal{O}^{(n)}(y; x_1, \ldots, x_n) = \langle \mathcal{V}(\sigma)\, \sigma(x_1) \ldots \sigma(x_n) \rangle_{\text{conn.}}.$$

With reasonable assumptions, but which have to be verified explicitly in examples, from the relation

$$\int d^d y\, W_\mathcal{O}^{(n)}(\lambda y; \lambda x_1, \ldots, \lambda x_n) = \lambda^{-d} \int d^d y\, W_\mathcal{O}^{(n)}(y; \lambda x_1, \ldots, \lambda x_n)$$

one infers, using the asymptotic behaviour (9.63),

$$\langle \mathcal{V}(\sigma) \sigma(\lambda x_1) \ldots \sigma(\lambda x_n) \rangle_{\text{conn.}} \propto \lambda^{d - nd_\sigma - d_\mathcal{O}}.$$

The right-hand side is the variation of the n-point function at first order when one adds a term proportional to $\mathcal{V}$ to the Hamiltonian:

$$\int [d\sigma] \sigma(x_1) \ldots \sigma(x_n)\, e^{-\mathcal{H} - \theta \mathcal{V}} = \int [d\sigma] \sigma(x_1) \ldots \sigma(x_n)\, e^{-\mathcal{H}} \left(1 - \theta \mathcal{V} + O(\theta^2)\right).$$

Comparing this term with the term of order zero, one observes that the perturbation $\mathcal{V}$ has generated an additional factor $\lambda^{d - d_\mathcal{O}}$ in the large-distance behaviour. This factor thus is related to the eigenvalue $\ell_\mathcal{V}$ associated with the eigen-perturbation $\mathcal{V}$. We have thus established a general relation between the dimension of a local eigen-operator and the eigenvalue of the corresponding perturbation (obtained by integrating over space) to the Hamiltonian:

$$\ell_\mathcal{V} + d_\mathcal{O} = d. \tag{9.64}$$

10 Perturbative renormalization group: Explicit calculations

In Chapter 9, we have shown that, at least in the vicinity of dimension 4, with some assumptions on the existence and properties of a renormalization group (RG) and the existence of a dimensional continuation, one can reduce the study of the flow of a general Hamiltonian in the vicinity of the Gaussian fixed point to the flow of the amplitude g of the relevant perturbation: $\int d^d x \, \sigma^4(x)$.

In this chapter, we assume that the analysis of Chapter 9 applies. Then, it is possible to determine the coefficient β_2 of the expansion of the function $\beta(g)$, which governs the flow of g (see equation (9.62)) and whose sign is crucial for the existence of a fixed point. Indeed, we show how β_2 can be inferred from a perturbative calculation of the two- and four-point vertex functions to first order in g beyond the quasi-Gaussian approximation. The calculation involves the Gaussian integration in the steepest descent method (see expression (6.31)) and thus corresponds to one-loop Feynman diagrams.

The value of β_2 being found to be positive in dimension $d = 4$, this implies the existence of a non-Gaussian fixed point for dimensions $\varepsilon = 4 - d \to 0$, as we have explained in Section 9.7. Having determined the fixed point, we then show how to determine the correlation exponent ν and the field dimension at leading order for $\varepsilon \to 0$ from a two-loop calculation of the two-point function.

In Chapter 13, we justify more generally that the perturbative expansion combined with the dimensional continuation, allows an extension of these calculations to all orders in g and in ε.

In the last part of the chapter, we generalize one-loop calculations to models with an $O(N)$ orthogonal symmetry and we conclude the chapter by displaying some numerical results derived from higher order calculations.

10.1 Critical Hamiltonian and perturbative expansion

We consider correlation functions of a real field $\sigma(x)$, $x \in \mathbb{R}^d$, associated to the measure $e^{-\mathcal{H}(\sigma)}/\mathcal{Z}$ where

$$\mathcal{Z} = \int [d\sigma] \exp[-\mathcal{H}(\sigma)]$$

and the critical Hamiltonian $\mathcal{H}$ involves only the relevant perturbation to the Gaussian model in dimension $d = 4 - \varepsilon$, $0 < \varepsilon \ll 1$:

$$\mathcal{H}(\sigma) = \mathcal{H}_G(\sigma) + \int d^d x \left(\frac{1}{2} r_c(g) \sigma^2(x) + \frac{1}{4!} g \sigma^4(x) \right) \qquad (10.1)$$

with

$$\mathcal{H}_G(\sigma) = \frac{1}{2}\int d^d x \left(\sum_\mu (\partial_\mu \sigma(x))^2 + \sum_{k=1} u_{k+1} \sum_\mu \partial_\mu \sigma(x)(-\nabla_x^2)^k \partial_\mu \sigma(x)\right). \quad (10.2)$$

The parameter g characterizes the amplitude of the perturbation to the Gaussian model. The parameter

$$r_c(g) = \sum_{\ell=1}(r_c)_\ell g^\ell,$$

is determined, order by order in g, by the condition that $\tilde{\Gamma}^{(2)}$, the Fourier transform of the two-point vertex function, satisfies

$$\tilde{\Gamma}^{(2)}(p=0, g) = 0 \quad (10.3)$$

and, thus, that the model remains critical for all values of g. The introduction of the parameter r_c, called mass renormalization in quantum field theory, is required since, as we have already pointed out, the quartic perturbation modifies the critical temperature.

Finally, the Hamiltonian (10.2) differs from the Hamiltonian of the Gaussian fixed point (9.45) by the addition of terms quadratic in σ with more than two derivatives of the type $\sigma(x)\nabla^{2k}\sigma(x)$, with arbitrary coefficients u_k, which are irrelevant at large distance. In the continuum, they provide a necessary substitute to the lattice structure that bounds the arguments of the field Fourier components to a Brillouin zone.

In terms of the Fourier components of the field σ,

$$\sigma(x) = \int d^d p\, e^{ipx}\, \tilde{\sigma}(p),$$

$\mathcal{H}_G$ takes the form

$$\mathcal{H}_G(\tilde{\sigma}) = \frac{1}{2}(2\pi)^d \int d^d p\, \tilde{\sigma}(p)\tilde{\mathcal{H}}^{(2)}(p)\tilde{\sigma}(-p) \quad (10.4)$$

with

$$\tilde{\mathcal{H}}^{(2)}(p) = p^2 + \sum_{k=2} u_k p^{2k}. \quad (10.5)$$

A necessary condition for the field integral to exist in any dimension $d > 2$, is that the function $\tilde{\mathcal{H}}^{(2)}(p)$ is positive for all $p \neq 0$.

Short-distance or large-momentum behaviour: 'Ultraviolet' (UV) divergences. Let us add a few comments justifying the introduction of the irrelevant quadratic terms. As we have already pointed out at the end of Section 8.4.3, the propagator of the Gaussian fixed point,

$$\Delta(x) = \frac{1}{(2\pi)^d}\int d^d p\, \frac{e^{-ipx}}{p^2},$$

has short-distance singularities since
$$\langle \sigma(x)\sigma(y)\rangle_{\text{gaus.}} \underset{|x-y|\to 0}{\propto} |x-y|^{2-d}.$$
In the Fourier representation, this difficulty translates into large momenta divergences (also called ultraviolet for historical reasons related to quantum field theory):
$$\langle \sigma(x)\sigma(x)\rangle_{\text{gaus.}} = \frac{1}{(2\pi)^d}\int \frac{d^d p}{p^2}.$$
This problem is generated by the transition to the continuum since, on the lattice, distinct points are separated by at least one lattice spacing and, in the Fourier representation, momenta belong to a finite Brillouin zone.

Expectation values of local monomials in the fields can only be defined, at least perturbatively, if the limits of correlation functions of products of fields at coinciding points exist (see the discussion in Section 9.8). This implies that $\Delta(x-y)$, the Gaussian two-point function also called the *propagator*, must be sufficiently regular at short distance and, thus, its Fourier transform $\tilde{\Delta}(p)$ must decrease sufficiently fast at large momenta. The modified propagator must cut the contributions of large momenta, introducing a cut-off scale.

The necessity of introducing such modified propagators is another indication of the non-decoupling of physics on very different distance scales. It is impossible to ignore the microscopic scale and to use directly a propagator that has the asymptotic scaling form.

The quadratic Hamiltonian (10.2) corresponds to a two-point function of the form
$$\Delta(x) = \frac{1}{(2\pi)^d}\int d^d p\, e^{-ipx}\, \tilde{\Delta}(p) \qquad (10.6a)$$
$$\tilde{\Delta}^{-1}(p) = \tilde{\mathcal{H}}^{(2)}(p) = p^2 + \sum_{k=2} u_k p^{2k}. \qquad (10.6b)$$
For the perturbative calculations that follow, a propagator $\tilde{\Delta}(p)$ decaying like $1/p^6$ suffices. If one wants to construct an exact perturbative RG, it becomes necessary to choose Gaussian two-point functions regular for $p \neq 0$ real and decreasing faster than any power for $|p|\to\infty$ (see Section 16.2).

Perturbative, asymptotic RG. We now use the analysis of Section 9.7. Moreover, we express the RG equations (RGE) in the form (9.14), that is, in terms of Fourier transforms $\tilde{\Gamma}^{(n)}(p_1,\ldots,p_n)$ of vertex functions. We thus consider equation (9.14), in which we neglect in the left-hand side any dependence other than in the coefficient $g(\lambda)$ of σ^4. In the vicinity of the Gaussian fixed point and in the logic of the ε-expansion, this amounts to neglecting the contributions to the asymptotic behaviour of perturbations irrelevant relative to the Gaussian fixed point. These contributions decay faster, relative to the leading term, by at least a factor $1/\lambda^2$ up to powers of $\ln \lambda$, to any finite order in g and ε. Equation (9.14) then simplifies and becomes
$$\tilde{\Gamma}^{(n)}_\lambda(p_1,\ldots,p_n) \sim \tilde{\Gamma}^{(n)}(p_1,\ldots,p_n;g(\lambda))$$
$$\underset{|p_i|\to 0}{\sim} Z^{n/2}(\lambda)\lambda^d \tilde{\Gamma}^{(n)}(p_1/\lambda,\ldots,p_n/\lambda;g), \qquad (10.7)$$

where this equation is valid only in the asymptotic regime $p_i \to 0$ after a first dilatation $p_i \mapsto \lambda_1 p_i$, $\lambda_1 \gg 1$, which has eliminated the irrelevant terms (see the discussion of Section 9.7). Moreover (equation (9.37)),

$$-\lambda \frac{d \ln Z}{d\lambda} = 2 D_\sigma(g(\lambda)). \tag{10.8}$$

10.2 Feynman diagrams at one-loop order

In Exercise 6.1, we have derived the first corrections to the leading terms in the steepest descent method contributing to vertex functions (one-loop 1-irreducible diagrams) from the contribution (6.31) to the generating function. All quantities have been calculated on the lattice and we now describe the continuum version. In particular, expression (6.31) then takes the form (see also Section A1.4 and equation (12.26))

$$\Gamma_1(M) = \frac{1}{2} \operatorname{tr} \ln \frac{\delta^2 \mathcal{H}}{\delta M(x_1) \delta M(x_2)}. \tag{10.9}$$

We write the Hamiltonian symbolically as

$$\mathcal{H}(\sigma) = \frac{1}{2} \int d^d x \, d^d y \, \sigma(x) \mathcal{H}^{(2)}(x-y) \sigma(y) + \frac{g}{4!} \int d^d x \, \sigma^4(x).$$

Then,

$$\frac{\delta^2 \mathcal{H}}{\delta \sigma(x_1) \delta \sigma(x_2)} = \mathcal{H}^{(2)}(x_1 - x_2) + \tfrac{1}{2} g \sigma^2(x_2) \delta^{(d)}(x_1 - x_2)$$

$$= \int d^d y \, \mathcal{H}^{(2)}(x_1 - y) \left[\delta^{(d)}(y - x_2) + \tfrac{1}{2} g \Delta(y - x_2) \sigma^2(x_2) \right],$$

where Δ is the inverse of $\mathcal{H}^{(2)}$ in the operator sense (equation (10.6)):

$$\int d^d y \, \mathcal{H}^{(2)}(x_1 - y) \Delta(y - x_2) = \delta^{(d)}(x_1 - x_2).$$

Thus,

$$\det \frac{\delta^2 \mathcal{H}}{\delta \sigma(x_1) \delta \sigma(x_2)} = \det \mathcal{H}^{(2)} \det(1 + \tfrac{1}{2} g \mathcal{K}),$$

where $\mathcal{K}$ is the operator associated to the kernel

$$\mathcal{K}(x_1, x_2) = \Delta(x_1 - x_2) \sigma^2(x_2).$$

Using the relation $\ln \det = \operatorname{tr} \ln$ (see Section A1.4) and expanding the logarithm, one infers

$$\ln \det(1 + \tfrac{1}{2} g \mathcal{K}) = \operatorname{tr} \ln(1 + \tfrac{1}{2} g \mathcal{K}) = -\sum_{n=1}^{\infty} \frac{1}{n} \left(-\frac{g}{2}\right)^n \operatorname{tr} \mathcal{K}^n.$$

More explicitly,
$$\operatorname{tr} \mathcal{K} = \int d^d x\, \mathcal{K}(x,x) = \Delta(0) \int d^d x\, \sigma^2(x),$$
$$\operatorname{tr} \mathcal{K}^2 = \int d^d x\, d^d y\, \mathcal{K}(x,y)\mathcal{K}(y,x) = \int d^d x\, d^d y\, \sigma^2(x)\Delta^2(x-y)\sigma^2(y).$$

Substituting $\sigma \mapsto M$, one concludes
$$\Gamma_1(M) - \Gamma_1(0) = \tfrac{1}{4}g\Delta(0) \int d^d x\, M^2(x)$$
$$- \tfrac{1}{16}g^2 \int d^d x\, d^d y\, M^2(x)\Delta^2(x-y)M^2(y) + O(g^3).$$

From this expression, one infers the one-loop contributions to the two- and four--point functions by successive functional differentiations with respect to $M(x)$. For example, setting
$$G_4 = \int d^d x\, d^d y\, M^2(x)\Delta^2(x-y)M^2(y),$$
one obtains successively,
$$\frac{\delta G_4}{\delta M(x_1)} = 4 \int d^d y\, M(x_1)\Delta^2(x_1-y)M^2(y),$$
$$\frac{\delta^2 G_4}{\delta M(x_1)\delta M(x_2)} = 8M(x_1)\Delta^2(x_1-x_2)M(x_2)$$
$$+ 4\delta^{(d)}(x_1-x_2) \int d^d y\, \Delta^2(x_1-y)M^2(y),$$
$$\frac{\delta^3 G_4}{\delta M(x_1)\delta M(x_2)\delta M(x_3)} = 8\Delta^2(x_1-x_2)M(x_1)\delta^{(d)}(x_2-x_3) + \text{two terms}$$
that are obtained by permuting x_1, x_2, x_3.

Completing the calculation, one finds
$$\Gamma_1^{(2)}(x_1,x_2) = \left.\frac{\delta^2\Gamma_1}{\delta M(x_1)\delta M(x_2)}\right|_{M=0} = \tfrac{1}{2}g\Delta(0)\delta^{(d)}(x_1-x_2),$$
$$\Gamma_1^{(4)}(x_1,\ldots,x_4) = -\tfrac{1}{2}g^2\left[\Delta^2(x_1-x_2)\delta^{(d)}(x_1-x_3)\delta^{(d)}(x_2-x_4) + 2\text{ terms}\right].$$

After Fourier transformation (see also equation (6.38a)),
$$\tilde{\Gamma}_1^{(2)}(p) = \int d^d x\, e^{ipx}\, \Gamma_1^{(2)}(x,0) = \tfrac{1}{2}g\Delta(0) = \tfrac{1}{2}g \int d^d p\, \tilde{\Delta}(p), \qquad (10.10)$$

where we have inverted the relation (10.6a). Similarly (see equations (6.14) and (6.38b)),
$$\tilde{\Gamma}_1^{(4)}(p_i) = \int \prod_{i=1}^{3} d^d x_i\, e^{i(p_1 x_1 + p_2 x_2 + p_3 x_3)}\, \Gamma_1^{(4)}(x_1,x_2,x_3,0)$$
$$= -\tfrac{1}{2}g^2 \int d^d x \left[e^{i(p_1+p_3)x} + e^{i(p_1+p_4)x} + e^{i(p_1+p_2)x}\right]\Delta^2(x)$$
$$= -\tfrac{1}{2}g^2\left[B(p_1+p_2) + B(p_1+p_3) + B(p_1+p_4)\right] \qquad (10.11)$$

with the definition

$$B(p) = \frac{1}{(2\pi)^d} \int d^d q\, \tilde{\Delta}(q) \tilde{\Delta}(p-q).$$

Dimensional continuation. Dimensional continuation can be defined using Gaussian integrals (see Section 12.5). Below we need only the property

$$\int d^d x\, e^{-x^2} = \pi^{d/2}.$$

Then, introducing radial and angular variables, one finds

$$\int d^d x\, e^{-x^2} = A_d \int_0^\infty dr\, r^{d-1} e^{-r^2} = \tfrac{1}{2} A_d \Gamma(d/2),$$

from which one infers the surface of the sphere S_{d-1}: $A_d = 2\pi^{d/2}/\Gamma(d/2)$.

10.3 Fixed point and critical behaviour

Since only two unknown functions have to be determined, $\beta(g)$ governing the flow of $g(\lambda)$ and

$$\eta(g) = 2D_\sigma(g) - d + 2$$

(introduced in (9.60)) related to $Z(\lambda)$, it suffices to calculate two vertex functions. It is especially convenient to choose the two- and four-point functions. More details concerning perturbative calculations can be found in Chapter 12.

10.3.1 *The two-point function*

The calculation of the vertex two-point function of Section 8.9 remains valid (see equation (8.74)), provided one uses the propagator $\tilde{\Delta}(q)$ of the continuum and one integrates freely over the Fourier variables (momenta). Another form of the calculation on the lattice has been presented in Section 6.5 (equation (6.38a)). Directly in the continuum, the first correction to the Gaussian form has been calculated in Section 10.2, and is given by equation (10.10). The function at order g follows:

$$\tilde{\Gamma}^{(2)}(p; g) = \tilde{\mathcal{H}}^{(2)}(p) + r_c(g) + \frac{g}{2}\frac{1}{(2\pi)^d}\int d^d q\, \tilde{\Delta}(q) + O(g^2)$$

$$= p^2 + O(p^4) + g(r_c)_1 + \frac{g}{2}\frac{1}{(2\pi)^d}\int d^d q\, \tilde{\Delta}(q) + O(g^2). \quad (10.12)$$

The first correction to the vertex two-point function $\tilde{\Gamma}^{(2)}$ is a constant. The criticality condition (10.3) determines the value of $(r_c)_1$:

$$(r_c)_1 = -\frac{1}{2(2\pi)^d}\int d^d q\, \tilde{\Delta}(q). \quad (10.13)$$

The two-point function becomes
$$\tilde{\Gamma}^{(2)}(p;g) = p^2 + O(p^4) + O(g^2)$$
and, thus, keeps the Gaussian behaviour for $p \to 0$ at order g. Using, for $p \to 0$, the RGE (10.7) for $n=2$,
$$\tilde{\Gamma}^{(2)}(p;g(\lambda)) \sim Z(\lambda)\lambda^d \tilde{\Gamma}^{(2)}(p/\lambda;g),$$
which yields at this order
$$p^2 = Z(\lambda)\lambda^d p^2/\lambda^2 + O(g^2),$$
one finds that the renormalization $Z(\lambda)$ keeps its Gaussian form,
$$Z(\lambda) = \lambda^{2-d} + O(g^2),$$
and, thus, using equation (10.8),
$$2D_\sigma(g) = d - 2 + \eta(g) = -\lambda \frac{\mathrm{d}}{\mathrm{d}\lambda} \ln Z(\lambda) = d - 2 + O(g^2).$$
The function $\eta(g)$ is at least of order g^2.

10.3.2 The four-point function

The algebraic part of the calculations leading to the first correction to the quasi-Gaussian, or mean-field, approximation, of the four-point vertex function has been presented in Section 6.5 on the lattice (equation (6.38b)) and in Section 10.2 directly in the continuum (equation (10.11)). In the Fourier representation,
$$\tilde{\Gamma}^{(4)}(p_1, p_2, p_3, p_4) = g - \tfrac{1}{2}g^2 \big(B(p_1+p_2) + B(p_1+p_3) + B(p_1+p_4)\big) + O(g^3) \quad (10.14)$$
with
$$B(p) = \frac{1}{(2\pi)^d} \int d^d q\, \tilde{\Delta}(q)\tilde{\Delta}(p-q). \tag{10.15}$$
This expression can also be obtained by expanding the continuum version of equation (6.31) (see Section 6.5).

At T_c for $q \to 0$, $\tilde{\Delta}(q)$, which is the Fourier transform of the Gaussian two-point function, behaves like $1/q^2$. Therefore, for $d \le 4$, the function $B(p)$ diverges for $p \to 0$.

In Section 10.3.3, we need the quantity $\lambda \partial B(p/\lambda)/\partial \lambda$ for $\lambda \to +\infty$. We thus decompose
$$B(p/\lambda) = B_+(p/\lambda) + B_-(p/\lambda)$$
with
$$B_+(p/\lambda) = \frac{1}{(2\pi)^d} \int_{|q|>\delta} d^d q\, \tilde{\Delta}(q)\tilde{\Delta}(q+p/\lambda),$$
$$B_-(p/\lambda) = \frac{1}{(2\pi)^d} \int_{|q|\le\delta} d^d q\, \tilde{\Delta}(q)\tilde{\Delta}(q+p/\lambda),$$

where δ is fixed, but can be chosen arbitrarily small.

For B_+, the integrand is analytic in the integration domain, the integral can be expanded in powers of $1/\lambda$. The leading term is a constant and thus

$$\lim_{\lambda \to +\infty} \lambda \frac{\partial B_+(p/\lambda)}{\partial \lambda} = 0.$$

For B_-, one can replace $\tilde{\Delta}(q)$ by its asymptotic form for $|q| \to 0$,

$$\tilde{\Delta}(q) \sim 1/q^2,$$

and thus

$$B_-(p/\lambda) \sim \frac{1}{(2\pi)^d} \int_{|q| \le \delta} \frac{d^d q}{q^2 (q + p/\lambda)^2}.$$

After the change of variables $q \mapsto q' = \lambda q$, one obtains

$$B_-(p/\lambda) \sim \lambda^\varepsilon \frac{1}{(2\pi)^d} \int_{|q| \le \lambda \delta} \frac{d^d q}{q^2 (q + p)^2},$$

where we have set

$$\varepsilon = 4 - d.$$

Then,

$$\lambda \frac{\partial B_-(p/\lambda)}{\partial \lambda} \sim \varepsilon \lambda^\varepsilon \frac{1}{(2\pi)^d} \int_{|q| \le \lambda \delta} \frac{d^d q}{q^2 (q + p)^2} + \lambda^\varepsilon \lambda \frac{\partial}{\partial \lambda} \frac{1}{(2\pi)^d} \int_{|q| \le \lambda \delta} \frac{d^d q}{q^2 (q + p)^2}. \tag{10.16}$$

In the second term, the λ-dependence appears only in the bound on the integration variables. For $\lambda \to +\infty$,

$$\lambda \frac{\partial}{\partial \lambda} \int_{|q| \le \lambda \delta} \frac{d^d q}{q^2 (q + p)^2} \sim \lambda \frac{\partial}{\partial \lambda} \int_{1 < |q| \le \lambda \delta} \frac{d^d q}{q^4}$$

$$\sim A_d \lambda \frac{\partial}{\partial \lambda} \int_{1 < q \le \lambda \delta} dq\, q^{d-3} \sim \frac{A_d}{(\lambda \delta)^\varepsilon}$$

where we have introduced radial and angular variables. The integral over angular variables yields the measure A_d of the surface of the sphere S_{d-1} in d dimensions which, by the rules of dimensional continuation (see Section 12.5.1), is given by

$$A_d = \frac{2\pi^{d/2}}{\Gamma(d/2)} = 2\pi^2 + O(\varepsilon).$$

In the limit $\varepsilon \to 0$, the first term in (10.16) goes to zero and the second term yields

$$\lim_{\varepsilon \to 0} \lim_{\lambda \to +\infty} \lambda \frac{\partial B_-(p/\lambda)}{\partial \lambda} = \frac{1}{8\pi^2}.$$

Therefore,

$$\lim_{\lambda \to +\infty} \lambda \frac{\partial B(p/\lambda)}{\partial \lambda} \underset{\varepsilon \to 0}{=} \frac{1}{8\pi^2} + O(\varepsilon), \tag{10.17}$$

a result independent of the explicit form of $\tilde{\Delta}$.

10.3.3 Fixed point

We now apply to the four-point function the relation (10.7) between vertex functions corresponding to the initial Hamiltonian and to the effective Hamiltonian at scale λ:

$$\tilde{\Gamma}^{(4)}(p_1,\ldots,p_4;g(\lambda)) \sim Z^2(\lambda)\lambda^d \tilde{\Gamma}^{(4)}(p_1/\lambda,\ldots,p_4/\lambda;g), \qquad (10.18)$$

where the expansion of $\tilde{\Gamma}^{(4)}$ is given by expression (10.14):

$$\tilde{\Gamma}^{(4)}(p_1,p_2,p_3,p_4;g) = g - \tfrac{1}{2}g^2\left[B(p_1+p_2) + B(p_1+p_3) + B(p_1+p_4)\right] + O(g^3).$$

As we have seen (Section 10.3.1), $Z(\lambda) = \lambda^{2-d} + O(g^2)$ and thus

$$\tilde{\Gamma}^{(4)}(p_1,\ldots,p_4;g(\lambda)) \sim \lambda^\varepsilon \tilde{\Gamma}^{(4)}(p_1/\lambda,\ldots,p_4/\lambda,g) + O(g^3), \qquad (10.19)$$

($\varepsilon = 4-d$) where the neglected term is of order g^3 since the function $\tilde{\Gamma}^{(4)}$ is of order g.

Differentiating both members with respect to λ, one finds

$$\lambda\frac{dg}{d\lambda}\frac{\partial}{\partial g}\tilde{\Gamma}^{(4)}(p_1,\ldots,p_4;g(\lambda)) = \varepsilon\lambda^\varepsilon \tilde{\Gamma}^{(4)}(p_1/\lambda,\ldots,p_4/\lambda,g)$$
$$+ \lambda^\varepsilon \lambda\frac{d}{d\lambda}\tilde{\Gamma}^{(4)}(p_1/\lambda,\ldots,p_4/\lambda,g) + O(g^3).$$

One then expands both members in g and ε with $g = O(\varepsilon)$. In the right-hand side, the first term is of order εg and the second of order g^2. Since $\partial\tilde{\Gamma}^{(4)}/\partial g$ is of order 1, $\lambda dg/d\lambda$ is of order g^2 or εg. One can thus omit all higher order terms. At leading order,

$$\lambda\frac{dg}{d\lambda} = \varepsilon g - \tfrac{1}{2}g^2 \lambda\frac{d}{d\lambda}\left[B((p_1+p_2)/\lambda) + B((p_1+p_3)/\lambda) + B((p_1+p_4)/\lambda)\right]$$
$$+ O(g^3, g^2\varepsilon).$$

Then, using the evaluation (10.17), one finds

$$\lambda\frac{dg}{d\lambda} = \varepsilon g - \frac{3}{16\pi^2}g^2 + O(g^3, g^2\varepsilon).$$

The possible dependence in the variables p_i, thus, has disappeared.

At leading order, the equation implies that the derivative of $g(\lambda)$ vanishes and thus

$$g(\lambda) = g(1) = g + O(g^2, g\varepsilon).$$

One infers

$$\lambda\frac{dg(\lambda)}{d\lambda} = \varepsilon g(\lambda) - \frac{3}{16\pi^2}g^2(\lambda) + O(g^3(\lambda), g^2(\lambda)\varepsilon).$$

In particular, one verifies that the right-hand side depends on λ only through $g(\lambda)$, a result consistent with the assumption (9.23).

Introducing the traditional notation (9.62), one defines the function
$$\beta(g(\lambda)) = -\lambda \frac{dg(\lambda)}{d\lambda}. \tag{10.20}$$

Then,
$$\beta(g) = -\varepsilon g + \frac{3}{16\pi^2} g^2 + O(g^3, g^2\varepsilon). \tag{10.21}$$

It is thus the case (iii) envisaged in Section 9.7.1 that is realized here: the coefficient of g^2 is positive. For $d < 4$, there exists a non-Gaussian stable fixed point, corresponding to the zero of order ε of the function $\beta(g)$:
$$g^* = \frac{16\pi^2 \varepsilon}{3} + O(\varepsilon^2). \tag{10.22}$$

The existence of this fixed point can be confirmed to all orders in ε, and allows a proof of universality properties of correlation and thermodynamic functions. Universal quantities can be calculated in the form of an expansion in powers of ε inferred from the perturbative expansion around the Gaussian model, following the method introduced by Wilson and Fisher (see Chapter 13).

10.3.4 The field dimension to order ε^2

The first correction to the dimension d_σ of the field σ requires a calculation of the two-point function at order g^2 (two loops in the sense of Feynman diagrams, Figure 2.6 of Section 2.5.1). The two-point function at next order on the lattice is then given by expression (6.38a). In the notation of the chapter and adjusting $r_c(g)$ at order g^2, one finds (see also expression (12.18))
$$\tilde{\Gamma}^{(2)}(p; g) = \tilde{\mathcal{H}}^{(2)}(p) - \tfrac{1}{6} g^2 [C(p) - C(0)] + O(g^3),$$

where $C(p)$, which corresponds to diagram (e) of Figure 2.6, is given by equation (6.39c):
$$C(p) = \frac{1}{(2\pi)^{2d}} \int d^d k_1 \, d^d k_2 \, \tilde{\Delta}(k_1) \tilde{\Delta}(k_2) \tilde{\Delta}(p - k_1 - k_2).$$

It is then necessary to evaluate $C(p) - C(0)$ for $p \to 0$ and $d \to 4$. One finds (cf. equation (12.38))
$$C(p) - C(0) = -K p^2 \ln(\Lambda/p) + o(1) \quad \text{with} \quad K = \frac{1}{4(8\pi^2)^2},$$

where the constant Λ depends on the specific form of Δ.

The RGE (10.7) for $n = 2$, in the limit $p \to 0$, reads at this order
$$p^2 + \tfrac{1}{6} K g^2(\lambda) p^2 \ln(\Lambda/p) = Z(\lambda) \lambda^d \left[p^2/\lambda^2 + \tfrac{1}{6} K g^2 \ln(\Lambda\lambda/p) p^2/\lambda^2 \right].$$

One infers
$$Z(\lambda) = \lambda^{2-d} \left(1 - \tfrac{1}{6} K g^2 \ln \lambda \right) + O(g^3),$$

and, thus,
$$\lambda \frac{\partial \ln Z(\lambda)}{\partial \lambda} = 2 - d - \tfrac{1}{6} K g^2(\lambda) + O(g^3).$$

Quite generally, we have defined (equations (9.24) and (9.60))
$$\lambda \frac{\partial \ln Z(\lambda)}{\partial \lambda} = -2 D_\sigma(g(\lambda)) = 2 - d - \eta(g(\lambda)).$$

Thus,
$$\eta(g) = \frac{1}{6} \frac{1}{(4\pi)^4} g^2 + O(g^3).$$

Field dimension: Correction to the Gaussian value. In the limit $\lambda \to \infty$, $g(\lambda)$ tends toward g^*. The deviation from the Gaussian value is then characterized by the value of the exponent
$$\eta \equiv \eta(g^*) = \tfrac{1}{54} \varepsilon^2 + O(\varepsilon^3).$$

One infers the asymptotic behaviour
$$Z(\lambda) \propto \lambda^{-2 d_\sigma}$$

with
$$d_\sigma \equiv D_\sigma(g^*) = \tfrac{1}{2}(d - 2 + \eta) = 1 - \tfrac{1}{2}\varepsilon + \tfrac{1}{108}\varepsilon^2 + O(\varepsilon^3).$$

The application of this result to relation (10.7) determines the large-distance behaviour of vertex functions:
$$\tilde{\Gamma}^{(n)}(p_1/\lambda, \ldots, p_n/\lambda; g) \sim \lambda^{-d + n d_\sigma} \tilde{\Gamma}^{(n)}(p_1, \ldots, p_n; g^*).$$

In the special case of the two-point function, one obtains
$$\tilde{\Gamma}^{(2)}(p/\lambda; g) \sim \lambda^{-2+\eta} \tilde{\Gamma}^{(2)}(p; g^*) \Rightarrow \tilde{\Gamma}^{(2)}(p; g) \underset{|p| \to 0}{\propto} p^{2-\eta}, \qquad (10.23)$$

which is consistent with the behaviour of the connected correlation function (see equation (9.21))
$$W^{(2)}(x) = \frac{1}{(2\pi)^d} \int d^d p \, e^{-ipx} \left[\tilde{\Gamma}^{(2)}(p; g) \right]^{-1} \underset{|x| \to \infty}{\propto} \int d^d p \, e^{-ipx} \, p^{\eta - 2} \propto \frac{1}{|x|^{2 d_\sigma}}. \qquad (10.24)$$

10.4 Critical domain

In the general RG discussion, we have emphasized that if fixed points exist, universality properties are expected not only for the critical Hamiltonian but also for critical Hamiltonians infinitesimally perturbed by relevant operators. This defines the critical domain.

In the ferromagnetic systems that we study here, the two relevant operators correspond to $\int \sigma(x) d^d x$ that is coupled to the magnetic field and breaks the reflection symmetry $\sigma \mapsto -\sigma$, and to the operator $\int \sigma^2(x) d^d x$ that is coupled to the temperature. We consider here the latter perturbation and add to the critical Hamiltonian (10.1) a term proportional to $\int \sigma^2(x) d^d x$:

$$\mathcal{H}_t(\sigma) = \mathcal{H}(\sigma) + \frac{t}{2} \int \sigma^2(x) d^d x, \qquad (10.25)$$

where the parameter t characterizes the deviation from the critical temperature. We choose t infinitesimal in a sense that we have already explained: its initial value is adjusted in such a way that, after a large dilatation λ, its effective value $t(\lambda)$ is of order 1. In what follows, we choose $t \propto T - T_c$ positive, which corresponds to the high-temperature phase.

10.4.1 Two-point function

The Fourier transform of the Gaussian two-point function $\tilde{\Delta}(q,t)$ is now given by

$$1/\tilde{\Delta}(q,t) = t + \tilde{\mathcal{H}}^{(2)}(q) = t + q^2 + O(q^4),$$

where $\tilde{\Delta}(q,0) \equiv \tilde{\Delta}(q)$, the function defined by equation (10.6b).

At order g, the calculation is almost identical, up to the parametrization, to the calculation of Section 8.9. The vertex two-point function becomes

$$\tilde{\Gamma}^{(2)}(p) = t + p^2 + O(p^4) + \frac{g}{2(2\pi)^d} \int d^d q \left(\tilde{\Delta}(q,t) - \tilde{\Delta}(q,0) \right) + O(g^2),$$

where $(r_c)_1$ has been replaced by its value given by equation (10.13).

The useful quantity is the derivative (see Section 8.9.2)

$$\frac{\partial \tilde{\Gamma}^{(2)}(p)}{\partial t} = 1 - \frac{g}{2} \int \frac{d^d q}{(2\pi)^d} \tilde{\Delta}^2(q,t) + O(g^2). \qquad (10.26)$$

Also, in what follows, we need the function

$$F(t,\lambda) = \int \frac{d^d q}{(2\pi)^d} \left[\tilde{\Delta}^2(q,t) - \tilde{\Delta}^2(q,\lambda^2 t) \right]$$

with $\lambda > 0$, in the limit $t \to 0$, $d \to 4$.

In the integral, we change variables $q = q'\sqrt{t}$ and let t go to zero. Then,

$$1/\tilde{\Delta}(q\sqrt{t},t) = t(1+q^2) + O(t^2).$$

The integral obtained by retaining only the leading term converges for $|q| \to \infty$ and thus one can take the limit $t \to 0$ in the integrand:

$$F(t,\lambda) \sim t^{-\varepsilon/2} \int \frac{d^d q}{(2\pi)^d} \left[\frac{1}{(q^2+1)^2} - \frac{1}{(q^2+\lambda^2)^2} \right].$$

This expression has the limit, for $d \to 4$,

$$F(t,\lambda) \sim \int \frac{d^4 q}{(2\pi)^4} \left[\frac{1}{(q^2+1)^2} - \frac{1}{(q^2+\lambda^2)^2} \right]$$

$$= \int_0^\infty \frac{q^3 dq}{8\pi^2} \left[\frac{1}{(q^2+1)^2} - \frac{1}{(q^2+\lambda^2)^2} \right] = \frac{1}{8\pi^2} \ln \lambda.$$

We conclude

$$\int \frac{d^4 q}{(2\pi)^4} \left[\tilde{\Delta}^2(q,t) - \tilde{\Delta}^2(q,t\lambda^2) \right] \underset{t \to 0}{=} \frac{1}{8\pi^2} \ln \lambda. \tag{10.27}$$

10.4.2 Renormalization group

At leading order in ε, one must now take into account the flow of g and t and thus equation (9.15) becomes

$$\tilde{\Gamma}_\lambda^{(2)}(p) \equiv \tilde{\Gamma}^{(2)}(p;t(\lambda);g(\lambda)) = Z(\lambda) \lambda^d \tilde{\Gamma}^{(2)}(p/\lambda;t;g). \tag{10.28}$$

At order g, $Z(\lambda)$ has the Gaussian form and the RGE reduces to

$$\tilde{\Gamma}^{(2)}(p;t(\lambda);g(\lambda)) = \lambda^2 \tilde{\Gamma}^{(2)}(p/\lambda;t;g) + O(g^2). \tag{10.29}$$

At order g^0 (the Gaussian approximation),

$$t(\lambda) = t\lambda^2.$$

This result agrees with the analysis of Section 9.6.

Let us point out that, in agreement with the general analysis of Section 9.4.4, we assume here that t goes to zero before λ goes to infinity in such a way that $t(\lambda)$ goes to zero.

Differentiating equation (10.29) with respect to t, one finds

$$\frac{\partial t(\lambda)}{\partial t} \partial_t \tilde{\Gamma}^{(2)}(p;t(\lambda);g(\lambda)) = \lambda^2 \partial_t \tilde{\Gamma}^{(2)}(p/\lambda;t;g) + O(g^2).$$

At order g, for $p = 0$, the relation becomes (equation (10.26))

$$\frac{\partial t(\lambda)}{\partial t} \left[1 - \frac{g(\lambda)}{2} \int \frac{d^d q}{(2\pi)^d} \tilde{\Delta}^2(q,t\lambda^2) \right] = \lambda^2 \left[1 - \frac{g}{2} \int \frac{d^d q}{(2\pi)^d} \tilde{\Delta}^2(q,t) \right] + O(g^2).$$

From this equation, one infers $\partial t(\lambda)/\partial t$. Since at this order $g(\lambda) = g$, one obtains

$$\frac{\partial t(\lambda)}{\partial t} = \lambda^2 \left\{ 1 - \frac{g}{2} \int \frac{\mathrm{d}^d q}{(2\pi)^d} \left[\tilde{\Delta}^2(q,t) - \tilde{\Delta}^2(q, t\lambda^2) \right] \right\} + O(g^2).$$

Using equation (10.27), one finds

$$\frac{\partial t(\lambda)}{\partial t} = \lambda^2 - \frac{g}{16\pi^2} \lambda^2 \ln \lambda + O(g^2, g\varepsilon).$$

In the right-hand side the dependence on t has cancelled. Integrating, one obtains

$$t(\lambda) = \lambda^2 t \left(1 - \frac{g}{16\pi^2} \ln \lambda \right) + O(g^2, g\varepsilon).$$

In differential form, the flow equation can be written as

$$\lambda \frac{\mathrm{d}}{\mathrm{d}\lambda} t(\lambda) = t(\lambda) \left(2 - \frac{g(\lambda)}{16\pi^2} + O(g^2(\lambda)) \right).$$

In this form, the right-hand side thus depends on λ only through the functions $g(\lambda)$ and $t(\lambda)$, a result consistent with the assumption (9.23).

Note that the flow equation is linear in t because t is the coefficient of a relevant perturbation and we consider only infinitesimal values of t.

Introducing the traditional notation

$$\lambda \frac{\mathrm{d}}{\mathrm{d}\lambda} t(\lambda) = t(\lambda) \left[2 + \eta_2\bigl(g(\lambda)\bigr) \right],$$

one finds

$$\eta_2(g) = -\frac{g}{16\pi^2} + O(g^2). \tag{10.30}$$

For $\lambda \to \infty$, $g(\lambda)$ tends toward g^* and thus

$$t(\lambda) \propto t\lambda^{d_t}, \tag{10.31}$$

where d_t is the eigenvalue corresponding to the eigen-operator $\int \mathrm{d}^d x\, \sigma^2(x)$ (at leading order) associated with a deviation from the critical temperature:

$$d_t = 2 - \tfrac{1}{3}\varepsilon + O(\varepsilon^2). \tag{10.32}$$

One observes that the result is *universal*, in the sense that it is independent of all parameters of the Hamiltonian like the coefficients u_k in expression (10.2) and g.

From the general relation (9.64), one can derive the dimension of the operator $\sigma^2(x)$:

$$d_{\sigma^2} = d - d_t = 2 - \tfrac{2}{3}\varepsilon + O(\varepsilon^2). \tag{10.33}$$

One notes, in particular,

$$d_{\sigma^2} - 2d_\sigma = \tfrac{1}{3}\varepsilon + O(\varepsilon^2) \neq 0,$$

which confirms that, for a non-Gaussian fixed point, the dimensions of local polynomials in the fields have no simple relation with the dimensions of the fields (see Section 9.8).

10.4.3 Two-point function: Scaling behaviour in the critical domain

The result (10.31) combined with the general expected behaviour (9.20),

$$Z(\lambda) \propto \lambda^{-2d_\sigma} = \lambda^{-(d-2+\eta)},$$

and RGE (10.28) for $\tilde{\Gamma}^{(2)}$, implies

$$\tilde{\Gamma}^{(2)}(p; t(\lambda); g^*) \propto \lambda^{2-\eta} \tilde{\Gamma}^{(2)}(p/\lambda; t; g)$$

or, changing p into λp,

$$\tilde{\Gamma}^{(2)}(p; t; g) \propto \lambda^{\eta-2} \tilde{\Gamma}^{(2)}(\lambda p; t(\lambda); g^*).$$

Choosing the dilatation parameter such that

$$t(\lambda) = 1 \;\Rightarrow\; \lambda \propto t^{-1/d_t} \tag{10.34}$$

(equation (10.31)), in agreement with the strategy explained in Section 9.4.4, one obtains the relation

$$\tilde{\Gamma}^{(2)}(p; t; g) \underset{t\to 0;\,|p|\to 0}{\propto} t^\gamma F^{(2)}(pt^{-\nu}), \tag{10.35}$$

where

$$F^{(2)}(p) = \tilde{\Gamma}^{(2)}(\text{const.} \times p; 1; g^*),$$

which involves the exponents ν and γ:

$$\nu = 1/d_t = \tfrac{1}{2}(1 + \varepsilon/6) + O(\varepsilon^2), \quad \gamma = \nu(2-\eta) = 1 + \tfrac{1}{6}\varepsilon + O(\varepsilon^2).$$

One obtains the remarkable result that a function of three variables p, t, g can be expressed in terms of a function of only one variable. One calls this property *scaling behaviour*.

Since $\tilde{\Gamma}^{(2)}(p=0; t; g)$ is the inverse of the magnetic susceptibility, γ is the magnetic susceptibility exponent defined by (7.35).

Finally, the form (10.35) implies for the connected correlation function

$$W^{(2)}(x, t) = \frac{1}{(2\pi)^d} \int \frac{d^d p \, e^{ipx}}{\tilde{\Gamma}^{(2)}(p; t; g)}$$

$$\underset{|x|\to\infty}{\propto} t^{-\gamma} \int \frac{d^d p \, e^{ipx}}{F^{(2)}(pt^{-\nu})} = t^{\nu d_\sigma} \int \frac{d^d p \, e^{ipxt^\nu}}{F^{(2)}(p)}.$$

It depends on the distance x only through the product xt^ν. The function $t^{-\nu}$ is thus proportional to the correlation length $\xi(t)$:

$$\xi(t) \propto t^{-\nu}.$$

According to the definition (8.26), this divergence of the correlation length implies that ν is the correlation exponent.

In particular, one verifies that the maximum dilatation (10.34) is of order of the correlation length, in agreement with intuitive arguments.

10.5 Models with $O(N)$ orthogonal symmetry

The preceding analysis generalizes directly to models where the field or order parameter has N components, σ_α, $\alpha = 1, \ldots, N$, and which are invariant under the $O(N)$ orthogonal group (rotation–reflection group in N space dimensions). Indeed, the $O(N)$ symmetry implies that on the critical surface $T = T_c$ there exists only one relevant operator in dimension $d = 4 - \varepsilon$.

Properties of $O(N)$ symmetric models have already been studied in the framework of the quasi-Gaussian approximation in Section 7.8.

The effective Hamiltonian with $O(N)$ symmetry has the form

$$\mathcal{H}(\sigma) = \int \left\{ \frac{1}{2} \sum_\mu [\partial_\mu \sigma(x)]^2 + \frac{1}{2}(r_c + t)\sigma^2(x) + \frac{1}{4!} g \left(\sigma^2\right)^2(x) \right\} d^d x, \quad (10.36)$$

where quadratic terms with more than two derivatives are implicit and r_c is determined by a generalized condition (10.3). Indeed, in the case of a vanishing magnetization ($T > T_c, H = 0$), the two-point vertex function in the critical domain takes the form

$$\tilde{\Gamma}^{(2)}_{\alpha\beta}(p) = \delta_{\alpha\beta} \tilde{\Gamma}^{(2)}(p).$$

The RGE then have exactly the same form as for the $N = 1$ Ising type models. Explicit results differ from the case $N = 1$ only by polynomials in N multiplying the various perturbative contributions. At order g, one finds

$$\tilde{\Gamma}^{(2)}(p) = t + p^2 + O(p^4) + \frac{N+2}{6} g \frac{1}{(2\pi)^d} \int d^d q (\tilde{\Delta}(q,t) - \tilde{\Delta}(q,0)) + O(g^2),$$

and thus $Z(\lambda)$ keeps its Gaussian form at this order.

The critical four-point function with four equal indices reads

$$\tilde{\Gamma}^{(4)}_{\alpha\alpha\alpha\alpha}(p_1, p_2, p_3, p_4) = g - \frac{N+8}{18} g^2 (B(p_1+p_2) + B(p_1+p_3) + B(p_1+p_4)) + O(g^3).$$

β-function and fixed point. In the calculations of Section 10.3.3, it suffices to change the coefficient of g^2. One then infers the β-function that describes the flow of the parameter $g(\lambda)$ at order g^2:

$$\beta(g) = -\varepsilon g + \frac{N+8}{48\pi^2} g^2 + O\left(g^3, g^2 \varepsilon\right). \quad (10.37)$$

At first order in ε, the non-trivial zero of the function $\beta(g)$ is then

$$g^* = \frac{48\pi^2}{N+8} \varepsilon + O\left(\varepsilon^2\right) \Rightarrow \omega \equiv \beta'(g^*) = \varepsilon + O\left(\varepsilon^2\right) > 0. \quad (10.38)$$

The corresponding fixed point is stable. Universality properties for all $O(N)$ symmetric models follow.

Critical domain. Again, the calculations of the $N = 1$ case generalize. The coefficient of the term of order g is multiplied by an N-dependent factor. The flow of the coefficient t of the relevant perturbation $\int d^d x\, \sigma^2(x)$ is described by the function

$$\frac{1}{\nu(g)} = 2 + \eta_2(g) = 2 - \frac{(N+2)}{48\pi^2}g + O(g^2)$$

and, thus,

$$d_t = \frac{1}{\nu} = 2 - \frac{N+2}{N+8}\varepsilon + O(\varepsilon^2) \;\Rightarrow\; d_{\sigma^2} = d - d_t = 2 - \frac{6}{N+8}\varepsilon + O(\varepsilon^2).$$

Field dimension: Correction to the Gaussian value. The diagram (e) of Figure 2.6 (Section 2.5.1), which determines the deviation from the Gaussian value, is now multiplied by a factor $(N+2)/18$. From its behaviour for $p \to 0$ and $d \to 4$, one infers

$$\eta(g) = \frac{N+2}{18}\frac{1}{(4\pi)^4}g^2 + O(g^3) \tag{10.39}$$

and, thus,

$$\eta \equiv \eta(g^*) = \frac{N+2}{2(N+8)^2}\varepsilon^2 + O(\varepsilon^3),$$

$$d_\sigma = \tfrac{1}{2}(d - 2 + \eta) = 1 - \tfrac{1}{2}\varepsilon + \frac{N+2}{4(N+8)^2}\varepsilon^2 + O(\varepsilon^3).$$

10.6 RG near dimension 4

A more systematic analysis requires the methods of quantum field theory that we briefly describe in Chapters 12 and 13. These methods allow us, in the framework of the $\varepsilon = 4 - d$ expansion, itself based on an expansion around the Gaussian model, to prove asymptotic RGE. Solutions of the RGE confirm the existence of a stable non-Gaussian fixed point beyond the first-order calculation that we have presented here. A set of universal properties follow, such as scaling relations in the critical domain (e.g., the form (10.35) of the two-point function) or relations between critical exponents that had been conjectured in the framework of the phenomenological scaling theory: all the critical exponents that had been introduced previously as $\alpha, \beta, \gamma, \delta, \eta, \nu$ are functions only of two of them, for example,

$$\alpha = 2 - d\nu, \quad \beta = \tfrac{1}{2}\nu(d - 2 + \eta), \quad \gamma = \nu(2 - \eta), \quad \beta(\delta - 1) = \gamma,$$

relations valid except for the Gaussian fixed point, thus for $d < 4$.

In this section, within the framework of the double expansion in powers of g, the coefficient of $\int d^d x\, \sigma^4(x)$, and ε, we briefly describe the RGE satisfied by connected correlation functions, asymptotic forms at large distance of the general RGE (9.38). More details are given in Chapter 13.

10.6.1 Critical Hamiltonian and RGE

One first defines a perturbative expansion around the Gaussian theory, based on the Hamiltonian (10.1),

$$\mathcal{H}(\sigma) = \mathcal{H}_G(\sigma) + \int d^d x \left(\frac{1}{2} r_c(g) \sigma^2(x) + \frac{1}{4!} g \sigma^4(x) \right),$$

that is, an expansion in powers of the parameter g, in generic non-integer dimension. For the critical Hamiltonian, one then proves, to all orders in a double expansion in powers of g and $\varepsilon = 4-d$, equations of the form (9.38), but with only one parameter g that describes the trajectory that interpolates between the Gaussian fixed point and the stable fixed point g^*:

$$\left[\sum_\ell x_\ell \cdot \frac{\partial}{\partial x_\ell} + \beta(g) \frac{\partial}{\partial g} + \frac{n}{2} D_\sigma(g) \right] W^{(n)}(g; x_1, \ldots, x_n) \sim 0, \qquad (10.40)$$

where the symbol $\sim$ indicates that contributions that decay faster at large distance have been neglected.

The integrated RGE (9.9), the parameter $g(\lambda)$ and the renormalization $Z(\lambda)$ at scale λ solutions of

$$\beta(g(\lambda)) = -\lambda \frac{d}{d\lambda} g(\lambda), \qquad (10.41)$$

$$-2 D_\sigma(g(\lambda)) = \lambda \frac{d}{d\lambda} \ln Z(\lambda), \qquad (10.42)$$

then appear in the solution of equation (10.40) by the method of characteristics (see Section 13.3.1).

The solution of the equation $\beta(g^*) = 0$ gives the fixed-point value and $d_\sigma = D_\sigma(g^*)$ the dimension of the order parameter.

In the case of an attractive fixed point ($\omega = \beta'(g^*) > 0$), the asymptotic behaviour of correlation functions is determined by the functions of the fixed-point theory $g = g^*$, which satisfy

$$\left(\sum_\ell x_\ell \cdot \frac{\partial}{\partial x_\ell} + n d_\sigma \right) W^{(n)}(g^*; x_1, \ldots, x_n) = 0$$

and, thus,

$$W^{(n)}(g^*; \lambda x_1, \ldots, \lambda x_n) = \frac{1}{\lambda^{n d_\sigma}} W^{(n)}(g^*; x_1, \ldots, x_n).$$

10.6.2 Critical domain

As we have already pointed out, in the whole critical domain $\xi \gg 1$ universality properties are expected. Universal quantities can be derived from the Hamiltonian

$$\mathcal{H}_t(\sigma) = \mathcal{H}(\sigma) + \frac{t}{2} \int d^d x\, \sigma^2(x), \tag{10.43}$$

where the constant $|t| \ll 1$ parametrizes the deviation from the critical temperature. One then proves the more general equation

$$D_{\mathrm{RG}} W^{(n)}(g, t; x_1, \ldots, x_n) \sim 0, \tag{10.44}$$

$$D_{\mathrm{RG}} = \sum_\ell x_\ell \cdot \frac{\partial}{\partial x_\ell} + \beta(g) \frac{\partial}{\partial g} - (2 + \eta_2(g)) t \frac{\partial}{\partial t} + \frac{n}{2} D_\sigma(g), \tag{10.45}$$

where the additional function $\eta_2(g)$ is directly related to the flow of the coefficient $t(\lambda)$ of $\int d^d x\, \sigma^2(x)$. At the fixed point $g = g^*$, the equation simplifies since

$$D_{\mathrm{RG}} = \sum_\ell x_\ell \cdot \frac{\partial}{\partial x_\ell} - d_t t \frac{\partial}{\partial t} + n d_\sigma$$

with

$$d_t = 2 + \eta_2(g^*) \equiv \frac{1}{\nu}.$$

The more general scaling property

$$W^{(n)}(g^*, t\lambda^{d_t}; \lambda x_1, \ldots, \lambda x_n) = \frac{1}{\lambda^{n d_\sigma}} W^{(n)}(g^*, t; x_1, \ldots, x_n)$$

follows. Choosing λ such that $t\lambda^{d_t} = 1$, one can rewrite the relation as

$$W^{(n)}(g^*, t; x_1, \ldots, x_n) = t^{n \nu d_\sigma} W^{(n)}(g^*, 1; t^\nu x_1, \ldots, t^\nu x_n).$$

This relation shows that connected correlation functions at the fixed point and, thus, all connected correlation functions asymptotically in the critical domain, do not depend on the x_i and the parameter t independently but only through the product xt^ν. This is the general form of a scaling relation that we have already exhibited in the case of the two-point function:

$$W^{(2)}(g^*, t; x) = t^{2\nu d_\sigma} W^{(2)}(g^*, 1; t^\nu x),$$

as a direct consequence of equation (10.35). In the Fourier representation,

$$\tilde{W}^{(2)}(g, t; p) \sim t^{-\gamma} \tilde{W}^{(2)}(g^*, 1; p/t^\nu)$$

with

$$\gamma = \nu(2 - \eta).$$

Above T_c, the two-point function decays exponentially and the scaling relation implies that the correlation length ξ is proportional to $t^{-\nu}$. $\xi(t)$ thus diverges at T_c like

$$\xi(t) \underset{t \propto T - T_c \to 0_+}{\propto} t^{-\nu}.$$

Equations in a field or below T_c. It is further possible to include the effect of a weak magnetic field, the relevant operator that breaks the reflection symmetry, or of a finite magnetization. The corresponding equations allow then a continuous transition from the ordered to the disordered phases, and a proof of the same scaling relations below T_c. At fixed magnetization M, one finds

$$D_{\mathrm{RG}} = \sum_\ell x_\ell \cdot \frac{\partial}{\partial x_\ell} + \beta(g)\frac{\partial}{\partial g} - (2+\eta_2(g))t\frac{\partial}{\partial t} + \frac{1}{2}D_\sigma(g)\left(n + M\frac{\partial}{\partial M}\right).$$

Specific heat. A qualitative difference with the quasi-Gaussian model is the behaviour of the specific heat $\mathcal{C}$, which is here the sum of a regular non-universal part and a singular universal part

$$\mathcal{C}_{\mathrm{sing.}} \sim A^\pm |t|^{-\alpha}, \qquad (10.46)$$

where $\alpha = 2 - d\nu$ is the specific-heat exponent.

10.7 Universal quantities: Numerical results

In addition to leading to a proof of general universality properties, the field theory methods and the corresponding RG allow also a calculation of universal quantities. For the $O(N)$ symmetric models, the simplest ones, like critical exponents, some amplitude ratios, and the equation of state in the scaling limit (but the latter only for the models with Ising symmetry) have been precisely determined.

It is important to stress, here, that the most precise results do not come from the ε-expansion, but from the *perturbative expansion in the Callan–Symanzik (CS) formalism* described in Section 13.5 directly in $d = 3$ (following a suggestion of Parisi (1973)), because the evaluation of the successive terms of the perturbative expansion is easier. For example, the functions $\beta_{\mathrm{r}}(g_{\mathrm{r}})$, $\eta_{\mathrm{r}}(g_{\mathrm{r}})$ and $\eta_{\mathrm{r},2}(g_{\mathrm{r}})$, where g_{r} is the renormalized coupling, are known up to order g_{r}^7 in $d = 3$, whereas the exponents are known only up to order ε^5 in the ε-expansion. However, the method requires to first solve the fixed-point equation $\beta_{\mathrm{r}}(g_{\mathrm{r}}) = 0$ numerically.

In all cases, it is necessary to confront the problem of the divergence of the perturbative expansion in field theory (see Table 10.1), a generic problem of expansions generated by the steepest descent method. One calls such perturbative expansions *asymptotic* expansions. One notes that the partial sums seem first to converge,

Table 10.1

Partial sums of the ε-expansion for the exponents γ and η for $\varepsilon = 1$, $N = 1$.

k	0	1	2	3	4	5
γ	1.000	1.1667	1.2438	1.1948	1.3384	0.8918
η	0.0...	0.0...	0.0185	0.0372	0.0289	0.0545

Table 10.2

Successive results obtained by Borel transformation and conformal mapping: The zero g_r^* of the function $\beta_r(g_r)$ and the exponents γ and ν for $d = 3$, $N = 1$ in the CS formalism.

k	2	3	4	5	6	7
g_r^*	31.456	25.359	23.707	23.637	23.630	23.633
ν	0.6338	0.6328	0.62966	0.6302	0.6302	0.6302
γ	1.2257	1.2370	1.2386	1.2398	1.2398	1.2398

before diverging with increasing oscillations, a feature typical for asymptotic expansions.

Summation methods are required to derive from these divergent expansions convergent sequences. The most precise results have been obtained by applying a Borel transformation on the expansion, followed by an analytic continuation based on a conformal transformation (see Table 10.2).

Tables 10.2 and 10.3 are based on the perturbative expansion at fixed dimension $d = 3$ in the CS formalism (Section 13.5). In the first line we display the successive values of g_r^* obtained from a numerical solution of the equation $\beta_r(g_r) = 0$. In the second line of Table 10.3, we give also the values of the combination

$$\tilde{g}_{\text{Ni}}^* = (N+8)g/(48\pi),$$

which is defined in such a way that the two first coefficients of the expansion of the β-function are equal, to give a better idea of the real size of the expansion parameter. In Table 10.3, we then report the values of critical exponents obtained by field theory methods as obtained by Guida and Zinn-Justin (1998), improving the previous results of Le Guillou and Zinn-Justin (1977–1980).

The integer N corresponds to models with different $O(N)$ symmetries. Let us mention here a few examples of models or physical systems corresponding to the different values of N. For example, $N = 0$ corresponds to statistical properties of polymers (from a more theoretical viewpoint, self-avoiding random walk or SAW), $N = 1$ to the liquid–vapour transition, separation transitions of binary mixtures or uniaxal antiferromagnetic transitions. The most significant representative of the $N = 2$ class is helium's superfluid transition. Finally, $N = 3$ corresponds to ferromagnetic transitions.

Table 10.3

Critical exponents of the $O(N)$ model, $d = 3$, obtained from the $(\sigma^2)^2_{d=3}$ field theory in the CS formalism.

N	0	1	2	3
g_r^*	26.63 ± 0.11	23.64 ± 0.07	21.16 ± 0.05	19.06 ± 0.05
$\tilde{g}_{\text{Ni}}^*$	1.413 ± 0.006	1.411 ± 0.004	1.403 ± 0.003	1.390 ± 0.004
γ	1.1596 ± 0.0020	1.2396 ± 0.0013	1.3169 ± 0.0020	1.3895 ± 0.0050
ν	0.5882 ± 0.0011	0.6304 ± 0.0013	0.6703 ± 0.0015	0.7073 ± 0.0035
η	0.0284 ± 0.0025	0.0335 ± 0.0025	0.0354 ± 0.0025	0.0355 ± 0.0025
β	0.3024 ± 0.0008	0.3258 ± 0.0014	0.3470 ± 0.0016	0.3662 ± 0.0025
α	0.235 ± 0.003	0.109 ± 0.004	-0.011 ± 0.004	-0.122 ± 0.010
ω	0.812 ± 0.016	0.799 ± 0.011	0.789 ± 0.011	0.782 ± 0.0013
$\theta = \omega\nu$	0.478 ± 0.010	0.504 ± 0.008	0.529 ± 0.009	0.553 ± 0.012

For a comparison, we display also in Table 10.4 the results for critical exponents derived from the ε-expansion after summation (Le Guillou and Zinn-Justin, 1985-1986, Guida and Zinn-Justin, 1998). The two versions 'free' and 'const.' correspond to a free summation, and a summation where information about the exact behaviour when $d \to 2$, derived from the non-linear σ model (see Chapter 15) has been explicitly incorporated.

A comparison with the values obtained by other numerical methods (high-temperature series, Monte Carlo type simulations in lattice models), or with experimental results, shows an excellent general agreement. As an illustration, we thus display in Table 10.5 a short compilation of published numerical results obtained in lattice models.

Finally, let us quote a few especially precise experimental results:

Based on a study of long polymeric chains, which correspond to the $N = 0$ class, the value

$$\nu = 0.586 \pm 0.004,$$

has been reported.

Experiments on transitions in binary mixtures ($N = 1$) have led to the estimates, for example,

$$\nu = 0.625 \pm 0.010, \quad \beta = 0.325 \pm 0.005.$$

Table 10.4

Critical exponents of the $O(N)$ model, $d = 3$, obtained from the ε-expansion.

N	0	1	2	3
γ (free)	1.1575 ± 0.0060	1.2355 ± 0.0050	1.3110 ± 0.0070	1.3820 ± 0.0090
γ (const.)	1.1571 ± 0.0030	1.2380 ± 0.0050	1.317	1.392
ν (free)	0.5875 ± 0.0025	0.6290 ± 0.0025	0.6680 ± 0.0035	0.7045 ± 0.0055
ν (const.)	0.5878 ± 0.0011	0.6305 ± 0.0025	0.671	0.708
η (free)	0.0300 ± 0.0050	0.0360 ± 0.0050	0.0380 ± 0.0050	0.0375 ± 0.0045
η (const.)	0.0315 ± 0.0035	0.0365 ± 0.0050	0.0370	0.0355
β (free)	0.3025 ± 0.0025	0.3257 ± 0.0025	0.3465 ± 0.0035	0.3655 ± 0.0035
β (const.)	0.3032 ± 0.0014	0.3265 ± 0.0015		
ω	0.828 ± 0.023	0.814 ± 0.018	0.802 ± 0.018	0.794 ± 0.018
θ	0.486 ± 0.016	0.512 ± 0.013	0.536 ± 0.015	0.559 ± 0.017

Table 10.5

Critical exponents of the $O(N)$ model evaluated by lattice methods.

N	0	1	2	3
γ	1.1575 ± 0.0006	1.2385 ± 0.0025	1.322 ± 0.005	1.400 ± 0.006
ν	0.5877 ± 0.0006	0.631 ± 0.002	0.674 ± 0.003	0.710 ± 0.006
α	0.237 ± 0.002	0.103 ± 0.005	-0.022 ± 0.009	-0.133 ± 0.018
β	0.3028 ± 0.0012	0.329 ± 0.009	0.350 ± 0.007	0.365 ± 0.012
θ	0.56 ± 0.03	0.53 ± 0.04	0.60 ± 0.08	0.54 ± 0.10

However, the most precise results are provided by the helium's superfluid transition (in micro-gravity experiments), which belongs to the $N = 2$ class, for example,

$$\nu = 0.6807 \pm 0.0005\,.$$

Universal ratios of critical amplitudes. In Table 10.6, we display a few universal ratios of critical amplitudes for $T - T_{\rm c} \to 0$, in the case of the universality class of the Ising model ($N = 1$) in three dimensions (Guida and Zinn-Justin 1996), and a comparison with other available results. The various definitions are given in (7.35) and (10.46) and in Sections 8.6 and 10.6.1. Moreover, $R_{\rm c} = \alpha A^+ C^+/M_0^2$.

The two first lines correspond to calculations in field theory. High-temperature (HT) series are obtained from lattice models. The last three lines correspond to experimental results, transitions in binary mixtures, in liquid–vapour and in anisotropic magnetic systems.

Table 10.6

Amplitude ratios for $N = 1$, the Ising model class.

	A^+/A^-	C^+/C^-	R_c	R_χ
ε-exp.,	0.527 ± 0.037	4.73 ± 0.16	0.0569 ± 0.0035	1.648 ± 0.036
$d = 3$	0.537 ± 0.019	4.79 ± 0.10	0.0574 ± 0.0020	1.669 ± 0.018
HT series	0.523 ± 0.009	4.95 ± 0.15	0.0581 ± 0.0010	1.75
	0.560 ± 0.010	4.75 ± 0.03		
bin. mixt.	0.56 ± 0.02	4.3 ± 0.3	0.050 ± 0.015	1.75 ± 0.30
liqu.–vap.	0.48–0.53	4.8–5.2	0.047 ± 0.010	1.69 ± 0.14
magn. syst.	0.49–0.54	4.9 ± 0.5		

11 Renormalization group: N-component fields

In this chapter, we study more general models with an N-component field (or order parameter), from the viewpoint of the renormalization group (RG). Indeed, one can find interesting physical systems for which the Hamiltonian does not have the $O(N)$ orthogonal symmetry of the models studied in Section 10.5.

A first family of such models is characterized by the presence of several independent correlation lengths. This happens typically when the quadratic part of the Hamiltonian involves several unrelated parameters. Generically, the different correlation lengths then diverge for different values of the temperature. The components of the fields that are not critical decouple and can be ignored in the study of the asymptotic large-distance behaviour (in the sense of the field integral, they can be integrated out). To classify the possible types of critical behaviour, one can thus restrict the discussion to models that, like the $O(N)$ model, have only one correlation length in the disordered phase.

Models with only one correlation length. Generic models with only one correlation length all correspond to Hamiltonians invariant under some symmetry group G acting on the field, a subgroup of the $O(N)$ group, which admits only one quadratic invariant. This implies, in particular, that the field transforms under an irreducible representation of the group G.

As a consequence, the two-point correlation function in the disordered phase is proportional to the identity matrix in component space. Denoting now by σ_α, the N-component field, we can express this condition as

$$\langle \sigma_\alpha(x)\sigma_\beta(y)\rangle = \frac{1}{N}\delta_{\alpha\beta}\sum_{\gamma=1}^{N}\langle \sigma_\gamma(x)\sigma_\gamma(y)\rangle. \tag{11.1}$$

Moreover, we assume that the group G contains the reflection group $\mathbb{Z}_2$, $\boldsymbol{\sigma}\mapsto-\boldsymbol{\sigma}$ as a subgroup and admits several, linearly independent, quartic invariants of σ^4 type, as the example of the cubic anisotropy studied in Section 11.3 will illustrate.

For this class of models, the effective Hamiltonians thus have the same quadratic terms as the Hamiltonian (10.36), but differ by the quartic contributions: they contain several independent terms of $\int d^d x\, \sigma^4(x)$ type, one of them always being the isotropic term present in Hamiltonian (10.36).

In this chapter, we discuss rather generally the RG flow of the relevant parameters at T_c in dimension $d = 4 - \varepsilon$.

11.1 RG: General remarks

We first establish a few general properties of field theories of σ^4 type in which the field is an N-component vector $\boldsymbol{\sigma}(x)$, which satisfy the condition (11.1) and thus admit a unique correlation length.

Hamiltonian flow. We assume that the Hamiltonian contains $p > 1$, linearly independent, quartic terms of $\sigma^4(x)$ type, with coefficients g_a, $a = 1, \ldots, p$. The flow equations that replace the unique flow equation (9.62), then take the general form

$$\lambda \frac{\mathrm{d}g_a}{\mathrm{d}\lambda} = -\beta_a(g(\lambda)). \tag{11.2}$$

With this convention, a dilatation $\lambda \to \infty$ corresponds to the large-distance behaviour.

The vector tangent to an RG trajectory at a point g_a is proportional to $\beta_a(g)$. It is thus unique in each point where it is defined. RG trajectories can intersect only at points that are solutions of $\beta_a(g) = 0$, that is, at fixed points.

In the framework of the ε-expansion with $g_a = O(\varepsilon)$, at leading order the β-functions can be written as

$$\beta_a(g) = -\varepsilon g_a + B_a(g), \tag{11.3}$$

where $B_a(g)$ is a homogeneous, second-degree, polynomial. It thus satisfies the identities

$$B_a(\rho g) = \rho^2 B_a(g), \tag{11.4a}$$

$$\Rightarrow \left. \frac{\mathrm{d} B_a(\rho g)}{\mathrm{d}\rho} \right|_{\rho=1} = \sum_{b=1}^{p} g_b \frac{\partial B_a(g)}{\partial g_b} = 2 B_a(g). \tag{11.4b}$$

Fixed points and stability. In the case of p parameters g_a, the maximum number of solutions g_a^* of the fixed-point equations

$$\beta_a(g^*) \equiv -\varepsilon g_a^* + B_a(g^*) = 0, \tag{11.5}$$

is 2^p. The local stability of these fixed points can be studied by linearizing equations (11.2):

$$\lambda \frac{\mathrm{d}}{\mathrm{d}\lambda}(g_a - g_a^*) = \sum_b L_{ab}^*(g_b - g_b^*)$$

with

$$L_{ab}^* = -\frac{\partial \beta_a(g^*)}{\partial g_b} = \varepsilon \delta_{ab} - \frac{\partial B_a(g^*)}{\partial g_b}.$$

The local stability properties thus depend on the sign of the eigenvalues (one proves that they are real, see Section 11.2.2) of the matrix $\mathbf{L}^*$ of partial derivatives. If all eigenvalues of $\mathbf{L}^*$ are negative, the fixed point is locally stable. Global properties depend on the complete solutions of equation (11.2), which determine, in the space of parameters g_a, the domain of attraction of each stable fixed point.

Eigenvalue $-\varepsilon$. From the homogeneity property (11.4b) and from the fixed-point equation (11.5), one infers

$$\sum_b L^*_{ab} g^*_b = \varepsilon g^*_a - 2 B_a(g^*) = -\varepsilon g^*_a.$$

One concludes that all non-Gaussian fixed points have at least one direction of stability corresponding to the eigenvector g^*_a, with eigenvalue $-\varepsilon + O(\varepsilon^2)$.

Special RG trajectories. One can prove an even more general result. One looks for special solutions of equation (11.2) of the form

$$g_a(\lambda) = \rho(\lambda) g^*_a, \quad g^* \neq 0, \quad \rho \geq 0.$$

Introducing the Ansatz into equation (11.2), using the fixed-point equation (11.5) and the homogeneity property (11.4a), one infers

$$g^*_a \lambda \frac{\mathrm{d}}{\mathrm{d}\lambda} \rho(\lambda) = \varepsilon \rho(\lambda) g^*_a - B_a\bigl(g^* \rho(\lambda)\bigr) = \varepsilon \rho(\lambda) g^*_a - \rho^2(\lambda) B_a(g^*)$$
$$= \varepsilon \rho(\lambda) g^*_a - \varepsilon g^*_a \rho^2(\lambda).$$

The flow equation is thus compatible with the Ansatz and the function $\rho(\lambda)$ is a solution of

$$\lambda \frac{\mathrm{d}}{\mathrm{d}\lambda} \rho(\lambda) = \varepsilon \rho(\lambda) \bigl(1 - \rho(\lambda)\bigr).$$

In the approximation (11.3), the half-lines joining the Gaussian fixed point to other fixed points are RG trajectories and on these trajectories non-Gaussian fixed points are stable.

11.2 Gradient flow

It has been verified up to order ε^5 (i.e., all known orders) that the functions β of the general σ^4 models can be written as (see Section 11.4.3 for the leading order)

$$\beta_a(g) = \sum_b T_{ab}(g) \frac{\partial U}{\partial g_b}, \qquad (11.6)$$

where the matrix $\mathbf{T}$ with elements T_{ab} is a symmetric positive matrix, and a regular function of the g_a. Equation (11.2) then defines a gradient flow. Let us point out, however, that the regularity and positivity properties of the matrix $\mathbf{T}$ can only be verified, in this framework, in the vicinity of the Gaussian fixed point $g_a = 0$.

11.2.1 Change of parametrization

The general form (11.6) is the only form of a gradient flow consistent with the transformation properties under reparametrization in the space of the coefficients g_a (diffeomorphisms). Indeed, let us introduce new parameters γ_a and change variables, $g_a = g_a(\gamma)$, in the flow equations. The matrix $\partial \gamma_b / \partial g_a$ must be invertible at $g = 0$ for the mapping $\gamma_a \mapsto g_a$ to be invertible within the framework of the ε-expansion.

The chain rule for partial derivatives then leads to

$$\lambda \frac{\mathrm{d}}{\mathrm{d}\lambda} \gamma_a = \sum_b \frac{\partial \gamma_a}{\partial g_b} \lambda \frac{\mathrm{d}}{\mathrm{d}\lambda} g_b, \quad \frac{\partial U}{\partial g_a} = \sum_b \frac{\partial \gamma_b}{\partial g_a} \frac{\partial U}{\partial \gamma_b}.$$

Then,

$$\lambda \frac{\mathrm{d}}{\mathrm{d}\lambda} \gamma_a = -\sum_b T'_{ab} \frac{\partial U}{\partial \gamma_b}$$

with

$$T'_{ab} = \sum_{c,d} \frac{\partial \gamma_a}{\partial g_d} T_{dc} \frac{\partial \gamma_b}{\partial g_c}. \tag{11.7}$$

One verifies that if the matrix $\mathbf{T}$ is symmetric and positive, the transformed matrix $\mathbf{T}'$ with elements T'_{ab} is also symmetric and positive.

In particular, even if the matrix $\mathbf{T}$ is proportional to the identity matrix in one special parametrization, this is in general no longer true in a different parametrization.

Finally, since the matrix $\mathbf{T}$ is positive and transforms under reparametrization as shown in (11.7), its inverse has the properties of a metric tensor.

11.2.2 Flow and potential

The property of gradient flow has important consequences:

(i) The potential decreases along an RG trajectory and thus fixed points are extrema of the potential, stable fixed points being local minima.

(ii) The eigenvalues of the matrix of first-order partial derivatives at a fixed point are *real*.

Derivation. The variation of the potential U along a trajectory satisfies

$$\lambda \frac{\mathrm{d}}{\mathrm{d}\lambda} U(g(\lambda)) = \sum_a \frac{\partial U}{\partial g_a} \lambda \frac{\mathrm{d} g_a}{\mathrm{d}\lambda} = -\sum_{a,b} \frac{\partial U}{\partial g_a} T_{ab} \frac{\partial U}{\partial g_b}.$$

Since the matrix $\mathbf{T}$ is positive, the right-hand side, which is the expectation value of a negative matrix, is negative. Thus, the potential decreases along a trajectory. The fixed points g^* are extrema of the function U:

$$\beta_a(g^*) = 0 \Leftrightarrow \frac{\partial U(g^*)}{\partial g_a} = 0.$$

A stable fixed point is a local minimum of $U(g)$.

At a fixed point, the elements of the matrix $\mathbf{L}^*$ of the derivatives of the functions $-\beta$ are given by

$$L^*_{ab} = -\frac{\partial \beta_a(g^*)}{\partial g_b} = -\sum_c T_{ac}(g^*)\frac{\partial^2 U(g^*)}{\partial g_c \partial g_b}, \qquad (11.8)$$

a relation that can be written more symbolically as

$$\mathbf{L}^* = -\mathbf{T}\mathbf{U}''. \qquad (11.9)$$

Since the matrix $\mathbf{T}$ is positive, it can be written as the square of a matrix $\mathbf{X}$, also symmetric and positive:

$$\mathbf{T} = \mathbf{X}^2, \quad \mathbf{X} > 0.$$

The matrix

$$\mathbf{M} = \mathbf{X}^{-1}\mathbf{L}^*\mathbf{X} = -\mathbf{X}\mathbf{U}''\mathbf{X},$$

has the same eigenvalues as $\mathbf{L}^*$, but since the matrices $\mathbf{U}''$ and $\mathbf{X}$ are symmetric, it is a symmetric matrix. The matrix $\mathbf{L}^*$, which has the same eigenvalues as a real symmetric matrix, thus has also *real* eigenvalues.

The relation also shows that if the matrix $\mathbf{U}''$ is positive, the matrix $\mathbf{X}\mathbf{U}''\mathbf{X}$ is positive (and conversely), and the corresponding fixed point thus is locally stable.

11.2.3 Fixed points and stability

In the framework of the ε-expansion, we now prove two other consequences of the property of gradient flow: there exists at most one stable fixed point; the stable fixed point corresponds to the lowest value of the potential.

Indeed, let us assume the existence of two fixed points corresponding to the parameters g^* and g'^*. We then consider the parameters g of the form

$$g(s) = sg^* + (1-s)g'^*, \quad 0 \le s \le 1,$$

and the corresponding potential

$$u(s) = U(g(s)).$$

Note that the positivity condition (11.12), which is verified by all fixed points (see Section 11.4.1) and thus for $s = 0$ and $s = 1$, is then also verified for all parameters $g(s)$ such that $0 \le s \le 1$.

As the explicit form (11.28) shows, at leading order $u(s)$ is a third-degree polynomial. The derivative

$$u'(s) = \sum_a g'_a(s)\frac{\partial U}{\partial g_a} = \sum_a (g^*_a - g'^*_a)\frac{\partial U}{\partial g_a} = \sum_{a,b}(g^*_a - g'^*_a)T^{-1}_{ab}\beta_b(g(s))$$

vanishes due to the fixed-point conditions at $s=0$ and $s=1$:
$$u'(0) = u'(1) = 0.$$

Since $u'(s)$ is a second-degree polynomial, it then necessarily has the form
$$u'(s) = As(1-s).$$

The second derivative $u''(s)$ is given in terms of the matrix of second partial derivatives of U and, thus, the partial derivatives of the functions β, by
$$u''(s) = \sum_{a,b}(g_a^* - g_a'^*)\frac{\partial^2 U(g(s))}{\partial g_a \partial g_b}(g_b^* - g_b'^*) = A(1-2s).$$

In particular, for $s=0$ and $s=1$,
$$A = \sum_{a,b}(g_a^* - g_a'^*)\frac{\partial^2 U(g'^*)}{\partial g_a \partial g_b}(g_b^* - g_b'^*), \quad -A = \sum_{a,b}(g_a^* - g_a'^*)\frac{\partial^2 U(g^*)}{\partial g_a \partial g_b}(g_b^* - g_b'^*).$$

As we have shown in Section 11.2.2, at a stable fixed point the matrix $\mathbf{U}''$ of partial second derivatives of U is positive. Thus, if g^* and g'^* are stable fixed points, A and $-A$ are both given by the expectation value of a positive matrix and thus are both positive, which is contradictory: the two fixed points cannot both be stable.

More generally, the sign of A characterizes, in some sense, the relative stability of these two fixed points. Let us assume, for example, that $A<0$ which is consistent with the assumption that g^* is stable. Then $u'(s) < 0$ in $[0,1]$ and $U(g(s))$ is a decreasing function. Thus,
$$U(g^*) < U(g'^*).$$

In particular, if g^* is a stable fixed point, it corresponds, among all fixed points, to the lowest value of the potential.

11.3 Model with cubic anisotropy

To illustrate the preceding results and before studying other properties of general models with an N-component order parameter, we examine a model that is simple but with interesting physics properties.

We consider an N-component field σ_α, $\alpha = 1, \ldots, N$ and a Hamiltonian invariant under the cubic group, the finite group of transformations generated by

$$\sigma_\alpha \mapsto -\sigma_\alpha, \quad \sigma_\alpha \leftrightarrow \sigma_\beta \quad \text{for all } \alpha \text{ and } \beta.$$

One hopes to describe by such models the critical properties of classical spin systems in which the interactions are modulated by an underlying cubic lattice (see Section 3.3.7).

The cubic symmetry group admits a unique quadratic invariant but two independent quartic invariants of σ^4 type. Using these symmetry properties, one verifies that a critical Hamiltonian in continuum space, truncated at order σ^4 as justified by the analysis of relevant operators near dimension 4, has the general form

$$\mathcal{H}_c(\sigma) = \int d^d x \left\{ \tfrac{1}{2} \sum_\alpha [\nabla \sigma_\alpha(x)]^2 + \cdots + \tfrac{1}{2} r_c(g,h) \sum_\alpha \sigma_\alpha^2(x) \right.$$
$$\left. + \frac{g}{24} \left(\sum_\alpha \sigma_\alpha^2(x) \right)^2 + \frac{h}{24} \sum_\alpha \sigma_\alpha^4(x) \right\}. \tag{11.10}$$

Since the model has a unique quadratic invariant, the condition (11.1) is satisfied, the two-point function in the disordered phase is proportional to the identity matrix and r_c is determined by the condition (10.3): $\tilde{\Gamma}^{(2)}(p=0) = 0$.

The appearance of two quartic terms implies that, on the critical surface, the RG in dimension $d = 4 - \varepsilon$ now involves two parameters g, h corresponding to two quartic relevant operators.

Positivity. Note that the two constants g, h must satisfy the two conditions $g + h \geq 0$, $Ng + h \geq 0$ to ensure that the Hamiltonian is positive for $\sigma \to \infty$ and, thus, that the transition is second order. The first condition is obtained by choosing all σ_α zero but one, the second by taking them all equal.

These conditions imply, in particular, that if the RG flow leads to parameters $g(\lambda), h(\lambda)$ outside this domain, the terms of higher degree in the expansion of the thermodynamic potential, *a priori* thought to be negligible, become important and the transition, in disagreement with the predictions of the quasi-Gaussian or mean field approximation, is generically weak first order.

11.3.1 RG and fixed points

The flow equations have the general form

$$\lambda \frac{dg}{d\lambda} = -\beta_g(g(\lambda), h(\lambda)),$$
$$\lambda \frac{dh}{d\lambda} = -\beta_h(g(\lambda), h(\lambda)).$$

A simple calculation, analogous to the calculation presented in Section 10.3, determines the two β-functions at leading order for $g, h = O(\varepsilon)$, $\varepsilon \to 0$:

$$\beta_g(g, h) = -\varepsilon g + \frac{1}{8\pi^2}\left(\frac{N+8}{6} g^2 + gh\right),$$
$$\beta_h(g, h) = -\varepsilon h + \frac{1}{8\pi^2}\left(2gh + \frac{3}{2}h^2\right).$$

One has then to study the flow of the parameters g and h as a function of the dilatation parameter λ. One first looks for the various fixed points for $\varepsilon = 4 - d > 0$, and discusses their stability for $\lambda \to \infty$ as a function of the integer N. One then tries to determine the nature of the transition as a function of the initial values of g and h. It is also interesting to determine the symmetry of the Hamiltonians corresponding to the various fixed points, in particular, to the stable fixed point

Fixed points. The equations $\beta_g = \beta_h = 0$ both factorize into two linear equations. Combining them in the four possible ways, one finds:

(i) The Gaussian fixed point
$$g = h = 0.$$

(ii) The decoupled fixed point
$$g = 0, \quad h = 16\varepsilon\pi^2/3,$$
which corresponds to N identical and decoupled copies of an Ising type model with a $\mathbb{Z}_2$ reflection symmetry.

(iii) The isotropic fixed point
$$h = 0, \quad g = 48\varepsilon\pi^2/(N+8),$$
which has an $O(N)$ symmetry, more extended than the cubic symmetry of the initial Hamiltonian.

(iv) Finally, the last fixed point
$$g = \frac{16\pi^2 \varepsilon}{N}, \quad h = \frac{16\pi^2 (N-4)\varepsilon}{3N},$$
is new and is called the *cubic* fixed point.

All fixed points belong to the half-plane $g \geq 0$. Only the cubic fixed point for $N < 4$ is such that $h < 0$. However, for $N \geq 1$, it satisfies the positivity condition $g + h \geq 0$ (and thus also $Ng + h \geq 0$). Thus, all fixed points satisfy the positivity condition (11.12), in agreement with the general result proved in Section 11.4.1.

11.3.2 Linearized flow and eigenvalues

The local stability properties of the four fixed points are determined by the eigenvalues of the matrix $\mathbf{L}^*$ of the partial derivatives, with respect to g and h, of the functions $-\beta_g, -\beta_h$. One finds

$$\mathbf{L}^* = \varepsilon \mathbf{1} - \frac{1}{8\pi^2} \begin{pmatrix} \frac{N+8}{3}g + h & g \\ 2h & 2g + 3h \end{pmatrix}.$$

For the different fixed points, the corresponding eigenvalues of the matrix $\mathbf{L}^*$ are

$$\begin{aligned}
\text{Gaussian fixed point:} &\quad \varepsilon, &&\varepsilon, \\
\text{Decoupled (Ising) fixed point:} &\quad \tfrac{1}{3}\varepsilon, &&-\varepsilon, \\
\text{Isotropic fixed point:} &\quad \frac{N-4}{N+8}\varepsilon, &&-\varepsilon, \\
\text{Cubic fixed point:} &\quad \frac{4-N}{3N}\varepsilon, &&-\varepsilon.
\end{aligned}$$

The existence of the eigenvalue $-\varepsilon$ is a general property of all non-Gaussian fixed points, which all have at least one stable direction (see Section 11.1).

The Gaussian fixed point is unstable in all directions. The decoupled (Ising type) fixed point always has one direction of instability.

For the isotropic fixed point, one finds a special example of a general result (see Section 11.4.2): the isotropic fixed point is stable for $N < N_c$ with $N_c = 4 + O(\varepsilon)$. Finally, the cubic fixed point is stable only if $N > N_c$. At $N = N_c$ the two fixed points merge and then exchange roles.

Remark. For $N < N_c$, the stable fixed point has an $O(N)$ symmetry. The asymptotic behaviour of correlation functions in the critical domain, the singularities of thermodynamic quantities at T_c, thus exhibit more symmetry than the initial microscopic model.

We have already met a similar phenomenon: the cubic symmetry of the lattice generates a continuous $O(d)$ spatial symmetry at large distance or in the critical domain (see Sections 3.3.9 and 8.2). Only a study of the corrections to the asymptotic critical behaviour reveals the more restricted symmetry of the microscopic model.

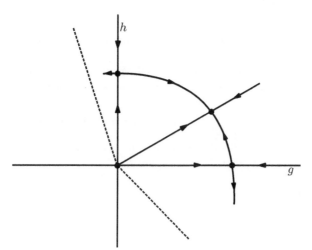

Fig. 11.1 Cubic anisotropy: RG flow for $N > 4$.

The RG flow. The RG trajectories can intersect only at a fixed point. The lines $h = 0$ and $g = 0$ and the half-line joining the origin to the cubic fixed point are stable lines for the RG, a special case of a general property proved in Section 11.1, and thus cannot be crossed. In the planar case, these conditions fix completely the topology of the RG trajectories (Figure 11.1).

In particular, one can find initial parameters g, h such that the RG trajectories can reach no fixed point but, instead, evolve toward unphysical regions, such as $g < 0$ for all N and $h < 0$ for $N > 4$ or $h < (N - 4)/3g$ for $N < 4$. These parameters, generically, correspond to weak first-order transitions: the correlation length remains finite when $T \to T_{c+}$ and, in microscopic units, takes a value that is of the order of the maximum dilatation parameter such that the parameters g, h are still in the allowed region.

11.4 Explicit general expressions: RG analysis

A general Hamiltonian satisfying all assumptions can be written as

$$\mathcal{H}(\sigma) = \int d^d x \left\{ \frac{1}{2} \sum_{i=1}^{N} \left[\sum_{\mu=1}^{d} (\partial_\mu \sigma_i)^2 + (r_c + t) \sigma_i^2 \right] + \frac{1}{4!} \sum_{i,j,k,l=1}^{N} g_{ijkl} \sigma_i \sigma_j \sigma_k \sigma_l \right\}, \quad (11.11)$$

where g_{ijkl} is a tensor symmetric in its four indices $ijkl$ that satisfies several conditions:

The Hamiltonian $\mathcal{H}$ must be positive for $|\sigma| \to \infty$ (for the phase transition to be continuous). This implies the condition

$$\sum_{i,j,k,l} g_{ijkl} \sigma_i \sigma_j \sigma_k \sigma_l > 0 \quad \forall \sigma_i \text{ such that } |\sigma| = 1. \quad (11.12)$$

The positivity condition gives to the space of admissible parameters g the structure of a convex cone.

As a consequence of the condition (11.1), which expresses that the connected two-point correlation function in the disordered phase is proportional to the identity matrix,

$$\langle \sigma_i(x) \sigma_j(y) \rangle_{\text{conn.}} = W_{ij}^{(2)}(x, y) = \delta_{ij} W^{(2)}(x, y), \quad (11.13)$$

the tensor g_{ijkl} has special properties that take the form of successive constraints on the tensor g_{ijkl} in the perturbative expansion.

11.4.1 RG functions

The flow equation for the parameters $g_{ijkl}(\lambda)$ in the Hamiltonian (11.11) now reads

$$\lambda \frac{d}{d\lambda} g_{ijkl}(\lambda) = -\beta_{ijkl}(g(\lambda)). \quad (11.14)$$

The large-distance behaviour of the field theory is governed by fixed points. These are solutions of the equation

$$\beta_{ijkl}(g^*) = 0. \quad (11.15)$$

The local stability properties of fixed points are related to the eigenvalues of the matrix

$$L^*_{ijkl, i'j'k'l'} = -\frac{\partial \beta_{ijkl}(g^*)}{\partial g_{i'j'k'l'}}. \quad (11.16)$$

RG functions as ε-expansions. It is not difficult to calculate the RG functions corresponding to a general σ^4 type theory. As in the example of the $O(N)$ symmetry, calculations differ from the case $N = 1$ only by some geometric factors.

The function β, at leading order, is given by

$$\beta_{ijkl}(g) = -\varepsilon g_{ijkl} + \frac{1}{16\pi^2} \sum_{m,n} \left(g_{ijmn}g_{mnkl} + g_{ikmn}g_{mnjl} + g_{ilmn}g_{mnkj} \right)$$
$$+ O\left(g^3\right). \tag{11.17}$$

The field dimension can be inferred from the function

$$\eta(g) = \frac{1}{6N(4\pi)^4} \sum_{i,j,k,l} g_{ijkl}g_{ijkl} + O\left(g^3\right), \tag{11.18}$$

a result that is, in particular, consistent with the general result of quantum field theory $\eta \geq 0$.

The flow equation for the deviation t of the critical temperature can be written as

$$\lambda \frac{\mathrm{d}}{\mathrm{d}\lambda} \ln t(\lambda) = \frac{1}{\nu(g(\lambda))}$$

with

$$\frac{1}{\nu(g)} = 2 - \frac{1}{16\pi^2 N} \sum_{i,j} g_{iijj} + O\left(g^2\right). \tag{11.19}$$

In the two equations (11.18) and (11.19), we have used explicitly the condition (11.13), which implies

$$\sum_k g_{ijkk} = \frac{\delta_{ij}}{N} \sum_{k,l} g_{kkll}, \quad \sum_{k,l,m} g_{iklm}g_{jklm} = \frac{\delta_{ij}}{N} \sum_{k,l,m,n} g_{klmn}g_{klmn}. \tag{11.20}$$

Moreover, at a fixed point

$$\varepsilon \sum_{i,j,k,l} g_{ijkl}\sigma_i\sigma_j\sigma_k\sigma_l = \frac{3}{16\pi^2} \sum_{i,j,k,l,m,n} \sigma_i\sigma_j g_{ijmn}g_{mnkl}\sigma_k\sigma_l$$

$$= \frac{3}{16\pi^2} \sum_{m,n} \left(\sum_{i,j} \sigma_i\sigma_j g_{ijmn} \right)^2.$$

Thus, any non-Gaussian fixed point satisfies the positivity condition (11.12).

11.4.2 Stability of the isotropic fixed point

Among the possible fixed points, one always finds, in addition to the Gaussian fixed point, the fixed point corresponding to the $O(N)$ symmetric Hamiltonian. It is possible to study its local stability at leading order in ε.

One can first specialize the expressions (11.17)–(11.19) to the example of the $(\sigma^2)^2$ field theory with $O(N)$ symmetry. This amounts to substituting in these equations

$$g_{ijkl} = \frac{g}{3}\left(\delta_{ij}\delta_{kl} + \delta_{ik}\delta_{jl} + \delta_{il}\delta_{jk}\right). \tag{11.21}$$

After a short calculation, one recovers the expressions (10.37) and (10.39) of the $\beta(g)$ and $\eta(g)$ functions and the corresponding value of g^*.

The stability conditions are given by the eigenvalues of the matrix L^* (equation (11.16)). Setting
$$g_{ijkl} = g^*_{ijkl} + s_{ijkl}, \qquad (11.22)$$
at leading order one finds
$$(L^*s)_{ijkl} = \varepsilon s_{ijkl} - \frac{\varepsilon}{N+8}\left(\delta_{ij}\sum_m s_{mmkl} + 5 \text{ terms} + 12 s_{ijkl}\right), \qquad (11.23)$$
where the five terms are obtained by permutation of the indices i, j, k, l.

Taking s_{ijkl} proportional to g^*_{ijkl}, one recovers the exponent $\omega = \beta'(g^*)$ of the isotropic model. More generally, the eigenvectors can be classified according to their trace properties. We thus parametrize s_{ijkl} in the form
$$s_{ijkl} = u g^*_{ijkl} + (v_{ij}\delta_{kl} + 5 \text{ terms}) + w_{ijkl}, \qquad (11.24)$$
where the tensors v_{ij} and w_{ijkl} are traceless:
$$\sum_i v_{ii} = 0, \quad \sum_k w_{ijkk} = 0. \qquad (11.25)$$

The three eigenvalues corresponding to the components u, w, v are, respectively, $-\omega$, $-\omega_{\text{anis.}}$ and $-\omega'$ with
$$\omega = \varepsilon + O\left(\varepsilon^2\right), \quad \omega_{\text{anis.}} = \varepsilon\frac{4-N}{N+8} + O\left(\varepsilon^2\right), \quad \omega' = \frac{8\varepsilon}{N+8} + O\left(\varepsilon^2\right). \qquad (11.26)$$

The perturbation proportional to v_{ij} does not satisfy the first condition (11.20). It lifts the degeneracy between the correlation lengths of the various field components. This induces a crossover to a situation where several components of the field decouple. However, one easily verifies that the corresponding eigenvalue ω' produces, for ε small, sub-leading effects compared to the eigenvalue corresponding to the quadratic operator $\sigma_i\sigma_j$.

For the set of models satisfying the condition (11.20), the leading eigenvalue is $\omega_{\text{anis.}}$. One then finds the following interesting result (which generalizes a result obtained in the example of the cubic anisotropy): the $O(N)$ symmetric fixed point is stable with respect to all perturbations for N smaller than a value N_{c}. The calculation of $\omega_{\text{anis.}}$ at order ε yields
$$N_{\text{c}} = 4 - 2\varepsilon + O\left(\varepsilon^2\right). \qquad (11.27)$$

This is a new example of a symmetry generated dynamically: for $N < N_{\text{c}}$, in the critical domain, correlation functions have a large-distance behaviour that exhibits more symmetry than the initial microscopic theory.

11.4.3 Gradient flow: Fixed points, stability and field dimension

One verifies that the expression (11.17) of the β-function derives from a potential. Indeed,
$$\beta_{ijkl}(g) = \frac{\partial U(g)}{\partial g_{ijkl}}$$
with
$$U(g) = -\frac{\varepsilon}{2}\sum_{i,j,k,l} g_{ijkl}g_{ijkl} + \frac{1}{(4\pi)^2}\sum_{i,j,k,l,m,n} g_{ijkl}g_{klmn}g_{mnij}. \tag{11.28}$$

Fixed-point stability and exponent η. In the framework of the ε-expansion, we now show that the *stable fixed point* (or at least the stablest) corresponds to the largest value of the exponent η and thus to the correlation functions that have the fastest decay at large distance.

For any fixed point g^*, the equations $\beta = 0$ imply the relation
$$\sum_{i,j,k,l} g^*_{ijkl}\beta_{ijkl} = 0 \Rightarrow \varepsilon \sum_{i,j,k,l} g^*_{ijkl}g^*_{ijkl} = \frac{3}{(4\pi)^2}\sum_{i,j,k,l,m,n} g^*_{ijkl}g^*_{klmn}g^*_{mnij}$$
and, thus,
$$U(g^*) = -\tfrac{1}{6}\varepsilon \sum_{i,j,k,l} g^*_{ijkl}g^*_{ijkl} + O(g^4),$$
a negative value and thus lower than the Gaussian fixed-point value
$$g^* \neq 0 \Rightarrow U(g^*) < U(0),$$
in agreement with the analysis of Section 11.2.3.

Moreover, for a system with only one correlation length, at leading order the exponent η is given by equation (11.18) and thus
$$\eta = \frac{1}{6N}\frac{1}{(4\pi)^4}\sum_{i,j,k,l} g^*_{ijkl}g^*_{ijkl} = -\frac{1}{N\varepsilon}\frac{1}{(4\pi)^4}U(g^*). \tag{11.29}$$

As we have shown, the stable fixed point corresponds to the lowest value of U. It thus corresponds also to the largest value of the exponent η, that is, of the dimension d_σ of the field σ: therefore, the correlation functions corresponding to the stable fixed point have the fastest large-distance decay.

The validity of this result, beyond the ε-expansion, remains a conjecture.

A bound on the exponent η. Let us derive, as an exercise, a general bound on the coefficient of the leading order ε^2 contribution to η.

The first condition (11.20) can be written as
$$\sum_k g_{ijkk} = 8\pi^2 G\delta_{ij}, \quad G = \frac{1}{8\pi^2 N}\sum_{j,k} g_{jjkk}. \tag{11.30}$$

The function $\nu(g)$ can be expressed, at this order, in terms of G (equation (11.19)):

$$\frac{1}{\nu(g)} = 2 - \frac{1}{2}G + O(g^2). \tag{11.31}$$

From a Gaussian expectation value with the measure $e^{-\sum \sigma_i^2/2}$ of inequality (11.12), one infers (using Wick's theorem)

$$\sum_{i,j,k,l} g_{ijkl} \langle \sigma_i \sigma_j \sigma_k \sigma_l \rangle = 3 \sum_{i,k} g_{ikik} > 0$$

and, thus,

$$G > 0. \tag{11.32}$$

Moreover, η is given by equation (11.18):

$$\eta = \frac{1}{6N(4\pi)^4} \sum_{i,j,k,l} (g_{ijkl})^2. \tag{11.33}$$

From the fixed-point equation $\beta_{ijkl}(g^*) = 0$, where β_{ijkl} is given at leading order by equation (11.17), one infers

$$0 = \frac{1}{8\pi^2} \sum_k \beta_{ijkk}(g) = -\varepsilon G \delta_{ij} + \frac{1}{2} G^2 \delta_{ij} + \frac{1}{(8\pi^2)^2} \sum_{k,l,m} g_{iklm} g_{jklm}. \tag{11.34}$$

At this order, the exponent η can thus also be expressed only in terms of G (and thus of ν):

$$\eta = \frac{1}{24} \left(\varepsilon G^* - \tfrac{1}{2} G^{*2} \right) + O(\varepsilon^3).$$

This expression implies, in particular, that the coefficient of order ε of G^* is bounded, $0 < G^* < 2\varepsilon$. One can improve this bound by using the decomposition

$$g_{ijkl}^* = \frac{8\pi^2 G^*}{N+2} (\delta_{ij}\delta_{kl} + \delta_{ik}\delta_{jl} + \delta_{il}\delta_{kj}) + \tilde{g}_{ijkl}^*,$$

where $\sum_i \tilde{g}_{iikl}^* = 0$. Then,

$$\sum_{i,j,k,l} g_{ijkl}^* g_{ijkl}^* = \frac{3N(8\pi^2)^2}{N+2} G^{*2} + \sum_{i,j,k,l} \tilde{g}_{ijkl}^* \tilde{g}_{ijkl}^*,$$

and combining this expression with equation (11.34):

$$\varepsilon G^* - \frac{1}{2} G^{*2} - \frac{3}{N+2} G^{*2} \geq 0.$$

One infers

$$0 < G^* \leq \frac{2(N+2)}{N+8} \varepsilon \leq 2\varepsilon. \tag{11.35}$$

The upper bound corresponds to a fixed point such that $\tilde{g}^*_{ijkl} = 0$ and thus to a fixed point with $O(N)$ symmetry.

Finally, expression (11.33) shows that η is bounded by its value for $G^* = \varepsilon$, if this value of G^* is allowed, and otherwise by the value corresponding to the largest value of G^*. One concludes

$$\begin{cases} \eta \leq \frac{N+8}{2(N+8)^2}\varepsilon^2 & \text{for } N \leq 4, \\ \eta \leq \frac{\varepsilon^2}{48} & \text{for } N \geq 4. \end{cases}$$

The bound for $N \leq 4$ corresponds to the isotropic (i.e. $O(N)$-symmetric) fixed point. It follows also directly from the stability properties of the isotropic fixed point (Section 11.4.2) and the correspondence between the stable fixed point and the largest value of η.

11.5 Exercise: General model with two parameters

To illustrate the preceding results in a simpler situation, we consider a general model with two parameters, which generalizes the example of the model with cubic anisotropy examined in Section 11.3. Moreover, we assume that two non-Gaussian fixed points g_1^* and g_2^*, have been found, one of them being necessarily the isotropic fixed point with $O(N)$ symmetry, for example g_1^*. The space of allowed parameters g being linear, g can take all values contained in the plane $sg_1^* + tg_2^*$. The positivity condition is satisfied at least in the sector $s, t \geq 0$.

At order ε^3, the potential $U(g(s,t))$ depends *a priori* on seven parameters. The fixed-point conditions for $s = 1, t = 0$ and $s = 0, t = 1$,

$$\partial_s U(s=1, t=0) = \partial_t U(s=1, t=0) = 0,$$
$$\partial_s U(s=0, t=1) = \partial_t U(s=0, t=1) = 0,$$

determine four parameters. The potential then takes the general form

$$U(s,t) = -\left(\tfrac{1}{2}as^2 + bst + \tfrac{1}{2}ct^2\right) + \tfrac{1}{3}as^3 + bst(s+t) + \tfrac{1}{3}ct^3, \tag{11.36}$$

where a, b, c are three constants that, due to the positivity of

$$\sum_{i,j,k,l} g_{ijkl} g_{ijkl} > 0 \quad \text{for } g_{ijkl} \neq 0,$$

satisfy the conditions

$$a > 0, \quad c > 0, \quad b^2 - ac < 0.$$

For example, in the model with cubic anisotropy, one finds

$$a = (8\pi^2)^2 \frac{12N(N+2)\varepsilon^3}{(N+8)^2}, \quad b = (8\pi^2)^2 \frac{4N\varepsilon^3}{N+8}, \quad c = (8\pi^2)^2 \frac{4N\varepsilon^3}{9}$$

and, thus,
$$ac - b^2 = (8\pi^2)^4 \frac{16N^2(N-1)\varepsilon^6}{3(N+8)^2}.$$

Expressing that, at the Gaussian fixed point, the degenerate eigenvalue is ε, one obtains the matrix $\mathbf{T}$, which here is a constant matrix because the transformation $g \mapsto \{s,t\}$ is linear and in the parametrization (11.11) $\mathbf{T} = \mathbf{1}$. Here,
$$\mathbf{T} = \frac{1}{ac-b^2} \begin{pmatrix} c & -b \\ -b & a \end{pmatrix},$$
which, indeed, is a positive matrix. Its action on the vector $(\partial_s U, \partial_t U)$ leads to the flow equations
$$\frac{1}{\varepsilon}\lambda \frac{ds}{d\lambda} = s - s^2 + \frac{2b(b-c)}{ac-b^2} st,$$
$$\frac{1}{\varepsilon}\lambda \frac{dt}{d\lambda} = t - t^2 + \frac{2b(b-a)}{ac-b^2} st.$$

We set
$$\alpha_1 = \frac{2b(c-b)}{ac-b^2}, \quad \alpha_2 = \frac{2b(a-b)}{ac-b^2}.$$

In the model with cubic anisotropy
$$\alpha_1 = \frac{2}{3}, \quad \alpha_2 = \frac{12}{N+8}.$$

One then finds a fourth fixed point
$$s = s_4 \equiv \frac{1-\alpha_1}{1-\alpha_1\alpha_2}, \quad t = t_4 \equiv \frac{1-\alpha_2}{1-\alpha_1\alpha_2}.$$

The matrix of partial derivatives of the functions $-\beta$ is
$$\mathbf{L}^* = \varepsilon \begin{pmatrix} 1 - 2s - \alpha_1 t & -\alpha_1 s \\ -\alpha_2 t & 1 - 2t - \alpha_2 s \end{pmatrix}.$$

The eigenvalues at the various fixed points are
$$\varepsilon \times \begin{cases} 1, 1 & \text{for } s=t=0 \\ -1, 1-\alpha_2 & \text{for } s=1, t=0 \\ -1, 1-\alpha_1 & \text{for } s=0, t=1 \\ -1, 1-\alpha_3 & \text{for } s=s_4, t=t_4, \end{cases}$$

where we have set
$$1 - \alpha_3 = -\frac{(1-\alpha_1)(1-\alpha_2)}{1-\alpha_1\alpha_2} \Leftrightarrow 2 - \alpha_1 - \alpha_2 - \alpha_3 + \alpha_1\alpha_2\alpha_3 = 0. \tag{11.37}$$

One notices that, from the algebraic viewpoint, the three fixed points play a completely symmetric role.

The stability condition for a fixed point is

$$\alpha_i > 1.$$

Let us verify that, in agreement with the general result, the conditions $\alpha_1 > 1$ and $\alpha_2 > 1$ are incompatible. For example, $\alpha_1 > 1$ implies

$$2b(c-b) > ac - b^2 \Rightarrow c(c-a) > (b-c)^2 \Rightarrow c > a.$$

It is clear that the condition $\alpha_2 > 1$ implies $a > c$, which is incompatible.

Moreover, the values of the potential at the fixed points g_1^* and g_2^*, respectively, are $-a/6$ and $-c/6$ and, thus, the stable fixed point, indeed, corresponds to the lowest potential value.

We now use the property that the isotropic fixed point is always present. We identify it with g_1^*:

$$(g_1^*)_{ijkl} = \frac{16\pi^2 \varepsilon}{N+8} (\delta_{ij}\delta_{kl} + \delta_{ik}\delta_{jl} + \delta_{il}\delta_{kj}).$$

The value of a is the same as in the model with cubic anisotropy. The parameter b is given by

$$b = \varepsilon \sum_{i,j,k,l} (g_1^*)_{ijkl} (g_2^*)_{ijkl} = (8\pi^2)^2 \frac{6\varepsilon^2}{N+8} NG_2^*,$$

and, finally,

$$c = \varepsilon \sum_{i,j,k,l} (g_2^*)_{ijkl} (g_2^*)_{ijkl} = (8\pi^2)^2 N\varepsilon \left(\varepsilon G_2^* - \tfrac{1}{2} G_2^{*2}\right).$$

At this order, all parameters can be expressed in terms of G_2^*. The parameters α_i become

$$\alpha_1 = G_2^*/\varepsilon, \quad \alpha_2 = \frac{12}{N+8}.$$

One sees that α_1 and α_2 are positive and, as a consequence of the bound (11.35),

$$\alpha_1 + \alpha_2 < 2.$$

Then, equation (11.37) implies

$$\alpha_3 > 0, \quad \alpha_i + \alpha_j < 2 \text{ for } i \neq j.$$

Finally, equation (11.37) shows conversely that $0 < \alpha_1, \alpha_2 < 1$ implies $\alpha_3 > 1$. Thus, there always exists a stable fixed point. Like in the example of the model with cubic anisotropy, if the fourth fixed point is stable, $s_4, t_4 > 0$.

Finally, notice that most of these considerations generalize to the next order in ε.

Exercises

Exercise 11.1

Cubic anisotropy. Using the explicit expressions of the β-functions given in Section 11.3, find the matrices $\mathbf{M}$ such that one can write

$$\frac{\partial U}{\partial g} = M_{11}\beta_g + M_{12}\beta_h$$

$$\frac{\partial U}{\partial h} = M_{21}\beta_g + M_{22}\beta_h,$$

and determine the corresponding potential function U. Calculate the values of the potential for the different fixed points.

Solution. All matrices are proportional to

$$\mathbf{M} = \begin{pmatrix} \frac{1}{3}(N+2) & 1 \\ 1 & 1 \end{pmatrix},$$

which is indeed a positive matrix for $N > 1$, and with this choice

$$8\pi^2 U(g,h) = -8\pi^2 \varepsilon \left[\tfrac{1}{6}(N+2)g^2 + gh + \tfrac{1}{2}h^2\right] + \tfrac{1}{54}(N+8)(N+2)g^3$$
$$+ \tfrac{1}{6}(N+8)g^2h + \tfrac{3}{2}gh^2 + \tfrac{1}{2}h^3.$$

The values are

$$U_{\text{gaus.}} = 0, \quad U_{\text{Is.}} = -(8\pi^2)^2 \frac{2}{27}\varepsilon^3, \quad U_{O(N)} = -(8\pi^2)^2 \frac{2(N+2)}{(N+8)^2}\varepsilon^3,$$

$$U_{\text{cub.}} = -(8\pi^2)^2 \frac{2(N-1)(N+2)}{27N^2}\varepsilon^3.$$

In particular, one verifies that

$$U_{O(N)} - U_{\text{cub.}} = (8\pi^2)^2 \frac{2(N+2)(N-4)^3}{27N^2(N+8)^2}\varepsilon^3$$

changes sign at $N = 4$, in correspondence with the change in fixed-point stability.

12 Statistical field theory: Perturbative expansion

An analysis of the leading corrections to the quasi-Gaussian (or mean-field) approximation, as well as renormalization group (RG) arguments, have shown that at least in dimension $d = 4 - \varepsilon$, that is, in an infinitesimal neighbourhood of dimension 4, the universal properties of second-order phase transitions can be entirely described by a statistical field theory in continuum space. In Chapter 10, we have used this idea to determine the RG functions to the first non-trivial order in ε and in the σ^4 coefficient. The calculation were based on assumptions that require some justification. We thus give in this chapter, devoted to perturbative calculations, and in the following where we relate renormalization theory to RG equations, a short outline of the field theory methods that allow, in the framework of the ε-expansion, that is, in the sense of formal series, proving RG results and calculating universal quantities.

In this chapter, we discuss the perturbative calculation of correlation or vertex functions expressed in terms of field (functional) integrals. The successive contributions to the perturbative expansion are Gaussian expectation values which can be calculated with the help, for example, of Wick's theorem and which have a representation in the form of Feynman diagrams (defined in Section 2.4). We illustrate diagrammatically the relations between the first connected correlation functions and the corresponding vertex functions.

Of course, these calculations have an algebraic structure which is analogous to the calculations presented in Sections 2.3 and 2.5 for ordinary integrals, and in Section 5.3 in the case of path integrals.

We also show that the calculation of a field integral by the steepest descent method organizes the perturbative expansion as an expansion in the number of loops in the Feynman diagram representation.

Finally, we have already introduced the idea of dimensional continuation of Feynman diagrams. We define here more generally dimensional continuation and introduce dimensional regularization.

12.1 Generating functionals

In Chapter 6, on the lattice, and in Sections 9.1.1 and 9.2 in the continuum, we have already introduced the generating functional of correlation functions. For convenience, we just recall here the definitions.

Let $\sigma(x)$ be a random classical field endowed with a probability distribution, a normalized positive measure in the space of fields $[d\sigma]\, e^{-\mathcal{H}(\sigma)}/\mathcal{Z}$, where $\mathcal{H}(\sigma)$ is the Hamiltonian and $\mathcal{Z}$ the associated partition function.

286   Statistical field theory: Perturbative expansion

The n-point correlation function,

$$\langle \sigma(x_1)\sigma(x_2)\ldots\sigma(x_n)\rangle = \frac{1}{\mathcal{Z}} \int [\mathrm{d}\sigma]\sigma(x_1)\sigma(x_2)\ldots\sigma(x_n)\,\mathrm{e}^{-\mathcal{H}(\sigma)}, \qquad (12.1)$$

can be inferred from the generating functional of correlation functions

$$\mathcal{Z}(H) = \int [\mathrm{d}\sigma]\exp\left[-\mathcal{H}(\sigma) + \int \mathrm{d}^d x\, H(x)\sigma(x)\right] \qquad (12.2a)$$

$$= \mathcal{Z}(0)\left\langle \exp\int \mathrm{d}^d x\, \sigma(x) H(x)\right\rangle \qquad (12.2b)$$

($\mathcal{Z}$ in (12.1) is identical to $\mathcal{Z}(0)$) by functional differentiation. Indeed,

$$\langle \sigma(x_1)\sigma(x_2)\ldots\sigma(x_n)\rangle = \frac{1}{\mathcal{Z}(0)}\left(\prod_i \frac{\delta}{\delta H(y_i)}\right)\mathcal{Z}(H)\bigg|_{H=0}. \qquad (12.3)$$

The functional

$$\mathcal{W}(H) = \ln \mathcal{Z}(H), \qquad (12.4)$$

generates the connected correlation functions:

$$\mathcal{W}(H) = \sum_{n=0}^{\infty} \frac{1}{n!}\int \mathrm{d}^d x_1 \ldots \mathrm{d}^d x_n\, W^{(n)}(x_1,\ldots,x_n)H(x_1)\ldots H(x_n).$$

In local field theories (continuum limits of statistical systems with short-range interactions) connected correlation functions satisfy the cluster property (see Section 6.2.1): $W^{(n)}(x_1,\ldots,x_n)$ decays, algebraically or exponentially, when the points $x_1, x_2, \ldots, x_n$ belong to two largely separated non-empty subsets.

Generating functional of vertex functions. In Section 9.2, we have also defined the generating functional $\Gamma(M)$ of *vertex functions* (see also Section 6.3):

$$\Gamma(M) = \sum_{n=0}^{\infty} \frac{1}{n!}\int \mathrm{d}^d x_1 \ldots \mathrm{d}^d x_n\, \Gamma^{(n)}(x_1,\ldots,x_n)M(x_1)\ldots M(x_n).$$

$\Gamma(M)$ is the Legendre transform of $\mathcal{W}(H)$ (equations (9.7)):

$$\mathcal{W}(H) + \Gamma(M) = \int \mathrm{d}^d x\, H(x)M(x), \quad M(x) = \frac{\delta \mathcal{W}(H)}{\delta H(x)}.$$

12.2 Gaussian field theory. Wick's theorem

In field theory, as in other stochastic processes, the simplest measure is the Gaussian measure. An example has been discussed in Section 8.4.

Gaussian field theory. In the Gaussian case, a translation-invariant Hamiltonian can, quite generally, be written as

$$\mathcal{H}_G(\sigma) = \frac{1}{2} \int d^d x \, d^d y \, \sigma(x) \mathcal{H}^{(2)}(x-y) \sigma(y). \tag{12.5}$$

The kernel $\mathcal{H}^{(2)}(x-y)$ is symmetric, positive and local. In the simplest examples, $\mathcal{H}^{(2)}(x-y)$ is a differential operator, polynomial in the Laplacian ∇_x^2:

$$\mathcal{H}^{(2)}(x-y) \equiv \mathcal{K}(-\nabla_x^2) \delta^{(d)}(x-y).$$

More precisely, one can write expression (12.5) in the form (see expression (10.2))

$$\int d^d x \, d^d y \, \sigma(x) \mathcal{H}^{(2)}(x-y) \sigma(y)$$

$$\equiv \int d^d x \left\{ \sum_{\mu=1}^d \partial_\mu \sigma(x) \left(1 + \sum_{k=1} u_{k+1}(-\nabla_x^2)^k \right) \partial_\mu \sigma(x) + m^2 \sigma^2(x) \right\}.$$

The limit $m = 0$ corresponds to a critical theory (or massless theory in the language of quantum field theory). For $m > 0$ (massive theory), the correlation length ξ is finite: $\xi \propto 1/m$.

The field integral,

$$\mathcal{Z}_G(H) = \int [d\sigma] \exp[-\mathcal{H}_G(\sigma)] \exp\left[\int d^d x \, \sigma(x) H(x)\right]$$

$$= \int [d\sigma] \exp\left[-\mathcal{H}_G(\sigma) + \int d^d x \, \sigma(x) H(x)\right],$$

a functional of the external field $H(x)$, is proportional to the generating functional of correlation functions corresponding to the measure $e^{-\mathcal{H}_G(\sigma)}/\mathcal{Z}_G(0)$,

$$\left\langle \exp\left[\int d^d x \, \sigma(x) H(x)\right] \right\rangle_{\mathcal{H}_G} = \mathcal{Z}_G(H)/\mathcal{Z}_G(0).$$

Calculation of the integral. The calculation of the field integral is a simple generalization of the calculation presented in Section 8.4. Denoting by Δ the inverse of $\mathcal{H}^{(2)}$,

$$\int d^d z \, \Delta(x-z) \mathcal{H}^{(2)}(z-y) = \delta^{(d)}(x-y), \tag{12.6}$$

one change variables $\sigma(x) \mapsto \sigma'(x)$ with

$$\sigma(x) = \sigma'(x) + \int d^d y\, \Delta(x-y) H(y). \tag{12.7}$$

This shift of $\sigma(x)$ eliminates the term linear in σ in the exponential. The measure is invariant and the integrand becomes

$$\mathcal{Z}_G(H) = \exp\left[\frac{1}{2}\int d^d y\, H(x)\Delta(x-y)H(y)\right]\int [d\sigma']\exp[-\mathcal{H}_G(\sigma')].$$

The dependence in H is now explicit. The residual integral yields only a normalization. Its calculation may be difficult but it cancels in the calculation of correlation functions. Indeed, the measure must be normalized in such a way that $\langle 1 \rangle_G = 1$, where $\langle \bullet \rangle_G$ means Gaussian expectation value (or free field in the context of quantum field theory). One concludes

$$\left\langle \exp\left[\int d^d x\, \sigma(x) H(x)\right]\right\rangle_G = \mathcal{Z}_G(H)/\mathcal{Z}_G(0)$$

$$= \exp\left[\frac{1}{2}\int d^d x\, d^d y\, H(x)\Delta(x-y)H(y)\right]. \tag{12.8}$$

The kernel Δ, the inverse of $\mathcal{H}^{(2)}$, is the Gaussian two-point function, and is also called the *propagator*. In a translation-invariant theory, it is convenient to introduce the Fourier representation:

$$\mathcal{H}^{(2)}(x) = \frac{1}{(2\pi)^d}\int d^d p\, e^{-ip\cdot x}\tilde{\mathcal{H}}^{(2)}(p), \quad \tilde{\mathcal{H}}^{(2)}(p) = \int d^d x\, e^{ip\cdot x}\mathcal{H}^{(2)}(x), \tag{12.9a}$$

$$\Delta(x) = \frac{1}{(2\pi)^d}\int d^d p\, e^{ip\cdot x}\tilde{\Delta}(p), \quad \tilde{\Delta}(p) = \int d^d x\, e^{ip\cdot x}\Delta(x), \tag{12.9b}$$

and thus

$$\tilde{\Delta}(p)\tilde{\mathcal{H}}^{(2)}(p) = 1.$$

In order for the field integral to exist, $\tilde{\mathcal{H}}^{(2)}(p)$ must be positive for $p \neq 0$. If $\mathcal{H}^{(2)}(x)$ is rotation-invariant (the special orthogonal group $SO(d)$), the function $\Delta(x)$ is a function only of $|x|$, and $\tilde{\Delta}(p)$ only of $|p|$.

Connected correlation functions. The generating functional of connected correlation functions $\mathcal{W}_G = \ln \mathcal{Z}_G$ then reduces to a simple quadratic form:

$$\mathcal{W}_G(H) - \mathcal{W}_G(0) = \frac{1}{2}\int d^d x\, d^d y\, H(x)\Delta(x-y)H(y).$$

In the Gaussian case, connected functions with more than two points vanish.

Vertex functions. Finally, the Legendre transform $\Gamma_G(M)$ is directly related to the Hamiltonian (Section 6.1) since

$$\Gamma_G(M) = \Gamma_G(0) + \mathcal{H}_G(M).$$

Wick's theorem. Expression (12.8) combined with the arguments of Section 2.2, leads to an immediate generalization of equations (2.15)–(2.20) or (5.21), which expresses Wick's theorem in field theory:

$$\left\langle \prod_1^{2s} \sigma(z_i) \right\rangle_G = \left[\prod_{i=1}^{2s} \frac{\delta}{\delta H(z_i)} \exp\left[\mathcal{W}_G(H) - \mathcal{W}_G(0)\right] \right]\Bigg|_{H \equiv 0}$$

$$= \sum_{\substack{\text{all pairings} \\ \text{of } \{1,2,\ldots,2s\}}} \Delta(z_{i_1} - z_{i_2}) \ldots \Delta(z_{i_{2s-1}} - z_{i_{2s}}). \quad (12.10)$$

12.3 Perturbative expansion

We now consider a more general Hamiltonian of the form

$$\mathcal{H}(\sigma) = \mathcal{H}_G(\sigma) + \mathcal{V}_I(\sigma), \quad (12.11)$$

where $\mathcal{H}_G(\sigma)$ is the quadratic form (12.5) and $\mathcal{V}_I(\sigma)$ is a polynomial in the field, which, in the context of quantum field theory, is called an *interaction*. In a local field theory, that is, in the class that we have introduced in the preceding chapters, $\mathcal{V}_I(\sigma)$ is the space integral of a function of the field and its derivatives:

$$\mathcal{V}_I(\sigma) = \int d^d x \, V_I[\sigma(x), \partial_\mu \sigma(x), \ldots]. \quad (12.12)$$

Note that in quantum field theory, the Hamiltonian $\mathcal{H}(\sigma)$ is sometimes also called the *Euclidean action* in the sense that, in simple examples, it can be formally obtained from a classical action by a continuation to imaginary times.

Although most results presented in this chapter will be illustrated only by Hamiltonians of type (10.1), these results apply to more general theories.

The generating functional of correlation functions is proportional to

$$\mathcal{Z}(H) = \int [d\sigma] \exp\left[-\mathcal{H}(\sigma) + \int d^d x \, H(x) \sigma(x)\right]. \quad (12.13)$$

12.3.1 Perturbative expansion

The perturbative expansion of correlation functions is obtained by expanding expression (12.13) in powers of H and of $\mathcal{V}_I$, keeping only the quadratic term $\mathcal{H}^{(2)}$ in the exponential. The interaction (12.12) is a sum of monomials called *interaction vertices*. The expansion then reduces to the calculation of Gaussian expectation values of products of fields of the form

$$\left\langle \sigma(x_1) \cdots \sigma(x_n) \int d^d y_1 \, \sigma^{p_1}(y_1) \int d^d y_2 \, \sigma^{p_2}(y_2) \cdots \int d^d y_k \, \sigma^{p_k}(y_k) \right\rangle_G \quad (12.14)$$

(to simplify, we have omitted possible derivatives) and, thus, to Wick's theorem. Wick's theorem involves the Gaussian two-point function or propagator Δ (equation (12.6)). Each contribution takes the form of a product of propagators integrated over all points corresponding to interaction vertices and has a graphical representation in terms of Feynman diagrams (see Section 2.4).

The perturbative expansion has a formal global representation. Using the property (9.1),

$$\frac{\delta}{\delta H(x)} \exp\left[\int d^d y\, H(y)\sigma(y)\right] = \sigma(x) \exp\left[\int d^d y\, H(y)\sigma(y)\right],$$

one can express $\mathcal{Z}(H)$ in terms of the Gaussian functional $\mathcal{Z}_G(H)$ in the form

$$\mathcal{Z}(H) = \exp\left[-\mathcal{V}_I\left(\frac{\delta}{\delta H}\right)\right] \mathcal{Z}_G(H) = \exp\left[-\mathcal{V}_I\left(\frac{\delta}{\delta H}\right)\right] \exp\left[\mathcal{W}_G(H)\right]$$

$$= \mathcal{Z}_G(0) \exp\left[-\mathcal{V}_I\left(\frac{\delta}{\delta H}\right)\right] \exp\left[\frac{1}{2}\int d^d x\, d^d y\, H(x)\Delta(x-y)H(y)\right]. \quad (12.15)$$

Combining identities (12.3) and (12.15), one can calculate correlation functions of the field σ as formal series in powers of the *interaction* $\mathcal{V}_I$, to use quantum field theory language. To each monomial contributing to $\mathcal{V}_I$ there corresponds a differential operator: a product of derivatives $\delta/\delta H$ that generates a product of propagators Δ.

Remark. As we have already noted in Chapter 10, the kernel $\mathcal{H}^{(2)}(x-y)$ cannot be reduced to the Gaussian fixed point form $-\nabla_x^2 \delta^{(d)}(x-y)$ because the perturbative expansion then contains short-distance divergences. This is again a manifestation of a coupling between the different physical scales. These divergences have no physical meaning in the theory of phase transitions since the fixed-point two-point function is only the asymptotic form at large distance, and the lattice, or more generally the microscopic structure, regularizes the theory at short distance.

In the continuum, one must instead add to the Hamiltonian irrelevant terms with higher order derivatives, to generate a propagator $\Delta(x-y)$ sufficiently regular for $|x-y| \to 0$, in contrast with the Gaussian two-point function (Section 8.4.3). This substitution is called a *regularization*.

In the Fourier representation, short-distance divergences become large-momentum or *ultraviolet (UV)* divergences (in quantum field theory the arguments of the Fourier transform are momenta or energies). On the lattice, these divergences are absent since momenta vary in a bounded domain, a Brillouin zone. Finiteness of the perturbative expansion requires, in the continuum, that $\tilde{\Delta}(p)$ decays fast enough for $p \to \infty$. We thus assume, in what follows, that the large momentum decay of the propagator, in the Fourier representation, is sufficiently fast to render the perturbative expansion finite to all orders.

12.3.2 Feynman diagrams: Loops

Feynman diagrams have already been defined in Section 2.4. To each monomial contributing to the interaction, one associates a vertex, a point from which originates a number of lines equal to the degree of the monomial. A propagator is represented by a line that joins the points that correspond to its arguments. These points are either vertices, or points corresponding to arguments of a correlation function. In what follows we call an *internal line* a line that connects two vertices. By contrast, an *external* line joins a vertex to a point of a correlation function.

In a local Hamiltonian, a vertex corresponds to a space integral of a product of fields and their derivatives (representation (12.12)). Each vertex, in a diagram, thus corresponds to an argument on which one integrates.

After Fourier transformation, to each line is attached a momentum, the argument of the propagator in the Fourier representation. This assumes an orientation of the lines: changing the orientation changes the sign of the momentum attached to the line. In the Fourier representation, due to translation invariance, at each vertex the sum of the entering momenta vanishes: one finds Kirchoff's laws for current intensities in an electric circuit. Finally, one integrates over all free momenta.

Remark. For any connected diagram, one proves the relation between the number of *loops* L, the number I of internal lines, or propagators relating vertices, and the number of vertices n:

$$L = I - n + 1. \qquad (12.16)$$

A method for establishing this relation is the following:

(i) When one cuts an internal line, one also suppresses one loop, thus $L - I$ is a constant.

(ii) A diagram without a loop is a tree. Each time one suppresses one vertex on the boundary of a tree, one transforms an internal line into an external line, thus $I - n$ is a constant. Finally, any diagram is reduced in this way to a vertex, thus if $L = I = 0$ then $n = 1$.

In the Fourier representation, L is also the number of free momenta over which one integrates. Indeed, this number is equal to the number of propagators minus the number of vertices, due to momentum conservation at each vertex, plus one because the conservation of the total momentum entering in a diagram is then automatically satisfied.

Remark. Local interactions can also contain derivatives of the field $\sigma(x)$. Then, the evaluation of expression (12.14) involves also derivatives of the propagator. The representation in terms of Feynman diagrams, as they have been defined so far, is no longer faithful since the presence of derivatives is not indicated. One can construct a more faithful representation by splitting vertices and by placing arrows on lines.

12.3.3 Connected and 1-irreducible diagrams

In the Feynman diagram representation, the perturbative expansion of $\mathcal{Z}(H)$ contains non-connected contributions in the sense of graphs. It follows from the arguments of Section 2.3 that $\mathcal{W}(H)$, by contrast, is the sum of connected contributions.

Finally, the functional $\Gamma(M)$ has the simplest perturbative properties: indeed, one proves that its expansion contains only 1-irreducible diagrams, that is, diagrams that cannot be decomposed into several connected components by cutting only one line. These are the diagrams that are directly involved in the renormalization theory.

For illustration, we give a graphical representation of the first relations between connected and vertex functions.

Fig. 12.1 Representations of the connected two-point correlation function and the n-point vertex function.

In Figure 12.1, we define the graphical representation of $W^{(2)}$ and $\Gamma^{(n)}$. In the representation of $\Gamma^{(n)}$, we have emphasized the property that no propagator is attached to the points of the boundary of the graph, in contrast with the diagrams contributing to connected functions.

The relation between two-point functions is

$$\int d^d z \, W^{(2)}(x-z) \Gamma^{(2)}(z-y) = \delta^{(d)}(x-y).$$

It is convenient to set

$$\Gamma^{(2)}(x-y) = \mathcal{H}^{(2)}(x-y) + \Sigma(x-y),$$

where we have separated the Gaussian contribution $\mathcal{H}^{(2)}$ from the sum Σ of contributions generated by the interactions $\mathcal{V}_\mathrm{I}$, also called the *mass operator*. In terms of Σ, the perturbative expansion of $W^{(2)}$ can be organized as a geometrical series:

$$W^{(2)}(x-y) = \Delta(x-y) - \int d^d z_1 \, d^d z_2 \, \Delta(x-z_1) \Sigma(z_1-z_2) \Delta(z_2-y) + \cdots .$$

Figure 12.2 displays one term of the sum expressed in terms of Σ, that is, of 1-irreducible components.

Fig. 12.2 Contribution to the connected two-point function $W^{(2)}$.

The graphical representations of the correlation functions $W^{(3)}$ and $W^{(4)}$, for example, in terms of the corresponding vertex functions and of $W^{(2)}$ are then given in Figures 12.3 and 12.4, respectively.

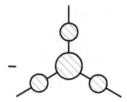

Fig. 12.3 The connected three-point function $W^{(3)}$.

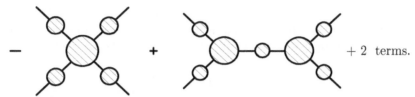

Fig. 12.4 The connected four-point function $W^{(4)}$.

12.3.4 Example: The σ^4 interaction

We now consider the example, specially useful from the viewpoint of critical phenomena, of the quartic interaction

$$\mathcal{V}_{\mathrm{I}}(\sigma) \equiv \frac{1}{4!} g \int \mathrm{d}^d x \, \sigma^4(x). \qquad (12.17)$$

The first non-trivial order in g of the RG functions can be derived from the expansion of the two- and four-point functions up to order g^2. One finds expressions whose algebraic structure is identical to expressions (2.30) and (2.35) (see also Section 5.4.2).

Two-point function. The two-point function at order g^2 is given by

$$\langle \sigma(x_1)\sigma(x_2) \rangle = (\mathrm{a}) - \tfrac{1}{2} g \, (\mathrm{b}) + \tfrac{1}{4} g^2 \, (\mathrm{c}) + \tfrac{1}{4} g^2 \, (\mathrm{d}) + \tfrac{1}{6} g^2 \, (\mathrm{e}) + O\left(g^3\right).$$

Note that three additional contributions that factorize in

$$\langle \sigma(x_1)\sigma(x_2) \rangle_{\mathrm{G}} \langle \sigma^4(y) \rangle_{\mathrm{G}}, \quad \langle \sigma(x_1)\sigma(x_2)\sigma^4(y_1) \rangle_{\mathrm{G}} \langle \sigma^4(y_2) \rangle_{\mathrm{G}} \text{ and}$$
$$\langle \sigma(x_1)\sigma(x_2) \rangle_{\mathrm{G}} \langle \sigma^4(y_1)\sigma^4(y_2) \rangle_{\mathrm{G}},$$

cancel after division by the partition function $\mathcal{Z}$. In quantum field theory, the connected diagrams contributing to $\ln \mathcal{Z}$ are also called vacuum diagrams, because in the quantum context they contribute to the ground state (the vacuum) energy (see equation (4.49)).

In the expansion, (a) is the propagator and (b) the Feynman diagram that appears at order g, and both are displayed in Figure 2.5 (Section 2.5.1). The diagrams (c), (d), (e) are displayed in Figure 2.6 (Section 2.5.1). More explicitly, one thus finds (here $W^{(2)} = Z^{(2)}$) (see equation (2.30))

$$W^{(2)}(x_1 - x_2) = \Delta(x_1 - x_2) - \tfrac{1}{2}g \int d^d y\, \Delta(x_1 - y)\Delta(0)\Delta(y - x_2)$$

$$+ g^2 \int d^d y_1\, d^d y_2 \left[\tfrac{1}{4}\Delta^2(0)\Delta(x_1 - y_1)\Delta(y_1 - y_2)\Delta(y_2 - x_2)\right.$$
$$+ \tfrac{1}{4}\Delta(0)\Delta^2(y_1 - y_2)\Delta(x_1 - y_2)\Delta(x_2 - y_2)$$
$$\left. + \tfrac{1}{6}\Delta(x_1 - y_1)\Delta^3(y_1 - y_2)\Delta(y_2 - x_2)\right] + O(g^3).$$

Only the diagram (c) is 1-reducible and it cancels in the Legendre transformation. Also, the external propagators are removed. One then finds (see Section 6.5)

$$\Gamma^{(2)}(x_1 - x_2) = \mathcal{H}^{(2)}(x_1 - x_2) + \tfrac{1}{2}g\delta^{(d)}(x_1 - x_2)\Delta(0)$$
$$- \tfrac{1}{4}g^2\Delta(0) \int d^d y\, \Delta^2(y)\delta^{(d)}(x_1 - x_2) - \tfrac{1}{6}g^2\Delta^3(x_1 - x_2) + O(g^3),$$

which is a simpler expression.

Fourier transformation. As we have already pointed out, in a translation-invariant theory the relations between correlation and vertex functions take simpler forms in the Fourier representation. We thus introduce the functions defined in (9.6), (9.8) and the representations (12.9) of the propagator and of its inverse. With this notation, the vertex two-point function becomes

$$\tilde{\Gamma}^{(2)}(p) = \tilde{\mathcal{H}}^{(2)}(p) + \frac{g}{2}\int \frac{d^d q}{(2\pi)^d} \tilde{\Delta}(q) - \frac{g^2}{4}\int \frac{d^d q_1}{(2\pi)^d} \tilde{\Delta}(q_1) \int \frac{d^d q_2}{(2\pi)^d} \tilde{\Delta}^2(q_2)$$
$$- \frac{g^2}{6}\int \frac{d^d q_1}{(2\pi)^d}\frac{d^d q_2}{(2\pi)^d} \tilde{\Delta}(q_1)\tilde{\Delta}(q_2)\tilde{\Delta}(p - q_1 - q_2) + O(g^3). \qquad (12.18)$$

The connected two-point function is then obtained by expanding the relation

$$\widetilde{W}^{(2)}(p)\tilde{\Gamma}^{(2)}(p) = 1.$$

Four-point function. At order g^2, the four-point function is given by (see equation (2.35))

$$\langle \sigma(x_1)\sigma(x_2)\sigma(x_3)\sigma(x_4)\rangle$$
$$= [(a)_{12}\,(a)_{34} + 2\,\text{terms}] - \frac{g}{2}[(a)_{12}\,(b)_{34} + 5\,\text{terms}] - g\,(f)$$
$$+ g^2\left\{(a)_{12}\left[\tfrac{1}{4}((c)_{34} + (d)_{34}) + \tfrac{1}{6}(e)_{34}\right] + 5\,\text{terms}\right\}$$
$$+ \frac{g^2}{4}[(b)_{12}\,(b)_{34} + 2\,\text{terms}] + \frac{g^2}{2}[(g) + 3\,\text{terms}] + \frac{g^2}{2}[(h) + 2\,\text{terms}]$$
$$+ O(g^3).$$

The diagrams (f), (g), (h) are displayed in Figure 2.7 (Section 2.5.2). The notation $(a)_{12}$, for example, means diagram (a), contributing to the two-point function, with arguments x_1 and x_2. Finally, the terms that must be added in order to restore the permutation symmetry of the four-point function are obtained by exchanging the external arguments.

Diagrams such as $a_{12}a_{34}$, which are expressed in terms of two-point function contributions, are not connected and factorize into a product of functions that depend on disjoint subsets of variables. The origin of this property has already been indicated in Section 2.3.

Again, as in the case of the two-point function, we have omitted non-connected diagrams in which one factor has no external arguments. These diagrams are cancelled by the perturbative contributions of the partition function $\mathcal{Z}$ in expression (12.3).

The connected four-point function, in a more explicit notation, reduces to

$$W^{(4)}(x_1, x_2, x_3, x_4)$$
$$= -g \int d^d y\, \Delta(x_1 - y)\Delta(y - x_2)\Delta(x_3 - y)\Delta(x_4 - y)$$
$$+ \tfrac{1}{2}g^2 \int d^d y_1 d^d y_2\, \Delta(x_1 - y_1)\Delta(x_2 - y_1)\Delta(x_3 - y_2)\Delta(x_4 - y_2)\Delta^2(y_1 - y_2)$$
$$+ \text{ 2 terms}$$
$$+ \tfrac{1}{2}g^2 \int d^d y_1 d^d y_2\, \Delta(y_1 - y_1)\Delta(y_1 - y_2)\Delta(x_1 - y_1)\Delta(x_2 - y_2)\Delta(x_3 - y_2)$$
$$\times \Delta(x_4 - y_2) + \text{ 3 terms} + O(g^3).$$

The Legendre transformation is simple also for the four-point function in this theory: here it suffices to remove the contributions of the two-point functions on the external lines, an operation called an *amputation*, and to change the sign. One finds (see Section 6.5)

$$\Gamma^{(4)}(x_1, x_2, x_3, x_4) = g\delta^{(d)}(x_1 - x_2)\delta^{(d)}(x_1 - x_3)\delta^{(d)}(x_1 - x_4) - \tfrac{1}{2}g^2\delta^{(d)}(x_1 - x_2)$$
$$\times \delta^{(d)}(x_3 - x_4)\Delta^2(x_1 - x_3) + \text{ 2 terms } + O(g^3).$$

The respective Fourier transforms are then given by

$$\widetilde{W}^{(4)}(p_1, p_2, p_3, p_4) = \tilde{\Delta}(p_1)\tilde{\Delta}(p_2)\tilde{\Delta}(p_3)\tilde{\Delta}(p_4)$$
$$\times \left[-g + \frac{g^2}{2}\int \frac{d^d q}{(2\pi)^d} \tilde{\Delta}(p_1 + p_2 - q)\tilde{\Delta}(q) + \text{ 2 terms} \right.$$
$$\left. + \frac{g^2}{2}\tilde{\Delta}(p_1)\int \frac{d^d q}{(2\pi)^d}\tilde{\Delta}(q) + \text{ 3 terms} \right] + O(g^3),$$

and

$$\tilde{\Gamma}^{(4)}(p_1, p_2, p_3, p_4) = g - \frac{g^2}{2}\int \frac{d^d q}{(2\pi)^d}\tilde{\Delta}(p_1 + p_2 - q)\tilde{\Delta}(q) + \text{ 2 terms} + O(g^3). \quad (12.19)$$

12.4 Loop expansion

The perturbative expansion can be organized in a different way, based on a calculation by the steepest descent method of the field integral (see Section 2.6). In the example of the integral (12.2a), the saddle point σ_c is given by the minimum of the functional

$$\mathcal{H}(\sigma, H) = \mathcal{H}(\sigma) - \int d^d x\, H(x)\sigma(x),$$

and thus is solution of the equation

$$\left.\frac{\delta \mathcal{H}(\sigma, H)}{\delta \sigma(x)}\right|_{\sigma=\sigma_c} = 0.$$

After the change of variables $\sigma \mapsto \chi$,

$$\sigma(x) = \sigma_c(H; x) + \chi(x),$$

one expands $\mathcal{H}(\sigma, H)$ in powers of χ:

$$\mathcal{H}(\sigma, H) = \mathcal{H}(\sigma_c, H) + \frac{1}{2!}\int d^d x_1 d^d x_2\, \chi(x_1) \left.\frac{\delta^2 \mathcal{H}(\sigma)}{\delta\sigma(x_1)\delta\sigma(x_2)}\right|_{\sigma=\sigma_c} \chi(x_2) + O(\chi^3).$$

One keeps the term quadratic in χ in the exponential and expands the terms of higher order in χ. This reduces the calculation of each term to a Gaussian expectation value. The two first terms of such an expansion have already been given, in a formalism with discrete variables, in Section 6.4.

12.4.1 Leading order: Tree diagrams

Approximating the field integral by its value at the saddle point, one finds the leading contribution to the connected functional

$$\mathcal{W}_0(H) = -\mathcal{H}(H, \sigma_c) = -\mathcal{H}(\sigma_c) + \int d^d x\, H(x)\sigma_c(x). \tag{12.20}$$

One then verifies that the Legendre transform Γ_0 of $\mathcal{W}_0(H)$ is given by (Section 6.4)

$$\Gamma_0(M) = \mathcal{H}(M). \tag{12.21}$$

The direct perturbative expansion is obtained by expanding the solution $\sigma_c(H)$ in powers of H. The diagrams generated in this way are tree diagrams (without loops). In the example of the Hamiltonian (12.11), the expansion takes the form

$$\sigma_c(x) = \int d^d y\, \Delta(x-y) H(y) - \int d^d y\, \Delta(x-y)\frac{\delta \mathcal{V}_I(\sigma_c)}{\delta \sigma(y)}$$
$$= \int d^d y\, \Delta(x-y) H(y) - \int d^d y\, \Delta(x-y)\frac{\delta \mathcal{V}_I}{\delta \sigma_c(y)}(\Delta H) + \cdots, \tag{12.22}$$

where the argument ΔH of $\mathcal{V}_{\mathrm{I}}$ represents the substitution

$$\sigma_{\mathrm{c}}(x) \mapsto \Delta H \equiv \int \mathrm{d}^d y\, \Delta(x-y) H(y).$$

If, for example,

$$\mathcal{V}_{\mathrm{I}}(\sigma) = \frac{g}{4!} \int \mathrm{d}^d x\, \sigma^4(x),$$

the expansion of σ_{c} in powers of H takes the diagrammatic form shown in Figure 12.5.

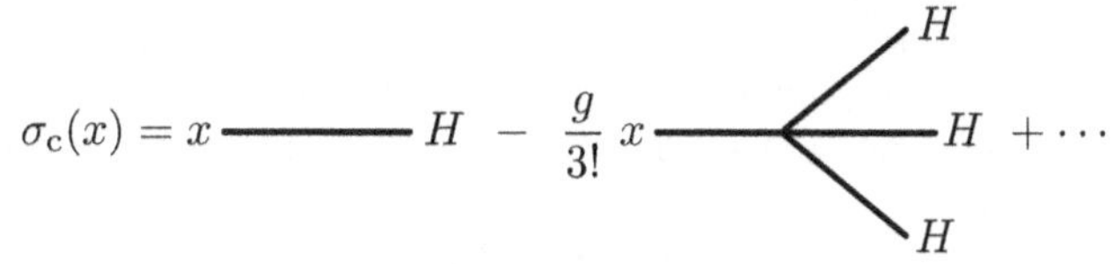

$$\sigma_{\mathrm{c}}(x) = x \!-\!\!\!-\!\!\!-\!\!\!- H \;-\; \frac{g}{3!} \, x \!-\!\!\!-\!\!\!-\!\!\!<\!\!\!\begin{array}{c} H \\ H \\ H \end{array} + \cdots$$

Fig. 12.5 Expansion of σ_{c} in powers of H.

12.4.2 Next order: One-loop diagrams

Keeping only the term quadratic in χ in the expansion of $\mathcal{H}(\sigma)$, one calculates the Gaussian integral and one finds

$$\mathcal{Z}(H) \propto \mathcal{Z}_0(H) \left[\det \frac{\delta^2 \mathcal{H}}{\delta \sigma_{\mathrm{c}}(x_1) \delta \sigma_{\mathrm{c}}(x_2)} \right]^{-1/2}, \tag{12.23}$$

where the normalization, which is independent of H, depends on the continuum limit of a specific lattice regularization. Setting

$$\mathcal{W}(H) = \mathcal{W}_0(H) + \mathcal{W}_1(H) + \cdots, \tag{12.24}$$

one finds the next contribution to the connected functional (see expression (2.51))

$$\mathcal{W}_1(H) = -\frac{1}{2} \operatorname{tr} \ln \frac{\delta^2 \mathcal{H}}{\delta \sigma_{\mathrm{c}}(x_1) \delta \sigma_{\mathrm{c}}(x_2)}, \tag{12.25}$$

where the identity $\ln \det = \operatorname{tr} \ln$ has been used. The functional $\mathcal{W}_1(H)$, expanded in powers of H, generates all one-loop connected diagrams. Figure 12.6 exhibits a typical contribution to $\mathcal{W}_1(H)$.

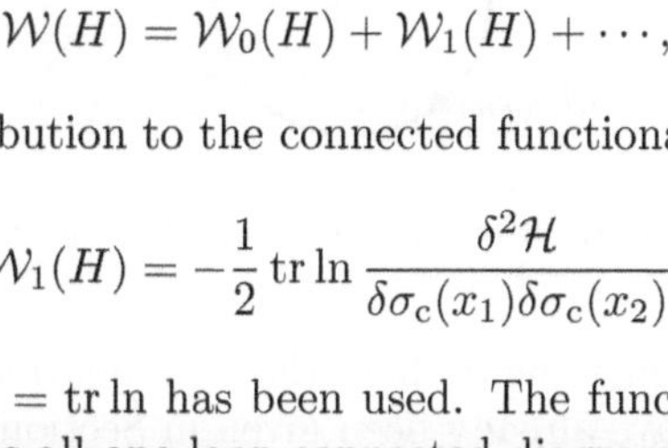

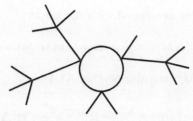

Fig. 12.6 Example of a connected one-loop contribution.

298　　　　　　　　Statistical field theory: Perturbative expansion

Finally, as a consequence of property (6.20), the leading corrections to $\mathcal{W}$ and Γ are opposite. The one-loop contributions to the functional Γ are thus generated by (see equations (6.31), (10.9))

$$\Gamma_1(M) = \frac{1}{2}\operatorname{tr}\ln\frac{\delta^2\mathcal{H}}{\delta M(x_1)\delta M(x_2)}, \qquad (12.26)$$

where M is the local magnetization.

More generally, the successive terms generated by the steepest descent method correspond to Feynman diagrams with an increasing number of loops. The expansion is also called the loop expansion.

To establish this property, one introduces a parameter $\hbar$ and replaces $\mathcal{H}(\sigma, H)$ by $\mathcal{H}(\sigma, H)/\hbar$. The steepest descent method generates an expansion in powers of $\hbar$. It is convenient to define

$$\mathcal{W} = \hbar \ln \mathcal{Z} \qquad (12.27)$$

in such a way that the tree contributions to $\mathcal{W}$ and Γ are of order $\hbar^0$. The introduction of the parameter $\hbar$ has the effect of replacing the propagator Δ by $\hbar\Delta$ (it is the inverse of the coefficient of $\sigma\sigma$) and to divide all vertices by $\hbar$. For a 1-irreducible diagram, calling I the number of propagators and n the total number of vertices, one thus finds a factor $\hbar^{I-n+1}$, the last factor $\hbar$ coming from the normalization (12.27) of $\mathcal{W}$. Using relation (12.16), one recognizes the factor $\hbar^L$ where L is the number of loops.

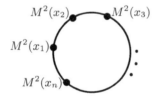

Fig. 12.7 Contribution to the 1-irreducible functional at one-loop.

Example. Again, we consider the σ^4 field theory (equation (12.17)). The two first terms of the expansion have already been given in Section 10.2. Then, in symbolic notation,

$$\Gamma_1(M) - \Gamma_1(0) = \tfrac{1}{2}\operatorname{tr}\ln\left(1 + \tfrac{1}{2}g\Delta M^2\right),$$

where ΔM^2 is the operator associated with the kernel

$$[\Delta M^2](x, y) = \Delta(x - y)M^2(y).$$

Expanding in powers of M^2 (see also Section A1.4),

$$\Gamma_1(M) - \Gamma_1(0) = \sum_{n=1}^{\infty}(-1)^{n+1}\frac{g^n}{n\,2^{n+1}}\operatorname{tr}(\Delta M^2)^n,$$

one generates the one-loop, 1-irreducible diagrams of Figure 12.7:

$$\operatorname{tr}(\Delta M^2)^n = \int \prod_{i=1}^{n} d^d x_i \, M^2(x_i) \Delta(x_i - x_{i-1}) \quad \text{with} \quad x_n = x_0. \quad (12.28)$$

Introducing the Fourier representations (12.9b) and

$$M(x) = \int d^d p \, e^{ipx} \, \tilde{M}(p),$$

one can rewrite the expansion as

$$\operatorname{tr}(\Delta M^2)^n$$

$$= \int \left(\prod_{i=1}^{2n} d^d p_i \, \tilde{M}(p_i) \right) \left(\prod_{j=1}^{n} d^d x_j \frac{d^d q_j}{(2\pi)^d} \, e^{i(x_j(p_{2j}+p_{2j-1})} \, e^{iq_j(x_{j-1}-x_j)} \, \tilde{\Delta}(q_j) \right)$$

$$= \int \left(\prod_{i=1}^{2n} d^d p_i \, \tilde{M}(p_i) \right) \delta^{(d)}\left(\sum_i p_i\right) \int d^d q \prod_{j=1}^{n} \tilde{\Delta}(q_j) \quad (12.29)$$

with

$$q_1 = q, \quad q_{j+1} - q_j = p_{2j-1} + p_{2j}, \quad q_{n+1} = q_1.$$

12.5 Dimensional continuation and dimensional regularization

A technique is specially useful in the study both of quantum field theory as applied to the theory of fundamental interactions, and of critical phenomena, the dimensional continuation of Feynman diagrams. It leads to a definition, but which is purely perturbative, of a field theory in generic, non-integer, dimension d. Dimensional regularization and, thus, minimal subtraction are based on it, but also the ε-expansion that we have introduced in Chapter 10.

12.5.1 Dimensional continuation

The main idea that allows defining a dimensional continuation of Feynman diagrams consists in representing the propagator as a Laplace transform of the form

$$\tilde{\Delta}(p) = \int_0^\infty dt \, \varrho(t\Lambda^2) \, e^{-t(p^2+m^2)}, \quad (12.30)$$

where Λ is a parameter characterizing the short-distance scale and the function $\varrho(t)$ is positive and satisfies the condition

$$|1 - \varrho(t)| < C e^{-\sigma t} \quad (\sigma > 0) \text{ for } t \to +\infty.$$

A simple example is
$$\begin{cases} \varrho(t) = 0 & \text{for } t < 1, \\ \varrho(t) = 1 & \text{for } t \geq 1. \end{cases}$$

In the limit $\Lambda \to \infty$, one recovers the two-point function of the Gaussian fixed point for $m = 0$ and, in general, the Gaussian two-point function in the critical domain corresponding to the Hamiltonian (8.31).

Moreover, one assumes that, for $|p| \to \infty$, the propagator $\tilde{\Delta}(p)$ decays sufficiently fast (i.e., faster than any power) to render all Feynman diagrams finite for all dimensions for $m \neq 0$ and above some lower dimension $d_{\min} > 2$ for the critical theory. This implies that the function $\varrho(t)$ vanishes faster than any power for $t \to 0$.

In Feynman diagrams, the loop integrations over free momenta then become Gaussian integrations. The loop integrals can thus be calculated in any dimension d. The dependence in the variable d then becomes explicit.

As a consequence of the condition of fast decay of the propagator, one obtains meromorphic functions of d, analytic for $\operatorname{Re} d > d_{\min}$ for critical theories ($m = 0$) and entire if the correlation length is finite (massive theories).

12.5.2 Dimensional regularization

After dimensional continuation, one chooses d complex, $\operatorname{Re} d < d_c$ such that naively all initial integrals converge in the power counting sense. Then, the integrals have limits, for $\Lambda \to \infty$, which are *independent of the function* $\varrho(t)$ and define meromorphic functions of the dimension. These functions have simple or multiple poles on the real axis, corresponding to divergences of diagrams with a propagator of the form $1/(p^2 + m^2)$. This construction is called the *dimensional regularization*.

Dimensional continuation and regularization satisfy the following important formal properties:

(i) $\quad \int d^d p\, F(p+q) = \int d^d p\, F(p) \qquad\qquad$ translation

(ii) $\quad \int d^d p\, F(\lambda p) = |\lambda|^{-d} \int d^d p\, F(p) \qquad$ dilatation

(iii) $\int d^d p\, d^{d'} q\, f(p) g(q) = \int d^d p\, f(p) \int d^{d'} q\, g(q)\ $ factorization.

The two first conditions imply invariance under affine changes of variables. This invariance plays an essential role in the preservation of the symmetries in some field theories.

In the example of the σ^4 theory in dimension d, dimensional regularization eliminates, in particular, automatically all contributions generating deviations from the critical point. It also eliminates all irrelevant contributions. But it introduces divergences in the form of simple or multiple poles when $\varepsilon = 4 - d \to 0$, which replace contributions that diverge logarithmically when $\Lambda \to \infty$, and that must be eliminated by a *renormalization*, a topic discussed in Chapter 13.

A surprising consequence. One sometimes meets the integral

$$\frac{1}{(2\pi)^d} \int d^d k\, k^{-2\alpha},$$

in particular, in massless theories. The rules of dimensional regularization (invariance under dilatation) imply

$$\lambda^{2\alpha-d}\int d^d k\, k^{-2\alpha} = 0 \quad \Rightarrow \quad \text{for } \alpha \ne d/2 \int d^d k\, k^{-2\alpha} \equiv 0.$$

(For $\alpha = d/2$, the integral is undefined.) Let us examine the integral with a cut-off function $\varrho(t)$, which renders it initially convergent:

$$D_\alpha(\Lambda, d) = \frac{1}{(2\pi)^d}\frac{1}{\Gamma(\alpha)} \int d^d k \int_0^\infty dt\, \varrho(t\Lambda^2) t^{\alpha-1} e^{-tk^2}.$$

By replacing $\varrho(t)$ by 1 and after integration over t, one recovers the initial integral. Integrating over k first, one finds

$$D_\alpha(\Lambda, d) = \frac{1}{(4\pi)^{d/2}}\frac{1}{\Gamma(\alpha)} \int_0^\infty dt\, \varrho(t\Lambda^2) t^{\alpha-1-d/2}.$$

The integral now converges for $\operatorname{Re} d > 2\alpha$, the domain in which $D_\alpha(\Lambda, d)$ is an analytic function of d. Changing $t\Lambda^2$ into t, one obtains

$$D_\alpha(\Lambda, d) = \Lambda^{d-2\alpha}\frac{1}{(4\pi)^{d/2}}\frac{1}{\Gamma(\alpha)} \int_0^\infty dt\, \varrho(t) t^{\alpha-1-d/2}.$$

The singularity of the integral is due to the behaviour for $t \to \infty$. One can separate the different contributions

$$D_\alpha(\Lambda, d) = \frac{\Lambda^{d-2\alpha}}{(4\pi)^{d/2}}\frac{1}{\Gamma(\alpha)}\left[\int_0^1 dt\, \varrho(t) t^{\alpha-1-d/2} + \int_1^\infty dt\, \bigl(\varrho(t)-1\bigr) t^{\alpha-1-d/2}\right]$$

$$+ \frac{\Lambda^{d-2\alpha}}{(4\pi)^{d/2}}\frac{1}{\Gamma(\alpha)}\frac{2}{d-2\alpha}. \qquad (12.31)$$

The two integrals are entire functions of d and, thus, this representation defines the function in the entire complex plane. The only singularity is the simple pole at $d = 2\alpha$. Following the procedure outlined above, we choose d complex, $\operatorname{Re} d < 2\alpha$ and take the $\Lambda \to \infty$ limit: the limit then vanishes.

A simple example illustrates this somewhat strange result, and allows verifying its consistency with dimensional regularization. The integral

$$I = \frac{1}{(2\pi)^d} \int \frac{d^d p}{p^2\,(p^2+1)}$$

can be calculated in two different ways. The first relies on the decomposition

$$\int d^d p \left(\frac{1}{p^2} - \frac{1}{p^2+1}\right) = -\int \frac{d^d p}{p^2+1}.$$

Then,
$$I = \frac{1}{(2\pi)^d} \int d^d p \int_0^\infty dt\, e^{-t(p^2+1)}$$
$$= \frac{1}{(4\pi)^{d/2}} \int_0^\infty dt\, t^{-d/2} e^{-t} = \frac{1}{(4\pi)^{d/2}} \Gamma(1 - d/2).$$

Alternatively, the first calculation of Section 12.5.3 shows how the same result can be obtained more directly.

Dimensional regularization and critical theories. This property must induce some caution in the use of dimensional regularization in critical (i.e., massless) theories. In such models (like the non-linear σ model of Chapter 15), one meets an integral of this type: $\int d^d k/k^2$. In dimension 2, this integral is both divergent for $k \to 0$ and $k \to \infty$. In this case, a naive usage of dimensional regularization may have undesirable consequences.

To understand this point, we first define the integral by dimensional continuation for $d > 2$ with a cut-off function of type (12.30). From expression (12.31), one infers
$$\int \frac{d^d k}{k^2} \propto \frac{\Lambda^{d-2}}{d-2}.$$

When $d \to 2$, the integral diverges. In models with continuous symmetries, this divergence due to the singularity at $k = 0$ has a physical meaning: it forbids the presence of Goldstone bosons (which are massless fields) in two dimensions and, thus, the spontaneous breaking of continuous symmetries (Section 7.7).

By contrast, in a massive theory with mass m, one can first define the theory for $\operatorname{Re} d < 2$ by dimensional regularization. For $d \to 2$, one finds
$$\int^\Lambda \frac{d^d k}{k^2 + m^2} = A_d \frac{\pi}{2 \sin(\pi d/2)} m^{d-2} \underset{d \to 2}{\sim} -A_d \frac{m^{d-2}}{d-2}, \quad A_d = \frac{2\pi^{d/2}}{\Gamma(d/2)}.$$

Here, the pole at $d = 2$ is a divergence for $k \to \infty$ and the residue has a sign opposite to the preceding IR pole. In dimension 2, a naive dimensional regularization yields a vanishing result that results from a cancellation between a UV divergence ($k \to \infty$) and an IR divergence ($k \to 0$), something that is physically unreasonable. In a critical theory, the use of dimensional regularization is sensible only for dimensions where no IR (i.e., zero momentum) divergences appear.

12.5.3 Examples

Two-point function at one-loop. The only one-loop diagram that contributes to the two-point vertex function in the σ^4 field theory is the diagram (b) of Figure 2.5 (Section 2.5.1):
$$(b) \equiv \Omega_d(m) = \frac{1}{(2\pi)^d} \int \frac{d^d k}{k^2 + m^2}. \tag{12.32}$$

From this diagram, we subtract its value at $m = 0$ to ensure that the theory remains critical for $m = 0$. Moreover, we modify the propagator at short distance as indicated in (12.30):

$$\Omega_d(m) - \Omega_d(0) = \frac{1}{(2\pi)^d} \int_0^\infty dt\, \varrho(t\Lambda^2)\, e^{-tk^2} \left(e^{-tm^2} - 1\right). \tag{12.33}$$

The integration over k yields the dimensional continuation

$$\Omega_d(m) - \Omega_d(0) = \frac{1}{(4\pi)^{d/2}} \int_0^\infty dt\, \varrho(t\Lambda^2) t^{-d/2} \left(e^{-tm^2} - 1\right).$$

The integral converges for $\operatorname{Re} d > 2$ where it defines an analytic function. We want to investigate the behaviour in the limit $m \to 0$, which is also the behaviour $\Lambda \to \infty$, for $d = 4 - \varepsilon$, ε positive and infinitesimal. For $d < 4$ fixed, the limit $\Lambda \to \infty$ exists and one can thus replace ϱ by 1. The leading correction is of order $\Lambda^{-\varepsilon}$, which cannot be neglected in the limit $\varepsilon \to 0$. One thus needs the first two terms of the expansion:

$$\Omega_d(m) - \Omega_d(0) = \frac{1}{(4\pi)^{d/2}} \int_0^\infty dt\, t^{-d/2} \left(e^{-tm^2} - 1\right)$$
$$+ \frac{1}{(4\pi)^{d/2}} \int_0^\infty dt\, (\varrho(t\Lambda^2) - 1) t^{-d/2} \left(e^{-tm^2} - 1\right).$$

The first integral converges for $2 < \operatorname{Re} d < 4$ and, after an integration by parts, yields a Γ-function. The second integral is a function of m^2 analytic at $m = 0$. Only the first term of the expansion is needed and, thus,

$$\Omega_d(m) - \Omega_d(0) = \frac{\Gamma(1 - d/2)}{(4\pi)^{d/2}} m^{d-2} + a(d) m^2 \Lambda^{d-4} + O(m^4 \Lambda^{d-6}) \tag{12.34}$$

with

$$a(d) = \frac{1}{(4\pi)^{d/2}} \int_0^\infty dt\, (1 - \varrho(t)) t^{1-d/2}. \tag{12.35}$$

The leading term for $m \to 0$ does not depend on the function ϱ and thus of the short-distance structure. The function $a(d)$ depends on the function ϱ explicitly, but for $\varepsilon = 4 - d \to 0$ has a pole whose residue is independent of ϱ and which cancels the pole of the coefficient of m^{d-2}:

$$a(d) \underset{d \to 4}{\sim} \frac{1}{8\pi^2 \varepsilon}.$$

In an expansion for $\varepsilon \to 0$, one finds

$$\Omega_d(m) - \Omega_d(0) \underset{m \to 0, d \to 4}{\sim} \frac{N_d m^2}{\varepsilon} \left(\Lambda^{-\varepsilon} - m^{-\varepsilon} + \text{const.}\right)$$
$$\underset{m \to 0, d \to 4}{\sim} N_d m^2 \ln(m/\Lambda) + O(\varepsilon) + O(m^4),$$

where
$$N_d = \frac{2}{(4\pi)^{d/2}\Gamma(d/2)} = \frac{1}{8\pi^2} + O(\varepsilon),$$

is a loop factor, an entire function of d, which one generally does not expand. Indeed, the ratio N_d/N_4 can be absorbed into a redefinition of the expansion parameter.

The factor $\ln(\Lambda/m)$ is directly related to the factor $\ln\lambda$ of equation (10.27).

Finally, within dimensional regularization the expression reduces to
$$\Omega_d(m) \equiv \Omega_d(m) - \Omega_d(0) = \frac{\Gamma(1-d/2)}{(4\pi)^{d/2}} m^{d-2}. \qquad (12.36)$$

Since the Γ-function is meromorphic, the expression is a meromorphic function of d with a simple pole at $d=2$ corresponding to the subtraction of the term with a critical propagator, and poles at $d=4,6,\ldots$ that are direct consequences of the large-momentum (ultraviolet, UV), or short-distance divergence of the Feynman diagram in the limit $\Lambda\to\infty$.

In the expansion in ε, the term $m^2\ln m$ is not affected, but a term singular in m^2/ε appears, which must be cancelled by *renormalization*.

Four-point function at one loop. We now consider the one-loop diagram contributing to the vertex four-point function in the σ^4 theory (diagram (g) of Figure 2.7, Section 2.5.2). In dimensional regularization, one finds
$$B_d(p) = \frac{1}{(2\pi)^d}\int \frac{d^d q}{(q^2+m^2)[(p-q)^2+m^2]}$$
$$= \frac{1}{(2\pi)^d}\int d^d q \int_0^\infty dt_1\, dt_2\, \exp\left[-t_1(q^2+m^2) - t_2((p-q)^2+m^2)\right]$$
$$= \frac{1}{(4\pi)^{d/2}}\int_0^\infty \frac{dt_1\, dt_2}{(t_1+t_2)^{d/2}} \exp\left[-p^2 t_1 t_2/(t_1+t_2) - m^2(t_1+t_2)\right].$$

The dependence on d is now explicit and the integrand is defined for all d. The integral over the remaining variables, by contrast, converges only for $\mathrm{Re}\,d<4$ in the absence of a cut-off.

After the change of variables $t_1,t_2 \mapsto t,s$:
$$t_1 = ts, \quad t_2 = (1-t)s,$$
the expression becomes
$$B_d(p) = \frac{1}{(4\pi)^{d/2}}\int_0^1 dt \int_0^\infty ds\, s^{1-d/2} e^{-st(1-t)p^2 - sm^2}.$$

The integral over s yields
$$B_d(p) = \frac{1}{(4\pi)^{d/2}}\Gamma(2-d/2) \int_0^1 dt\, \left[p^2 t(1-t) + m^2\right]^{(d/2)-2}.$$

For $m = 0$, the last integration over t can be carried out for $\operatorname{Re} d > 2$ and leads to the explicit result

$$B_d(p) = \frac{1}{(4\pi)^{d/2}} \Gamma(2 - d/2) \frac{\Gamma^2(d/2 - 1)}{\Gamma(d - 2)} (\mathbf{p}^2)^{(d/2)-2}. \tag{12.37}$$

Again, the expression is a meromorphic function of d with a pole at $d = 2$ corresponding to zero-momentum or large-distance singularities because we have specialized the calculation to the critical theory $m = 0$, and poles at $d = 4, 6, \ldots$ that are direct consequences of the large-momentum behaviour of the Feynman diagram.

The expansion in ε leads to

$$B_d(p) \sim N_d \left(\frac{1}{\varepsilon} - \ln p \right) + O(\varepsilon).$$

Again, the simple dimensional continuation with a modified propagator leads to a cancellation of the pole in ε and a substitution by $\ln \Lambda$, up to an additive numerical constant that depends on ϱ.

The result obtained here allows recovering the result (10.17).

Two-point function at two loops: Dimensional regularization. To determine the leading correction to the field dimension σ, one needs the diagram (e) of Figure 2.6 (Section 2.5.1), contributing to the two-point function $\tilde{\Gamma}^{(2)}(p)$, in the critical theory:

$$(e) = \frac{1}{(2\pi)^{2d}} \int \frac{d^d q_1 d^d q_2}{q_1^2 q_2^2 (p - q_1 - q_2)^2}.$$

The Feynman diagram can be calculated more easily using the initial space variables. In dimension d, as a function of the space variables, the propagator is given by expression (8.40):

$$\Delta(x) = \frac{2^{d-2}}{(4\pi)^{d/2}} \Gamma(d/2 - 1) |x|^{2-d}.$$

Moreover,

$$(e) = \int e^{ipx} \Delta^3(x) d^d x.$$

This reduces the calculation to an integral of the general form

$$\Phi(p) = \int d^d x \, \frac{e^{ipx}}{|x|^\alpha}$$

with here $\alpha = 3d - 6$. One uses the representation

$$\frac{1}{|x|^\alpha} = \frac{1}{\Gamma(\alpha/2)} \int_0^\infty dt \, t^{\alpha/2 - 1} e^{-tx^2}.$$

Statistical field theory: Perturbative expansion

After, successively, an integration over x and the change of variables $s = p^2/4t$, which reduces the integral to a representation of the Γ-function, one finds

$$\Phi(p) = \frac{\pi^{d/2}}{\Gamma(\alpha/2)} \int_0^\infty dt\, t^{\alpha/2-1-d/2} e^{-p^2/4t} = \frac{\pi^{d/2}}{\Gamma(\alpha/2)} \Gamma((d-\alpha)/2) \left(\frac{p}{2}\right)^{\alpha-d}.$$

The final result is

$$(e) = -N_d^2 \frac{3\Gamma(d/2-1)\Gamma(2-d)}{2\Gamma(3d/2-2)} |p|^{2d-6} = N_d^2 p^2 \left(-\frac{1}{8\varepsilon} + \frac{1}{4}\ln p\right) + O(\varepsilon). \quad (12.38)$$

The diagram is a meromorphic function of d, with a double pole at $d = 2$, corresponding to large-distance singularities, and with simple poles at $d = 3, 4, \ldots$ corresponding to short-distance singularities.

Cut-off. In the presence of a cut-off function ϱ, $1/\varepsilon$ is replaced by $\ln\Lambda$ plus a non-universal constant.

Exercises

Exercise 12.1

Recover the result (12.37) by using the method explained in the case of the diagram (e).

13 The σ^4 field theory near dimension 4

In Chapter 10, we have shown how perturbative calculations based on a local statistical field theory of the $\int \mathrm{d}^d x\, \sigma^4(x)$ type, combined with the assumption of the existence of a renormalization group (RG) of the form postulated in Section 9.7, allow determining some universal properties of critical statistical models, at least in the vicinity of dimension 4. Using the idea of the expansion in $\varepsilon = 4 - d$, where d is the dimension of space, introduced by Wilson and Fisher, we have exhibited a non-Gaussian fixed point and calculated the corresponding critical exponents to first order in the deviation $\varepsilon > 0$ from dimension 4.

In this chapter, we explain how the asymptotic renormalization group equations (RGE), introduced without too much justification in Section 10.6, can be proved within the framework of statistical (or quantum) field theory, to all orders in a double expansion in the coefficient of $\int \mathrm{d}^d x\, \sigma^4(x)$ and $\varepsilon = 4 - d$. The proof is based on the methods of perturbative statistical field theory introduced in Chapter 12, and a few assumptions that it is thus possible to clarify.

RGE, which are satisfied by correlation functions of statistical field theory, are a particular form of the general RGE (9.38) introduced in Section 9.1, within an asymptotic limit in which the Hamiltonian can be described by a small number of parameters. They appear as a consequence of the renormalizability of certain local field theories.

The renormalizability of a field theory expresses, in a slightly formal way, its relative insensitivity to the short-distance structure in the vicinity of the Gaussian fixed point and is, therefore, quite directly related to the universality of critical phenomena. It depends on the nature of the divergences of Feynman diagrams calculated with the critical propagator. Dimension 4 finds a simple interpretation within this framework. Indeed, for $d < 4$, the Gaussian fixed point is stable from the viewpoint of large-momentum (ultraviolet) behaviour. This phenomenon ceases at $d = 4$, the dimension where, by definition, the field theory σ^4 is *renormalizable*.

When an IR stable fixed point exists, the RGE of field theory allow proving universal properties of critical phenomena, determining the large-distance behaviour of connected correlation functions or the singularities of thermodynamic functions.

In Chapter 10, postulating the existence of a RG, we have determined the non-trivial RG fixed point at first order in ε.

The solution of these RGE confirms these results and enables generalizing them to all orders in the ε-expansion, and then leads to the calculation of universal quantities in the form of an ε-expansion.

In this chapter, we restrict the discussion to theories with an Ising type symmetry and the field σ has only one component. Generalization to models with N-component fields and $O(N)$ symmetry is simple.

Remarks.

(i) The limitation of the methods used to derive universality properties resides in their perturbative character: they are applicable only to situations where a fixed point is, in a sense that will become gradually clearer, close to the Gaussian fixed point.

(ii) The physics that one tries to describe corresponds to integer values of ε, $\varepsilon = 1, 2$. Although one can show the validity of the RG results only within the framework of the ε-expansion, one thus assumes their validity beyond the infinitesimal vicinity of dimension 4. One is then confronted with another difficulty: the ε-expansion is divergent for any value of ε. It is necessary to introduce suitable summation methods to extract from it precise estimates of universal quantities. Alternatively, it is also possible to sum the perturbative expansion at fixed dimension.

A comparison between the results obtained after summation and experimental or numerical data for a broad class of statistical models (see Section 10.7) then provides a crucial test for the theory.

13.1 Effective Hamiltonian. Renormalization

We thus continue the study of the statistical field theory (SFT) whose Hamiltonian takes the form (10.1) of Section 10.1 and, more generally, the form (10.25) of Section 10.4. This SFT is *local*, that is, the Hamiltonian depends only on the field and its derivatives, as a consequence of the assumption of short-range interactions.

We first describe the properties of renormalizability of such a theory, which are at the basis of the perturbative derivation of the RGE.

13.1.1 Effective Hamiltonian

The effective Hamiltonian has the general structure (12.11),

$$\mathcal{H}(\sigma) = \mathcal{H}_G(\sigma) + \mathcal{V}_I(\sigma), \tag{13.1}$$

but with the more specific forms

$$\mathcal{H}_G(\sigma) = \frac{1}{2} \int d^d x \left[\sum_{\mu=1}^{d} \partial_\mu \sigma(x) \left(1 + \sum_{k=1}^{} u_{k+1}(-\nabla_x^2)^k \right) \partial_\mu \sigma(x) + u_0 \sigma^2(x) \right],$$

$$\mathcal{V}_I(\sigma) = \frac{1}{4!} g \int d^d x \, \sigma^4(x).$$

It is the sum of the Hamiltonian of the Gaussian fixed point, of two relevant, even perturbations to the Gaussian model and of terms quadratic in the field with a sufficient number of derivatives, in order to regularize all large-momentum (or short-distance) divergences (called the Landau–Ginzburg–Wilson Hamiltonian). These divergences are not present in lattice models because momenta are bounded by the Brillouin zone. The parameter u_0 is here the sum of the contribution $r_c(g)$ that ensures that the Hamiltonian remains critical, and a relevant infinitesimal deviation

$t \propto T - T_c$ from the critical theory. Based on the analysis of the preceding chapters, we assume that all parameters u_k and g are *regular* functions of the temperature for T close to T_c.

At leading order, such a Hamiltonian reproduces all the results of the quasi-Gaussian or mean-field approximations. Beyond, it leads to a double, perturbative and dimensional expansion of thermodynamic quantities. In particular, correlation functions have a finite *expansion to all orders in* $\varepsilon = 4 - d$ *and* g except at T_c at zero momentum.

One also calls such a field theory an *effective field theory* to emphasize that it is not a microscopic model, but only a model that reproduces correctly the asymptotic behaviour at large distance (this denomination has nowadays become almost a pleonasm in the sense that almost all quantum or statistical field theories that are encountered in physics have such an interpretation).

13.1.2 Gaussian renormalization

In Hamiltonian (13.1), it is convenient to rescale distances,

$$x = \Lambda x', \tag{13.2}$$

where Λ is a parameter that has a momentum dimension. Through this change, we want to take as a reference scale the macroscopic scale relevant for large-distance physics, rather than the initial microscopic scale. The latter is now characterized by the parameter $1/\Lambda$ (related, for example, to the lattice spacing of an initial statistical model). Instead of studying the large-distance limit, one then studies the limit $\Lambda \to \infty$.

We have seen that the dimension of $\sigma(x)$ in the Gaussian model is $(d-2)/2$. We thus change the field normalization correspondingly (the Gaussian renormalization):

$$\sigma(x) \mapsto \sigma'(x) = \Lambda^{(d-2)/2} \sigma(\Lambda x) \Rightarrow \frac{\partial \sigma'(x)}{\partial x_\mu} = \Lambda^{d/2} \frac{\partial \sigma(\Lambda x)}{\partial x_\mu}, \tag{13.3}$$

in such a way that the Hamiltonian of the Gaussian fixed point remains unchanged. One recognizes the RG transformations (9.46) that lead to the Gaussian fixed point.

In these new variables, the Hamiltonian then reads (now omitting primes)

$$\mathcal{H}(\sigma) = \mathcal{H}_G(\sigma) + \frac{1}{4!} g \Lambda^{4-d} \int d^d x \, \sigma^4(x), \tag{13.4}$$

$$\mathcal{H}_G(\sigma) = \frac{1}{2} \int d^d x \sum_\mu (\partial_\mu \sigma(x))^2 + \frac{1}{2} \Lambda^2 u_0 \int d^d x \, \sigma^2(x)$$

$$+ \frac{1}{2} \int d^d x \sum_{k=1} u_{k+1} \Lambda^{-2k} \sum_\mu \partial_\mu \sigma(x)(-\nabla^2)^k \partial_\mu \sigma(x). \tag{13.5}$$

In the context of quantum field theory, the parameter Λ, which reflects the initial microscopic structure, is called the *cut-off*.

Actually, this operation corresponds to the first step in the construction of Section 9.7.1, where a first dilatation suppresses (at leading order) the terms that are irrelevant from the viewpoint of the fixed Gaussian point, since, for $g \ll 1$, the RG flow is first dominated by the local flow near the Gaussian fixed point.

13.1.3 Dimensional analysis and critical dimension

After the introduction of the Λ parameter, all quantities get a dimension in units of Λ. Momenta have dimension 1, space coordinates dimension -1, the field σ has dimension $\frac{1}{2}(d-2)$, which is equal to its dimension from the viewpoint of the Gaussian fixed point. More generally, all monomials in the Hamiltonian have as dimensions the eigenvalues of the corresponding eigenvectors of the RG flow in the linear approximation near the Gaussian fixed point.

It is easy to verify that this property generalizes to interactions with higher powers of σ and more derivatives, which we have omitted, since the transformations (13.2), (13.3) are analogous to the transformations (9.46). With the introduction of the Λ parameter, we have reduced the calculation of the dimension of the field and the analysis of the perturbations at the Gaussian fixed point to *dimensional analysis*. In particular, the Gaussian fixed point is stable and the quasi-Gaussian model is valid if the coefficients of all terms in the Hamiltonian, but for the first two in expression (13.5), go to zero for $\Lambda \to \infty$. For the quadratic terms with more than two derivatives, this condition is always satisfied.

Moreover, we set
$$\Lambda^2 u_0 = \Lambda^2 u_{0c}(g) + t,$$
where $u_{0c}(g)$ corresponds to the value of u_0 for which the theory is critical ($T = T_c$), that is, at which the correlation length diverges or the mass vanishes:
$$u_0 = u_{0c}(g) \Leftrightarrow \tilde{\Gamma}^{(2)}(p = 0) = 0.$$

The parametrization
$$u_0 - u_{0c} \mapsto t/\Lambda^2,$$
corresponds to a renormalization of the deviation to the critical temperature adapted to the vicinity of the Gaussian fixed point. Indeed, the σ^2 term is relevant and has Gaussian dimension 2 and this renormalization thus cancels the scaling factor generated by the RG in the linear approximation.

After this change, correlation functions have a finite limit when $\Lambda \to \infty$ at $\{p_i, t\}$ fixed if the Gaussian fixed point is stable. This property generalizes to correlation functions at fixed magnetization. For $d > 4$, one can verify this property with a perturbative calculation. As an example, we calculate the diagram (b) of Figure 2.5 (Section 2.5.1) to which its value at $t = 0$ has been subtracted (equation (12.33)). For $d > 4$, the leading contribution is now linear in t (see also equation (12.34)):

$$\Omega_d(\sqrt{t}) - \Omega_d(0) = -\frac{t\Lambda^{d-4}}{(4\pi)^{d/2}} \int_0^\infty ds\, \varrho(s) s^{1-d/2} + O\left(t^{d/2-1}\right).$$

This contribution is multiplied by $\frac{1}{2}g\Lambda^{4-d}$ and thus (equation (12.18))

$$\tilde{\Gamma}^{(2)}(p) \underset{\Lambda \to \infty}{=} p^2 + t - gt \frac{1}{2(4\pi)^{d/2}} \int_0^\infty ds\, \varrho(s) s^{1-d/2} + O(g^2).$$

Similarly, one verifies that the contribution of order g^2 to the four-point function has a finite limit. For $d > 4$, the effect of the quartic interaction is to modify (renormalize) the parameters of the quasi-Gaussian approximation and to generate corrections due to the irrelevant operators encountered in the local RG analysis.

For $d < 4$, by contrast, one does not expect a finite limit since the dimensions of the field as well as $u_0 - u_{0c}$ change. This is what one verifies here since, for example, $\Omega_d(\sqrt{t}) - \Omega_d(0)$ has a finite limit when $\Lambda \to \infty$, but the diagram is multiplied by $g\Lambda^{4-d}$, which diverges.

In four dimensions (the critical dimension) in an expansion in powers of g and, more generally, in dimension $d < 4$, in a double expansion in powers of g and $\varepsilon = 4 - d$, these divergences take the form of powers of $\ln \Lambda$, the degree increasing with the order in g and in ε.

Vertex functions. We also need the dimensions of the Fourier components of vertex functions. The calculation follows from a few remarks. The dimension $[\tilde\sigma]$ of the Fourier components $\tilde\sigma$ of the field can be inferred from the relation

$$\sigma(x) = \int d^d p \, e^{ipx} \, \tilde\sigma(p) \quad \Rightarrow \quad [\tilde\sigma] = -(d+2)/2 \,.$$

The vertex functions $\tilde\Gamma^{(n)}$ are obtained by expanding the thermodynamic potential in powers of the local magnetization $M(x) = \langle \sigma(x) \rangle$ or rather its Fourier transform $\tilde M(p)$:

$$\Gamma(M) = \sum_n \frac{1}{n!} \int d^d p_1 \ldots d^d p_n \, \delta^{(d)}\left(\sum_i p_i\right) \tilde M(p_1) \ldots \tilde M(p_n) \tilde\Gamma^{(n)}(p_1, \ldots, p_n).$$

One infers the dimension

$$[\tilde\Gamma^{(n)}] = -nd + n(d+2)/2 + d = d - n(d-2)/2 \,. \tag{13.6}$$

Comparing this result with equation (9.14) for $Z(\lambda) = \lambda^{2-d}$, the Gaussian form, one again verifies that this dimensional analysis generates the dilatation factors of the RG transformations adapted to the Gaussian fixed point.

13.1.4 Renormalization theorem

First, we state explicitly the renormalization theorem for the critical theory $t = 0$. It applies, in the sense of formal series, to the double series expansion in powers of the coefficient g of the quartic term (the interaction) and of $\varepsilon = 4 - d$.

To formulate the renormalization theorem, one introduces a momentum μ, called the renormalization scale, and a parameter g_r characterizing the effective interaction at scale μ, called the renormalized interaction. One can then find two functions $Z(\Lambda/\mu, g_r)$ and $Z_g(\Lambda/\mu, g_r)$, which satisfy

$$\Lambda^{4-d} g = \mu^{4-d} Z_g(\Lambda/\mu, g_r) g_r = \mu^{4-d} g_r + O(g_r^2), \tag{13.7}$$
$$Z(\Lambda/\mu, g_r) = 1 + O(g_r), \tag{13.8}$$

calculable order by order in a double series expansion in powers of $g_{\rm r}$ and ε, such that all vertex functions

$$\tilde\Gamma_{\rm r}^{(n)}(p_i;g_{\rm r},\mu,\Lambda)=Z^{n/2}(g_{\rm r},\Lambda/\mu)\tilde\Gamma^{(n)}(p_i;g,\Lambda),\tag{13.9}$$

called *renormalized*, have, order by order, finite limits $\tilde\Gamma_{\rm r}^{(n)}(p_i;g_{\rm r},\mu)$ when $\Lambda\to\infty$ at $p_i,\mu,g_{\rm r}$ fixed.

Remarks.

(i) There is some arbitrariness in the choice of the *renormalization constants* Z and Z_g since they can be multiplied by arbitrary finite functions of $g_{\rm r}$. The constants can be completely determined by *renormalization conditions*, for example,

$$\frac{\rm d}{{\rm d}p^2}\tilde\Gamma_{\rm r}^{(2)}(p=\mu,\mu,g_{\rm r})=1\,,$$
$$\tilde\Gamma_{\rm r}^{(4)}(p_i=\mu\theta_i,\mu,g_{\rm r})=\mu^{4-d}g_{\rm r}\,,$$

where the θ_i are four vectors with vanishing sum and such that

$$\theta_i\cdot\theta_j=\tfrac{4}{3}\delta_{ij}-\tfrac{1}{3}\,.$$

Then, one proves, order by order in an expansion in powers of g and ε, that the functions $\tilde\Gamma_{\rm r}^{(n)}$ are unique, in the sense that they are independent of the special choice of the coefficients of the quadratic terms irrelevant with respect to the Gaussian fixed point in $\mathcal H_{\rm G}(\sigma)$. This result can be generalized: one can prove, in the sense of formal series, that the contributions of all irrelevant perturbations, from the viewpoint of the Gaussian fixed point, go to zero.

(ii) In the language of quantum field theory, the parameters or correlation functions of the initial theory are called *bare* parameters or *bare* correlation functions.

(iii) Physically, the momentum μ is the inverse of the macroscopic distance scale and the parameter $g_{\rm r}$ plays the role of $g(\Lambda/\mu)$, where Λ/μ is the dilatation parameter (for more details, see Section 13.6).

(iv) Beyond the perturbative expansion, the interpretation of renormalized correlation functions is subtle. In formal renormalization theory, the parameters of the renormalized theory are fixed, and the initial parameters, which are the coefficients of all operators relevant or marginal with respect to the Gaussian fixed point, are adjustable parameters and thus vary when the scale factor Λ/μ varies. This corresponds to applying, in the vicinity of the Gaussian fixed point and for $d=4-\varepsilon$, a generalization of the strategy that has already been used for the coefficient of σ^2 and which has allowed deriving universal properties in the critical domain.

In this sense, the existence of renormalized functions implies universal properties in the largest possible critical domain near the Gaussian fixed point.

However, the existence of such an extended critical domain beyond the perturbative expansion has to be investigated. The study of the RGE will provide some information about this question.

Finally, of course, the renormalization theorem has implications for the large-distance behaviour of statistical models only if the property of renormalizability remains true at fixed dimension.

Super-renormalizable field theories. At fixed dimension $d < 4$, the perturbative expansion has no meaning for $\Lambda \to \infty$, as indicated by the factor Λ^{4-d} multiplying the relevant σ^4 interaction. Actually, it can easily be shown that, even at Λ fixed, in the case of the critical or massless theory, the perturbative expansion does not exist due to IR (zero momentum) divergences. For example, for $d = 3$, the contribution of order g^4 to the two-point function diverges. In the framework of dimensional continuation, for any $\varepsilon = 4 - d > 0$, one can find an order $k = O(1/\varepsilon)$ in the expansion in powers of g such that some diagrams diverge.

By contrast, for $t > 0$ and $g_0 = g\Lambda^{4-d}/t^{2-d/2} \sim g\Lambda^{4-d}\xi^{4-d}$ fixed (ξ is the correlation length), correlation functions have a limit for $\Lambda \to \infty$. This condition implies decreasing g as a function of the dilatation parameter with a power appropriate to the Gaussian fixed point in order to cancel the dilatation factor, as we have done systematically for the deviation to the critical theory due to the σ^2 operator. One then obtains a finite perturbative expansion, and the corresponding SFT is called *super-renormalizable*.

The interesting theory, from the viewpoint of statistical physics, then corresponds, in general, to the limit $g_0 \to \infty$. However, there exists exceptional situations where such a SFT has a physics application, for example, in the problem of cold and very diluted quantum gases, where the parameter g is naturally very small and simultaneously the coefficients of all interaction terms of higher degree in the field are even more negligible.

Renormalization and dimensional regularization. In the framework of dimensional regularization, the field renormalization as well as the relations between initial (bare) and renormalized parameters have a finite limit for $\Lambda \to \infty$. The corresponding renormalization constants become meromorphic functions of the dimension, singular at the dimensions corresponding to logarithmic divergences in Λ (Section 12.5.2).

13.2 RG equations

In the framework of the ε-expansion, that is, in a double series expansion in powers of the parameter g and of ε, the critical behaviour differs from the quasi-Gaussian behaviour only by powers of logarithms. These logarithms are organized by the RGE.

13.2.1 RGE for the critical theory

From equation (13.9) and the existence of a limit $\Lambda \to \infty$, there follows a new equation obtained by differentiation of the equation with respect to Λ at μ, g_r fixed:

$$\Lambda \frac{\partial}{\partial \Lambda}\bigg|_{g_r, \mu \text{ fixed}} Z^{n/2}(g, \Lambda/\mu)\tilde{\Gamma}^{(n)}(p_i; g, \Lambda) = o(\Lambda^{-2+\upsilon}), \quad \upsilon > 0, \tag{13.10}$$

where the renormalization factor Z has been expressed in terms of g instead of g_r.

In agreement with the perturbative philosophy, one first neglects all contributions that, order by order, decay as powers of Λ. Thus, we define asymptotic functions $\tilde{\Gamma}^{(n)}_{\text{as.}}(p_i; g, \Lambda)$ and $Z_{\text{as.}}(g, \Lambda/\mu)$ as sums of the perturbative contributions to the functions $\tilde{\Gamma}^{(n)}(p_i; g, \Lambda)$ and $Z(g, \Lambda/\mu)$, respectively, that do not go to zero when $\Lambda \to \infty$. One can show that such correlation functions can also be obtained by adding to the Hamiltonian all possible irrelevant terms and adjusting order by order their amplitudes as functions of g in order to eliminate systematically the contributions that go to zero. This corresponds to the strategy suggested at the end of Section 9.7, which leads to the unique equation (9.62).

The asymptotic functions then satisfy equation (13.10) with a right-hand side that vanishes exactly. Using the chain rule, one derives from equation (13.10)

$$\left[\Lambda\frac{\partial}{\partial\Lambda} + \beta(g,\Lambda/\mu)\frac{\partial}{\partial g} - \frac{n}{2}\eta(g,\Lambda/\mu)\right]\tilde{\Gamma}^{(n)}_{\text{as.}}(p_i; g, \Lambda) = 0, \tag{13.11}$$

where the functions β and η are defined by

$$\beta(g,\Lambda/\mu) = \Lambda\frac{\partial}{\partial\Lambda}\bigg|_{g_r,\mu} g, \tag{13.12a}$$

$$\eta(g,\Lambda/\mu) = -\Lambda\frac{\partial}{\partial\Lambda}\bigg|_{g_r,\mu} \ln Z_{\text{as.}}(g,\Lambda/\mu). \tag{13.12b}$$

Since the two functions are dimensionless, they can depend on g, Λ, μ only through the dimensionless combinations g and Λ/μ. Moreover, the functions β and η can be calculated directly, by using equation (13.11), in terms of the initial vertex functions, which do not depend on μ. Thus, the functions β and η cannot depend on the Λ/μ. Equation (13.11) then takes the simpler form (Zinn-Justin 1973)

$$\left(\Lambda\frac{\partial}{\partial\Lambda} + \beta(g)\frac{\partial}{\partial g} - \frac{n}{2}\eta(g)\right)\tilde{\Gamma}^{(n)}_{\text{as.}}(p_i; g, \Lambda) = 0. \tag{13.13}$$

Formulated within the formalism with cut-off Λ, the fundamental idea of the RG becomes: it is possible to modify the parameter Λ and in a correlated way the normalization of the field σ and the coefficients of all interactions in a way that leaves all correlation functions invariant.

Equation (13.13) is satisfied by the functions $\tilde{\Gamma}^{(n)}$ asymptotically in the limit $|p_i| \ll \Lambda$. It is possible to verify to all orders in the expansion in g and ε that the neglected terms are of the form $(\ln \Lambda)^L/\Lambda^2$, where the degree L increases with the order.

Notation. In what follows, the vertex functions are implicitly the functions $\tilde{\Gamma}^{(n)}_{\text{as.}}$ without corrections, which satisfy the RGE exactly and *we omit the subscript 'as.'*.

Remark. Equation (13.13) is equivalent to equation (10.40), which we have postulated in Chapter 10. Indeed, the dimensional relation (13.6) implies

$$\Lambda\frac{\partial}{\partial\Lambda} + \sum_i p_i\frac{\partial}{\partial p_i} = d - \frac{n}{2}(d-2)$$

and, thus,

$$\left(d - \sum_i p_i \frac{\partial}{\partial p_i} + \beta(g)\frac{\partial}{\partial g} - \frac{n}{2}[d-2+\eta(g)]\right)\tilde{\Gamma}^{(n)}(p_i; g, \Lambda) = 0.$$

After Fourier and Legendre transformations, one recovers equation (10.40).

13.2.2 Perturbative solution of the RGE

To illustrate the preceding discussion, it is instructive to solve perturbatively RGE (13.13) for the two-point vertex function $\tilde{\Gamma}^{(2)}(p)$, at $d = 4$ for simplicity. It is convenient to introduce the function

$$\zeta(g) = \exp\left[\int_0^g dg' \frac{\eta(g')}{\beta(g')}\right].$$

As we show later, the functions η and β are of order g^2 and thus this function can be expanded in powers of g.

We then set

$$\tilde{\Gamma}^{(2)}(p) = p^2 \zeta(g) \mathcal{G}^{(2)}(g, p/\Lambda).$$

The function $\mathcal{G}^{(2)}$ satisfies the equation

$$\left(\Lambda \frac{\partial}{\partial \Lambda} + \beta(g)\frac{\partial}{\partial g}\right) \mathcal{G}^{(2)}(g, p/\Lambda) = 0, \qquad (13.14)$$

which expresses that it is an RG invariant. We expand $\mathcal{G}^{(2)}$ and $\beta(g)$ in powers of g, setting

$$\mathcal{G}^{(2)}(g, p/\Lambda) = 1 + \sum_1^\infty g^n \gamma_n(p/\Lambda), \quad \beta(g) = \sum_2^\infty \beta_n g^n. \qquad (13.15)$$

Introducing these expansions into equation (13.14), one finds for $n \geq 2$,

$$-z\gamma_n'(z) + \sum_{m=1}^{n-1} m\beta_{n-m+1}\gamma_m(z) = 0. \qquad (13.16)$$

The equation $n = 1$ is special:

$$-z\gamma_1'(z) = 0 \quad \Rightarrow \quad \gamma_1(z) = C_1.$$

Let us examine the case $n = 2$:

$$-z\gamma_2'(z) + C_1\beta_2 = 0 \Rightarrow \gamma_2(z) = C_1\beta_2 \ln z + C_2.$$

This example exhibits the general structure. $\gamma_n(z)$ is a polynomial of degree $(n-1)$ in $\ln z$,

$$\gamma_n(z) = P_{n-1}(\ln z),$$

which satisfies the recursion relation

$$P'_{n-1}(x) = \sum_{m=1}^{n-1} m P_{m-1}(x) \beta_{n-m+1}. \quad (13.17)$$

Let us emphasize, in particular, that the new information specific to order n reduces to two constants, β_n that is the coefficient of $\ln z$, and C_n that is an integration constant (to which one must add a coefficient of $\eta(g)$, which is hidden in the function $\zeta(g)$). Moreover, the term of highest degree of P_n is determined by a one-loop calculation, the following term by a two-loop calculation, and so on. Finally, $\tilde{\Gamma}^{(2)}(p)$ is entirely determined by the functions $\beta(g)$ and $\eta(g)$ and, for example, $\tilde{\Gamma}^{(2)}(1,g)/\Lambda^2$, which is a third function of g.

From this analysis, one infers that $\tilde{\Gamma}^{(2)}(p)$ has a limit for $p = 0$:

$$\tilde{\Gamma}^{(2)}(p=0) = 0, \quad (13.18)$$

but not its derivative $\partial \tilde{\Gamma}^{(2)}/\partial p^2$, which diverges at $p = 0$. More generally, all other vertex functions diverge at zero momentum.

Renormalized correlation functions and RG. The renormalized correlation functions (13.9) and the initial physical functions are by construction proportional and, therefore, have the same behaviour for $|p_i| \to 0$. One can thus study asymptotic behaviours by using the renormalized functions. Moreover, the calculation of renormalized functions, within the framework of dimensional regularization and the renormalization by minimal subtraction of poles in $1/\varepsilon$, is easier. Renormalized functions also satisfy RGE derived from equation (13.9) (in which the asymptotic correlation functions of the initial theory and the renormalized functions play a symmetric role) by differentiating with respect to μ at Λ and g fixed (see Section 13.5). But studying only renormalized functions does not allow determining the behaviour of the parameters of the renormalized theory as functions of the initial parameters, or the nature of all corrections to the asymptotic behaviour.

13.3 Solution of RG equations: The ε-expansion

The solution of equation (13.13), combined with one-loop perturbative calculations, allows proving the universality of the large-distance asymptotic behaviour of correlation functions and determining this behaviour to first order in ε.

13.3.1 General solution

Equation (13.13) can be solved, for example, by the method of characteristics. One introduces a dilatation parameter λ and one looks for two functions $g(\lambda)$ and $Z(\lambda)$ such that

$$\lambda \frac{d}{d\lambda} \left[Z^{-n/2}(\lambda) \tilde{\Gamma}^{(n)}(p_i; g(\lambda), \Lambda/\lambda) \right] = 0. \quad (13.19)$$

By differentiating explicitly with respect to λ, one finds that equation (13.19) is consistent with equation (13.13) if

$$\lambda \frac{\mathrm{d}}{\mathrm{d}\lambda} g(\lambda) = -\beta(g(\lambda)), \qquad g(1) = g; \tag{13.20a}$$

$$\lambda \frac{\mathrm{d}}{\mathrm{d}\lambda} \ln Z(\lambda) = -\eta(g(\lambda)), \qquad Z(1) = 1. \tag{13.20b}$$

Equations (13.19) and (13.20) then imply

$$\tilde{\Gamma}^{(n)}(p_i; g, \Lambda) = Z^{-n/2}(\lambda)\tilde{\Gamma}^{(n)}(p_i; g(\lambda), \Lambda/\lambda). \tag{13.21}$$

Remark. This equation becomes quite similar to equation (13.9) if one chooses $\lambda = \Lambda/\mu$ and if one identifies g_r with $g(\Lambda/\mu)$, that is, the effective interaction at scale μ, and $Z(\Lambda/\mu)$ with the renormalization of the field. One may then wonder why it was necessary to introduce the partial differential equations. The main reason is the following: it allows showing that the coefficients of the RGE do not depend on the ratio μ/Λ, in contrast with the renormalization constants. This implies properties of renormalization constants that are contained in equation (13.9) only implicitly. This also proves that the flow equations (13.20) are independent of λ, a property that we have postulated for the general RGE (9.23), (9.24).

Other forms. First, it is convenient to multiply Λ by a factor λ in equation (13.21):

$$\tilde{\Gamma}^{(n)}(p_i; g, \Lambda\lambda) = Z^{-n/2}(\lambda)\tilde{\Gamma}^{(n)}(p_i; g(\lambda), \Lambda). \tag{13.22}$$

The dimensional considerations of Section 13.1.3, in particular the relation (13.6), imply

$$\tilde{\Gamma}^{(n)}(p_i; g, \Lambda\lambda) = \lambda^{d-(n/2)(d-2)}\tilde{\Gamma}^{(n)}(p_i/\lambda; g, \Lambda). \tag{13.23}$$

Then, equation (13.22) can be rewritten as

$$\tilde{\Gamma}^{(n)}(p_i/\lambda; g, \Lambda) = \lambda^{-d+(n/2)(d-2)} Z^{-n/2}(\lambda)\tilde{\Gamma}^{(n)}(p_i; g(\lambda), \Lambda), \tag{13.24}$$

which is identical to equation (9.14) after the *substitution* $Z \mapsto Z\lambda^{2-d}$.

Equations (13.20) and (13.24) realize asymptotically (because terms sub-leading by powers of Λ have been neglected) the general ideas of the RG. The parameter $g(\lambda)$ characterizes the effective Hamiltonian $\mathcal{H}_\lambda$ at scale λ and, therefore, equation (13.20a) governs the flow of Hamiltonians. However, the function $Z(\lambda)$ defined in (13.20b) differs from the function (9.24) by a factor λ^{d-2}. This corresponds to the initial renormalization of the field σ performed in Section 13.1.2, which has the form of the field RG transformation adapted to the Gaussian fixed point. The function defined here corresponds to the ratio between the general renormalization (9.24) and the Gaussian renormalization. This factorization is well adapted to a situation where one operator is marginal or weakly relevant (from the viewpoint of the Gaussian fixed point), as one assumes in the framework of the ε-expansion.

Discussion. Integrating equations (13.20), one finds

$$\int_g^{g(\lambda)} \frac{dg'}{\beta(g')} = -\ln\lambda, \qquad (13.25a)$$

$$\int_1^\lambda \frac{ds}{s} \eta(g(s)) = -\ln Z(\lambda). \qquad (13.25b)$$

Equation (13.13) is the RGE in differential form. Equations (13.24) and (13.25) are the integrated RGE. In what follows, we assume explicitly that the RG functions, $\beta(g)$ and $\eta(g)$, are *regular* functions of g, for $g > 0$.

In equation (13.22), one notes that it is equivalent to increase Λ or λ. To study the limit $\Lambda \to \infty$, one must thus determine the behaviour of the amplitude $g(\lambda)$ of the effective interaction for $\lambda \to \infty$. Equation (13.25a) shows that $g(\lambda)$ increases if the β-function is negative, or decreases in the opposite case. Fixed points correspond to zeros g^* of the β-function which, therefore, play an essential role in the study of the critical behaviour. Those for which the function β has a negative slope are repulsive fixed points in the IR: $g(\lambda)$ moves away from such zeros, except if initially $g(1) = g^*$. On the contrary, those for which the slope is positive are attractive fixed points from the viewpoint of the large-distance behaviour.

13.3.2 Calculations to one-loop order: Fixed-point and scaling properties

β- and η-functions. The RG functions β and η can be calculated perturbatively by expressing that the two- and four-point functions satisfy the RGE (it suffices to calculate the diagrams (e) and (g) in Figures 2.6 and 2.7, Section 2.5). The first correction to the two-point function is a constant and induces only a modification of the critical parameter u_{0c} in such a way that

$$\tilde{\Gamma}^{(2)}(p) = p^2 + O(g^2) \;\Rightarrow\; \eta(g) = O(g^2). \qquad (13.26)$$

The combinatorial factors of the four-point vertex function have been explained in Chapter 12. With the normalizations of field theory, one finds (see equation (10.11))

$$\tilde{\Gamma}^{(4)}(p_1, p_2, p_3, p_4) = \Lambda^\varepsilon g - \tfrac{1}{2} g^2 \Lambda^{2\varepsilon} \left[B(p_1 + p_2) + B(p_1 + p_3) + B(p_1 + p_4) \right]$$
$$+ O(g^3)$$

with

$$B(p) = \frac{1}{(2\pi)^d} \int d^d q\, \tilde{\Delta}(q)\tilde{\Delta}(p - q).$$

To this order, we need only this function for $d = 4$. Then, in the relevant limit, it is convenient to express B in terms of position variables in the form

$$B(p) = \int d^4 x\, e^{ipx}\, \Delta^2(x).$$

In dimension 4, as a function of position variables, the propagator with a cut-off can be parametrized as
$$\Delta(x) = \frac{\vartheta(\Lambda x)}{4\pi^2 x^2},\qquad(13.27)$$
where the large-distance behaviour is determined by expression (8.40) and the cut-off ensures regularity at short distance:
$$\lim_{x\to\infty}\vartheta(x)=1,\qquad \vartheta(x)\underset{x\to 0}{\propto}x^2.$$

For $\Lambda\to\infty$,
$$B(p)\sim K_B\ln\Lambda$$
and the coefficient K_B is given by
$$K_B = \Lambda\frac{\partial}{\partial\Lambda}B(0) = \int d^4x\,\Lambda\frac{\partial}{\partial\Lambda}\Delta^2(x) = \frac{1}{16\pi^4}\int\frac{d^4x}{x^4}\Lambda\frac{\partial}{\partial\Lambda}\vartheta^2(\Lambda x)$$
$$=\frac{1}{16\pi^4}\int\frac{d^4x}{x^4}x\frac{\partial}{\partial x}\vartheta^2(\Lambda x) = \frac{1}{8\pi^2}\int_0^\infty dx\,\frac{\partial}{\partial x}\vartheta^2(\Lambda x)$$
$$=\frac{1}{8\pi^2}\left[\vartheta^2(\infty)-\vartheta^2(0)\right] = \frac{1}{8\pi^2}.$$

One infers
$$\tilde{\Gamma}^{(4)}(p_1,p_2,p_3,p_4)\underset{\Lambda\to\infty}{=}\Lambda^\varepsilon g - \frac{3g^2}{16\pi^2}\ln\Lambda + O(g^2\times 1,g^2\varepsilon)$$
and, thus,
$$\varepsilon\Lambda^\varepsilon g - \frac{3g^2}{16\pi^2} + \beta(g)\Lambda^\varepsilon = O(g^3,g^2\varepsilon),$$
a result consistent with expression (10.21):
$$\beta(g,\varepsilon) = -\varepsilon g + \frac{3g^2}{16\pi^2} + O\left(g^3,g^2\varepsilon\right).$$

Below four dimensions, if g is initially small, expression (10.21) shows that $g(\lambda)$ first increases as a consequence of the instability of the Gaussian fixed point. But, within the framework of the ε-expansion, $\beta(g)$ has another zero g^* (equation (10.22)):
$$\beta(g^*)=0\quad\text{for}\quad g^* = \frac{16\pi^2}{3}\varepsilon + O\left(\varepsilon^2\right).$$

Then, equation (13.25a) shows that the asymptotic limit of $g(\lambda)$ is g^*. We call ω the slope of $\beta(g)$ at the fixed point g^*:
$$\omega=\beta'(g^*)=\varepsilon+O\left(\varepsilon^2\right)>0,\qquad(13.28)$$

(equation (9.59)) and linearize equation (13.25a) at the fixed point:

$$\int_g^{g(\lambda)} \frac{dg'}{\omega(g'-g^*)} \sim -\ln\lambda. \tag{13.29}$$

Integrating, one finds

$$|g(\lambda) - g^*| \underset{\lambda\to\infty}{\propto} \lambda^{-\omega}. \tag{13.30}$$

Below four dimensions, at least for ε small enough, this non-Gaussian fixed point is IR stable. In four dimensions, this fixed point merges with the Gaussian fixed point and the eigenvalue ω vanishes indicating the presence of a marginal operator.

From equation (13.25b), one also infers the behaviour of $Z(\lambda)$ for $\lambda \to \infty$. The integral in the right-hand side of the equation is dominated by the vicinity of $s = 0$. It follows that

$$\ln Z(\lambda) \underset{\lambda\to\infty}{\sim} -\eta \ln \lambda, \tag{13.31}$$

where we have set

$$\eta = \eta(g^*).$$

Correlation functions. In the framework of the ε-expansion, $\tilde{\Gamma}^{(n)}(g^*)$ is finite. We assume that this property remains valid for some finite range of values of ε. Equation (13.24) then determines the behaviour of $\tilde{\Gamma}^{(n)}(p_i; g, \Lambda)$ for $\Lambda \to \infty$ or $p_i \to 0$:

$$\tilde{\Gamma}^{(n)}(p_i/\lambda; g, \Lambda) \underset{\lambda\to\infty}{\sim} \lambda^{-d+(n/2)(d-2+\eta)} \tilde{\Gamma}^{(n)}(p_i; g^*, \Lambda). \tag{13.32}$$

The equation shows that critical correlation functions have an asymptotic power law behaviour that does not depend on the initial value of the coefficient g of σ^4.

In particular, for $n = 2$, equation (13.32) yields the behaviour of the inverse of the connected two-point function $\widetilde{W}^{(2)}(p)$. Inverting, one infers

$$\widetilde{W}^{(2)}(p) \equiv \left[\tilde{\Gamma}^{(2)}(p)\right]^{-1} \underset{|p|\to 0}{\propto} 1/p^{2-\eta}. \tag{13.33}$$

One verifies that the definition of the critical exponent η in equation (13.31) coincides with the usual definition (8.22). The spectral representation of the two-point function implies that $\eta > 0$.

13.3.3 The $\eta(g)$ function at two loops and the exponent η

To determine the first correction to the dimension of the field σ, it is necessary to calculate the two-point function $\tilde{\Gamma}^{(2)}(p)$ up to order g^2 for $d = 4$ and, thus, the diagram (e) in Figure 2.6 (Section 2.5.1):

$$(\text{e}) = \frac{1}{(2\pi)^8} \int d^4q_1\, d^4q_2\, \tilde{\Delta}(q_1)\tilde{\Delta}(q_2)\tilde{\Delta}(p - q_1 - q_2).$$

(Some elements of the calculation can be found in Section 12.5.3.) The Feynman diagram can be calculated more easily in position variables where it takes the form

$$(e) = \int e^{ipx} \Delta^3(x) d^4x.$$

In dimension 4, as a function of position variables, we represent the propagator with a cut-off in the form (13.27).

The condition $\tilde\Gamma^{(2)}(0) = 0$ determines u_{0c} at order g^2. The vertex function then takes the form

$$\tilde\Gamma^{(2)}(p) = p^2 - \frac{g^2}{6} K p^2 \ln(\Lambda/p) + O(g^2 \times 1, g^2\varepsilon).$$

The coefficient K is given by

$$K = \frac{\partial}{\partial p^2} \Lambda \frac{\partial}{\partial \Lambda} \int e^{ipx} \Delta^3(x) d^4x \bigg|_{p=0}.$$

Notice the identity

$$\sum_{\mu=1}^{4} \left(\frac{\partial}{\partial p_\mu}\right)^2 \Phi(p^2) = 8\Phi'(p^2) + 4p^2 \Phi''(p^2).$$

Thus,

$$K = -\frac{1}{8(4\pi^2)^3} \int \frac{d^4x}{x^4} \Lambda \frac{\partial}{\partial \Lambda} \vartheta^3(\Lambda x) = -\frac{1}{(4\pi)^4} \int_0^\infty dx \frac{\partial}{\partial x} \vartheta^3(\Lambda x).$$

The integrand is an explicit derivative. Only $x = \infty$ contributes and the result thus is independent of the function ϑ. One finds

$$K = -\frac{1}{(4\pi)^4}.$$

One infers

$$\tilde\Gamma^{(2)}(p) = p^2 + \frac{1}{24} \frac{g^2}{(8\pi^2)^2} p^2 \ln(\Lambda/p) + O(g^2 \times 1, g^2\varepsilon).$$

RGE (13.13) then implies

$$\frac{g^2}{6(4\pi)^4} p^2 - \eta(g)p^2 = 0,$$

and thus

$$\eta(g) = \frac{1}{6(4\pi)^4} g^2 + O(g^3),$$

in agreement with expression (10.39) for $N = 1$.

Substituting the value of g^*, one obtains the first correction to the Gaussian value of the exponent η (see also Section 10.3.4):

$$\eta = \frac{\varepsilon^2}{54} + O\left(\varepsilon^3\right). \tag{13.34}$$

The field $\sigma(x)$, which at the Gaussian fixed point has the *canonical* dimension $(d-2)/2$, now has the so-called *anomalous* dimension

$$d_\sigma = \tfrac{1}{2}(d - 2 + \eta).$$

Universality. These results call for a few comments. In the framework of the ε-expansion, one shows that all correlation functions have, for $d < 4$, a large-distance behaviour different from the one predicted by the quasi-Gaussian approximation. Moreover, this critical behaviour does not depend on the initial value of the coefficient g of σ^4. At least for $\varepsilon \ll 1$, one can hope that the analysis of leading IR singularities remains valid and, as a consequence, the critical behaviour does not depend on any other parameter in the Hamiltonian. The critical behaviour thus is *universal*, though less universal than in the quasi-Gaussian approximation or the mean-field approximations: it depends on a limited number of qualitative characteristic properties of the statistical system.

Moreover, the correlation functions obtained by neglecting, in the double expansion in g and in ε, the power law corrections when the cut-off is large, and which satisfy the RGE (13.13) exactly, define implicitly a one-parameter family of critical Hamiltonians. They correspond to a RG trajectory that goes from the vicinity of the Gaussian fixed point $g = 0$, which is IR unstable for dimensions smaller than 4, to the non-trivial IR stable fixed point g^*.

Finally, the consistency of this approach, based on the ε-expansion, relies on the following observation: the divergences found in the perturbative expansion at fixed dimension are generated by an expansion around a wrong, since repulsive, fixed point. The ε-expansion enables exchanging limits and to follow perturbatively the stable fixed point.

13.4 The critical domain above T_c

We now study the critical domain $t \neq 0$. We thus modify the Hamiltonian:

$$\mathcal{H}(\sigma) \mapsto \mathcal{H}(\sigma) + \frac{t}{2} \int d^d x \, \sigma^2(x),$$

where t, the coefficient of σ^2, characterizes the deviation from the critical temperature: $t \propto T - T_c$.

The renormalization theorem generalizes to correlation functions of $\sigma(x)$ and $\sigma^2(x)$, and leads to the appearance of a new renormalization factor $Z_2(\Lambda/\mu, g_r)$

associated with the parameter t. By the same arguments as in the critical situation, one derives a more general RGE of the form

$$\left[\Lambda\frac{\partial}{\partial\Lambda}+\beta(g)\frac{\partial}{\partial g}-\frac{n}{2}\eta(g)-\eta_2(g)t\frac{\partial}{\partial t}\right]\tilde{\Gamma}^{(n)}(p_i;t,g,\Lambda)=0\,,\qquad(13.35)$$

where a new function $\eta_2(g)$ appears. The additional term is proportional to t since it must vanish for $t=0$. The dimensionless function η_2 may still have a regular dependence in the ratio t/Λ^2, but we have neglected such a possible dependence for the same reason that we have already neglected all other contributions of the same order.

To determine $\eta_2(g)$, one can calculate the two-point function and apply the RGE. At order g, one finds

$$\tilde{\Gamma}^{(2)}(p)=p^2+t+\frac{g}{32\pi^4}\int^\Lambda d^4q\left(\frac{1}{q^2+t}-\frac{1}{q^2}\right)+\cdots.$$

Using

$$\frac{1}{32\pi^2}\int^\Lambda d^4q\left(\frac{1}{q^2+t}-\frac{1}{q^2}\right)\sim -\frac{t}{16\pi^2}\ln(\Lambda/\sqrt{t}),$$

and applying the RGE, one finds

$$-\frac{gt}{16\pi^2}-\eta_2(g)t=0$$

and, thus (see equation (10.30))

$$\eta_2(g)=-\frac{g}{16\pi^2}\,.\qquad(13.36)$$

13.4.1 Solution of RGE

To study the critical behaviour above T_c, we integrate equation (13.35) by the method of characteristics, as we have done for previous RGE. In addition to the functions $g(\lambda)$ and $Z(\lambda)$ of equations (13.25), one must now introduce a function $t(\lambda)$ that is determined by imposing that equation (13.35) is consistent with equation

$$\lambda\frac{d}{d\lambda}\left[Z^{-n/2}(\lambda)\tilde{\Gamma}^{(n)}(p_i;t(\lambda),g(\lambda),\Lambda/\lambda)\right]=0\,.\qquad(13.37)$$

The consistency condition is equivalent to the set of equations

$$\lambda\frac{d}{d\lambda}g(\lambda)=-\beta(g(\lambda)),\qquad g(1)=g\,,\qquad(13.38)$$

$$\lambda\frac{d}{d\lambda}\ln t(\lambda)=\eta_2(g(\lambda)),\qquad t(1)=t\,,\qquad(13.39)$$

$$\lambda\frac{d}{d\lambda}\ln Z(\lambda)=-\eta(g(\lambda)),\qquad Z(1)=1\,.\qquad(13.40)$$

Dimensional analysis (see Section 13.1.3) implies

$$\tilde{\Gamma}^{(n)}(p_i; t(\lambda), g(\lambda), \Lambda/\lambda) = (\Lambda/\lambda)^{d-n(d-2)/2} \tilde{\Gamma}^{(n)}(\lambda p_i/\Lambda; \lambda^2 t(\lambda)/\Lambda^2, g(\lambda), 1). \tag{13.41}$$

The critical region is defined in particular by $|t| \ll \Lambda^2$, and this is the source of the IR singular behaviour that is present in the perturbative expansion. We now assume that the equation

$$t(\lambda) = \Lambda^2/\lambda^2, \tag{13.42}$$

has a solution in λ. Then, after a dilatation of parameter λ, the correlation length is of order unity. This also corresponds, by relating the dilatation parameter λ and t, to choosing the initial value (for $\lambda = 1$) of the coefficient of the relevant term σ^2 in such a way that its value after dilatation is of order unity.

Combining equations (13.37)–(13.42), one finds

$$\tilde{\Gamma}^{(n)}(p_i; t, g, \Lambda) = Z^{-n/2}(\lambda) m^{(d-n(d-2)/2)} \tilde{\Gamma}^{(n)}(p_i/m; 1, g(\lambda), 1), \tag{13.43}$$

where we have introduced the notation

$$m = \Lambda/\lambda. \tag{13.44}$$

The solution of equation (13.39) can be written as

$$t(\lambda) = t \exp\left[\int_1^\lambda \frac{d\sigma}{\sigma} \eta_2(g(\sigma))\right]. \tag{13.45}$$

Substituting the relation into equation (13.42), one obtains

$$\ln(t/\Lambda^2) = -\int_1^\lambda \frac{d\sigma}{\sigma} \frac{1}{\nu(g(\sigma))}, \tag{13.46}$$

where, to relate the function $\eta_2(g)$ to standard exponents, we have introduced the function

$$\nu(g) = [\eta_2(g) + 2]^{-1}. \tag{13.47}$$

One looks for a solution λ in the limit $t/\Lambda^2 \ll 1$. Since $\nu(g)$ is a positive function, at least for g small enough, as equations (13.36), (13.47) show, equation (13.46) implies that the parameter λ diverges and, thus, that $g(\lambda)$ converges toward the IR fixed point g^*. In this limit, $\nu(g(\sigma))$ can be replaced, at leading order, by the critical exponent

$$\nu = \nu(g^*) = \tfrac{1}{2}(1 + \tfrac{1}{6}\varepsilon) + O(\varepsilon^2), \tag{13.48}$$

(using the expansion (13.36)). Equation (13.46) implies

$$\ln t/\Lambda^2 \sim -\frac{1}{\nu} \ln \lambda. \tag{13.49}$$

The asymptotic form of $Z(\lambda)$ is given by equation (13.31):

$$Z(\lambda) \propto \lambda^{-\eta}. \qquad (13.50)$$

Finally, in the limit $\Lambda \to \infty$, or equivalently $\lambda \to \infty$, using equations (13.49), (13.50) together with (13.43), one obtains an asymptotic behaviour of the form

$$\tilde{\Gamma}^{(n)}(p_i; t, g, \Lambda = 1) \underset{\substack{t \ll 1 \\ |p_i| \ll 1}}{\sim} m^{d-n(d-2+\eta)/2} F_+^{(n)}(p_i/m), \qquad (13.51)$$

where

$$m(\Lambda = 1) = \xi^{-1} \propto t^{\nu}. \qquad (13.52)$$

In addition to the general scaling behaviour, another result has been obtained. From equation (13.51), one infers that the quantity m is proportional to the physical mass or the inverse correlation length. Equation (13.52) then shows that the divergence of the correlation length $\xi = m^{-1}$ at T_c is characterized by the exponent ν.

For $t \neq 0$, the correlation functions are finite at zero momentum and behave like

$$\tilde{\Gamma}^{(n)}(0; t, g, \Lambda) \propto t^{\nu(d-n(d-2+\eta)/2)}. \qquad (13.53)$$

In particular, for $n = 2$, one obtains the inverse of the magnetic susceptibility χ:

$$\chi^{-1} = \tilde{\Gamma}^{(2)}(p = 0; t, g, \Lambda) \propto t^{\nu(2-\eta)}. \qquad (13.54)$$

The divergence of χ is characterized by an exponent that is usually denoted by γ (see equation (7.35)). Equation (13.53) establishes the relation between exponents

$$\gamma = \nu(2 - \eta). \qquad (13.55)$$

Finally, since we have assumed that the critical theory exists in the $t = 0$ limit, the different powers of t must cancel in equation (13.51). From these remarks, one recovers equation (13.32) in the form

$$\tilde{\Gamma}^{(n)}(p_i/\lambda; t, g, \Lambda = 1) \underset{t^{\nu} \ll |p_i|/\lambda \ll 1}{\propto} \lambda^{-d+n(d-2+\eta)/2}. \qquad (13.56)$$

13.4.2 Functions at fixed magnetization or below T_c

It is easy to generalize RGE to correlation functions in a fixed, external, uniform, magnetic field H or to vertex functions at fixed magnetization $M = \langle \sigma(x) \rangle$. In particular, the external field H, the magnetization M and the deviation from the critical temperature t are related by the equation of state

$$H = H(M, t, g, \Lambda).$$

In the critical domain, $H(M, t, g, \Lambda)$ satisfies the RGE

$$\left[\Lambda\frac{\partial}{\partial\Lambda} + \beta(g)\frac{\partial}{\partial g} - \frac{1}{2}\eta(g)\left(1 + M\frac{\partial}{\partial M}\right) - \eta_2(g)t\frac{\partial}{\partial t}\right]H(M,t,g,\Lambda) = 0. \quad (13.57)$$

The equation can be solved by the same methods. Since it involves no new RG function, the solution can be entirely expressed in terms of the functions (13.38), (13.39) and (13.40). Asymptotically in the limit $M, t, H \ll 1$, one can replace g by g^*. Moreover, dimensional analysis shows that H has dimension $\frac{1}{2}(d+2)$ and, thus,

$$\Lambda\frac{\partial}{\partial\Lambda} = \frac{d+2}{2} - 2t\frac{\partial}{\partial t} - \frac{d-2}{2}M\frac{\partial}{\partial M},$$

in such a way that, asymptotically, $H(M,t)$ is solution of

$$\left[(2+\eta_2)t\frac{\partial}{\partial t} + \frac{1}{2}(d-2+\eta)M\frac{\partial}{\partial M}\right]H(M,t) = \frac{1}{2}(d+2-\eta)H(M,t).$$

This equation implies the asymptotic universal scaling property of the equation of state

$$H = M^\delta f(t/M^{1/\beta}),$$

where the function $f(x)$ is universal, up to normalizations of f and x, and the exponents β and δ are related to η and ν by

$$\delta = \frac{d+2-\eta}{d-2+\eta} = \frac{d}{d_\sigma} - 1, \quad \beta = \nu d_\sigma = \frac{1}{2}\nu(d-2+\eta).$$

By varying the external field H, one can then relate continuously the disordered phase $(T > T_c, H = 0)$ to the ordered phase $(T < T_c, H = 0)$. The RGE satisfied by vertex functions,

$$\left[\Lambda\frac{\partial}{\partial\Lambda} + \beta(g)\frac{\partial}{\partial g} - \frac{1}{2}\eta(g)\left(n + M\frac{\partial}{\partial M}\right) - \eta_2(g)t\frac{\partial}{\partial t}\right]\tilde{\Gamma}^{(n)}(p_i; t, M, g, \Lambda) = 0,$$
(13.58)
allow then proving scaling relations in a field or at given magnetization and, thus, in the limit $H \to 0$, $T < T_c$, scaling relations in the ordered phase.

13.5 RG equations for renormalized vertex functions

RGE can also be written for renormalized correlation functions, at the price of making the discussion of corrections to scaling more indirect. Indeed, one can rewrite equation (13.9) as

$$Z^{-n/2}(g_r, \Lambda/\mu)\tilde{\Gamma}_r^{(n)}(p_i; g_r, \mu, \Lambda) = \tilde{\Gamma}^{(n)}(p_i; g, \Lambda), \quad (13.59)$$

and express now that the right hand side does not depend on the renormalization scale μ:

$$\mu \frac{\mathrm{d}}{\mathrm{d}\mu}\bigg|_{g,\Lambda \text{ fixed}} \tilde{\Gamma}^{(n)}(p_i; g, \Lambda) = 0.$$

The role of g, Λ and g_r, μ are thus interchanged and by the same algebraic manipulations as in Section 13.2.1, in the large Λ limit, one obtains an equation of the form

$$\left(\mu \frac{\partial}{\partial \mu} + \beta_r(g_r) \frac{\partial}{\partial g_r} - \frac{n}{2} \eta_r(g_r) \right) \tilde{\Gamma}_r^{(n)}(p_i; g_r, \mu) = 0. \tag{13.60}$$

After the limit $\Lambda \to \infty$ has been taken, the equation is exact. Note that the functions β_r and η_r differ from the functions β and η defined in (13.12) but coïncide at leading order for $g \to 0$.

In the renormalized scheme, functions can be calculated using dimensional regularization and, instead of introducing renormalization conditions, the renormalized theory can also be defined by minimal subtraction (or modified minimal subtraction), that is, by subtracting from perturbative contributions the simple or multiple poles in $1/\varepsilon$. RG functions then have a simple dependence in ε (see, e.g., Section 11.3 of [9]).

Again a renormalized deviation t_r from the critical theory can be introduced and the analogue of equation (13.35) is obtained:

$$\left[\mu \frac{\partial}{\partial \mu} + \beta_r(g_r) \frac{\partial}{\partial g_r} - \frac{n}{2} \eta_r(g_r) - \eta_{2,r}(g_r) t_r \frac{\partial}{\partial t_r} \right] \tilde{\Gamma}_r^{(n)}(p_i; t_r, g_r, \mu) = 0. \tag{13.61}$$

The solutions of these RGE then follows immediately from the corresponding solutions of equations (13.35).

Callan–Symanzik equations. Callan–Symanzik (CS) equations [22] are useful, inhomogeneous, variants of the renormalized RGE (13.61). One considers only a non-critical theory, but in the critical domain, above T_c. The renormalized vertex functions are then be defined by the conditions

$$\tilde{\Gamma}_r^{(2)}(p, m, g_r) = m^2 + p^2 + O(p^4), \quad \tilde{\Gamma}_r^{(4)}(0,0,0,0) = m^\varepsilon g_r,$$

where $m \ll \Lambda$ is proportional to the inverse of the correlation length. Moreover, one introduces the vertex (1PI) function $\Gamma^{(1,n)}$ associated to the correlation function (see also Section 13.4)

$$\langle \tfrac{1}{2}\sigma^2(y)\sigma(x_1)\ldots\sigma(x_n) \rangle.$$

Its renormalization is determined by the condition

$$\tilde{\Gamma}_r^{(1,2)}(0;0,0;g_r,m) = 1.$$

Dimensional analysis then implies

$$\tilde{\Gamma}_r^{(n)}(p_i; g_r, m) = m^{d-n(d-2)/2} F_+^{(n)}(p_i/m),$$

a form consistent with equation (13.51), since the missing factor $m^{-n\eta/2}$ is provided by the field renormalization (13.59).

One then proves the RGE in the CS form

$$\left(m\frac{\partial}{\partial m} + \beta_{\rm r}(g_{\rm r})\frac{\partial}{\partial g_{\rm r}} - \frac{n}{2}\eta_{\rm r}(g_{\rm r})\right)\tilde{\Gamma}_{\rm r}^{(n)}(p_i;g_{\rm r},m) = (2-\eta_{\rm r}(g_{\rm r}))m^2\tilde{\Gamma}_{\rm r}^{(1,n)}(0;p_i;g_{\rm r},m),$$
(13.62)

where the right hand side refers to the vertex function involving the expectation value of n fields σ and $\frac{1}{2}\int d^dx\,\sigma^2(x)$. The advantage of the CS equations is that their *derivation below four dimensions no longer relies on the ε-expansion*. However, they are predictive only if the right hand side becomes negligible for large momenta $|p_i| \gg m$ (but still $|p_i| \ll \Lambda$), a property that can be explicitly verified only within the framework of the ε-expansion. Then,

$$\left(m\frac{\partial}{\partial m} + \beta_{\rm r}(g_{\rm r})\frac{\partial}{\partial g_{\rm r}} - \frac{n}{2}\eta_{\rm r}(g_{\rm r})\right)\tilde{\Gamma}_{\rm r}^{(n)}(p_i;g_{\rm r},m) \underset{|p_i|\gg m}{\to} 0.$$
(13.63)

The analysis of Section 13.6 applies here also, the role of μ being played by m. Therefore, in the limit $\Lambda/m \to \infty$, the renormalized coupling $g_{\rm r}$ tends toward an IR fixed point (a zero of the $\beta_{\rm r}$ function with a positive slope) if such a fixed point exists. Numerically, such a fixed point $g_{\rm r}^*$ is found (Table 10.3) and, thus, for $\Lambda \gg |p_i| \gg m$, the equation reduces at leading order to

$$\left(m\frac{\partial}{\partial m} - \frac{n}{2}\eta_{\rm r}(g_{\rm r}^*)\right)\tilde{\Gamma}_{\rm r}^{(n)}(p_i;g_{\rm r}^*,m) = 0.$$
(13.64)

From the solution of the asymptotic homogeneous equation combined with dimensional analysis, one recovers then the behaviour (13.56).

The corresponding perturbative expansion of the RG functions at $d=3$ fixed has led to the most precise field theory critical exponents as reported in Table 10.3.

13.6 Effective and renormalized interactions

We have stated that the renormalized parameter $g_{\rm r}$ (equation (13.7)) characterizes the strength of the effective interaction at scale μ (or m in the case of the CS equations (13.62)). We now give a more precise meaning to this remark. We write equation (13.7) in the form

$$g_{\rm r} = G(\Lambda/\mu, g).$$
(13.65)

Then, differentiating with respect to Λ at $g_{\rm r}, \mu$ fixed, one finds

$$\left(\Lambda\frac{\partial}{\partial \Lambda} + \beta(g)\frac{\partial}{\partial g}\right)G(\Lambda/\mu,g) = 0,$$

where the same contributions as in the case of correlation functions have been neglected: in the expansion of $G(\Lambda/\mu, g)$ in powers of g and ε, one keeps only the contributions that diverge or remain finite when $\Lambda \to \infty$.

Introducing a dilatation parameter λ, one infers
$$G(\Lambda/\mu, g) = G(\Lambda/\mu\lambda, g(\lambda)).$$
For $\lambda = \Lambda/\mu$, the relation becomes
$$g_\mathrm{r} = G(\Lambda/\mu, g) = G(1, g(\lambda))$$
with, for $g \to 0$,
$$g_\mathrm{r} = g(\lambda) + O(g^2).$$
Thus, g_r is not identical to $g(\lambda)$ except for $g \to 0$, but it is a function of $g(\lambda)$.

In the limit $\Lambda/\mu \to \infty$, λ diverges and $g(\lambda)$ converges toward g^*. One concludes
$$\Lambda/\mu \to \infty \ \Rightarrow \ g_\mathrm{r} = G(\Lambda/\mu, g) \to G(1, g^*).$$
For any fixed initial value of $g > 0$, the renormalized parameter converges toward the same fixed point value. More precisely,
$$g_\mathrm{r} - g_\mathrm{r}^* = O\left((\mu/\Lambda)^\omega\right).$$
In particular, in four dimensions, the fixed point remains Gaussian and, thus, $g_\mathrm{r} \to 0$ when $\Lambda \to \infty$. This leads to the so-called *triviality* problem in the framework of the corresponding quantum field theory.

Finally, since at g^* the theory is scale invariant, one infers from the RGE (13.60) (or (13.63)) of the renormalized theory that g_r is also solution of
$$\beta_\mathrm{r}(g_\mathrm{r}^*) = 0 \ \Leftrightarrow \ g_\mathrm{r}^* = G(1, g^*).$$

Fixed renormalized parameter. Conversely, one may wonder whether, beyond perturbation theory, it is always possible to find a value of g for any given value of g_r. The question can be studied by rewriting equation (13.25a) as
$$\int_{g(\mu/\Lambda)}^{g(1)} \frac{dg'}{\beta(g')} = \ln(\mu/\Lambda).$$
At g_r fixed, the equation implies that for $\Lambda/\mu \to \infty$, the initial parameter g moves away from the IR stable fixed point. If a UV fixed point exists, that is, a zero of $\beta(g)$ with a negative slope, g may converge toward such a fixed point.

If initially $g(1) < g^*$, $g(\mu/\Lambda)$ converges toward $g = 0$. This is not surprising since this amounts to cancelling the effect of dilatations for a relevant parameter in the vicinity of the Gaussian fixed point:
$$g(\mu/\Lambda) \propto (\mu/\Lambda)^\varepsilon.$$
Note that $g(1) < g^*$ also implies $g_\mathrm{r} < g_\mathrm{r}^*$. Instead, if initially $g_\mathrm{r} > g_\mathrm{r}^*$, $g(\mu/\Lambda)$ increases and the existence of a solution eventually depends on the possibility of still using an asymptotic RG with only one parameter and on the existence of an additional UV fixed point. Empirical arguments comforted by numerical calculations seem to exclude such a fixed point and, thus, one is tempted to conclude that g_r is always bounded by g_r^*. In particular, in four dimensions this seems to exclude a solution to the triviality problem ($g_\mathrm{r} = 0$) based on the existence of a UV fixed point value of g.

14 The $O(N)$ symmetric $(\phi^2)^2$ field theory in the large N limit

In Chapter 13, we have explained how one can derive, by the methods of statistical, or quantum, field theory, universal properties of critical systems in the framework of the formal $\varepsilon = 4 - d$ expansion where d is the dimension of space. It is encouraging to be able to verify by analytic methods, at least in some special limit, that the results obtained in this way remain valid even when ε is no longer infinitesimal. In this chapter, we study a statistical field theory with an $O(N)$ orthogonal symmetry and a $(\phi^2)^2$ interaction (we denote here by $\phi = (\phi_1, \ldots, \phi_N)$ the N-component field rather than σ, in contrast with previous chapters), at fixed dimension, in the framework of another approximation scheme, the $N \to \infty$ limit. The results confirm the universal properties derived in the framework of the formal ε-expansion. More generally, but this will not be done here, all the properties derived by means of the renormalization group (RG) can also be proved order by order in an expansion in $1/N$.

Finally, note that it is also possible to study N-component spin models in the limit $N \to \infty$ directly on the lattice. The choice of continuum space is a technical simplification but does affect universal results.

General symmetric Hamiltonian. In a first step, we study a more general problem and consider a Hamiltonian of the form

$$\mathcal{H}(\phi) = \mathcal{H}^{(2)}(\phi) + N \int d^d x\, U(\phi^2(x)/N), \qquad (14.1)$$

where

$$\mathcal{H}^{(2)}(\phi) = \frac{1}{2} \int d^d x \sum_\mu \partial_\mu \phi(x) \cdot D(-\nabla_x^2/\Lambda^2) \partial_\mu \phi(x).$$

This Hamiltonian is invariant under the $O(N)$ group acting on the vector ϕ.

The function $D(s)$ satisfies the usual conditions. It is normalized: $D(0) = 1$, strictly positive and analytic for $s \geq 0$. Moreover, for $s \to +\infty$, it increases faster than $s^{(d-2)/2}$ in order for the perturbative expansion to be defined. The parameter Λ is the cut-off. Finally, $U(\rho)$ is a polynomial in ρ with, for $\rho \geq 0$, a unique minimum at $\rho \neq 0$ where its second derivative does not vanish.

In terms of the field Fourier components defined by

$$\phi(x) = \int d^d p\, e^{ipx}\, \tilde\phi(p),$$

$\mathcal{H}^{(2)}(\phi)$ can also be written as

$$\mathcal{H}^{(2)}(\phi) = (2\pi)^d \int d^d p\, \tilde\phi(p) D(p^2/\Lambda^2) \tilde\phi(-p). \qquad (14.2)$$

We want to determine the behaviour of the partition function

$$Z = \int [d\boldsymbol{\phi}(x)] \exp[-\mathcal{H}(\boldsymbol{\phi})] \tag{14.3}$$

and of correlation functions for $N \to \infty$ (at U fixed).

14.1 Algebraic preliminaries

For $N \to \infty$, the study of models specified by a Hamiltonian of type (14.1) relies on ideas of the central limit theorem (Section 3.1) or mean-field theory (Section 8.10) types. Indeed, one expects that quantities invariant under the $O(N)$ group like

$$\boldsymbol{\phi}^2(x) = \sum_{i=1}^{N} \phi_i^2(x),$$

self-average and, thus, have weak fluctuations. This implies, for example,

$$\langle \boldsymbol{\phi}^2(x)\boldsymbol{\phi}^2(y) \rangle \underset{N \to \infty}{\sim} \langle \boldsymbol{\phi}^2(x) \rangle \langle \boldsymbol{\phi}^2(y) \rangle.$$

Therefore, it seems more appropriate to take $\rho(x) = \boldsymbol{\phi}^2(x)$ as a dynamic variable rather than $\boldsymbol{\phi}(x)$. This idea can be implemented by the following method: one introduces two new fields $\lambda(x)$ and $\rho(x)$ (see also Section 8.10) and imposes the constraint $\rho(x) = \boldsymbol{\phi}^2(x)/N$ by an integral over $\lambda(x)$. At each point x in space, one uses the identities

$$\frac{N}{4\pi} \int d\rho \int_{-\infty}^{+\infty} d\lambda \, e^{i\lambda(\boldsymbol{\phi}^2 - N\rho)/2} = N \int d\rho \, \delta(\boldsymbol{\phi}^2 - N\rho) = 1,$$

where δ is Dirac's distribution. Insertion of this identity into integral (14.3) leads to the new representation of the partition function

$$Z = \int [d\boldsymbol{\phi}][d\rho][d\lambda] \exp[-\mathcal{H}(\boldsymbol{\phi}, \rho, \lambda)] \tag{14.4}$$

with

$$\mathcal{H}(\boldsymbol{\phi}, \rho, \lambda) = \mathcal{H}^{(2)}(\boldsymbol{\phi}) + \int \left[NU(\rho(x)) - \tfrac{1}{2}i\lambda(x)(\boldsymbol{\phi}^2(x) - N\rho(x)) \right] d^d x. \tag{14.5}$$

The field integral (14.4) is then Gaussian in $\boldsymbol{\phi}$. The integrals over the different components of $\boldsymbol{\phi}$ factorize and each integral yields the same result

$$\int [d\phi] \exp\left[-\mathcal{H}^{(2)}(\phi) + \tfrac{1}{2}i \int d^d x\, \lambda(x)\phi^2(x)\right] \propto e^{-w(\lambda)},$$

where formally (tr ln = ln det)

$$w(\lambda) = \tfrac{1}{2} \ln \det \left[-\nabla_x^2 D(-\nabla_x^2/\Lambda^2) - i\lambda(\bullet) \right] = \tfrac{1}{2} \operatorname{tr} \ln \left[-\nabla_x^2 D(-\nabla_x^2/\Lambda^2) - i\lambda(\bullet) \right].$$

The dependence on N of the partition function then becomes explicit.

Actually, it is convenient to separate the components of ϕ in one component σ, and $N-1$ components π, and to integrate only over π (for $T < T_c$ it may even be useful to integrate only over $N-2$ components). This modification does not affect the $N \to \infty$ limit but allows calculating directly correlation functions of the σ component of the field ϕ. For this purpose, we add to the Hamiltonian a term corresponding to the coupling of the σ component to an external field. The generating functional of correlation functions of the σ field is then given by

$$\mathcal{Z}(H) = \int [\mathrm{d}\sigma][\mathrm{d}\rho][\mathrm{d}\lambda] \exp\left[-\mathcal{H}_N(\sigma, \rho, \lambda) + \int \mathrm{d}^d x\, H(x)\sigma(x) \right] \qquad (14.6)$$

with

$$\mathcal{H}_N(\sigma, \rho, \lambda) = \mathcal{H}^{(2)}(\sigma) + \int \mathrm{d}^d x \left[NU(\rho) - \tfrac{1}{2} i\lambda(x)(\sigma^2(x) - N\rho(x)) \right] + (N-1)w(\lambda). \qquad (14.7)$$

Correlation functions of the field $\phi^2(x)$. In this formalism, it is easy to generate correlation functions involving also the field ρ, which, by construction, is proportional to ϕ^2. The field $\phi^2(x)$, which is coupled to the variation of the temperature, yields the leading contribution to the local energy.

14.2 Integration over the field ϕ: The determinant

In the calculation of Section 14.1, the integration over each component of ϕ has generated the determinant of a differential operator, which is not a simple quantity since $\lambda(x)$ is a fluctuating field. Fortunately, the calculations that follow require only a perturbative definition, which we give below, valid when $\lambda(x)$ fluctuates only weakly around some constant imaginary value.

14.2.1 The determinant: Perturbative definition

We assume that $\lambda(x)$ fluctuates weakly around the constant expectation value im^2, $m > 0$. We then set

$$\lambda(x) = im^2 + \mu(x), \quad m > 0$$

and define the determinant by (see Sections 10.2, 12.4)

$$w(\lambda) = \tfrac{1}{2} \operatorname{tr} \ln \left[-\nabla_x^2 D(-\nabla_x^2/\Lambda^2) - i\lambda(\bullet) \right]$$
$$= \tfrac{1}{2} \operatorname{tr} \ln \left[-\nabla_x^2 D(-\nabla_x^2/\Lambda^2) + m^2 \right] - \tfrac{1}{2} \sum_{n=1}^{\infty} \frac{i^n}{n} \operatorname{tr}(\Delta \mu)^n, \qquad (14.8)$$

where Δ is the propagator, the inverse of the operator $-\nabla_x^2 D(-\nabla_x^2/\Lambda^2) + m^2$:

$$[-\nabla_x^2 D(-\nabla_x^2/\Lambda^2) + m^2]\Delta(x-y) = \delta^{(d)}(x-y).$$

This equation can be solved by introducing the Fourier representation

$$\Delta(x) = \frac{1}{(2\pi)^d}\int d^d p\, e^{-ipx}\tilde{\Delta}(p).$$

One finds

$$\tilde{\Delta}(p) = \frac{1}{m^2 + p^2 D(p^2/\Lambda^2)}. \tag{14.9}$$

In the expansion (14.8), one recognizes a sum of one-loop diagrams of the type discussed in Section 12.4.2 and

$$\mathrm{tr}(\Delta\mu)^n \equiv \int \prod_{i=1}^n d^d x_i\, \mu(x_i)\Delta(x_i - x_{i-1}) \quad \text{with} \quad x_n = x_0.$$

Introducing the Fourier representation of the field

$$\mu(x) = \int d^d p\, e^{ipx}\tilde{\mu}(p),$$

one can rewrite the expression

$$\mathrm{tr}(\Delta\mu)^n = \int \left(\prod_{i=1}^n d^d p_i\, \tilde{\mu}(p_i)\right) \delta^{(d)}\left(\sum_i p_i\right) \int d^d q \prod_{i=1}^n \tilde{\Delta}(q_i) \tag{14.10}$$

with

$$q_{i+1} - q_i = p_i, \quad q_{n+1} = q_1 = q.$$

The first three terms of the expansion play a special role later. For $n = 1, 2$,

$$\mathrm{tr}\,\Delta\mu = \Delta(0)\int d^d x\, \mu(x) = (2\pi)^d \tilde{\mu}(0)\Omega_d(m),$$

$$\mathrm{tr}(\Delta\mu)^2 = \int d^d x\, d^d y\, \Delta^2(x-y)\mu(x)\mu(y) = (2\pi)^d \int d^d p\, \tilde{\mu}(p)\tilde{\mu}(-p) B_\Lambda(p,m),$$

where we have defined

$$\Omega_d(m) = \Delta(0) = \int \frac{d^d q}{(2\pi)^d}\tilde{\Delta}(q), \tag{14.11}$$

$$B_\Lambda(p,m) = \int \frac{d^d q}{(2\pi)^d}\tilde{\Delta}(q)\tilde{\Delta}(p-q). \tag{14.12}$$

The functions Ω_d correspond to the one-loop diagram (b) in Figure 2.5 in its 1-irreducible form. This diagram has already appeared in equations (6.39a) and

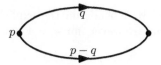

Fig. 14.1 The 'bubble' diagram $B_\Lambda(p,m)$.

(10.10), and has been discussed in an identical form, after the identification $m^2 \mapsto t$, in Section 10.4.

The function B_Λ involves the diagram (g) in Figure 2.7, under its 1-irreducible form, displayed more accurately in Figure 14.1. For $m = 0$, the function $B_\Lambda(p,m)$ reduces to the function $B(p)$ defined in (10.15) whose asymptotic behaviour for $p \to 0$ and $d \to 4$ is discussed in Section 10.3.2.

14.2.2 First one-loop diagrams: Discussion

For what follows, a few asymptotic properties of the functions $\Omega_d(m)$ and $B_\Lambda(p,m)$ are needed, which for $\Lambda = \infty$ have already been studied in the framework of the dimensional regularization in Section 12.5.3.

The function Ω_d. The function (14.11) can be rewritten more explicitly as

$$\Omega_d(m) = \frac{1}{(2\pi)^d} \int \frac{d^d q}{m^2 + q^2 D(q^2/\Lambda^2)}. \tag{14.13}$$

The change of variables $q/\Lambda \mapsto q$ shows that Ω_d is of the form

$$\Omega_d(m) = \Lambda^{d-2} \omega_d(m/\Lambda). \tag{14.14}$$

The first terms of the expansion of $\Omega_d(m)$ for $m^2 \to 0$ can be inferred from the expansion of $\omega_d(z)$ for $z \to 0$. For $d > 2$, the expansion has the form (for details see Appendix A4.1)

$$\omega_d(z) = \omega_d(0) - K(d) z^{d-2} + a(d) z^2 + O\left(z^4, z^d\right). \tag{14.15}$$

The constant $K(d)$ is universal, that is, independent of the cut-off function D:

$$N_d = \frac{2}{(4\pi)^{d/2} \Gamma(d/2)} \tag{14.16a}$$

$$K(d) = -\frac{\pi}{2 \sin(\pi d/2)} N_d = -\frac{1}{(4\pi)^{d/2}} \Gamma(1 - d/2), \tag{14.16b}$$

where, again, we have introduced the loop factor N_d.

The constant $a(d)$, which characterizes, for $d < 4$, the leading correction in equation (14.15), is given by

$$a(d) = N_d \int_0^\infty k^{d-5} dk \left(1 - \frac{1}{D^2(k^2)}\right) \tag{14.17}$$

(form valid for $2 < d < 4$). It depends on the explicit form of the propagator and thus of the regularization. However, for $\varepsilon = 4 - d \to 0$, it has the universal divergence

$$a(d) \underset{\varepsilon=4-d\to 0}{\sim} N_4 \int_1^\infty k^{d-5} dk \sim 1/(8\pi^2 \varepsilon). \tag{14.18}$$

For $d = 4$, the behaviour of $w_d(z)$ involves a logarithm:

$$w_d(z) - w_d(0) \sim \frac{1}{8\pi^2} z^2 \ln z. \tag{14.19}$$

The function $B_\Lambda(p, m)$. The expansion of $B_\Lambda(0, m)$ for $m \to 0$ can be derived from the expansion (14.15). Indeed,

$$B_\Lambda(0, m) = \int \frac{d^d q}{(2\pi)^d} \tilde{\Delta}^2(q)$$

$$= -\frac{\partial}{\partial m^2} \Omega_d(m) \underset{m \ll \Lambda}{=} (d/2 - 1) K(d) m^{-\varepsilon} - a(d) \Lambda^{-\varepsilon} + \cdots . \tag{14.20}$$

In the critical theory ($m = 0$ at this order), for $2 \le d \le 4$,

$$B_\Lambda(p, 0) \underset{2<d<4}{=} b(d) p^{-\varepsilon} - a(d) \Lambda^{-\varepsilon} + O\left(\Lambda^{d-6} p^2, \Lambda^{-2} p^{d-2}\right), \tag{14.21}$$

where

$$b(d) = -\frac{\pi}{\sin(\pi d/2)} \frac{\Gamma^2(d/2)}{\Gamma(d-1)} N_d. \tag{14.22}$$

For $d = 4$, the function has a logarithmic behaviour of the form (see Section 10.3.2)

$$B_\Lambda(p, 0) \underset{p \ll \Lambda}{\sim} \frac{1}{8\pi^2} \ln(\Lambda/p) + \text{constant}. \tag{14.23}$$

Contribution to the free energy. The first term in the expansion (14.8),

$$w(im^2) = \tfrac{1}{2} \operatorname{tr} \ln[-\nabla_x^2 D(-\nabla_x^2/\Lambda^2) + m^2],$$

is proportional to the volume and diverges when the volume becomes infinite. Moreover, it also diverges because the integral over the field ϕ has not been normalized. Normalizing the partition function corresponds to multiplying the determinant by a constant that must be chosen in such a way that the product is finite. This corresponds also to subtracting a divergent contribution to its logarithm. It is convenient to choose a normalization independent of m^2. We thus define

$$\mathcal{D}(m^2) = 2[w(im^2) - w(0)]/\text{volume}. \tag{14.24}$$

The variation of the determinant, when m^2 varies by δm^2, can be inferred from the second term in the expansion (14.8) by setting $\mu(x) = i\delta m^2$:

$$-i\operatorname{tr}\Delta\mu = \delta m^2 \Delta(0) \int d^d x = \delta m^2 \Delta(0) \times \text{volume}.$$

As a consequence,

$$\frac{\partial \mathcal{D}(m^2)}{\partial m^2} = \Delta(0) = \int \frac{d^d k}{(2\pi)^d} \tilde{\Delta}(k) = \Omega_d(m).$$

Integrating $\Omega_d(m)$ over m^2, one obtains the finite expression

$$\mathcal{D}(m^2) = \int_0^m 2s\,ds\,\Omega_d(s). \tag{14.25}$$

For $\Lambda \to \infty$, the function $\mathcal{D}$ can be evaluated by using the regularized expression (14.25), and expressing the expansion in terms of the quantities defined in (14.15). One finds

$$\mathcal{D}(m^2) = -2\frac{K(d)}{d}m^d + \Omega_d(0)m^2 + \frac{a(d)}{2}m^4 \Lambda^{4-d} + O(m^6 \Lambda^{d-6}, m^{d+2}\Lambda^{-2}). \tag{14.26}$$

14.3 The limit $N \to \infty$: The critical domain

We now study the limit $N \to \infty$, the function $U(\rho)$ being assumed independent of N. With the Ansatz $\sigma = O(N^{1/2})$, $\rho = O(1)$, $\lambda = O(1)$ all terms in $\mathcal{H}_N$ are of order N and the integral can be evaluated by the steepest descent method. One looks for a saddle point that corresponds to constant classical fields $\sigma(x), \rho(x), \lambda(x)$:

$$\sigma(x) = \sigma, \quad \rho(x) = \rho, \quad \lambda(x) = im^2, \tag{14.27}$$

where the parametrization stresses the positivity of $-i\lambda$ at the saddle point (see (14.9)).

For $H = 0$, the Hamiltonian density then reads

$$\mathcal{E} \underset{N \to \infty}{\sim} NU(\rho) + \tfrac{1}{2}m^2(\sigma^2 - N\rho) + \tfrac{1}{2}N\mathcal{D}(m^2), \tag{14.28}$$

where the representation (14.25) has been used. Differentiating $\mathcal{E}$ with respect to σ, ρ and m^2, one finds the saddle point equations (using the definition (14.11))

$$m^2 \sigma = 0, \tag{14.29a}$$
$$\tfrac{1}{2}m^2 = U'(\rho), \tag{14.29b}$$
$$\sigma^2/N - \rho + \Omega_d(m) = 0. \tag{14.29c}$$

Discussion. In expression (14.7), the term proportional to $w(\lambda)$ comes from the integration over the $N-1$ components $\boldsymbol{\pi}$. The expectation value im^2 of the field λ comes from the mass term of the field $\boldsymbol{\pi}$ and thus, at leading order, m is the mass of the $N-1$ components $\boldsymbol{\pi}$. Equation (14.29a) implies either $\sigma = 0$ or $m = 0$. This alternative is a manifestation of the Goldstone phenomenon (Section 7.8): if $\sigma = 0$, the $O(N)$ symmetry is explicit and the N components of the field have the same mass m; on the contrary, if $\sigma \neq 0$, the $O(N)$ symmetry is spontaneously broken, $m = 0$ and, thus, the mass of the field $\boldsymbol{\pi}$ vanishes, the $N-1$ components of $\boldsymbol{\pi}$ corresponding to the expected $N-1$ Goldstone modes.

We now consider the saddle point equations in the two phases.

(i) *Ordered phase (spontaneously broken symmetry)*. Note that equation (14.29c) is compatible with a solution $m = 0$ only for $d > 2$ since, for $d \leq 2$, the integral diverges at $m = 0$. This property is directly related to the Mermin–Wagner–Coleman theorem: in a statistical model involving only short-range forces, a continuous symmetry cannot be spontaneously broken for $d \leq 2$, in the sense that the expectation value σ of the order parameter necessarily vanishes (see also the discussion of Section 7.7). In the present context, the Goldstone modes are at the origin of this property: being massless, as general arguments imply and as the explicit calculation at leading order for $N \to \infty$ confirms, their presence would induce IR (zero momentum) divergences for $d \leq 2$.

The saddle point equations (14.29) then reduce to $(d > 2)$

$$U'(\rho) = 0, \quad \sigma^2/N - \rho + \Omega_d(0) = 0.$$

The first equation expresses that ρ is given by the minimum of U, which we have assumed to be unique, and the second equation then determines the expectation value of the field. A solution exists only if the value of ρ at the minimum of U satisfies

$$\rho > \rho_c = \Omega_d(0). \tag{14.30}$$

Then,

$$\sigma = \sqrt{N(\rho - \rho_c)}. \tag{14.31}$$

(ii) *Disordered (symmetric) phase*. In the case $\sigma = 0$, the saddle point equations (14.29) read

$$m^2 = 2U'(\rho), \quad \rho - \rho_c = \Omega_d(m) - \Omega_d(0). \tag{14.32}$$

The second equation implies $\rho \leq \rho_c$. The value $\rho = \rho_c$ thus corresponds to a transition between an ordered phase $\rho > \rho_c$ and a symmetric phase $\rho \leq \rho_c$. The condition

$$U'(\rho_c) = 0 \tag{14.33}$$

determines the critical interactions $U(\rho)$.

The parameter m now is the mass of the field ϕ and $\xi = 1/m$ the correlation length. The condition $m \ll \Lambda$, which is equivalent to $\xi \gg 1/\Lambda$, defines the critical domain. It implies that ρ is close to ρ_c by the second equation (14.32). The first equation implies that $U'(\rho)$ is small and thus ρ is close to the location of the minimum of $U(\rho)$. We set

$$U'(\rho_c) = \tfrac{1}{2}\tau, \quad |\tau| \ll 1. \tag{14.34}$$

We then parametrize the expansion of $U(\rho)$ at ρ_c as

$$U(\rho) = \tfrac{1}{2}\tau(\rho - \rho_c) + \tfrac{1}{2}U_c''(\rho - \rho_c)^2 + O((\rho - \rho_c)^3). \tag{14.35}$$

With this parametrization

$$m^2 = 2U_c''(\rho - \rho_c) + \tau + O((\rho - \rho_c)^2).$$

We have assumed that $U''(\rho)$ does not vanish at its minimum and, thus, U'' does not vanish in its vicinity. Since ρ is close to ρ_c, U_c'' is strictly positive. Then τ is positive in the symmetric phase and negative in the phase with symmetry breaking.

Without this assumption, the critical point could be a multicritical point, a situation we do not discuss in this context.

To determine the behaviour in the vicinity of the critical point, it is sufficient to expand U up to second order. We can thus reduce $U(\rho)$ to a quadratic polynomial, like in the quasi-Gaussian approximation, and the problem reduces to a discussion of the same $(\phi^2)^2$ field theory that, in the framework of the $\varepsilon = 4 - d$ expansion, describes universal critical properties.

14.4 The $(\phi^2)^2$ field theory for $N \to \infty$

The discussion that follows is now specific to the $(\phi^2)^2$ field theory. We parametrize U in the form

$$U(\rho) = \frac{1}{2}r\rho + \frac{u}{4!}\rho^2. \tag{14.36}$$

The Hamiltonian, expressed in terms of the initial N-component field ϕ, then reads

$$\mathcal{H}(\phi) = \mathcal{H}^{(2)}(\phi) + \int \left\{ \frac{1}{2}r\phi^2(x) + \frac{1}{4!}\frac{u}{N}\left[\phi^2(x)\right]^2 \right\} d^dx. \tag{14.37}$$

The $N \to \infty$ limit is taken at $U(\rho)$ fixed and this implies, with our conventions, that u, the coefficient of $(\phi^2)^2/N$, is fixed.

The potential U being quadratic (equation (14.36)), the integral over $\rho(x)$ in (14.6) is Gaussian and can thus be calculated explicitly. However, note that the field ρ has a more direct physical interpretation than the field λ.

The integration first involves translating $\rho(x)$ by its value at the minimum of the quadratic form. This value is solution of the field equation

$$\frac{\delta \mathcal{H}_N(\sigma, \rho, \lambda)}{\delta \rho(x)} = 0 \Rightarrow \lambda(x) = i\left[\tfrac{1}{6}u\rho(x) + r\right]. \tag{14.38}$$

After the translation, the remaining Gaussian integration yields only a constant factor. The integration thus corresponds to replacing, in the Hamiltonian, ρ by the solution of equation (14.38). One finds

$$\mathcal{H}_N(\sigma,\lambda) = \mathcal{H}^{(2)}(\sigma) - \int d^d x \left[\tfrac{1}{2}i\lambda(x)\sigma^2(x) + (3N/2u)(i\lambda(x)+r)^2\right] + (N-1)w(\lambda). \tag{14.39}$$

The Hamiltonian density at the saddle point (14.28) becomes

$$\mathcal{E} = \tfrac{1}{2}m^2\sigma^2 - (3N/2u)(r-m^2)^2 + \tfrac{1}{2}N\mathcal{D}(m^2). \tag{14.40}$$

Diagrammatic interpretation. In the $(\phi^2)^2$ field theory, when $N \to \infty$ (at u fixed), at each order in the perturbative expansion the leading terms come from chains of 'bubble' diagrams of the form displayed in Figure 14.2. Asymptotically for $N \to \infty$, these diagrams form a geometric series that the preceding algebraic techniques sum.

Fig. 14.2 Leading diagrams for $N \to \infty$.

Low-temperature phase in zero field. In the ordered phase $\rho > \rho_c$, equation (14.31) yields the non-vanishing expectation value of the field

$$\sigma = \sqrt{N(\rho - \rho_c)},$$

where $\rho_c = \Omega_d(0)$ (see equation (14.30)), and the symmetry $O(N)$ is broken spontaneously. The constant ρ is then given by equation (14.29b), which reduces to $U'(\rho) = 0$. The condition (14.33) determines the critical potential U:

$$U'(\rho_c) = 0 \Rightarrow r = r_c = -u\rho_c/6. \tag{14.41}$$

The field expectation value vanishes for $r = r_c$, which thus corresponds to the critical temperature T_c. Introducing the parameter τ that characterizes the deviation from the critical temperature (defined by equation (14.34)),

$$\tau = 2U'(\rho_c) = r - r_c,$$

one obtains

$$U'(\rho) = 0 \Rightarrow \rho - \rho_c = -(6/u)\tau. \tag{14.42}$$

The $O(N)$ symmetry is broken for $\tau < 0$, that is, at low temperature and equation (14.31) can then be rewritten as

$$\sigma^2 = -(6/u)\tau \propto (-\tau)^{2\beta} \quad \text{with } \beta = \tfrac{1}{2}. \tag{14.43}$$

This behaviour shows that, in the limit $N \to \infty$, the exponent β remains quasi-Gaussian in all dimensions.

High-temperature phase in zero field. For $\tau > 0$, that is, above T_c, σ vanishes. From expression (14.7) one infers the two-point function of the field σ, which in this phase is identical to the propagator (14.9). Asymptotically, it takes the form (neglecting the corrections coming from irrelevant terms)

$$\tilde{\Delta}_\sigma(p) = \tilde{\Delta}(p) \underset{|p|,m \ll \Lambda}{\sim} \frac{1}{p^2 + m^2}. \tag{14.44}$$

Therefore, m is the physical mass of the field σ (and thus of all components of the field ϕ) and its inverse

$$\xi = 1/m \tag{14.45}$$

is the correlation length. Equation (14.29c) implies that $\partial \rho / \partial m$ is negative and that r is an increasing function of m. The minimum of r, obtained for $m = 0$, is r_c. Using the definitions (14.34), (14.41) in equations (14.32), one then finds

$$m^2 = (u/6)(\rho - \rho_c) + \tau, \tag{14.46a}$$
$$\rho - \rho_c = \Omega_d(m) - \Omega_d(0). \tag{14.46b}$$

(i) For $d > 4$, the expansion (14.15) implies that the leading contribution to $\rho - \rho_c$ is proportional to m^2, like the left-hand side of equation (14.46a) and, thus,

$$m^2 = \xi^{-2} \sim \tau \quad \Rightarrow \quad \nu = \tfrac{1}{2}, \tag{14.47}$$

which is the Gaussian or mean-field result.

(ii) For $2 < d < 4$, by contrast, the leading term is of order m^{d-2} and thus

$$\rho - \rho_c \sim -K(d) m^{d-2}.$$

The leading contribution for $m \to 0$, in equation (14.46a) now comes from $\rho - \rho_c$. Keeping only the leading term in (14.15), one infers

$$m = \xi^{-1} \sim \tau^{1/(d-2)}, \tag{14.48}$$

which yields the non-Gaussian value

$$\nu = \frac{1}{d-2}, \tag{14.49}$$

of the correlation exponent.

(iii) For $d = 4$, the leading contribution in equation (14.46a) again comes from $\rho - \rho_c$ and, thus,

$$m^2 \sim \frac{48\pi^2}{u} \frac{\tau}{\ln(\Lambda/m)} \sim \frac{96\pi^2}{u} \frac{\tau}{\ln(\Lambda^2/\tau)}. \tag{14.50}$$

The correlation length no longer has a power-law behaviour but, instead, a Gaussian model behaviour modified by a logarithm. This behaviour is typical of a situation where the Gaussian fixed point is stable in the presence of a marginal operator.

(iv) If one examines equation (14.29b) for $\sigma = 0$ and $d = 2$, one finds that the correlation length can diverge only for $r \to -\infty$. This special situation will be studied in the framework of the non-linear σ model in Section 14.10.

Finally, in the critical limit $\tau = 0$, m vanishes and, thus, from the form (14.44) of the two-point function of the field σ, one derives that the critical exponent η remains Gaussian for all d:

$$\eta = 0 \Rightarrow d_\phi = \tfrac{1}{2}(d-2). \tag{14.51}$$

One also verifies that, for $d \le 4$, the exponents β, ν, η satisfy the scaling relations proved in the framework of the ε-expansion:

$$\beta = \tfrac{1}{2}\nu(d-2+\eta) = \nu d_\phi.$$

14.5 Singular part of the free energy and equation of state

In the presence of a constant magnetic field H in the σ direction, the free energy density $W(H)$ (at this order the opposite of the Hamiltonian density (14.40) in which the saddle point equations have been used) is given by

$$W(H) = \ln \mathcal{Z}/\text{volume} = -\mathcal{E}$$
$$= N\left[\frac{3}{2u}(m^2-r)^2 - \frac{1}{2}m^2\sigma^2/N + H\sigma/N - \frac{1}{2}\mathcal{D}(m^2)\right],$$

where ρ has been eliminated with the help of equation (14.29b) (or (14.38)) and the contribution of the determinant has been taken from equation (14.25). The saddle point values m^2, σ are given by equation (14.29c) and the modified saddle point equation (14.29a):

$$m^2\sigma = H. \tag{14.52}$$

The magnetization, expectation value of ϕ, is

$$M = \frac{\partial W}{\partial H} = \sigma,$$

since the partial derivatives of W with respect to m^2, σ vanish as a consequence of the saddle point equations. The thermodynamic potential density $\mathcal{G}(M)$, the Legendre transform of $W(H)$, is then

$$\mathcal{G}(M) = HM - W(H)$$
$$= N\left[-\frac{3}{2u}(m^2-r)^2 + \frac{1}{2}m^2M^2/N + \frac{1}{2}\mathcal{D}(m^2)\right].$$

The expansion of $\mathcal{D}(m^2)$ for $m \to 0$ is given by equation (14.26). The thermodynamic potential then takes the form

$$\mathcal{G}(M)/N = \frac{3}{2}\left(\frac{1}{u^*} - \frac{1}{u}\right)m^4 + \frac{3(r-r_c)}{u}m^2 + \frac{1}{2}m^2 M^2/N - \frac{K(d)}{d}m^d - \frac{3r^2}{2u} \qquad (14.53)$$

with the definition

$$u^* = \frac{6}{a(d)}\Lambda^\varepsilon. \qquad (14.54)$$

For $d < 4$, for $m \to 0$, the term proportional to m^4 is negligible with respect to the singular term m^d. Thus, at leading order in the critical domain

$$\mathcal{G}(M)/N = \frac{3}{u}\tau m^2 + \frac{1}{2}m^2 M^2/N - \frac{K(d)}{d}m^d + \text{const.}, \qquad (14.55)$$

where τ has been defined in (14.34).

As a consequence of the property (6.9) of the Legendre transformation, the saddle point equation corresponding to m^2 can also be obtained by expressing that the derivative of $\mathcal{G}$ with respect to m^2 vanishes. One finds

$$(6/u)\tau + M^2/N - K(d)m^{d-2} = 0,$$

whose solution is

$$m = \xi^{-1} = [K(d)]^{1/(2-d)}\left[(6/u)\tau + M^2/N\right]^{1/(d-2)},$$

an expression that one can compare with the quasi-Gaussian expression (8.61).

The leading contribution to the thermodynamic potential in the critical domain follows:

$$\frac{1}{N}\mathcal{G}(M) \sim \frac{d-2}{2d(K(d))^{2/(d-2)}}\left[(6/u)\tau + M^2/N\right]^{d/(d-2)}. \qquad (14.56)$$

Various quantities can be inferred from $\mathcal{G}(M)$. For example, by differentiating with respect to M, one obtains the equation of state in the scaling limit,

$$H = \frac{\partial \mathcal{G}}{\partial M} = h_0 M^\delta f\left(a_0 \tau / M^2\right), \qquad (14.57)$$

where h_0 and a_0 are two normalization constants. The exponent δ is given by

$$\delta = \frac{d+2}{d-2}, \qquad (14.58)$$

in agreement with the general relation between exponents $\delta = d/d_\phi - 1$, and the function $f(x)$ by

$$f(x) = (1+x)^{2/(d-2)}. \qquad (14.59)$$

The asymptotic form of $f(x)$ for $x \to +\infty$ implies $\gamma = 2/(d-2)$, which, again is in agreement with the scaling relation $\gamma = \nu(2-\eta)$. Finally, in agreement with general results, the equation of state can be expressed in a parametric form $\{M, \tau\} \mapsto \{R, \theta\}$:

$$M = (a_0)^{1/2} R^{1/2} \theta,$$
$$\tau = 3R(1 - \theta^2),$$
$$H = h_0 R^{\delta/2} \theta (3 - 2\theta^2)^{2/(d-2)}.$$

Specific heat exponent. Amplitude ratios. Differentiating $\mathcal{G}(M)$ twice with respect to τ, one obtains the singular part of the specific heat at fixed magnetization M:

$$\mathcal{C}_{\text{sing.}} \propto \left[(6/u)\tau + M^2/N\right]^{(4-d)/(d-2)}. \tag{14.60}$$

For $M = 0$, the singular part of the specific heat thus vanishes like

$$\mathcal{C}_{\text{sing.}} \propto |\tau|^{-\alpha}, \tag{14.61}$$

with the specific heat exponent

$$\alpha = \frac{d-4}{d-2}, \tag{14.62}$$

which is also equal to $2 - d\nu$, in agreement with RG predictions.

Note that, for N large, the leading contribution to the specific heat when $\tau \to 0$ is not the universal singular part, which vanishes, but the part regular in τ, which has a finite, non-universal, limit (e.g., it depends on u).

Among the standard critical amplitude ratios, one can calculate, for example, $R_\xi^+ = f_1^+ (\alpha A^+)^{1/d}$ and $R_c = \alpha A^+ C^+ / M_0^2$ (see equation (7.35) and Section 8.6)

$$(R_\xi^+)^d = \frac{4N}{(d-2)^3} \frac{\Gamma(3-d/2)}{(4\pi)^{d/2}}, \quad R_c = \frac{4-d}{(d-2)^2}. \tag{14.63}$$

Leading corrections to scaling properties. The term proportional to m^4 is the leading correction to the asymptotic scaling form. Comparing it with the leading term, one finds

$$m^4/m^d = O(\tau^{(4-d)/(d-2)}).$$

The exponent that characterizes the leading corrections to the asymptotic form in the temperature variable is $\omega\nu$ (ω is defined by equation (13.28)) and, thus,

$$\omega\nu = (4-d)/(d-2) \Rightarrow \omega = 4 - d. \tag{14.64}$$

Let us point out that for the special value $u = u^*$ this correction vanishes.

14.6 The $\langle\lambda\lambda\rangle$ and $\langle\phi^2\phi^2\rangle$ two-point functions

In the high-temperature phase, differentiating twice the Hamiltonian (14.39) with respect to $\lambda(x)$, and replacing the field $\lambda(x)$ by its expectation value im^2, one obtains the $\langle\lambda\lambda\rangle$ two-point function, which is also the propagator of the λ field in the $1/N$ expansion. For the derivative of $w(\lambda)$, one can use the term $n=2$ of expansion (14.8). The part quadratic in λ of the Hamiltonian is

$$\mathcal{H}_{\lambda\lambda} = \frac{3N}{2u}\int d^dx\, \lambda^2(x) + \frac{N}{4}\int d^dx\, d^dy\, \Delta^2(x-y)\lambda(x)\lambda(y).$$

In terms of the Fourier components of λ,

$$\lambda(x) = \int d^dp\, e^{ipx}\, \tilde{\lambda}(p),$$

the expression can be rewritten as

$$\mathcal{H}_{\lambda\lambda} = \frac{1}{2}N(2\pi)^d \int d^dp\, \tilde{\lambda}(p)\tilde{\lambda}(-p)\left(\frac{3}{u} + \frac{1}{2}B_\Lambda(p,m)\right), \tag{14.65}$$

where $B_\Lambda(p,m)$ is the function (14.12) corresponding to the diagram of Figure 14.1. One then finds

$$\langle\lambda(x)\lambda(0)\rangle = \frac{1}{(2\pi)^d}\int d^dp\, e^{ipx}\, \tilde{\Delta}_\lambda(p), \tag{14.66}$$

where $\tilde{\Delta}_\lambda$ follows from expression (14.65):

$$\tilde{\Delta}_\lambda(p) = \frac{2}{N}\left(\frac{6}{u} + B_\Lambda(p,m)\right)^{-1}. \tag{14.67}$$

Using the relation (14.38), one obtains the $\langle\rho\rho\rangle$ two-point function (in the Fourier representation, translation by a constant produces only a δ distribution at $p=0$) and thus, as pointed out in Section 14.4, the $\phi^2(x)$ two-point function (which is also the energy–energy correlation function) in the Fourier representation:

$$\int d^dx\, e^{ipx}\, \langle\phi^2(x)\phi^2(0)\rangle = N^2\int d^dx\, e^{ipx}\, \langle\rho(x)\rho(0)\rangle = -\frac{12N/u}{1+(u/6)B_\Lambda(p,m)}. \tag{14.68}$$

The value at zero momentum of the function yields the specific heat. The expansion for $m\to 0$ of $B_\Lambda(0,m)$ is given by equation (14.20). The singular part of the specific heat thus vanishes as m^{4-d}, in agreement with equation (14.60) for $M=0$.

In the critical theory ($m=0$ at this order), for $2\le d\le 4$, the expansion (14.21) shows that, when the momentum p goes to zero, the denominator is also dominated by the singular part of $B_\Lambda(p,0)$. One infers

$$\tilde{\Delta}_\lambda(p) \underset{p\to 0}{\sim} \frac{2}{Nb(d)}p^{4-d}. \tag{14.69}$$

Again, one verifies consistency with the scaling relations. Moreover, the dimension $[\lambda]$ of the field λ and, thus, of the energy operator, in the limit $N \to \infty$ is

$$[\lambda] = \tfrac{1}{2}[d + (4-d)] = 2, \qquad (14.70)$$

a result important for the perturbative expansion in $1/N$. The result is consistent with the properties (relations (9.64), (10.33) and (10.35))

$$[\lambda] = [\rho] = [\phi^2] = d - 1/\nu = d - (d-2).$$

Remarks.
(i) For $d = 4$, $B_\Lambda(p,0)$ has the logarithmic behaviour (14.23) and still gives the leading contribution to the two-point function, which becomes

$$\tilde{\Delta}_\lambda(p) \sim \frac{16\pi^2}{N \ln(\Lambda/p)}.$$

(ii) Therefore, for $d \leq 4$, the contributions generated by the term proportional to $\lambda^2(x)$ in (14.7) are always negligible in the critical domain.

The $\langle \sigma\sigma \rangle$ function at low temperature. In the phase with symmetry breaking, after a translation of the fields by their expectation values, the Hamiltonian contains a term proportional to $\sigma\lambda$ and thus the propagators of the fields σ and λ are elements of a 2×2 matrix $\mathbf{M}$. In the Fourier representation,

$$\mathbf{M}^{-1}(p) = \begin{pmatrix} p^2 & -i\sigma \\ -i\sigma & 3N/u + \tfrac{1}{2}NB_\Lambda(p,0) \end{pmatrix}, \qquad (14.71)$$

where $\sigma = \langle \sigma(x) \rangle$ and B_Λ is given by equation (14.21). For $d < 4$, asymptotically, the determinant is given by

$$1/\det \mathbf{M}(p) \sim N\left[b(d)p^{d-2} + 6\tau/u\right].$$

This expression defines a mass scale

$$m_{\mathrm{cr}} = (-\tau/u)^{1/(d-2)} \propto \Lambda\big((r_c - r)/\Lambda^2\big)^{1/(d-2)} = \Lambda\big((r_c - r)/\Lambda^2\big)^\nu, \qquad (14.72)$$

to which is associated a *crossover* between a behaviour dominated by Goldstone modes and the critical behaviour.

At $d = 4$, $B_\Lambda(p,0)$ is given by expression (14.23) and

$$m_{\mathrm{cr}}^2 \propto \frac{r_c - r}{\ln[\Lambda^2/(r_c - r)]}. \qquad (14.73)$$

Finally, for $d > 4$, $B_\Lambda(p,0)$ has a finite limit at $p = 0$ and thus

$$m_{\mathrm{cr}} \propto \sqrt{r_c - r}. \qquad (14.74)$$

Thus, in all dimensions, for $|r \to r_c| \to 0$, m_{cr} has the same scaling behaviour as the physical mass.

14.7 RG and corrections to scaling

To verify more directly the consistency of the $N \to \infty$ limit with the perturbative RG scheme, we calculate, at leading order for $N \to \infty$, the RG functions and the leading corrections to the scaling behaviour.

14.7.1 RG functions

We set (equation (14.54))

$$u = Ng\Lambda^{4-d}, \quad g^* = u^*\Lambda^{d-4}/N = 6/(Na), \qquad (14.75)$$

where the constant $a(d)$ is defined in (14.15) and diverges for $\varepsilon = 4 - d \to 0$ like $1/(8\pi^2\varepsilon)$ (equation (14.18)).

The mass m is a solution to equations (14.46), which imply

$$m^2 = \tfrac{1}{6}N\Lambda^{4-d}g\left[\Omega_d(m) - \Omega_d(0)\right] + \tau.$$

For $\Lambda \to \infty$, in the equation we keep the leading term and the correction that becomes marginal at $d = 4$ (equations (14.14), (14.15), (14.75)). The equation becomes

$$E(m,\tau,g,\Lambda) \equiv m^2(1 - g/g^*) + \tfrac{1}{6}NgK(d)m^{d-2}\Lambda^{4-d} - \tau = 0. \qquad (14.76)$$

Differentiating this equation with respect to Λ, g, τ, one infers

$$\Lambda\frac{\partial m}{\partial \Lambda}\frac{\partial E}{\partial m} + (4-d)\frac{1}{6}NgK(d)m^{d-2}\Lambda^{4-d} = 0,$$

$$g\frac{\partial m}{\partial g}\frac{\partial E}{\partial m} - m^2\frac{g}{g^*} + \frac{1}{6}NgK(d)m^{d-2}\Lambda^{4-d} = 0,$$

$$-\tau\frac{\partial m}{\partial \tau}\frac{\partial E}{\partial m} + m^2\left(1 - \frac{g}{g^*}\right) + \frac{1}{6}NgK(d)m^{d-2}\Lambda^{4-d} = 0,$$

where, in the last equation, τ has been eliminated by using the equation $E = 0$. Taking a linear combination of the three equations with coefficients $1, -\varepsilon(1 - g/g^*), -\varepsilon g/g^*$, respectively, one verifies that, at this order, m satisfies an equation that expresses that it is an RG invariant:

$$\left(\Lambda\frac{\partial}{\partial\Lambda} + \beta(g)\frac{\partial}{\partial g} - \eta_2(g)\tau\frac{\partial}{\partial\tau}\right)m(\tau,g,\Lambda) = 0, \qquad (14.77)$$

where, in the right-hand side, the contributions of order $1/\Lambda^2$ have thus been neglected. The RG functions $\beta(g)$ and $\eta_2(g)$ are then

$$\beta(g) = -\varepsilon g(1 - g/g^*), \qquad (14.78)$$

$$\eta_2(g) = -\varepsilon g/g^* \;\Rightarrow\; \nu^{-1}(g) = 2 + \eta_2(g) = 2 - \varepsilon g/g^*. \qquad (14.79)$$

For the simplest regularizations, $a(d)$ is positive in all dimensions and the fixed-point equation $\beta(g) = 0$ has a non-trivial solution $g = g^*$ corresponding, for $d \leq 4$, to an IR fixed point. Integrating the flow equation

$$\lambda \frac{dg(\lambda)}{d\lambda} = -\beta(g(\lambda)),$$

one obtains

$$\left| \frac{g^*}{g(\lambda)} - 1 \right| = \lambda^{d-4}. \tag{14.80}$$

One then finds the exponents $\nu^{-1} = d-2$, this result being consistent with equations (14.64) and (14.49), and $\omega = \beta'(g^*) = \varepsilon$. In the framework of the ε-expansion, ω is associated with the leading corrections to the asymptotic scaling behaviour. In the limit $N \to \infty$, ω remains smaller than 2 for $\varepsilon < 2$. Therefore, it remains associated with the leading corrections, for N large, for all dimensions $2 \leq d \leq 4$.

Finally, applying the RG equations to the propagator (14.44), one finds

$$\eta(g) = 0, \tag{14.81}$$

a result consistent with the value (14.51) of the exponent η.

Let us point out, however, that one can find regularizations such that $a(d)$ is negative. In the latter case, which we discuss briefly later, the RG method is confronted with a serious difficulty.

14.7.2 Leading corrections to scaling relations

From the RG general analysis, one expects that the leading corrections to the scaling behaviour vanish for $u = u^*$. This property has already been verified for the free energy. We again consider the correlation length or the mass m given by equations (14.46). Keeping the leading correction to the integral for $m \to 0$, one obtains equation (14.76), which can be rewritten as

$$1 - \frac{u}{u^*} + (u/6)K(d)m^{d-4} + O\left(m^{d-2}\Lambda^{-2}\right) = \frac{\tau}{m^2}.$$

One notes that for the special value $u = u^*$ ($g = g^*$), the leading correction to scaling, which becomes marginal or logarithmic at $d = 4$, indeed vanishes.

One verifies that the corrections, present when $g \neq g^*$, are proportional to $(u - u^*)\tau^{\varepsilon/(2-\varepsilon)}$. One infers that $\omega\nu = \varepsilon/(2-\varepsilon)$, a result consistent with equations (14.64) and (14.49).

The $\langle\lambda\lambda\rangle$ two-point function. Similarly, if one keeps the leading corrections to the $\langle\lambda\lambda\rangle$ function in the critical theory (equations (14.67), (14.21)), one finds

$$\Delta_\lambda(p) = \frac{2}{N}\left[\frac{6}{u} - \frac{6}{u^*} + b(d)p^{-\varepsilon}\right]^{-1}, \tag{14.82}$$

where corrections of order Λ^{-2} have been neglected. The leading corrections to scaling again vanish identically for $u = u^*$, as expected.

Discussion.

(i) One can show that a perturbation generated by irrelevant operators is equivalent, at leading order in the critical region, to a modification of the coefficient of $(\phi^2)^2$. This can be explicitly verified here. The amplitude of the leading correction to scaling is, as we have noted, proportional to $6/u - a(d)\Lambda^{-\varepsilon}$ where the value of $a(d)$ depends on the form of the cut-off function and, thus, on contributions of irrelevant terms. Let us denote by u' the coefficient of $(\phi^2)^2$ in another schema where a is replaced by a'. Identifying the leading corrections to scaling, one finds the relation ($g' = u'\Lambda^{d-4}$)

$$\frac{1}{g} - \frac{a(d)}{6} = \frac{1}{g'} - \frac{a'(d)}{6},$$

which is a homographic relation consistent with the special form (14.78) of the β-function. Indeed, using the flow equation (14.80) between two values λ_1 and λ_2, one finds

$$\frac{1}{g(\lambda_1)} - \frac{\lambda_1^{d-4}}{g^*} = \frac{1}{g(\lambda_2)} - \frac{\lambda_2^{d-4}}{g^*}.$$

(ii) *The sign of $a(d)$.* One generally assumes that $a(d)$ is positive. This is indeed what is found for the simplest regularization schemes. For example, expression (14.17) shows that a sufficient condition is that $D(k)$ in expression (14.9) is a monotonically increasing function. Moreover, $a(d)$ is always positive near dimension 4 where it diverges like

$$a(d) \underset{d \to 4}{\sim} \frac{1}{8\pi^2 \varepsilon}.$$

Then, in the limit $N \to \infty$, the β-function has indeed a non-trivial zero $u = u^*$, which corresponds to an IR fixed point and governs large-distance physics. For this value of u, the leading corrections to the asymptotic scaling forms vanish.

However, at d fixed, $d < 4$, this is not the general situation. It is simple to find regularizations for which, for example, $a(d)$ is negative at $d = 3$.

If $a(d)$ is negative, the RG method, at least in its asymptotic form where it involves only one parameter, is confronted with a serious difficulty for $N \to \infty$. Indeed, the effective parameter $u(\lambda)$ at scale λ increases for $\lambda \to \infty$ until the validity of the $1/N$ expansion ceases. The interpretation of this phenomenon is unclear. It could be a real effect: the initial Hamiltonians of this type do not belong to the basin of attraction of the IR fixed point. It could also be an RG problem: the RG must be modified to ensure convergence. Finally, it could also simply be a pathology of the $N \to \infty$ limit.

Another way of formulating the problem consists in examining directly the relation between bare and renormalized parameters. Denoting by $g_r m^{4-d}$ the four-point function renormalized at zero momentum, one finds

$$m^{4-d} g_r = \frac{\Lambda^{4-d} g}{1 + \Lambda^{4-d} g N B_\Lambda(0, m)/6}. \tag{14.83}$$

In the limit $m \ll \Lambda$, the relation can be written as
$$\frac{1}{g_r} = \frac{(d-2)NK(d)}{12} + \left(\frac{m}{\Lambda}\right)^{4-d}\left(\frac{1}{g} - \frac{Na(d)}{6}\right). \tag{14.84}$$

As a consequence, for $a(d) < 0$, the renormalized fixed-point value cannot be reached, by varying $g > 0$, for any finite value of m/Λ. Similarly, the leading corrections to scaling behaviour cannot be cancelled.

14.8 The $1/N$ expansion

In the calculation of the field integral by the steepest descent method (14.6), the higher order terms generate an expansion in powers of $1/N$. To study the properties of the expansion, it is useful to first rewrite the Hamiltonian (14.7) differently, by translating the field $\lambda(x)$ by its expectation value im^2 (equation (14.27)), $\lambda(x) \mapsto im^2 + \lambda(x)$. One finds

$$\mathcal{H}_N(\sigma,\lambda) = \mathcal{H}^{(2)}(\sigma) + \frac{1}{2}\int \mathrm{d}^d x \left[m^2\sigma^2 - i\lambda\sigma^2 + \frac{3N}{u}\lambda^2 + i\frac{6N}{u}(m^2 - r)\lambda\right]$$
$$+ (N-1)w(im^2 + \lambda). \tag{14.85}$$

14.8.1 Dimensional analysis

One can then analyse the Hamiltonian (14.85) from the viewpoint of the local RG near the fixed point. The dimension of the field $\sigma(x)$ is $(d-2)/2$. The quadratic terms in σ with more than two derivatives are always irrelevant. The dimension $[\lambda]$ of the field $\lambda(x)$, inferred from the critical behaviour (14.69) of the propagator, is given by equation (14.70):

$$[\lambda] = 2.$$

Therefore, λ^2 has a dimension $4 > d$ and the perturbation $\int \mathrm{d}^d x\, \lambda^2(x)$, which corresponds to the eigenvalue $d - 4$, is irrelevant, in agreement with the analysis of Section 14.7.2. The interaction term $\int \lambda(x)\sigma^2(x)\mathrm{d}^d x$ has dimension zero. One can verify that the non-local interactions that depend on the field λ and arise from the expansion (14.8) of the determinant, also have all dimension zero when one keeps in the propagator the leading terms at large distance:

$$\left[\mathrm{tr}(\lambda\Delta)^k\right] = k[\lambda] - 2k = 0.$$

Therefore, in the framework of the $1/N$ expansion, the theory has, in the sense of dimensional analysis, the properties of a renormalizable theory for all dimensions $2 < d \le 4$. By contrast with the usual perturbation theory, the $1/N$ expansion generates only logarithmic corrections to the leading long-distance scaling behaviour. The situation is thus similar to the one encountered in the case of the ε-expansion at a fixed point and one expects to be able to calculate universal quantities, such as critical exponents, for example, as series in powers of $1/N$. However, since some interactions are non-local, the proof of this property is not immediate because the results of standard renormalization theory do not apply. Nevertheless, it is possible to construct an alternative, quasi-local, field theory that has the same asymptotic large-distance properties and to which renormalization theory applies.

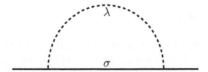

Fig. 14.3 Diagram contributing to $\tilde{\Gamma}^{(2)}_{\sigma\sigma}$ at order $1/N$.

14.8.2 Application: Perturbative expansion, IR singularities and large-momentum behaviour

A study of the $1/N$ correction to the two-point function gives some indication about the structure of IR divergences in the perturbative expansion of the massless (critical) theory at fixed dimension $d < 4$.

We have seen that, in the framework of the $1/N$ expansion, one can calculate at fixed dimension $d < 4$ in the critical limit $(T = T_c, m^2 = 0)$. One thus expects that the terms in the $1/N$ expansion cannot be expanded in a Taylor series in the coefficient g of $(\phi^2)^2$.

Note that the Gaussian fixed point is a stable fixed point for the large-momentum (UV) behaviour of renormalized correlation functions (i.e., large but still small compared with the cut-off). The behaviour for $g \to 0$ is thus related to the large--momentum behaviour.

To understand how the critical two-point function can be IR divergent in the perturbative expansion but finite in the $1/N$ expansion, we study the two-point function $\langle\sigma\sigma\rangle$ at order $1/N$. At this order, only one diagram contributes (Figure 14.3), containing two $\lambda^2\sigma$ vertices. In the $\Lambda \to \infty$ limit, and after a mass renormalization (to ensure $\tilde{\Gamma}^{(2)}_{\sigma\sigma}(p=0) = 0$), one finds

$$\tilde{\Gamma}^{(2)}_{\sigma\sigma}(p) = p^2 + \frac{2}{N(2\pi)^d} \int \frac{d^d q}{(6/u) + b(d)q^{-\varepsilon}} \left(\frac{1}{(p+q)^2} - \frac{1}{q^2} \right) + O\left(\frac{1}{N^2}\right). \quad (14.86)$$

An analytic study of the integral shows that it has an expansion of the form

$$\sum_{k \geq 1} \alpha_k(\varepsilon) u^k p^{2-k\varepsilon} + \beta_k(\varepsilon) u^{(2+2k)/\varepsilon} p^{-2k}. \quad (14.87)$$

The coefficients α_k, β_k can be calculated from the Mellin transform of the integral considered as a function of u (see Appendix A4.2). The terms with integer powers of u correspond to the perturbative expansion, which exists for ε small enough in dimensional regularization. α_k has poles at $\varepsilon = (2l+2)/k$ at which the corresponding powers of p^2 is $-l$, that is, an integer. One verifies that β_l has poles for the same value of ε and that the singular contributions cancel in the sum. For the dimensions corresponding to poles, logarithms of u appear in the expansion.

14.9 The exponent η at order $1/N$

To illustrate the preceding discussion, we calculate the exponent η at order $1/N$ (Figure 14.3). This requires the evaluation of the σ two-point function at order $1/N$, at T_c and at the IR fixed point. In the limit $p \to 0$, the leading contribution to expression (14.86) reduces to

$$\tilde{\Gamma}^{(2)}_{\sigma\sigma}(p) = p^2 + \frac{2}{Nb(\varepsilon)} \int \frac{d^d q}{(2\pi)^d} q^{4-d} \left[\frac{1}{(p+q)^2} - \frac{1}{q^2} \right] + O\left(\frac{1}{N^2}\right). \tag{14.88}$$

From the dimensional analysis of Section 14.8, one expects that after mass renormalization, the integral over q diverges logarithmically with the cut-off for all dimensions, a property one verifies here. One thus introduces a cut-off Λ whose specific form is not important at this order. The integral then behaves for Λ large, or small momenta, like

$$\int^\Lambda \frac{d^d q}{(2\pi)^d} q^\varepsilon \left[(p+q)^{-2} - q^{-2} \right] \sim Ap^2 \ln(\Lambda/p). \tag{14.89}$$

Comparing with the expansion of the two-point function at the IR fixed point,

$$\tilde{\Gamma}^2_{\sigma\sigma}(p) \propto p^{2-\eta} = p^2 \left(1 - \eta \ln p + O\left(1/N^2\right)\right), \tag{14.90}$$

one infers

$$\eta = 2A/(Nb(\varepsilon)).$$

Expression (14.89) shows that the coefficient of $\ln p$ can be inferred from the divergent part of the integral. The latter corresponds to large values of the integration variables q in (14.88) and is a regular function of the momentum p. One can thus expand it for $p \to 0$. One finds

$$\int_1^\Lambda \frac{d^d q}{(2\pi)^d} q^{4-d} \left[(p+q)^{-2} - q^{-2} \right] \sim p^2 \int_1^\Lambda \frac{d^d q}{q^d (2\pi)^d} \left(\frac{4(p \cdot q)^2}{p^2 q^2} - 1 \right). \tag{14.91}$$

Using rotation invariance, one can substitute

$$\frac{q_\beta q_\beta}{q^2} \mapsto \frac{1}{d}\delta_{\alpha\beta} \quad \text{and thus} \quad \frac{4(p \cdot q)^2}{p^2 q^2} \mapsto \frac{4}{d}.$$

The divergent part of the integral thus is

$$\int_1^\Lambda \frac{d^d q}{(2\pi)^d} q^{4-d} \left[(p+q)^{-2} - q^{-2} \right] \sim \frac{2(4-d)}{d} \frac{1}{(4\pi)^{d/2}\Gamma(d/2)} p^2 \ln \Lambda. \tag{14.92}$$

Using the definition (14.22) of $b(\varepsilon)$, one infers the value of the exponent η. It is convenient to set

$$X_1 = -\frac{2\Gamma(d-1)}{\Gamma^3(d/2)\Gamma(1-d/2)} = \frac{2\sin(\pi\varepsilon/2)\Gamma(3-\varepsilon)}{\pi\Gamma^2(2-\varepsilon/2)}. \tag{14.93}$$

After some algebra, one obtains

$$\eta = \frac{4-d}{Nd} X_1 + O\left(\frac{1}{N^2}\right). \tag{14.94}$$

14.10 The non-linear σ-model

We have emphasized that the term proportional to $\int \lambda^2(x)\mathrm{d}^d x$ is negligible in the critical domain for $d \le 4$ and can thus be omitted at leading order. In Section 14.8, we have related this observation to the dimension of the monomial $\lambda^2(x)$ in the framework of the $1/N$ expansion: λ^2 has dimension 4 in all space dimensions and thus is irrelevant for $d < 4$. For $d = 4$, the situation is more subtle, because it is marginal. However, even in this case, it generates only logarithmic corrections to the leading behaviour. Technically, the constant part in the inverse propagator, as it appears in equation (14.82), plays the role of a large-momentum cut-off. We thus consider the Hamiltonian (14.85) without the λ^2 term.

If one reintroduces the initial field ϕ and one integrates over $\lambda(x)$, one finds

$$\mathcal{Z} = \int [\mathrm{d}\phi(x)]\, \delta\left[\phi^2(x) - \phi_0^2\right] \exp\left[-\mathcal{H}^{(2)}(\phi)\right] \qquad (14.95)$$

with

$$\phi_0^2 = 6\left(m^2 - r\right)/u.$$

In this form, one recognizes the partition function of a field theory model called the non-linear σ-model, in an unusual parametrization. This model will be briefly discussed in Chapter 15. This observation leads to a remarkable correspondence: to all orders in the $1/N$ expansion, the connected correlation functions of the non-linear σ-model and of the $(\phi^2)^2$ field theory have the same asymptotic large-distance behaviour.

The $N \to \infty$ limit. The non-linear σ-model can be directly solved in the form of a $1/N$ expansion. The solution in a more natural parametrization will show more explicitly the correspondence with the parametrization of the $(\phi^2)^2$ field theory.

The partition function of the σ-model can be written as

$$\mathcal{Z} = \int [\mathrm{d}\sigma(x)\mathrm{d}\lambda(x)] \exp\left[-\mathcal{H}(\phi, \lambda)\right] \qquad (14.96)$$

with

$$\mathcal{H}(\phi, \lambda) = \frac{1}{T}\mathcal{H}^{(2)}(\phi) - \frac{i}{2T}\int \mathrm{d}^d x\, \lambda(x)\left(\phi^2(x) - 1\right), \qquad (14.97)$$

where the integration over λ implements the constraint $\phi^2 = 1$ and the parameter T is the temperature. Integrating, as we have done in Section 14.1, over $(N-1)$ components of ϕ and calling σ the remaining component, one obtains

$$\mathcal{Z} = \int [\mathrm{d}\sigma(x)\mathrm{d}\lambda(x)] \exp\left[-\mathcal{H}_N(\sigma, \lambda)\right], \qquad (14.98)$$

where

$$\mathcal{H}_N(\sigma, \lambda) = \frac{1}{T}\mathcal{H}^{(2)}(\sigma) - \frac{i}{2T}\int \mathrm{d}^d x\, (\sigma^2(x) - 1)\lambda(x) + (N-1)w(\lambda). \qquad (14.99)$$

One now takes the limit $N \to \infty$ at TN fixed. The saddle point equations that correspond to equations (14.29) are

$$m^2 \sigma = 0, \qquad (14.100a)$$
$$\sigma^2 = 1 - (N-1) T \Omega_d(m), \qquad (14.100b)$$

where we have set $im^2 = \langle \lambda(x) \rangle$ and Ω_d has been defined in (14.11). At low temperature, σ is different from zero for $d > 2$ and thus m, which is the mass of the π field, vanishes. Equation (14.100b) gives the spontaneous magnetization:

$$\sigma^2 = 1 - (N-1) T \Omega_d(0). \qquad (14.101)$$

For $T = T_c$, where we have defined

$$1/T_c = (N-1) \Omega_d(0),$$

σ vanishes. Therefore, equation (14.101) can be rewritten as

$$\sigma^2 = 1 - T/T_c, \qquad (14.102)$$

in agreement, for $T \to T_c$, with the result (14.43).

Above T_c, σ vanishes and m, which is now the mass common to all fields π and σ, is given for $2 < d < 4$ by (equation (14.15))

$$\frac{1}{T_c} - \frac{1}{T} = m^{d-2}(N-1)K(d) + O\left(m^2 \Lambda^{d-4}\right). \qquad (14.103)$$

For $T \to T_c$, one recovers the scaling form (14.48) of the correlation length $\xi = 1/m$.

We will see in Chapter 15 that the correlation functions of the σ-model satisfy RGE, which can be derived in a double series expansion in T and $\varepsilon = d - 2$. Combining the results (14.102), (14.103) with the general expressions (15.52), (15.51), one can derive the RG functions at leading order for $N \to \infty$:

$$\beta(T) = (d-2)T - \frac{(d-2)}{T_c} T^2, \qquad \zeta(T) = \frac{d-2}{T_c} T. \qquad (14.104)$$

The solution of the equation $\beta = 0$ is, in this context, the critical temperature T_c, $\beta'(T_c) = -1/\nu = 2 - d$ and $\zeta(T_c) = d_\phi$ (see Section 15.8).

The calculation of other physical quantities and the expansion in $1/N$ then follow from the arguments of Sections 14.1–14.8.

15 The non-linear σ-model

In this chapter, we discuss the non-linear σ-model, a field theory characterized by an orthogonal $O(N)$ symmetry acting non-linearly on the fields. The study has several motivations.

From the viewpoint of statistical physics, the model appears in the study of the large-distance properties, in the ordered phase at low temperature, of lattice spin models with $O(N)$ symmetry and short-range interactions. Indeed, in the case of continuous symmetries, the whole low-temperature phase has a non-trivial large-distance physics due to the presence of Goldstone modes with vanishing mass or infinite correlation length.

Moreover, the model possesses, in two dimensions, the property of *asymptotic freedom* (the Gaussian fixed point is marginally stable for the large-momentum or short-distance behaviour) and the spectrum is non-perturbative. These properties are shared, in dimension 4, by quantum chromodynamics (QCD), a non-Abelian gauge theory and a piece of the Standard Model of fundamental interactions describing physics at the microscopic scale.

From the viewpoint of statistical physics, the properties of the non-linear σ-model derived from renormalization group (RG) considerations provide additional information about the large-distance behaviour of spin models, in the whole low-temperature ordered phase. The RG allows proving universal properties at fixed temperature below T_c. These properties are specific to models with continuous symmetries, since they are direct consequences of the interactions between Goldstone modes.

Perhaps more surprisingly, when the $SO(N)$ group is non-Abelian, that is, for $N > 2$, the σ-model allows studying the critical behaviour near two dimensions. The results are derived from an analysis of the instability of the ordered low-temperature phase near two dimensions, again due to the interactions between Goldstone modes.

The nature of these results can more easily be understood in the framework of the $N \to \infty$ limit in Section 14.10: to all orders of an expansion in $1/N$, in the critical domain, the $(\phi^2)^2$ and non-linear σ-models are equivalent.

Finally, let us point out that, unlike the $(\phi^2)^2$ field theory, the non-linear σ-model can be defined perturbatively only using lattice or dimensional regularizations (Section 12.5.2).

15.1 The non-linear σ-model on the lattice

The non-linear σ-model is a statistical field theory invariant under the group $O(N)$ acting on an N-component field (or classical spin) $\hat{\phi}(x)$ belonging to the sphere S_{N-1}:

$$\hat{\phi}^2(x) = 1. \tag{15.1}$$

We first define a lattice model that, technically, can be considered as a regularized version of the corresponding statistical field theory. We introduce the finite differences

$$\mathrm{D}_\mu \hat{\phi}(x) = \frac{1}{a}\left[\hat{\phi}(x+\epsilon_\mu) - \hat{\phi}(x)\right],$$

where x belongs to the d-dimensional cubic lattice of points with integer coordinates and ϵ_μ is the unit vector in the μ direction. The configuration energy is chosen to be of the form

$$\mathcal{E}(\hat{\phi}) = \frac{1}{2}\sum_{x,\mu}\left(\mathrm{D}_\mu\hat{\phi}(x)\right)^2 = \sum_{x,\mu}\left[1 - \hat{\phi}(x+\epsilon_\mu)\cdot\hat{\phi}(x)\right]. \tag{15.2}$$

The partition function is then given by

$$\mathcal{Z} = \int \prod_{x\in\mathbb{Z}^d}\delta\left(\hat{\phi}^2(x) - 1\right)\mathrm{d}\hat{\phi}(x)\,\exp\left[-\mathcal{E}(\hat{\phi})/T\right], \tag{15.3}$$

where T is the temperature of the statistical model. The measure and the integrand are explicitly invariant under the $O(N)$ group.

In expression (15.3), one recognizes the partition function of a classical spin model on a lattice with nearest-neighbour ferromagnetic interactions.

N-component ferromagnetic lattice models. The statistical model (15.3) provides a particular example (nearest-neighbour interactions) of a class of $O(N)$ symmetric lattice models: N-component classical spins with unit length interact through ferromagnetic, short-range pair interactions of the class defined in Section 8.1. In zero external field, the partition function of such models takes the general form

$$\mathcal{Z} = \int \prod_{i\in\mathbb{Z}^d}\mathrm{d}\mathbf{S}_i\,\delta\left(\mathbf{S}_i^2 - 1\right)\exp\left[-\mathcal{E}(\mathbf{S})/T\right], \tag{15.4}$$

where $\mathbf{S}_i$ is an N-component vector, i denotes lattice sites, and the configuration energy is determined by the pair interaction V_{ij}:

$$\mathcal{E}(\mathbf{S}) = -\sum_{i,j\in\mathbb{Z}^d} V_{ij}\mathbf{S}_i\cdot\mathbf{S}_j. \tag{15.5}$$

The critical properties of such models can be inferred from the RG analysis of a $(\phi^2)^2$ effective field theory. The correspondence between field theory and lattice models is established by the following method: one first finds an approximation for the partition function, the mean-field approximation, valid for high dimensions (see Section 8.10). One characterizes the critical properties of all physical quantities in the framework of this approximation scheme. One then shows that the mean--field approximation is the first term in a systematic expansion. Examining the first correction, one discovers the role of dimension 4 where the validity of mean-field (or quasi-Gaussian) theory ceases (see Section 8.9.2). Finally, for $d \leq 4$, a summation

The non-linear σ-model

to all orders in the expansion of the leading IR divergences leads to an effective ϕ^4 field theory whose critical properties can be analysed by RG methods for dimensions $4-\varepsilon$, that is, near the upper-critical dimension, as we have shown in Sections 13.1.1 and 13.2.1.

Here, by contrast, we are interested in properties of these spin models at low temperature in the ordered phase or in a weak uniform external field, and this leads to another type of expansion.

15.2 Low-temperature expansion

At low temperature, in zero external field, the leading configurations are those that minimize the configuration energy (15.2) and that, thus, satisfy

$$\left|D_\mu \hat{\phi}(x)\right| = 0 \quad \Rightarrow \quad \hat{\phi}(x) = \mathbf{n}, \quad \mathbf{n}^2 = 1.$$

The leading configurations correspond to all spins aligned. As a direct consequence of the symmetry, the configuration energy admits a continuous set of equivalent leading configurations related by transformations of the $O(N)$ group.

In the case of $O(N)$ invariant correlation functions, all minima give exactly the same contribution. The summation over all minima just gives a factor, the volume of the sphere S_{N-1}, which cancels with the normalization of the partition function.

However, in the case of non-$O(N)$-invariant correlation functions, a sum over all minima is equivalent to an $O(N)$ group average. Therefore, it would seem that only the $O(N)$-invariant correlation functions do not vanish. This question is directly related to the possibility of spontaneous symmetry breaking. In the disordered phase, one has to sum over all configurations and, thus, all leading configurations. In the ordered phase, by contrast, the statistical system is no longer ergodic and, therefore, one has to sum only on the subset of spin configurations that fluctuate around the direction of spontaneous magnetization. The leading configuration at low temperature is unique.

Any leading configuration is a starting point for a low-temperature expansion. Moreover, the form (15.3) shows that the low-temperature expansion is technically a loop expansion (see the argument of Section 12.4.2).

Finally, in the presence of a uniform magnetic field, the degeneracy is lifted, and the magnetization and the leading configuration at low temperature are aligned along the magnetic field.

15.2.1 Parametrization

We now choose as leading configuration the vector $\mathbf{n} = (1, 0)$. To generate the low-temperature expansion of the statistical model, it is necessary to parametrize the spin vector in terms of independent variables. A parametrization of the sphere (15.1) that is adapted to the low-temperature expansion is, for example,

$$\hat{\phi}(x) = (\sigma(x), \boldsymbol{\pi}(x)), \tag{15.6}$$

where σ is the component of $\hat{\phi}$ along $\mathbf{n}$ and $\boldsymbol{\pi}(x)$ is an $(N-1)$-component vector. The component $\sigma(x)$ then is a function of $\boldsymbol{\pi}(x)$ through equation (15.1). The equation can be solved locally. Since $1 - \sigma \ll 1$, the solution is

$$\sigma(x) = \left(1 - \boldsymbol{\pi}^2(x)\right)^{1/2}. \tag{15.7}$$

The consequences of the singularity of the parametrization will be discussed later.

Representation of the SO(N) group. In what follows, we discuss transformations of the $O(N)$ group close to the identity and thus belonging to the sub-group $SO(N)$ of matrices with determinant 1 (see Section A1.5).

The sub-group $SO(N-1)$ that leaves σ invariant acts linearly on $\boldsymbol{\pi}(x)$:

$$\pi_\alpha(x) \mapsto \sum_{\beta=1}^{N-1} O_{\alpha\beta} \pi_\beta(x),$$

where $O_{\alpha\beta}$ is an orthogonal matrix $(N-1) \times (N-1)$ with determinant 1.

One can then decompose the set of generators of the Lie algebra of $SO(N)$ into the set of generators of the Lie algebra of $SO(N-1)$ and the complementary set. This complementary set corresponds to infinitesimal group transformations that generate the variations

$$\boldsymbol{\pi} \mapsto \boldsymbol{\pi} + \mathcal{D}_\omega \boldsymbol{\pi} \quad \text{with} \quad [\mathcal{D}_\omega \boldsymbol{\pi}](x) = \boldsymbol{\omega}\left(1 - \boldsymbol{\pi}^2(x)\right)^{1/2}, \tag{15.8}$$

where the constants ω_α are the infinitesimal parameters of the transformation. Since the transform of the vector $\boldsymbol{\pi}$ is a non-linear function of $\boldsymbol{\pi}$, one speaks here of a non-linear representation of the $O(N)$ group.

The transformation of the field σ is then a consequence of the relation (15.7) and of the transformation (15.8) of the field $\boldsymbol{\pi}$:

$$\mathcal{D}_\omega \sigma(x) \equiv \mathcal{D}_\omega \left(1 - \boldsymbol{\pi}^2(x)\right)^{1/2} = -\left(1 - \boldsymbol{\pi}^2(x)\right)^{-1/2} \boldsymbol{\pi} \cdot [\mathcal{D}_\omega \boldsymbol{\pi}](x) = -\boldsymbol{\omega} \cdot \boldsymbol{\pi}(x). \tag{15.9}$$

Partition function and generating function. In the representation (15.6), the configuration energy (15.2) takes the form

$$\mathcal{E}(\hat{\phi}) = \frac{1}{2} \sum_{x,\mu} (D_\mu \hat{\phi}(x))^2 \equiv \mathcal{E}(\boldsymbol{\pi}, \sigma) = \frac{1}{2} \sum_{x,\mu} \left[(D_\mu \boldsymbol{\pi}(x))^2 + (D_\mu \sigma(x))^2\right]. \tag{15.10}$$

Using the relation (15.7), one finds

$$\sigma = (1 - \boldsymbol{\pi}^2)^{1/2} = 1 - \tfrac{1}{2}\boldsymbol{\pi}^2 + O(\boldsymbol{\pi}^4) \Rightarrow D_\mu \sigma(x) = -\tfrac{1}{2} D_\mu \boldsymbol{\pi}^2 + O(\boldsymbol{\pi}^4),$$

and thus

$$\mathcal{E}(\boldsymbol{\pi}, \sigma) = \frac{1}{2} \sum_{x,\mu} (D_\mu \boldsymbol{\pi}(x))^2 + O(\boldsymbol{\pi}^4). \tag{15.11}$$

The generating functional of correlation functions is given by

$$\mathcal{Z}(\mathbf{H}) = \int \prod_{x \in \mathbb{Z}^d} \frac{\mathrm{d}\boldsymbol{\pi}(x)}{(1-\boldsymbol{\pi}^2(x))^{1/2}} \exp\left[-\frac{1}{T}\left(\mathcal{E}(\boldsymbol{\pi},\sigma) - \sum_x \mathbf{H}(x)\cdot\hat{\boldsymbol{\phi}}(x)\right)\right], \quad (15.12)$$

where $\mathrm{d}\boldsymbol{\pi}(1-\boldsymbol{\pi}^2)^{-1/2}$ is the invariant measure on the sphere in the parametrization (15.6).

In the parametrization (15.6), the configuration energy is no longer polynomial, as it was in the field theory models studied so far. Moreover, the measure term

$$\prod_x (1-\boldsymbol{\pi}^2(x))^{-1/2} = \exp\left[-\frac{1}{2}\sum_x \ln\left(1-\boldsymbol{\pi}^2(x)\right)\right],$$

expanded in powers of $\boldsymbol{\pi}$, generates additional interactions that do not contribute at leading order for $T \to 0$.

Finally, we recall that the form (15.12) is valid only for $\sigma(x) > 0$, that is, in the phase ordered with magnetization along the vector $\mathbf{n}$.

Mathematical remark. The field $\hat{\boldsymbol{\phi}}$ can also be identified with an element of the homogeneous (symmetric) quotient space $O(N)/O(N-1)$.

15.2.2 Perturbative expansion

The parametrization (15.6) is adapted to an expansion starting from the configuration $\boldsymbol{\pi}(x) = 0$. In the field integral (15.3), the temperature T orders the perturbative expansion. For $T \to 0$, the fields $\boldsymbol{\pi}(x)$ that contribute to the partition function are then such that

$$|D_\mu \boldsymbol{\pi}(x)| = O\left(\sqrt{T}\right).$$

Combined with the choice of expanding around $\boldsymbol{\pi}(x) = 0$, this also implies

$$|\boldsymbol{\pi}(x)| = O\left(\sqrt{T}\right). \quad (15.13)$$

The dependence on T in the integrand is typical of integrals calculable by the steepest descent method (Sections 2.6 and 2.7.1) The values of $\boldsymbol{\pi}(x)$ of order 1 give exponentially small contributions to the field integral (of order $\exp(-\mathrm{const.}/T)$) which are negligible to all orders in the perturbative expansion. The integral can be expanded around the Gaussian approximation and the perturbative expansion reduces to the evaluation of Gaussian expectation values. To all orders in the expansion in powers of T, in the field integral one can integrate freely over $\boldsymbol{\pi}(x)$ from $+\infty$ to $-\infty$, the constraints generated by the parametrization (15.7) ($\sigma(x) > 0$) and by

$$|\boldsymbol{\pi}(x)| \leq 1,$$

again corresponding to exponentially small corrections.

Since π is of order $\sqrt{T}$, it can be convenient to introduce a different normalization of the field,
$$\pi \mapsto \pi \sqrt{T} . \qquad (15.14)$$

Then,
$$\sigma = (1 - T\pi^2)^{1/2} = 1 - \tfrac{1}{2}T\pi^2 + O(T^2\pi^4), \quad \mathcal{E}(\pi, \sigma) = \frac{1}{2}\sum_{x,\mu}(D_\mu\pi(x))^2 + O(T\pi^4).$$

This normalization shows that the expansion of expression (15.10) in powers of T generates an infinite number of different vertices with arbitrary even powers of π. However, in the initial normalization of π, using the relation between expansion in powers of T and loop expansion, it is easy to verify that to all finite orders in T and for a given correlation function, only a finite number of vertices contributes. Let us point out that the additional, derivative-free, vertices generated by the measure contain no $1/T$ factor and begin contributing only at one-loop order.

Introducing the Fourier representation of the field
$$\pi(x) = \int d^d p \, e^{ipx} \, \tilde{\pi}(p), \qquad (15.15)$$

one notes
$$D_\mu\pi(x) = \int d^d p \, e^{ipx} \left(e^{ip_\mu} - 1\right) \tilde{\pi}(p).$$

One infers
$$\sum_{x,\mu}(D_\mu\pi(x))^2 = 2(2\pi)^d \int d^d p \, \tilde{\pi}(p) \cdot \tilde{\pi}(-p) \sum_\mu (1 - \cos p_\mu).$$

In the Fourier representation, the propagator of the π field then reads
$$\langle \pi_\alpha(x)\pi_\beta(0)\rangle|_{T=0} = \frac{1}{(2\pi)^d}\int d^d p \, e^{-ipx} \tilde{\Delta}_{\alpha\beta}(p)$$

with
$$\tilde{\Delta}_{\alpha\beta}(p) = \frac{2\delta_{\alpha\beta}}{d - \sum_\mu \cos p_\mu} = \frac{\delta_{\alpha\beta}}{p^2 + O(p^4)}. \qquad (15.16)$$

At leading order, the π field thus is massless and the correlation length is infinite: the low-temperature expansion corresponds automatically to the situation where the symmetry $O(N)$ is spontaneously broken. The π field then corresponds to the Goldstone modes associated with the symmetry breaking. The massive partner of the π field in the linear realization, the component σ, has been eliminated by the constraint (15.1).

This property explains the existence of a continuum limit in the low-temperature phase that the expansion allows studying.

Let us point out here that these properties are independent of the particular choice (15.6) of the parametrization of $\hat{\phi}(x)$.

15.2.3 Gaussian fixed point and perturbations

The form of the propagator (15.16) shows that the field $\boldsymbol{\pi}$ has the usual Gaussian dimension of a scalar field $[\boldsymbol{\pi}] = \frac{1}{2}(d-2)$.

All vertices are multiplied by a factor of the form $d - \sum_\mu \cos p_\mu = \frac{1}{2}p^2 + O(p^4)$. We thus rescale distances and change the field normalization as indicated in Section 13.1.2:
$$x \mapsto \Lambda x \;\Rightarrow\; p \mapsto p/\Lambda, \quad \boldsymbol{\pi}(x) \mapsto \Lambda^{(2-d)/2}\boldsymbol{\pi}(x/\Lambda), \tag{15.17}$$
where $1/\Lambda$ is the lattice spacing in this new scale. We then extend the definition of $\boldsymbol{\pi}(x)$ by an infinitely differentiable interpolation to arbitrary real values of x. With this definition, $D_\mu \boldsymbol{\pi}$ admits an expansion in powers of Λ of the form
$$\Lambda^{(2-d)/2} D_\mu \boldsymbol{\pi}(x) = \Lambda^{(2-d)/2}\left[\boldsymbol{\pi}(x + \epsilon_\mu/\Lambda) - \boldsymbol{\pi}(x)\right]$$
$$= \sum_{r=1} \frac{1}{r!}\Lambda^{(2-d)/2-r}(\partial_\mu)^r \boldsymbol{\pi}(x).$$

The field σ has an expansion in powers of $\boldsymbol{\pi}^2$. A contribution to $D_\mu \sigma$ has the form
$$D_\mu \Lambda^{n(2-d)}\left(\boldsymbol{\pi}^2(x)\right)^n = \sum_{r=1} \frac{1}{r!}\Lambda^{n(2-d)-r}(\partial_\mu)^r\left(\boldsymbol{\pi}^2(x)\right)^n.$$

One then infers the form of the contributions to the configuration energy. Factorizing the volume element Λ^{-d}, one finds that a vertex containing n fields $\boldsymbol{\pi}$ and r powers of ∂, $n \geq 2$, $r \geq 2$, is affected by a factor $\Lambda^{l_{nr}}$ with
$$l_{nr} = d - n(d-2)/2 - r.$$

As we have shown in Section 13.1.3, the changes of scale and normalization reduce the analysis of perturbations to the Gaussian fixed point to dimensional analysis and the quantities l_{nr} are also the eigenvalues associated with the eigenvectors of the RG (see equation (9.50)). As a consequence:

(i) For $d > 2$, $l_{nr} > 0$ and thus all vertices are irrelevant. The large-distance behaviour for $T < T_c$, in the ordered phase, is given by a quasi-Gaussian theory.

(ii) For $d = 2$, the vertices with $r > 1$ are irrelevant and the vertices with $r = 1$ are marginal. It is thus necessary to study systematically the corrections to the Gaussian approximation.

(iii) For $d < 2$, the same vertices become relevant. This is not particularly surprising since we have shown that a phase transition is then impossible.

For $d > 2$, however, a problem arises: the measure generates derivative-free vertices, which thus seem relevant. In fact, these vertices maintain the symmetry $O(N)$ of the model and, in particular, cancel the mass corrections generated in the perturbative expansion by the other vertices. They play a role somewhat analogous to the $r_c \sigma^2$ term in Hamiltonian (10.1).

From this analysis emerges the special role of dimension 2. We will thus study more systematically the model at and near two dimensions.

15.3 Formal continuum limit

The divergence of the correlation length corresponding to the field $\boldsymbol{\pi}$ implies the existence of a non-trivial large-distance physics, and thus of a continuum limit. One can try to describe the continuum limit, by taking directly the formal continuum limit of the lattice model. We have shown that the leading corrections to the Gaussian model come from the vertices with two derivatives. The most general configuration energy, or Euclidean action, invariant under the $O(N)$ group, a function of $\hat{\boldsymbol{\phi}}$ and its derivatives, and which involves at most two derivatives (the leading terms at large distance) is

$$\mathcal{E}(\hat{\boldsymbol{\phi}}) = \frac{1}{2} \int d^d x \sum_\mu \partial_\mu \hat{\boldsymbol{\phi}}(x) \cdot \partial_\mu \hat{\boldsymbol{\phi}}(x), \tag{15.18}$$

up to a multiplicative constant. (Translation invariance and isotropy in $\mathbb{R}^d$ are implicit.) Indeed, each symmetric and derivative free term is a function of $(\hat{\boldsymbol{\phi}})^2$ and thus, due to the constraint (15.1), reduces to a constant, and $\hat{\boldsymbol{\phi}} \cdot \partial_\mu \hat{\boldsymbol{\phi}}$ vanishes.

The partition function then reads

$$\mathcal{Z} = \int [d\rho(\hat{\boldsymbol{\phi}})] \exp\left[-\mathcal{E}(\hat{\boldsymbol{\phi}})/T\right], \tag{15.19}$$

where $d\rho(\hat{\boldsymbol{\phi}})$ is the product over all points of the invariant measure on the sphere.

In the parametrization (15.6), adapted to perturbative expansion for $T \to 0$, the configuration energy (15.18), or Euclidean action, takes the geometric form

$$\mathcal{E}(\boldsymbol{\pi}, \sigma) = \frac{1}{2} \int d^d x \sum_{\alpha,\beta} G_{\alpha\beta}(\boldsymbol{\pi}(x)) \sum_{\mu=1}^{d} \partial_\mu \pi_\alpha(x) \partial_\mu \pi_\beta(x), \tag{15.20}$$

where

$$G_{\alpha\beta}(\boldsymbol{\pi}) = \delta_{\alpha\beta} + \frac{\pi_\alpha \pi_\beta}{1 - \boldsymbol{\pi}^2} \tag{15.21}$$

is the metric tensor on the sphere and T then characterizes the deviation of the statistical field theory from the Gaussian theory.

The geometric expression (15.20), which involves the metric tensor, is independent of the parametrization of the sphere.

Partition function and generating function. The generating functional of correlation functions is then given by

$$\mathcal{Z}(\mathbf{J}) = \int \left[\frac{d\boldsymbol{\pi}(x)}{(1 - \boldsymbol{\pi}^2(x))^{1/2}}\right] \exp\left[-\frac{1}{T}\left(\mathcal{E}(\boldsymbol{\pi}, \sigma) - \int d^d x \, \mathbf{J}(x) \cdot \boldsymbol{\pi}(x)\right)\right]. \tag{15.22}$$

A problem appears immediately: in continuum space the functional measure is not defined. From a formal viewpoint, the measure can be interpreted as a determinant and thus, since $\ln \det = \operatorname{tr} \ln$, its logarithm as a trace:

$$\ln \prod_x \left(1 - \boldsymbol{\pi}^2(x)\right)^{-1/2} = \exp\left[-\tfrac{1}{2} \operatorname{tr} \mathbf{K}\right]$$

with

$$\mathbf{K}(x,y) = \delta^d(x-y)\ln(1-\pi^2(x)) \Rightarrow \mathrm{tr}\,\mathbf{K} = {}`\delta^d(0)'\int d^d x\, \ln(1-\pi^2(x)). \quad (15.23)$$

One must thus find a regularization for what appears formally as an additional, infinite, contribution of order T to the effective Hamiltonian.

15.4 Regularization

Even without taking into account the contributions of the measure, the Feynman diagrams generated by the theory (15.22) have short-distance, or large-momentum, divergences. It is thus necessary to modify the theory to render all diagrams finite, an operation called regularization.

Derivative or Pauli–Villars regularizations. Any regularized version of the non-linear σ-model must remain invariant under the transformations of the $O(N)$ group. It is more complicated to implement this condition here than in the case of the $(\phi^2)^2$ field theory, for example, because the $O(N)$ symmetry relates the interaction terms in the expression (15.20) to the quadratic part. A simple method consists in starting again from a description of the model in terms of the constrained $\hat{\phi}(x)$ field because the configuration energy (15.18) is formally identical to a Gaussian critical Hamiltonian. We thus replace $\mathcal{E}(\hat{\phi})$ by

$$\mathcal{E}_\Lambda(\hat{\phi}) = \frac{1}{2}\int d^d x\, \hat{\phi}(x)\cdot\left(-\nabla_x^2 + \frac{\alpha_2}{\Lambda^2}\nabla_x^4 - \frac{\alpha_3}{\Lambda^4}\nabla_x^6 + \cdots\right)\hat{\phi}(x). \quad (15.24)$$

Expressing then $\hat{\phi}(x)$ in terms of $\boldsymbol{\pi}(x)$, one discovers that the large-momentum behaviour of the propagator is improved, but simultaneously new, more singular, interactions have been introduced. A power counting analysis then reveals that all diagrams can be regularized except one-loop diagrams, whose power counting has not changed. This property is not completely independent of another limitation of Pauli–Villars regularization: it does not regularize the divergent term (15.23) generated by the measure. Actually, one can verify that the remaining one-loop divergences generated by the new interactions are needed to cancel formally the divergent contributions coming from the measure.

Dimensional regularization. Dimensional regularization (Section 12.5.2) preserves the $O(N)$ symmetry of the configuration energy. Moreover, as a consequence of the rule

$$\delta^d(0) = \frac{1}{(2\pi)^d}\int d^d k = 0,$$

the term (15.23) coming from the measure can be ignored and, therefore, the perturbative expansion has no large-momentum divergences for $d < 2$. Due to its technical simplicity, this is the regularization that one uses generally in explicit calculations. A theoretical inconvenience is that the role of the measure term is then hidden. Therefore, for the theoretical discussion of the renormalization of the non-linear σ-model, it is useful to consider both dimensional and lattice regularizations. In two dimensions, one finds both short (UV) and large (IR) distance divergences and this is a source of another difficulty. To define the perturbative expansion by dimensional regularization, it is necessary to give a mass (a finite correlation length) to the field π and thus to break the $O(N)$ symmetry explicitly. A simple way of generating a mass consists in adding to the Hamiltonian a magnetic field term linearly coupled to σ.

Lattice regularization. Expression (15.3), which corresponds to the initial lattice model, provides, of course, a suitable lattice regularization and also allows using non-perturbative methods to study the non-linear σ-model. Moreover, this is the only regularization that allows a discussion of the role of the measure in the perturbative expansion. On the other hand, perturbative calculations are technically complicated and one can define only asymptotically a differential RG since the dilatation parameter cannot vary continuously.

15.5 Zero-momentum or IR divergences

Since in a massless field theory the propagator behaves like $1/p^2$, it is necessary to discuss the perturbative expansion from the viewpoint of zero-momentum or IR divergences. For example, let us calculate the expectation value of the field σ at one-loop order. In the initial normalization (15.12),

$$\langle \sigma(x) \rangle = 1 - \tfrac{1}{2} \langle \pi^2(x) \rangle + O\left(\pi^4\right)$$
$$= 1 - (N-1)T \int \frac{\mathrm{d}^d p}{p^2 + O(p^4)} + O\left(T^2\right).$$

The integral is finite for $d > 2$, but diverges at the critical dimension $d = 2$, the dimension at which the non-linear σ-model is renormalizable. This property is directly related to the physics of spontaneous breaking of continuous symmetries.

(i) In Section 7.7, we have shown that in the N-component model, for $d > 2$, the $O(N)$ symmetry is spontaneously broken at low temperature. Consistently, the perturbative expansion of the σ-model, which predicts also spontaneous symmetry breaking, is not divergent at zero momentum for $d > 2$. For $T < T_c$ fixed, the large-distance perturbative behaviour is dominated by massless excitations (Goldstone modes) also called spin waves. By contrast, the perturbative expansion gives no indication about the possible existence of a critical region $T \sim T_c$.

(ii) For $d \leq 2$, the Mermin–Wagner–Coleman theorem states that spontaneous symmetry breaking with order ($\langle \hat{\phi} \rangle \neq 0$) of a continuous symmetry is impossible in a model with short-range interactions, and this, again, is consistent with the appearance of IR divergences in the perturbative expansion (see the discussion of Section 7.7). For $d \leq 2$, the critical temperature T_c vanishes and the perturbative expansion is meaningful only in the presence of a zero-momentum (IR) regularization. As a consequence, the perturbative expansion gives no direct information about the large-distance properties of the model.

IR regularization. To be able to define a perturbative expansion in two dimensions, one must first introduce a zero-momentum regularization. In an infinite volume, this necessitates giving a non-zero mass (i.e., a finite correlation length) to the field $\boldsymbol{\pi}$. Since the vanishing mass is a consequence of the spontaneous breaking of the $O(N)$ symmetry, it is necessary to break the symmetry explicitly. One can, for example, add to expression (15.10) an explicit mass term. However, a study of the mechanism of symmetry breaking suggests a more convenient method that consists in introducing a uniform magnetic field:

$$\mathcal{E}(\boldsymbol{\pi},\sigma) \mapsto \mathcal{E}(\boldsymbol{\pi},\sigma;H) = \mathcal{E}(\boldsymbol{\pi},\sigma) - \frac{H}{T}\sum_x \sigma(x), \qquad H > 0. \qquad (15.25)$$

An immediate consequence of such a modification is that the minimum of the configuration energy is no longer degenerate. Indeed, the minimum is now obtained by maximizing also the magnetic field contribution and this implies $\boldsymbol{\pi} = 0$ at the minimum.

Moreover, an expansion of σ in powers of $\boldsymbol{\pi}$,

$$\sigma = \left(1 - \boldsymbol{\pi}^2\right)^{1/2} = 1 - \tfrac{1}{2}\boldsymbol{\pi}^2 + O\left((\boldsymbol{\pi}^2)^2\right), \qquad (15.26)$$

shows that the quadratic terms in $\mathcal{E}(\boldsymbol{\pi},\sigma;H)$ lead to the new propagator of the field $\boldsymbol{\pi}$,

$$\tilde{\Delta}_{\alpha\beta}(p) = \frac{\delta_{\alpha\beta}T}{p^2 + H + O(p^4)}. \qquad (15.27)$$

The linear σ term has thus generated a mass proportional to $H^{1/2}$ for the field $\boldsymbol{\pi}$. It has also generated new, derivative-free, interactions.

We recall that in the case of the $(\boldsymbol{\phi}^2)^2$ field theory, the breaking term $H\sigma$ is linear in the independent field component σ and, therefore, does not generate a new renormalization constant. The same result can be proved here, even though the component σ is a non-linear function of the field $\boldsymbol{\pi}$.

15.6 Renormalization group

We have shown that the large-distance physics in a model with $O(N)$ symmetry can be described, in dimension $d > 2$, below T_c, by the non-linear σ-model. One verifies that this field theory is renormalizable in two dimensions. One thus proceeds in a way formally analogous to the case of the $(\boldsymbol{\phi}^2)^2$ field theory, that is, one studies the theory in dimension $d = 2 + \varepsilon$ within the framework of a double series expansion in powers of the temperature T and of ε. In this way, the perturbative expansion is renormalizable and RG equations (RGE) follow.

15.6.1 Renormalization and RGE

In the formal continuum limit, in dimensional regularization, the partition function in a uniform magnetic field H reads

$$\mathcal{Z}(H) = \int \left[(1 - \pi^2(x))^{-1/2} \mathrm{d}\pi(x)\right] \exp\left[-\mathcal{E}(\pi, \sigma; H)/T\right], \tag{15.28}$$

where

$$\mathcal{E}(\pi, \sigma; H) = \int \mathrm{d}^d x \left\{ \frac{1}{2} \sum_\mu \left[(\partial_\mu \pi(x))^2 + \frac{(\pi \cdot \partial_\mu \pi(x))^2}{1 - \pi^2(x)} \right] - H\sqrt{1 - \pi^2(x)} \right\}. \tag{15.29}$$

In the field integral, one then substitutes

$$\pi(x) = Z^{1/2} \pi_{\mathrm{r}}(x), \quad \sigma(x) = Z^{1/2} \sigma_{\mathrm{r}}(x), \tag{15.30}$$

where Z is the field renormalization. The relation between π and σ becomes

$$\sigma_{\mathrm{r}}(x) = \left[Z^{-1} - \pi_{\mathrm{r}}^2(x)\right]^{1/2}. \tag{15.31}$$

One also introduces a mass scale μ, at which the renormalized theory is defined, and two parameters g, h, the renormalized temperature and magnetic field, which are also the effective parameters at scale μ. In some renormalization schemes, like minimal subtraction, the renormalization constants are independent of h. We adopt here one of these schemes. Let us point out that, in what follows, μ plays the role of an intermediate scale characteristic of the critical domain around the Gaussian fixed point. The ultimate large-distance behaviour corresponds to momenta small with respect to μ.

The relations between initial and renormalized parameters take the form

$$T = \mu^{2-d} g Z_g(g), \quad \mu^{d-2} \frac{h}{g} = Z^{1/2}(g) \frac{H}{T}, \tag{15.32}$$

where Z_g characterizes the temperature renormalization and we have assumed that the renormalization constants are chosen independent of h.

With our conventions, the parameter g, which is proportional to the temperature, is dimensionless. The relation between regularized and renormalized vertex functions is

$$Z^{n/2}(g) \tilde{\Gamma}^{(n)}(p_i; T, H) = \tilde{\Gamma}_{\mathrm{r}}^{(n)}(p_i; g, h, \mu), \tag{15.33}$$

where the functions $\tilde{\Gamma}_{\mathrm{r}}^{(n)}$ have, order by order in an expansion in powers of g, a finite limit when $\varepsilon \to 0$.

By differentiating with respect to μ, at fixed initial parameters T, H, one obtains the RGE

$$\left[\mu \frac{\partial}{\partial \mu} + \beta(g) \frac{\partial}{\partial g} - \frac{n}{2} \zeta(g) + \rho(g) h \frac{\partial}{\partial h} \right] \tilde{\Gamma}_{\mathrm{r}}^{(n)}(p_i; g, h, \mu) = 0, \tag{15.34}$$

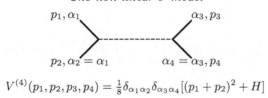

$$V^{(4)}(p_1, p_2, p_3, p_4) = \tfrac{1}{8}\delta_{\alpha_1\alpha_2}\delta_{\alpha_3\alpha_4}[(p_1 + p_2)^2 + H]$$

Fig. 15.1 π^4 vertex: Faithful representation, the dotted line does not correspond to a propagator but allows representing faithfully the flow of indices by full lines.

where the RG functions are defined by

$$\mu \frac{\partial}{\partial \mu}\bigg|_{T,H \text{ fixed}} g = \beta(g),$$

$$\mu \frac{\partial}{\partial \mu}\bigg|_{T,H \text{ fixed}} \ln Z = \zeta(g), \qquad (15.35)$$

$$\mu \frac{\partial}{\partial \mu}\bigg|_{T,H \text{ fixed}} \ln h = \rho(g).$$

Using the first relation (15.32) and the definition of Z, one relates the RG functions in the renormalized theory to the renormalization constants Z, Z_g:

$$\beta(g) = \varepsilon g \left(1 + g \frac{d \ln Z_g}{dg}\right)^{-1}, \qquad (15.36a)$$

$$\zeta(g) = \beta(g) \frac{d \ln Z}{dg}, \qquad (15.36b)$$

where one has set

$$d = 2 + \varepsilon. \qquad (15.37)$$

The coefficient of $\partial/\partial h$ can be inferred from the second equation (15.32), which implies (taking the logarithm of both sides)

$$0 = h^{-1}\mu \frac{\partial}{\partial \mu} h + d - 2 - \frac{1}{2}\zeta(g) - \frac{\beta(g)}{g}, \qquad (15.38)$$

and, therefore,

$$\rho(g) = 2 - d + \frac{1}{2}\zeta(g) + \frac{\beta(g)}{g}. \qquad (15.39)$$

To be able to discuss correlation functions involving the field σ, one also needs the RGE satisfied by the connected correlation functions $W_r^{(n)}$:

$$\left[\mu \frac{\partial}{\partial \mu} + \beta(g) \frac{\partial}{\partial g} + \frac{n}{2}\zeta(g) + \left(\frac{1}{2}\zeta(g) + \frac{\beta(g)}{g} - \varepsilon\right) h \frac{\partial}{\partial h}\right] W_r^{(n)} = 0. \qquad (15.40)$$

15.6.2 Calculations at one-loop order

The two RG functions can be inferred from the calculation of the two-point vertex function. At one-loop order, the calculation requires the vertex of order π^4:

$$\int \mathrm{d}^d x \sum_\mu [\boldsymbol{\pi}(x) \cdot \partial_\mu \boldsymbol{\pi}(x)]^2 = \frac{1}{4} \int \mathrm{d}^d x \sum_\mu [\partial_\mu (\boldsymbol{\pi}^2(x))]^2$$
$$= -\frac{1}{4} \int \mathrm{d}^d x\, \boldsymbol{\pi}^2(x) \nabla_x^2 (\boldsymbol{\pi}^2(x)).$$

In the Fourier representation (15.15),

$$-\frac{1}{4} \int \mathrm{d}^d x\, \boldsymbol{\pi}^2(x) \nabla_x^2 (\boldsymbol{\pi}^2(x))$$
$$= \frac{1}{4} \int \mathrm{d}^d x \int \mathrm{d}^d p_1\, \mathrm{d}^d p_2\, e^{ix(p_1+p_2)}\, \tilde{\boldsymbol{\pi}}(p_1) \cdot \tilde{\boldsymbol{\pi}}(p_2) \int \mathrm{d}^d p_3\, \mathrm{d}^d p_4\, (p_3 + p_4)^2$$
$$\times e^{ix(p_3+p_4)}\, \tilde{\boldsymbol{\pi}}(p_3) \cdot \tilde{\boldsymbol{\pi}}(p_4)$$
$$= \frac{1}{4}(2\pi)^d \int \mathrm{d}^d p_1\, \mathrm{d}^d p_2\, \mathrm{d}^d p_3\, \mathrm{d}^d p_4\, \delta\left(\sum p_i\right) (p_3 + p_4)^2\, \tilde{\boldsymbol{\pi}}(p_1) \cdot \tilde{\boldsymbol{\pi}}(p_2)$$
$$\times \tilde{\boldsymbol{\pi}}(p_3) \cdot \tilde{\boldsymbol{\pi}}(p_4).$$

The contribution of order π^4 coming from the expansion of $H\sigma(x)$ has the effect of replacing $(p_3 + p_4)^2$ by $(p_3 + p_4)^2 + H$ and leads to the vertex $V^{(4)}$ of Figure 15.1.

Fig. 15.2 Two-point function: One-loop contribution.

The calculation of the one-loop diagram of Figure 15.2 follows. The diagram decomposes, in a faithful representation, into the sum of the two diagrams of Figure 15.3, where the flow of group indices is explicit. It can be expressed in terms of the function Ω_d introduced in Section 12.5.3 (equation (12.32)).

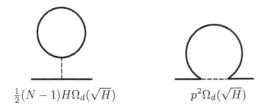

Fig. 15.3 Faithful one-loop diagrams.

The second diagram of Figure 15.3 is given by

$$\frac{1}{(2\pi)^d} \int \mathrm{d}^d q\, \frac{(p+q)^2 + H}{q^2 + H} = \frac{p^2}{(2\pi)^d} \int \frac{\mathrm{d}^d q}{q^2 + H} + \frac{1}{(2\pi)^d} \int \mathrm{d}^d q.$$

The contribution proportional to $\int d^d q$ is cancelled by the first contribution coming from the measure. Moreover, like the measure contribution, it vanishes in dimensional regularization.

For the two-point vertex function, one then finds

$$\tilde{\Gamma}^{(2)}(p) = \frac{1}{T}(p^2 + H) + [p^2 + \tfrac{1}{2}(N-1)H]\,\Omega_d(\sqrt{H}) + O(g), \qquad (15.41)$$

where (equation (12.32))

$$\Omega_d(m) = \frac{1}{(2\pi)^d} \int \frac{d^d q}{q^2 + m^2}.$$

Renormalized vertex function. At one-loop, the renormalized two-point function, defined in (15.33), and expressed in terms of the renormalized parameters (15.32), then reads

$$\tilde{\Gamma}^{(2)}_r(p) = \frac{\mu^\varepsilon}{g}\left(\frac{Zp^2}{Z_g} + hZ^{1/2}\right) + [p^2 + \tfrac{1}{2}(N-1)h]\,\Omega_d(\sqrt{h}) + O(g). \qquad (15.42)$$

In dimensional regularization, the function Ω_d is given by equation (12.36), which can be written as

$$\Omega_d(m) = \frac{\pi N_d}{2\sin(\pi d/2)} m^{d-2} \underset{\varepsilon \to 0}{\sim} -\frac{N_d}{\varepsilon}, \qquad (15.43)$$

where N_d is the usual loop factor:

$$N_d = \frac{2}{(4\pi)^{d/2}\Gamma(d/2)} = \frac{1}{2\pi} + O(\varepsilon).$$

Expressing that the two-point function is finite when $\varepsilon = d - 2 \to 0$, one derives from the expression of $\tilde{\Gamma}^{(2)}_r$ the renormalization constants at one-loop order. In the so-called *minimal subtraction* scheme, one finds

$$\begin{aligned}Z &= 1 + (N-1)\frac{N_d}{\varepsilon}g + O\left(g^2\right),\\ Z_g &= 1 + (N-2)\frac{N_d}{\varepsilon}g + O\left(g^2\right).\end{aligned} \qquad (15.44)$$

Using expressions (15.36), one infers the RG functions at one-loop order:

$$\beta(g) = \varepsilon g - \frac{(N-2)}{2\pi}g^2 + O\left(g^3\right), \qquad (15.45a)$$

$$\zeta(g) = \frac{(N-1)}{2\pi}g + O\left(g^2\right). \qquad (15.45b)$$

The two RG functions and the corresponding critical exponents have been calculated up to four-loop order.

15.7 Solution of the RGE. Fixed points

Near dimension 2, the fixed points correspond to the zeros of the function $\beta(g)$ given by equation (15.36a). Integrating then the RGE, one can derive, from the existence and stability of the fixed points, universality properties.

15.7.1 Fixed points

From the form of the function $\beta(g)$, one concludes immediately:

(i) For $d \leq 2$ ($\varepsilon \leq 0$), $g = 0$ is an unstable IR fixed point, the IR instability being a direct consequence of the IR divergences that massless Goldstone bosons would generate. The spectrum of the theory thus cannot be inferred from the perturbative expansion and the implicit perturbative assumption of spontaneous symmetry breaking at low temperature is inconsistent. As we have already mentioned, this result is in agreement with a more rigorous analysis.

Section 15.10.1 then contains a brief discussion of physics in two dimensions. Since the model then depends only on one marginal parameter, $g = 0$ is also a UV stable fixed point (a large-momentum property called *asymptotic freedom* in the literature).

Finally, there is a special case to which the analysis does not apply: $d = 2$, $N = 2$, where the β-function vanishes identically, and which must be examined separately (Section 15.10.2).

(ii) For $d > 2$, that is, $\varepsilon > 0$, $g = 0$ is a stable IR fixed point, the symmetry $O(N)$ is spontaneously broken at low temperature in zero field. The effective interaction that determines the large-distance behaviour, goes to zero for all effective temperatures $g < g^*$, g^* being the first non-trivial zero of $\beta(g)$. As a consequence, the large-distance properties of the model can be inferred from the low-temperature expansion and RG arguments, by replacing the parameters at scale μ by the effective parameters obtained by solving the RGE.

Critical temperature. Finally, one observes that, at least for $0 < \varepsilon \ll 1$ and $N > 2$, the RG function $\beta(g)$ has a non-trivial zero

$$g^* = \frac{2\pi\varepsilon}{N-2} + O\left(\varepsilon^2\right) \Rightarrow \beta(g^*) = 0 \quad \text{with} \quad \beta'(g^*) = -\varepsilon + O\left(\varepsilon^2\right), \quad (15.46)$$

which corresponds to an unstable IR fixed point. A value of the temperature that has such a property corresponds to a critical temperature. The consequences of this result will be discussed later.

Here, let us point out only that g^* is also a UV fixed point, that is, it governs the large-momentum behaviour of the renormalized theory. Therefore, the perturbative analysis that indicates that the theory is not renormalizable for $d > 2$ cannot be trusted. Indeed, correlation functions have, for large momenta, a non-perturbative behaviour. This opens the possibility that a quantum field theory consistent for all distance scales in the continuum could exist.

15.7.2 RGE integration: $d > 2$, $g < g^*$

We now present a more detailed discussion of the solutions of the RGE.

We first examine the implications of the RGE for the large-distance behaviour of correlation functions for $d > 2$ where $g = 0$ is an IR stable fixed point. Like in Section 13.3.1, we solve equation (15.34) by introducing a scale factor λ and by looking for a solution of the form

$$\tilde{\Gamma}_{\mathrm{r}}^{(n)}(p_i, g, h, \mu) = Z^{-n/2}(\lambda)\tilde{\Gamma}_{\mathrm{r}}^{(n)}(p_i, g(\lambda), h(\lambda), \mu/\lambda). \tag{15.47}$$

Compatibility with equation (15.34) implies

$$\ln \lambda = -\int_g^{g(\lambda)} \frac{dg'}{\beta(g')}, \tag{15.48a}$$

$$\ln Z(\lambda) = \int_g^{g(\lambda)} dg' \frac{\zeta(g')}{\beta(g')}, \tag{15.48b}$$

$$\ln\bigl(h(\lambda)/h\bigr) = \int_g^{g(\lambda)} dg' \left[\frac{\zeta(g')}{2\beta(g')} - \frac{d-2}{\beta(g')} + \frac{1}{g'}\right]. \tag{15.48c}$$

The latter equation can be integrated and yields

$$h(\lambda) = \lambda^{d-2} Z^{1/2}(\lambda) \frac{g(\lambda)}{g} h.$$

With our conventions, $\tilde{\Gamma}_{\mathrm{r}}^{(n)}$ has dimension d and h dimension 2. Taking into account dimensional analysis, one can then rewrite relation (15.47) as

$$\tilde{\Gamma}_{\mathrm{r}}^{(n)}(p_i, g, h, \mu) = Z^{-n/2}(\lambda)(\mu/\lambda)^d \tilde{\Gamma}_{\mathrm{r}}^{(n)}\bigl(\lambda p_i/\mu, g(\lambda), h(\lambda)\lambda^2/\mu^2, 1\bigr). \tag{15.49}$$

For $h = 0$, the perturbative expansion has large IR contributions. The choice of λ solution of the equation

$$h(\lambda) = (\mu/\lambda)^2, \tag{15.50}$$

ensures that the perturbative expansion in the effective theory at scale λ is not IR singular.

It is easy to verify that, at least in the vicinity of $g = 0$, for $h \ll \mu^2$ the equation implies $\lambda \to \infty$ and $g(\lambda)$ converge toward the IR stable Gaussian fixed point $g = 0$. We introduce the three functions of the temperature

$$M_0(g) = \exp\left[-\frac{1}{2}\int_0^g \frac{\zeta(g')}{\beta(g')} dg'\right] = Z^{-1/2}(g), \tag{15.51}$$

$$\xi(g) = \mu^{-1} g^{1/\varepsilon} \exp\left[\int_0^g \left(\frac{1}{\beta(g')} - \frac{1}{\varepsilon g'}\right) dg'\right], \tag{15.52}$$

$$K(g) = M_0(g)\left[\mu\xi(g)\right]^{d-2}/g = 1 + O(g). \tag{15.53}$$

The integral in equation (15.48c) has a limit when $g \to 0$ and thus $h(\lambda) \to hK(g)$. Solving equation (15.50), one finds

$$\lambda \sim \mu/\sqrt{K(g)h}. \tag{15.54}$$

Equations (15.48a, b) then determine the behaviours of $g(\lambda)$ and $Z(\lambda)$:

$$g(\lambda) \sim \lambda^{2-d}\left(\mu\xi(g)\right)^{d-2} \sim [K(g)]^{(d-2)/2}\left(\mu\xi(g)\right)^{d-2}(h/\mu^2)^{(d-2)/2}, \tag{15.55}$$
$$Z(\lambda) \sim M_0^2(g). \tag{15.56}$$

One infers

$$\tilde{\Gamma}_r^{(n)}(p_i, g, h, \mu) \sim M_0^{-n}(g)[K(g)h]^{d/2}\tilde{\Gamma}_r^{(n)}\left(p_i[K(g)h]^{-1/2}, g(\lambda), 1, 1\right). \tag{15.57}$$

Actually, it is simple to verify directly, using dimensional analysis in the form

$$\mu\frac{\partial}{\partial\mu} + 2h\frac{\partial}{\partial h} + p_i\frac{\partial}{\partial p_i} = d,$$

that equation (15.57) yields the general solution of equation (15.34).

Since the effective temperature $g(\lambda) \to 0$, the leading terms in the limit $h \to 0$ and $p_i \to 0$ can then be calculated perturbatively.

15.8 Correlation functions: Scaling form

It is convenient to rewrite equation (15.57) in the form

$$\tilde{\Gamma}_r^{(n)}(p_i; g, h, \mu) = \xi^{-d}(g)M_0^{-n}(g)F^{(n)}\left(p_i\xi(g), h/h_0(g)\right), \tag{15.58}$$

where

$$h_0(g) = \left[K(g)\xi^2(g)\right]^{-1}. \tag{15.59}$$

Equation (15.58) shows that $\xi(g)$ has, in zero field, the nature of a correlation length.

For connected correlation functions, the same analysis leads to

$$W_r^{(n)}(p_i; g, h, \mu) = \xi^{d(n-1)}(g)M_0^n(g)G^{(n)}\left(p_i\xi(g), h/h_0(g)\right). \tag{15.60}$$

The one-point function is the magnetization and thus

$$M = M_0(g)F^{(0)}\left(h/h_0(g)\right). \tag{15.61}$$

By inverting the relation, one obtains the scaling form of the equation of state

$$h = h_0(g)f\left(\frac{M}{M_0(g)}\right). \tag{15.62}$$

The vertex function can also be expressed in terms of the magnetization as
$$\tilde{\Gamma}_r^{(n)}(p_i, g, M, \mu) = \xi^{-d}(g) M_0^{-n}(g) F^{(n)}(p_i \xi(g), M/M_0(g)). \tag{15.63}$$
The scaling forms (15.62), (15.63) are consistent with the solutions of the RGE in the $(\phi^2)^2$ field theory (see Section 13.4): the appearance of two different functions $\xi(g)$ and $M_0(g)$ corresponds to the existence of two independent critical exponents ν, β, in the $(\phi^2)^2$ field theory. They extend the large-distance scaling form of correlation functions, valid in the critical region, to all temperatures below g^*. However, there is an important difference between the RGE of the $(\phi^2)^2$ theory and those of the σ-model: the $(\phi^2)^2$ theory depends on two parameters, the coefficient of ϕ^2 that plays the role of the temperature and the coefficient of $(\phi^2)^2$ that has no equivalent here. This coefficient allows interpolating between the Gaussian and the IR stable fixed points. The correlation functions of the continuum $(\phi^2)^2$ theory have the exact scaling form (15.63) only at the IR fixed point. By contrast, in the case of the σ-model, it has been possible to eliminate all corrections to scaling, related to irrelevant operators, order by order in the perturbative expansion. We are thus led to the remarkable conclusion: the correlation functions of the non-linear σ-model with $O(N)$ symmetry are identical to the correlation functions of the $(\phi^2)^2$ field theory at the IR stable fixed point. This conclusion is consistent with the analysis of the scaling behaviour presented in the framework of the $1/N$ expansion in Section 14.10.

Coexistence curve and spontaneous magnetization. The analysis can be applied to the determination of singularities on the coexistence curve, that is, at g fixed below the critical temperature when the magnetic field goes to zero.

With the same notation, the renormalized magnetization, which is the one-point function, satisfies
$$M(g, h, \mu) \equiv \langle \sigma_r(x) \rangle = Z^{1/2}(\lambda) M(g(\lambda), 1, 1). \tag{15.64}$$
To calculate M, we expand equation (15.31):
$$\langle \sigma_r(x) \rangle = Z^{-1/2}(g) \left[1 - \tfrac{1}{2} Z \pi^2(x) + O(\pi^4) \right].$$
At one-loop order, in a field, M is thus given by (equation (15.43))
$$M = Z^{-1/2} - \tfrac{1}{2}(N-1)\mu^{-\varepsilon} g \frac{1}{(2\pi)^d} \int \frac{d^d q}{q^2 + h} + O(g^2)$$
$$= 1 - (N-1)\frac{N_d}{2\varepsilon} g - (N-1)\frac{\pi N_d}{4\sin(\pi d/2)} g(h/\mu^2)^{(d-2)/2} + O(g^2).$$
Using then relation (15.64), one finds
$$M(g, h, \mu = 1) = M_0(g) \left[1 - (N-1) M_1(g) h^{(d-2)/2} + O\left(h, h^{d-2}\right) \right]$$
with
$$M_1(g) = N_d \left(\frac{1}{2(d-2)} + \frac{\pi}{4\sin(\pi d/2)} \right) [K(g)]^{(d-2)/2} \left(\mu \xi(g) \right)^{d-2}.$$
This result shows that $M_0(g)$ is the *spontaneous magnetization* and establishes the form of the leading singularity in h on the coexistence curve, that is, at $h = 0$, of the equation of state. For the physical dimension $d = 3$, one finds a $h^{1/2}$ singularity.

15.9 The critical domain: Critical exponents

We now study the behaviour of thermodynamic quantities when g approaches g^* (for $N > 2$). The correlation length $\xi(g)$ diverges like

$$\xi(g) \sim \mu^{-1}(g^* - g)^{1/\beta'(g^*)}. \tag{15.65}$$

One concludes that the correlation length exponent is given by

$$\nu = -\frac{1}{\beta'(g^*)}. \tag{15.66}$$

For $d \to 2_+$, the exponent ν thus behaves like

$$\nu \sim 1/(d-2). \tag{15.67}$$

The spontaneous magnetization $M_0(g)$ vanishes at g^*:

$$\ln M_0(g) \sim -\frac{1}{2}\frac{\zeta(g^*)}{\beta'(g^*)}\ln(g^* - g) \sim \beta \ln(g^* - g). \tag{15.68}$$

The magnetization exponent $\beta = \frac{1}{2}\nu d_\phi$ (not be confused with the β-function) follows, and thus also η, from the relation $d_\phi = d - 2 + \eta$:

$$d_\phi = \zeta(g^*), \quad \eta = \zeta(g^*) + 2 - d. \tag{15.69}$$

At leading order, one finds

$$d_\phi = \frac{N-1}{N-2}(d-2) + O\left((d-2)^2\right), \quad \eta = \frac{d-2}{N-2} + O\left((d-2)^2\right). \tag{15.70}$$

Finally, one verifies that the singularity of $\tilde{\Gamma}_r^{(n)}$ for $g = g^*$ coming from the factor $\xi^{-d}M_0^{-n}$ coincides with the result derived from the $(\phi^2)^2$ field theory.

The nature of the correlation length $\xi(g)$. The length $\xi(g)$ is a crossover scale between two different behaviours of correlation functions. For distances large with respect to $\xi(g)$, the behaviour of correlation functions is governed by Goldstone modes and can thus be inferred from the low-temperature perturbative expansion. By contrast, when g tends toward g^*, $\xi(g)$ diverges. There exists then distances large with respect to the microscopic scale but small with respect to $\xi(g)$ for which correlation functions have the critical behaviour. In such a situation, one can define, in the continuum, a quantum field theory consistent on all scales, the critical behaviour being also the large-momentum behaviour of *renormalized* correlation functions.

General comments. Starting from the low-temperature expansion, we have been able to describe, for theories with a continuous symmetry, not only the complete structure of the low-temperature phase, as expected, but also in the non-Abelian case $N > 2$ (the rotation group $SO(N)$ is Abelian, that is, commutative for $N = 2$) the critical behaviour near two dimensions.

What is slightly surprising in this result is that the perturbative expansion is sensitive only to the local structure of the sphere $\hat{\phi}^2 = 1$ even though symmetry restoration seems to depend on the global structure of the sphere. This is confirmed by the peculiarity of the Abelian case $N = 2$: locally the circle cannot be distinguished from the non-compact straight line and, thus, the σ-model becomes a free field theory. For $N > 2$, by contrast, the sphere has a characteristic local curvature. One still faces the problem that different regular compact manifolds may have the same local metric and, thus, lead to the same perturbative expansion. They all lead to the same low-temperature physics. However, the preceding results concerning the critical behaviour are physically relevant only if they are still valid when ε is no longer infinitesimal and g is close to g^*, a condition that cannot be verified directly. In particular, the low-temperature expansion misses in general terms decreasing like $\exp(\text{const.}/g)$, which may in some cases be important for physics.

On the other hand, we recall that a direct relation between the $(\phi^2)^2$ and σ models is provided in continuum space by the large N expansion (Section 14.10). The large N analysis suggests that the preceding considerations are valid for the N-component model, at least for large enough values of N.

By contrast, the physics for $N = 2$ is not well reproduced. The Kosterlitz–Thouless phase transition, which relies on effects invisible in the low-temperature expansion, is not found. Cardy and Hamber have speculated on the RG flow for N close to 2 and for the dimension d close to 2, by incorporating in a phenomenological way the Kosterlitz–Thouless transition in the RG analysis.

15.10 Dimension 2

We now examine briefly the model in dimension 2, the dimension in which it is renormalizable, because it possesses some interesting properties.

15.10.1 The non-Abelian model

For the $N > 2$ non-Abelian case, the non-linear σ-model shares an important property with the model describing strong interactions in the theory of fundamental microscopic interactions: the σ-model is the simplest example of an *asymptotically free* (for large momenta) field theory, since the first coefficient in the expansion of the β-function is negative:

$$\beta(g) = -\frac{(N-2)}{2\pi}g^2\left(1 + g/2\pi\right) + O(g^4), \tag{15.71}$$

in contrast, for example, with the ϕ^4 field theory. Therefore, the large-momentum behaviour of universal correlation functions is entirely calculable from the perturbative expansion and RG arguments (the property of asymptotic freedom). On the other hand, the Gaussian fixed point is IR unstable and, thus, in zero field h, the spectrum of the theory is non-perturbative. This is consistent with the expected absence of symmetry breaking, which implies a spectrum composed of N degenerate massive states.

Defining now the function

$$\xi_0(g) = \mu^{-1} g^{1/(N-2)} e^{2\pi/[(N-2)g]}$$
$$\times \exp\left[\int_0^g dg' \left(\frac{1}{\beta(g')} + \frac{2\pi}{(N-2)g'^2} - \frac{1}{(N-2)g'}\right)\right], \quad (15.72)$$

one can integrate the RGE as before and one finds that $\xi_0(g)$ is proportional to the correlation length in zero field $\xi(g)$. Thus,

$$\xi^{-1}(g) = m(g) = K\mu g^{-1/(N-2)} e^{-2\pi/[(N-2)g]} (1 + O(g)). \quad (15.73)$$

However, the integration constant K, which relates the physical mass to the RG scale, cannot be calculated by perturbative techniques.

Finally, the scaling forms (15.58), (15.60) imply that the perturbative expansion at fixed magnetic field is valid, at small momenta or large distances, and for $h/h_0(g)$ large.

Elitzur's conjecture. The configuration energy (15.18) with $O(N)$ symmetry has a sphere of classical degenerate minima. To define the perturbative expansion, one is forced to add to the energy a linear term that breaks the symmetry and selects a particular classical minimum. We have stated, and this can be easily verified, that for $d \leq 2$ correlation functions have IR divergences when the parameter h goes to zero, a property that is consistent with the absence of spontaneous symmetry breaking for $d \leq 2$. However, perturbative calculations can be organized in a different way: one does not introduce a symmetry breaking term but, rather, a set of *collective coordinates* that parametrize the set of classical minima. In a finite volume with periodic boundary conditions, one then expands as usual starting from a fixed minimum, but only the modes of the field that do not correspond to a global rotation (the non-zero momentum modes) are taken into account in the perturbative expansion. Eventually, one sums exactly over all classical minima. Clearly, after this last summation, only $O(N)$ invariant correlation functions survive. Elitzur has conjectured, and David has proved, that in two dimensions $O(N)$ invariant correlation functions obtained by this procedure have a regular expansion at low temperature. This implies that if one calculates perturbatively $O(N)$ invariant correlation functions in a non-zero field and then one takes the limit $h = 0$, this limit is finite.

To get an idea of the cancellation mechanism that leads to an IR finite expansion, one can calculate the $O(N)$ invariant two-point function at one-loop order:

$$\langle \sigma(x)\sigma(0) + \boldsymbol{\pi}(x) \cdot \boldsymbol{\pi}(0)\rangle = 1 - \tfrac{1}{2}\left\langle (\boldsymbol{\pi}(x) - \boldsymbol{\pi}(0))^2 \right\rangle + O(\pi^4)$$
$$= 1 - (N-1)T \int \frac{d^d p}{(2\pi)^d} \left(\frac{1 - e^{ipx}}{p^2}\right) + O(T^2).$$

One observes that the numerator in the integrand vanishes at $p = 0$.

15.10.2 The Abelian model $N = 2$, $d > 2$

We have seen that the non-linear σ-model with $O(2)$ symmetry is special because the RG function β reduces, near two dimensions, in the low-temperature expansion, to the dimensional term $(d-2)g$. Therefore, the properties, from the RG viewpoint, are quite different from those of the generic non-linear σ-model with $O(N)$ symmetry, $N > 2$. For example, in two dimensions the $O(2)$ symmetric model is not asymptotically free.

Actually, if one parametrizes the spin $\hat{\phi}(x)$ in terms of an angle θ as

$$\hat{\phi}(x) = \{\cos\theta(x), \sin\theta(x)\},$$

the configuration energy of the non-linear σ-model in zero field becomes the Euclidean action of a free massless field, which is also the Hamiltonian of the Gaussian critical model. Indeed,

$$\partial_\mu \cos\theta(x) = -\partial_\mu\theta(x)\sin\theta(x), \qquad \partial_\mu \sin\theta(x) = \partial_\mu\theta(x)\cos\theta(x)$$

and, thus,

$$\mathcal{E}(\theta) = \frac{\mu^{d-2}}{2g} \int d^d x \sum_\mu [\partial_\mu \theta(x)]^2.$$

The β-function thus reduces to the dimensional term $(d-2)g$ to all orders.

The origin of this difference can be traced back to the local structure of the field manifold: for $N = 2$ the sphere S_{N-1} reduces to a circle, which is a flat manifold, and cannot be distinguished locally from a straight line. The action has a larger symmetry than the initial lattice action since it is invariant under any translation of the θ field by a constant.

Since θ is a free massless field, the correlation length remains infinite for all g. This property has to be contrasted with a simple high-temperature analysis of the corresponding spin model on a lattice, which shows that the correlation length is finite at high temperature and implies the existence of a phase transition at some finite temperature. The Hamiltonian of the σ-model thus cannot describe the large-distance properties of the lattice model for all temperatures.

The origin of this discrepancy can be found in the cyclic nature of the angular θ field, a property that is apparent in the regularized lattice model but is no longer present in the continuum σ-model, which has the larger, non-compact, translation symmetry. At low temperature, the lattice model and the continuum theory differ only by contributions exponentially small in $1/g$, which play no role. But at higher temperature, they drive the phase transition.

15.10.3 The Abelian model in two dimensions

A new problem then arises: as shown in Section 8.2.2, the massless two-point function of the θ field does not exist in two dimensions. However, here the physical correlation functions are not the correlations functions of the field θ but, instead, those of the periodic functions $\sin\theta$ or $\cos\theta$ or equivalently of $e^{\pm i\theta}$. As shown below, these correlation functions exist provided they have the symmetry of the action, that is, are invariant under constant translations of the θ field.

In two dimensions, the β-function then vanishes identically. But, even in a free field theory, composite fields must be renormalized and the $e^{\pm i\theta}$ fields have a non--Gaussian dimension for all temperatures g.

This property can be verified by an explicit calculation of the corresponding correlation functions. Since the massless propagator does not exist, it is convenient to calculate first in the massive theory corresponding to the Hamiltonian

$$\mathcal{E}(\theta) = \frac{1}{2g} \int d^2 x \left[\sum_\mu [\partial_\mu \theta(x)]^2 + m^2 \theta^2(x) \right].$$

All correlation functions of $\sin \theta$ and $\cos \theta$ can be obtained from the quantities

$$\left\langle \prod_{i=1}^n e^{i\epsilon_i \theta(x_i)} \right\rangle = \int [d\theta] \exp\left[-\mathcal{E}(\theta) + i \sum_i \epsilon_i \theta(x_i) \right] \qquad (15.74)$$

with $\epsilon_i = \pm 1$. Setting

$$J(x) = i \sum_i \epsilon_i \delta(x - x_i),$$

one can rewrite the integral as

$$\left\langle \prod_{i=1}^n e^{i\epsilon_i \theta(x_i)} \right\rangle = \int [d\theta] \exp\left[-\mathcal{E}(\theta) + \int d^2 x \, J(x) \theta(x) \right].$$

The integration yields

$$\left\langle \prod_{i=1}^n e^{i\epsilon_i \theta(x_i)} \right\rangle = \exp\left[\frac{g}{2} \int d^2 x \, d^2 y \, J(x) \Delta(x - y) J(y) \right]$$

$$= \exp\left[-\frac{g}{2} \sum_{i,j} \epsilon_i \epsilon_j \Delta(x_i - x_j) \right], \qquad (15.75)$$

where $\Delta(x)$ is the massive propagator

$$\Delta(x) = \frac{1}{(2\pi)^2} \int \frac{d^2 p}{p^2 + m^2} e^{ipx}.$$

In the limit $m \to 0$, the propagator diverges (see equation $(A7)$). The sum of all divergent contributions is

$$\frac{g}{4\pi} \ln m \left(\sum_i \epsilon_i \right)^2.$$

Thus, in the limit $m \to 0$ only the correlation functions such that $\sum_i \epsilon_i = 0$ do not vanish. These functions are invariant under the transformations of the $SO(2)$ symmetry group, which translate θ. One concludes that the $SO(2)$ symmetry is not broken at low temperature, even though the phase is massless (actually, it is the complete translation symmetry that is not broken).

The invariant two-point function has the limit

$$\left\langle e^{i\theta(x)-i\theta(0)} \right\rangle \propto x^{-g/2\pi}. \tag{15.76}$$

Despite the absence of symmetry breaking, the functions have an algebraic behaviour at low temperature with a variable exponent

$$\eta = g/2\pi \Rightarrow d_\phi = g/4\pi, \tag{15.77}$$

a behaviour generally associated with a line of fixed points.

Finally, let us point out that in two dimensions it is possible to construct another field theory that, at low temperature, reduces to a free theory, but which incorporates the property that θ is an angular variable. This leads to the study of the sine-Gordon model, which displays a phase transition, the famous Kosterlitz–Thouless (KT) transition. This transition separates a phase with an infinite correlation length, but *without order* (the phase of the low-temperature $O(2)$ model) from a phase with a finite correlation length.

15.11 The $(\phi^2)^2$ field theory at low temperature

Let us first emphasize that dimensional regularization (see Section 12.5.2) is assumed in the algebraic calculations that follow, but a lattice version requires only simple modifications.

At low temperature, that is, at T fixed, $T < T_c$, in a system in which a discrete symmetry is broken spontaneously, connected correlation functions decay exponentially at large distance. The situation is quite different when the symmetry is continuous: the presence of Goldstone modes induces a non-trivial large-distance physics for all temperatures in the ordered phase. This physics is described by the non-linear σ-model.

We now verify this property directly in continuum space, by studying the $(\phi^2)^2$ field theory, corresponding to the Hamiltonian

$$\mathcal{H}(\phi) = \int d^d x \left[\frac{1}{2} \sum_\mu (\partial_\mu \phi)^2 + \frac{1}{2} r \phi^2 + \frac{1}{4!} u \left(\phi^2\right)^2 \right], \tag{15.78}$$

in the low-temperature phase ($r < r_c$). The partition function is given by the field integral

$$\mathcal{Z} = \int [d\phi(x)] \exp[-\mathcal{H}(\phi)].$$

We change variables, setting
$$\phi(x) = \rho(x)\hat{\phi}(x) \quad \text{with} \quad \hat{\phi}^2(x) = 1. \tag{15.79}$$

The field integral becomes
$$\mathcal{Z} = \int \left[\rho^{N-1}(x)\mathrm{d}\rho(x)\right] \left[\mathrm{d}\hat{\phi}(x)\right] \exp\left[-\mathcal{H}(\rho,\hat{\phi})\right] \tag{15.80}$$

with
$$\mathcal{H}(\rho,\hat{\phi}) = \int \mathrm{d}^d x \left\{ \tfrac{1}{2}\rho^2(x)\left[\partial_\mu\hat{\phi}(x)\right]^2 + \tfrac{1}{2}[\partial_\mu\rho(x)]^2 + \tfrac{1}{2}r\rho^2 + \frac{1}{4!}u\rho^4 \right\}. \tag{15.81}$$

In the ordered phase, at T fixed below T_{c}, the field $\rho(x)$ has a non-vanishing expectation value and is massive (the correlation length is finite); its dynamics thus is not critical. The integration over the field $\rho(x)$ generates a local effective configuration energy $\mathcal{E}(\hat{\phi})$ for the field $\hat{\phi}$:

$$\exp\left[-\mathcal{E}(\hat{\phi})\right] = \int \left[\rho^{N-1}(x)\mathrm{d}\rho(x)\right] \exp\left[-\mathcal{E}(\rho,\hat{\phi})\right]. \tag{15.82}$$

Moreover, the field integral (15.82) can be calculated perturbatively. At leading order, the field $\rho(x)$ can be replaced by its expectation value M, which in the tree approximation is given by
$$r + \tfrac{1}{6}uM^2 = 0.$$

One then obtains
$$\mathcal{E}^{(0)}(\hat{\phi}) = \mathrm{const.} + \frac{1}{2}M^2 \int \mathrm{d}^d x \left[\partial_\mu\hat{\phi}(x)\right]^2. \tag{15.83}$$

One recognizes the configuration energy of the non-linear σ-model.
To evaluate loop corrections, one then sets
$$\rho(x) = M + \rho'(x) \tag{15.84}$$

In terms of ρ', the Hamiltonian (15.81) reads
$$\mathcal{H}(\rho',\hat{\phi}) = \int \mathrm{d}^d x \left\{ \tfrac{1}{2}\left(M^2 + 2M\rho' + \rho'^2\right)\left[\partial_\mu\hat{\phi}(x)\right]^2 + \tfrac{1}{2}[\partial_\mu\rho'(x)]^2 + \tfrac{1}{2}r(M+\rho')^2 \right.$$
$$\left. + \frac{1}{4!}u(M+\rho')^4 \right\}.$$

The loop corrections coming from the integration over ρ' generate additional interactions in $\hat{\phi}$. However, as long as one explores momenta much smaller than the mass of the field ρ or distances much larger than the corresponding correlation length, the effective configuration energy resulting from the integration over the field ρ can be expanded in local terms. The leading term at large distance is the term involving only two derivatives. Due to the $O(N)$ symmetry, it is proportional to $\mathcal{E}^{(0)}(\hat{\phi})$. The only effect of the integration, from the viewpoint of large-distance physics, is a renormalization of the coefficient M^2 that appears in equation (15.83). The other interactions involve at least four derivatives and correspond to irrelevant operators. Note that for a temperature T close to T_c, the domain in momentum space where these arguments apply is

$$|p_i| \ll (T_c - T)^\nu.$$

In such a limit, the non-linear σ-model (15.83) describes completely the properties at large distance of the $(\phi^2)^2$ field theory for T fixed, $T < T_c$. Moreover, the coefficient in front of $\mathcal{E}(\hat{\phi})$ becomes large at low temperature like in the lattice model.

16 Functional renormalization group

In this chapter, we briefly describe a general approach to the renormalization group (RG) close to ideas initially developed by Wegner and Wilson, and based on a partial integration over the large-momentum modes of fields. This RG takes the form of *functional renormalization group* (FRG) equations that express the equivalence between a change of a scale parameter related to microscopic physics and a change of the parameters of the Hamiltonian. Some forms of these renormalization group equations (RGE) are exact and one then also speaks of the *exact renormalization group*.

Polchinski has shown later that these FRG equations could also provide a different proof of the renormalizability of field theories.

More recently, several variants have been proposed that have led to new approximation schemes no longer based on the standard perturbative expansion but, for example, on a derivative expansion.

Technically, these FRG equations follow from identities that express the invariance of the partition function under a correlated change of the propagator and the other parameters of the Hamiltonian. We discuss these equations, in continuum space, in the framework of local statistical field theory. It is easy to verify that, except in the Gaussian case, these equations are closed only if all possible local interactions are included.

It is then possible to infer various RGE satisfied by correlation functions. Depending on the chosen form, these RGE are either exact or only exact at large distance or small momenta, up to corrections decaying faster than any power of the dilatation parameter.

In this chapter, we do not describe the various approximation schemes in which the RGE satisfied by the Hamiltonian have been solved but we show explicitly that, within the framework of the ε-expansion, these RGE also allow a perturbative calculation of RG functions as the β-function (whose definition has to be adapted to this more general framework), even if technically the usual quantum field theory methods are much more efficient.

16.1 Partial field integration and effective Hamiltonian

Using identities that involve only Gaussian integrations, we first prove equality between two partition functions corresponding to two different Hamiltonians. This relation can be interpreted as resulting from a partial integration over some components of the fields. We infer a sufficient condition for correlated modifications of the propagator and interactions, in a statistical field theory, to leave the partition function invariant.

In what follows, we assume, except if stated otherwise, that the field theory is translation invariant and thus that the propagators correspond to kernels of the form $\Delta(x-y)$.

16.1.1 Partial integration

We first establish a simple relation between partition functions corresponding to two Hamiltonians of the class considered in Section 9.1.1.

The first Hamiltonian depends on a field ϕ and we write it in the form

$$\mathcal{H}_1(\phi) = \frac{1}{2}\int dx\, dy\, \phi(x) K_1(x-y)\phi(y) + \mathcal{V}_1(\phi), \qquad (16.1)$$

where K_1 is a positive operator and the functional $\mathcal{V}_1(\phi)$ is expandable in powers of the field ϕ, local and translation invariant. To the explicit quadratic part is associated a Gaussian two-point function, or propagator, Δ_1 the inverse of K_1:

$$\int dz\, K_1(x-z)\Delta_1(z-y) = \delta(x-y).$$

The second Hamiltonian depends on two fields ϕ_1, ϕ_2 in the form

$$\mathcal{H}(\phi_1, \phi_2) = \frac{1}{2}\int dx\, dy\, [\phi_1(x) K_2(x-y)\phi_1(y) + \phi_2(x)\mathcal{K}(x-y)\phi_2(y)]$$
$$+ \mathcal{V}_1(\phi_1 + \phi_2). \qquad (16.2)$$

Again, we define

$$\int dz\, K_2(x-z)\Delta_2(z-y) = \delta(x-y), \quad \int dz\, \mathcal{K}(x-z)\mathcal{D}(z-y) = \delta(x-y).$$

The kernels K_1, K_2 and $\mathcal{K}$ are positive, which is a necessary condition for the field integrals to exist, at least in a perturbative sense. Moreover, the kernels Δ_1, Δ_2 and $\mathcal{D}$ (thus also positive) have the form of the regularized propagators (12.30) in order to ensure the existence of a formal expansion of the field integrals in powers of the interaction $\mathcal{V}_1$.

Then, if

$$\Delta_1 = \Delta_2 + \mathcal{D} \quad \Rightarrow \quad K_1 = K_2(K_2 + \mathcal{K})^{-1}\mathcal{K}, \qquad (16.3)$$

the ratio of the partition functions

$$\mathcal{Z}_1 = \int [d\phi]\, e^{-\mathcal{H}_1(\phi)}$$

and

$$\mathcal{Z}_2 = \int [d\phi_1\, d\phi_2]\, e^{-\mathcal{H}(\phi_1,\phi_2)} \qquad (16.4)$$

does not depend on $\mathcal{V}_1$:

$$\mathcal{Z}_2 = \left(\frac{\det(\mathcal{D}\Delta_2)}{\det \Delta_1}\right)^{1/2} \mathcal{Z}_1, \qquad (16.5)$$

where factors $\sqrt{2\pi}$ have been integrated in the integration measure and a regularization allowing a proper definition of the ratio of determinants is implicit.

Notation. In what follows, we use the compact notation

$$\int \mathrm{d}x\, \mathrm{d}y\, \phi(x) K(x-y) \phi(y) \equiv (\phi K \phi). \qquad (16.6)$$

Proof. After the change of variables $\{\phi_1, \phi_2\} \mapsto \{\phi_1, \phi = \phi_2 + \phi_1\}$, the integral (16.4) takes the form

$$\mathcal{Z}_2 = \int [\mathrm{d}\phi]\, \mathrm{e}^{-\mathcal{V}_1(\phi)}\, \mathcal{Z}(\phi)$$

with

$$\mathcal{Z}(\phi) = \int [\mathrm{d}\phi_1] \exp\left\{-\tfrac{1}{2}\left[(\phi_1 K_2 \phi_1) + ((\phi - \phi_1) \mathcal{K}(\phi - \phi_1))\right]\right\}$$

$$= \int [\mathrm{d}\phi_1] \exp\left\{-\tfrac{1}{2}\left[(\phi_1 (K_2 + \mathcal{K}) \phi_1) - 2(\phi \mathcal{K} \phi_1) + (\phi \mathcal{K} \phi)\right]\right\}.$$

After the change of variables

$$\phi_1 \mapsto \chi = \phi_1 - (K_2 + \mathcal{K})^{-1} \mathcal{K} \phi,$$

the Gaussian integration over χ yields

$$\mathcal{Z}(\phi) = (\det(K_2 + \mathcal{K}))^{-1/2} \exp\left[-\tfrac{1}{2}(\phi K_1 \phi)\right].$$

Another form of the identity. One can also define

$$\mathrm{e}^{-\mathcal{V}_2(\phi)} = (\det \mathcal{D})^{-1/2} \int [\mathrm{d}\varphi] \exp\left[-\tfrac{1}{2}(\varphi \mathcal{K} \varphi) - \mathcal{V}_1(\phi + \varphi)\right], \qquad (16.7)$$

as well as

$$\mathcal{H}_2(\phi) = \tfrac{1}{2}(\phi K_2 \phi) + \mathcal{V}_2(\phi).$$

Then, identity (16.5) can be rewritten as

$$\int [\mathrm{d}\phi]\, \mathrm{e}^{-\mathcal{H}_2(\phi)} = \left(\frac{\det \Delta_2}{\det \Delta_1}\right)^{1/2} \int [\mathrm{d}\phi]\, \mathrm{e}^{-\mathcal{H}_1(\phi)}. \qquad (16.8)$$

This identity can be interpreted as resulting from a partial integration over the field ϕ since the propagator $\mathcal{D}$ is positive and, thus, in the sense of operators, $\Delta_2 < \Delta_1$.

16.1.2 Differential form

We now assume that the propagator Δ is a function of a real parameter s: $\Delta \equiv \Delta(s)$. Moreover, $\Delta(s)$ is C^∞ with a negative derivative. We define

$$D(s) = \frac{d\Delta(s)}{ds} < 0, \tag{16.9}$$

where $D(s)$ is represented by the kernel $D(s; x - y)$.

For $s < s'$, we identify

$$\Delta_1 = \Delta(s), \quad \Delta_2 = \Delta(s') \quad \text{and thus} \quad \mathcal{D}(s,s') = \Delta(s) - \Delta(s') > 0. \tag{16.10}$$

Similarly,

$$K_1 = K(s) = \Delta^{-1}(s), \quad K_2 = K(s'), \quad \mathcal{K}(s,s') = [\mathcal{D}(s,s')]^{-1} > 0. \tag{16.11}$$

Since the kernels $K(s)$, $\mathcal{K}(s, s')$ are positive, all Gaussian integrals are defined.

Finally, we set

$$\mathcal{V}_1(\phi) = \mathcal{V}(\phi, s), \quad \mathcal{V}_2(\phi) = \mathcal{V}(\phi, s'), \quad \mathcal{H}_1(\phi) = \mathcal{H}(\phi, s), \quad \mathcal{H}_2(\phi) = \mathcal{H}(\phi, s').$$

Identity (16.8) then takes the form

$$\int [d\phi] \, e^{-\mathcal{H}(\phi, s')} = \left(\frac{\det \Delta(s')}{\det \Delta(s)} \right)^{1/2} \int [d\phi] \, e^{-\mathcal{H}(\phi, s)}.$$

Equation (16.7) becomes

$$e^{-\mathcal{V}(\phi, s')} = \left(\det \mathcal{D}(s, s') \right)^{-1/2} \int [d\varphi] \exp\left[-\tfrac{1}{2}(\varphi \mathcal{K}(s, s')\varphi) - \mathcal{V}(\phi + \varphi, s)\right]. \tag{16.12}$$

We set

$$s' = s + \sigma, \quad \sigma > 0,$$

and consider the $\sigma \to 0$ limit. Then,

$$\mathcal{K}^{-1}(s, s + \sigma) = \mathcal{D}(s, s + \sigma) = -\sigma D(s) + O(\sigma^2). \tag{16.13}$$

In the limit $\sigma \to 0$, in expression (16.12) the term quadratic in the field φ is multiplied by a factor $1/\sigma$ and, thus, the values of the field φ contributing to the integral are of order $\sqrt{\sigma}$. We thus expand $\mathcal{V}(\phi + \varphi, s)$ in powers of φ:

$$\mathcal{V}(\phi + \varphi, s) = \mathcal{V}(\phi, s) + \int dx \frac{\delta \mathcal{V}(\phi, s)}{\delta \phi(x)} \varphi(x)$$
$$+ \frac{1}{2} \int dx \, dy \frac{\delta^2 \mathcal{V}(\phi, s)}{\delta \phi(x) \delta \phi(y)} \varphi(x)\varphi(y) + O\left(\varphi^3\right).$$

We expand $e^{-\mathcal{V}(\phi+\varphi,s)}$ correspondingly. In relation (16.12), the expansion of the integral in powers of σ reduces to a sum of Gaussian integrals that can be evaluated using Wick's theorem. Multiplying the two sides by $e^{\mathcal{V}(\phi,s)}$, one infers from relation (16.12)

$$e^{\mathcal{V}(\phi,s)-\mathcal{V}(\phi,s+\sigma)}$$
$$= 1 + \frac{1}{2}\int dx\,dy \left[\frac{\delta\mathcal{V}}{\delta\phi(x)}\frac{\delta\mathcal{V}}{\delta\phi(y)} - \frac{\delta^2\mathcal{V}}{\delta\phi(x)\delta\phi(y)}\right]\langle\varphi(x)\varphi(y)\rangle + O(\sigma^2),$$

where $\langle\varphi(x)\varphi(y)\rangle$ is the Gaussian expectation value corresponding, at this order, to the propagator $-\sigma D(s)$:

$$\langle\varphi(x)\varphi(y)\rangle = -\sigma D(s; x-y) + O(\sigma^2).$$

Moreover,

$$e^{\mathcal{V}(\phi,s)-\mathcal{V}(\phi,s+\sigma)} = 1 - \sigma\frac{d}{ds}\mathcal{V}(\phi,s) + O(\sigma^2).$$

Identifying the terms of order σ, one obtains the fundamental equation

$$\frac{d}{ds}\mathcal{V}(\phi,s) = -\frac{1}{2}\int dx\,dy\,D(s;x-y)\left[\frac{\delta^2\mathcal{V}}{\delta\phi(x)\delta\phi(y)} - \frac{\delta\mathcal{V}}{\delta\phi(x)}\frac{\delta\mathcal{V}}{\delta\phi(y)}\right]. \qquad (16.14)$$

The equation expresses a *sufficient condition* for the partition function

$$\mathcal{Z}(s) = (\det\Delta(s))^{-1/2}\int[d\phi]\,e^{-\mathcal{H}(\phi,s)} \text{ with}$$
$$\mathcal{H}(\phi,s) = \tfrac{1}{2}(\phi K(s)\phi) + \mathcal{V}(\phi,s), \qquad (16.15)$$

to be independent of the parameter s.

This property relates a modification of the propagator to a modification of the interaction, completely in the spirit of the RG.

Remarks.
(i) If one sets

$$\mathcal{V}(\phi,s) = \mathcal{V}(\phi=0,s) + \mathcal{V}'(\phi,s),$$

the term $\mathcal{V}(\phi=0,s)$ decouples and the equation reduces to

$$\frac{d}{ds}\mathcal{V}'(\phi,s) = -\frac{1}{2}\int dx\,dy\,D(s;x-y)\left[\frac{\delta^2\mathcal{V}'}{\delta\phi(x)\delta\phi(y)} - \frac{\delta\mathcal{V}'}{\delta\phi(x)}\frac{\delta\mathcal{V}'}{\delta\phi(y)}\right] - (\phi=0). \qquad (16.16)$$

In what follows, this is the form we generally use in explicit calculations, the subtractions of $\mathcal{V}(\phi)$ and of the equation being implicit to keep the notation simple.

(ii) A sufficient condition for $\mathcal{Z}(s)$ to be independent of s, is that equation (16.14) is satisfied as an expectation value with the measure $e^{-\mathcal{H}(\phi,s)}$. It is thus possible to add to the equation contributions with vanishing expectation value to derive other sufficient conditions.

16.1.3 Hamiltonian evolution

From equation (16.14), one also infers the evolution of the Hamiltonian (16.15). First,
$$\frac{d}{ds}\mathcal{H}(\phi,s) = \frac{d}{ds}\mathcal{V}(\phi,s) - \frac{1}{2}(\phi\Delta^{-1}(s)D(s)\Delta^{-1}(s)\phi).$$

Moreover,
$$\frac{\delta\mathcal{V}}{\delta\phi(x)} = \frac{\delta\mathcal{H}}{\delta\phi(x)} - [\Delta^{-1}(s)\phi](x), \quad \frac{\delta^2\mathcal{V}}{\delta\phi(x)\delta\phi(y)} = \frac{\delta^2\mathcal{H}}{\delta\phi(x)\delta\phi(y)} - [\Delta^{-1}(s)](x-y).$$

One infers
$$\int dx\,dy\, D(s; x-y) \frac{\delta\mathcal{V}}{\delta\phi(x)}\frac{\delta\mathcal{V}}{\delta\phi(y)}$$
$$= \int dx\,dy\, \left[D(s; x-y) \frac{\delta\mathcal{H}}{\delta\phi(x)}\frac{\delta\mathcal{H}}{\delta\phi(y)} - 2\phi(x)L(s; x-y)\frac{\delta\mathcal{H}}{\delta\phi(y)} \right]$$
$$+ (\phi\Delta^{-1}(s)D(s)\Delta^{-1}(s)\phi),$$

where the operator $L(s)$, with kernel $L(s; x-y)$, is defined by
$$L(s) \equiv D(s)\Delta^{-1}(s) = \frac{d\ln\Delta(s)}{ds}. \tag{16.17}$$

Using these equations in equation (16.14), one verifies that the two terms quadratic in ϕ cancel. One then finds
$$\frac{d}{ds}\mathcal{H}(\phi,s) = -\frac{1}{2}\int dx\,dy\, D(s; x-y) \left[\frac{\delta^2\mathcal{H}}{\delta\phi(x)\delta\phi(y)} - \frac{\delta\mathcal{H}}{\delta\phi(x)}\frac{\delta\mathcal{H}}{\delta\phi(y)} \right]$$
$$- \int dx\,dy\, \phi(x) L(s; x-y) \frac{\delta\mathcal{H}}{\delta\phi(y)} + \frac{1}{2}\operatorname{tr} L(s). \tag{16.18}$$

16.1.4 Connected functional and formal solution

Conversely, we now solve equation (16.14), which is a first-order differential equation in s. The form of the functional $\mathcal{V}(\phi, s)$ for an initial value s_0 of the parameter s thus determines the solution for all $s \geq s_0$. From the proof of equation (16.14), one expects that the solution directly follows from equation (16.12):

$$e^{-\mathcal{V}(\phi,s)} = \left(\det \mathcal{D}(s_0, s)\right)^{-1/2} \int [d\varphi] \exp\left[-\tfrac{1}{2}(\varphi\mathcal{K}(s_0,s)\varphi) - \mathcal{V}(\phi+\varphi, s_0)\right]. \tag{16.19}$$

We now verify this property. As a byproduct, we give another proof of equation (16.14).

Connected functional. We change variables $\phi + \varphi \mapsto \varphi$ in the integral (16.19). The equation becomes

$$e^{-\mathcal{V}(\phi,s)} = \left(\det \mathcal{D}(s_0,s)\right)^{-1/2} e^{-\frac{1}{2}(\phi\mathcal{K}(s_0,s)\phi)}$$
$$\times \int [\mathrm{d}\varphi]\exp\left[-\tfrac{1}{2}(\varphi\mathcal{K}(s_0,s)\varphi) - \mathcal{V}(\varphi,s_0) + (\phi\mathcal{K}(s_0,s)\varphi)\right]. \quad (16.20)$$

The expression shows that $\mathcal{V}(\phi,s)$ is related to the generating functional $\mathcal{W}(H,s)$ of connected correlation functions of the field φ:

$$\mathcal{V}(\phi,s) = \tfrac{1}{2}\ln\det\mathcal{D}(s_0,s) + \tfrac{1}{2}(\phi\mathcal{K}(s_0,s)\phi) - \mathcal{W}(\mathcal{K}\phi,s) \quad (16.21)$$

with the notation $[\mathcal{K}\phi](x) \equiv \int \mathrm{d}y\, \mathcal{K}(s_0,s;x-y)\phi(y)$ and

$$e^{\mathcal{W}(H,s)} = \int [\mathrm{d}\varphi]\exp\left[-\tfrac{1}{2}(\varphi\mathcal{K}(s_0,s)\varphi) - \mathcal{V}(\varphi,s_0) + \int \mathrm{d}x\, H(x)\varphi(x)\right].$$

This representation suggests another, algebraically equivalent, proof of identity (16.14), but which does not reveal its deeper meaning.

Another derivation of equation (16.14). We start from the definition

$$e^{\mathcal{W}(H,s)} = \int [\mathrm{d}\varphi]\exp\left[-\mathcal{H}(\varphi,s) + \int \mathrm{d}x\, H(x)\varphi(x)\right] \quad \text{with}$$
$$\mathcal{H}(\varphi,s) = \tfrac{1}{2}(\varphi\mathcal{K}(s)\varphi) + \mathcal{V}(\varphi),$$

where we assume that the kernel $\mathcal{K}$ depends on a parameter s and is differentiable. Differentiating both sides with respect to s, one obtains

$$\frac{\mathrm{d}\mathcal{W}}{\mathrm{d}s} e^{\mathcal{W}(H,s)} = -\frac{1}{2}\int [\mathrm{d}\varphi]\,(\varphi\mathcal{K}'(s)\varphi)\, e^{-\mathcal{H}(\varphi,s)+\int \mathrm{d}x\, H(x)\varphi(x)}.$$

Using inside the integral the usual identity

$$\frac{\delta}{\delta H(x)}\exp\left[\int \mathrm{d}x\, H(x)\varphi(x)\right] = \varphi(x)\exp\left[\int \mathrm{d}x\, H(x)\varphi(x)\right], \quad (16.22)$$

one can rewrite the equation as

$$\frac{\mathrm{d}\mathcal{W}}{\mathrm{d}s}e^{\mathcal{W}(H,s)} = -\frac{1}{2}\int \mathrm{d}x\,\mathrm{d}y\,\frac{\delta}{\delta H(x)}\frac{\mathrm{d}\mathcal{K}(s;x-y)}{\mathrm{d}s}\frac{\delta}{\delta H(y)}e^{\mathcal{W}(H,s)},$$

and, thus, finally

$$\frac{\mathrm{d}\mathcal{W}}{\mathrm{d}s} = -\frac{1}{2}\int \mathrm{d}x\,\mathrm{d}y\,\frac{\mathrm{d}\mathcal{K}(s;x-y)}{\mathrm{d}s}\left[\frac{\delta^2\mathcal{W}}{\delta H(x)\delta H(y)} + \frac{\delta\mathcal{W}}{\delta H(x)}\frac{\delta\mathcal{W}}{\delta H(y)}\right]. \quad (16.23)$$

Using relation (16.21) with $\mathcal{K}(s) \equiv \mathcal{K}(s_0, s)$ to express $\mathcal{W}$ as a function of $\mathcal{V}$, after some algebra one recovers equation (16.14).

The functional $\mathcal{V}(\phi, s)$ thus follows directly from the calculation of the connected correlation functions corresponding to the propagator $\mathcal{D}(s_0, s)$ and the initial interaction $\mathcal{V}(\phi, s_0)$.

Moreover, this proof indicates that equation (16.14) is also a field equation, as the explicit calculations of Section 16.3 and the proofs of equations (16.14), (16.29) in Section A5.1 indeed confirm.

Solution of equation (16.23). We first linearize equation (16.23), setting

$$\mathcal{Z}(H, s) = e^{\mathcal{W}(H, s)}.$$

The equation becomes

$$\frac{d\mathcal{Z}}{ds} = -\frac{1}{2}\int dx\, dy\, \frac{d\mathcal{K}(s; x - y)}{ds}\, \frac{\delta^2 \mathcal{Z}}{\delta H(x)\delta H(y)}, \qquad (16.24)$$

a functional generalization of a diffusion or Fokker–Planck equation. In particular, one verifies that the kernel $d\mathcal{K}/ds$ is negative, so that the functional differential operator in the right-hand side is formally negative, like in a usual diffusion equation.

One may wonder why one does not work directly with this equation, which is much simpler since it is linear. The reason is the following: while it is possible to assume quite generally that $\mathcal{V}(\phi, s)$ and thus $\mathcal{W}(H, s)$, are functionals local in the field, this is not the case for $\mathcal{Z}(H, s)$ and, thus, it becomes difficult to implement the locality condition. We show later that this locality condition plays an essential role in the perturbative solution of equation (16.33).

The solution of the equation is based on a functional Fourier transformation. One sets

$$\mathcal{Z}(H, s) = \int [d\varphi]\, e^{-\mathcal{H}(\varphi, s) + \int dx\, H(x)\varphi(x)}.$$

Using the remark (16.22), one transforms equation (16.24) into an equation for $\mathcal{H}$:

$$\frac{d\mathcal{H}}{ds} = \frac{1}{2}\int dx\, dy\, \frac{d\mathcal{K}(s; x - y)}{ds}\, \varphi(x)\varphi(y).$$

The solution then is immediate:

$$\mathcal{H}(\varphi, s) = \tfrac{1}{2}\bigl(\varphi \mathcal{K}(s)\varphi\bigr) + \mathcal{V}(\varphi),$$

where $\mathcal{V}(\varphi)$ is independent of s and thus determined by the boundary condition. For the choice $\mathcal{K}(s) \equiv \mathcal{K}(s_0, s)$, one finds $\mathcal{V}(\varphi) \equiv \mathcal{V}(\varphi, s_0)$.

16.1.5 Correlation functions

Introducing an additional classical field $\chi(x)$, we substitute
$$\mathcal{V}_1(\phi) \mapsto \mathcal{V}_1(\phi + \chi)$$
in equation (16.8). The identity (16.7) then shows that the transform of $\mathcal{V}_1(\phi+\chi)$ is $\mathcal{V}_2(\phi+\chi)$. The corresponding Hamiltonian reads
$$\mathcal{H}_1(\phi,\chi) = \tfrac{1}{2}(\phi K_1 \phi) + \mathcal{V}_1(\phi + \chi).$$

After the change of variables $\phi + \chi \mapsto \phi$ in the field integral, one finds
$$\mathcal{H}_1(\phi,\chi) \mapsto \tfrac{1}{2}(\phi K_1 \phi) + \mathcal{V}_1(\phi) - (\phi K_1 \chi) + \tfrac{1}{2}(\chi K_1 \chi)$$
$$= \mathcal{H}_1(\phi) - (\phi K_1 \chi) + \tfrac{1}{2}(\chi K_1 \chi).$$

One then obtains a relation between generating functionals of correlation functions
$$\mathcal{Z}_1(\chi) = \int [\mathrm{d}\phi]\, e^{-\mathcal{H}_1(\phi)+(\phi K_1 \chi)-(\chi K_1 \chi)/2} = \left(\frac{\det \Delta_1}{\det \Delta_2}\right)^{1/2} \mathcal{Z}_2(\chi) \qquad (16.25)$$

with
$$\mathcal{Z}_2(\chi) = \int [\mathrm{d}\phi]\, e^{-\mathcal{H}_2(\phi)+(\phi K_2 \chi)-(\chi K_2 \chi)/2}.$$

Then, choosing
$$K_1 = K(s), \quad K_2 = K(s_0)$$
and setting
$$\chi(x) = [\Delta(s_0) H](x),$$
one verifies that the functional
$$\mathcal{Z}(H,s) = \left(\frac{\det \Delta(s_0)}{\det \Delta(s)}\right)^{1/2} \int [\mathrm{d}\phi] \exp\left[-\mathcal{H}(\phi,s) + (\phi K(s)\Delta(s_0) H)\right.$$
$$\left. - (H\Delta(s_0)(K(s)-K(s_0))\Delta(s_0)H)/2\right] \qquad (16.26)$$

is independent of s and satisfies the boundary condition
$$\mathcal{Z}(H,s_0) = \int [\mathrm{d}\phi] \exp\left[-\mathcal{H}(\phi,s_0) + \int \mathrm{d}x\, \phi(x) H(x)\right]. \qquad (16.27)$$

16.1.6 Field renormalization

In order to be able to find RG fixed-point solutions, it is necessary to introduce a field renormalization. To prove the corresponding identities, we set

$$\phi(x) = \sqrt{Z(s)}\phi'(x), \quad \mathcal{H}(\phi, s) = \mathcal{H}'(\phi', s),$$

where $Z(s)$ is an arbitrary differentiable function. Then, using the chain rule,

$$\frac{d}{ds}\mathcal{H}(\phi, s) = \frac{d}{ds}\mathcal{H}'(\phi', s) + \int dx \, \frac{\delta \mathcal{H}'(\phi')}{\delta \phi'(x)} \frac{\partial \phi'(x)}{\partial s}$$

$$= \frac{d}{ds}\mathcal{H}'(\phi', s) - \frac{1}{2}\eta(s) \int dx \, \phi'(x) \frac{\delta \mathcal{H}'(\phi')}{\delta \phi'(x)},$$

where we have defined

$$\eta(s) = \frac{d \ln Z(s)}{ds}. \tag{16.28}$$

Omitting now the primes, we infer from the flow equation (16.18) the new equation (omitting the constant term)

$$\frac{d}{ds}\mathcal{H}(\phi, s) = -\frac{1}{2} \int dx \, dy \, D(x-y) \left[\frac{\delta^2 \mathcal{H}}{\delta\phi(x)\delta\phi(y)} - \frac{\delta \mathcal{H}}{\delta\phi(x)} \frac{\delta \mathcal{H}}{\delta\phi(y)} \right]$$
$$- \int dx \, dy \, \phi(x) L(s; x-y) \frac{\delta \mathcal{H}}{\delta\phi(y)} + \frac{1}{2}\eta(s) \int dx \, \phi(x) \frac{\delta \mathcal{H}}{\delta\phi(x)}. \tag{16.29}$$

16.2 High-momentum mode integration and RG equations

The equivalence (16.8) or the equations (16.19) and (16.14) that follow, lead to the RGE. Indeed, as we shall verify, they can be applied to a situation where the partial integration over the field corresponds to a partial integration over its high--momentum modes, which in position space also corresponds to an integration over short-distance degrees of freedom.

In what follows, we specialize Δ to a critical propagator. A possible deviation to the critical theory is included in $\mathcal{V}(\phi)$. Moreover, we consider explicitly the case of d-dimensional space.

Cut-off parameter and propagator. In the preceding formalism, we now identify $s \equiv -\ln \Lambda$, where Λ is a large-momentum cut-off, which also represents the inverse of the microscopic scale. A variation of s then corresponds to a dilatation of the parameter Λ.

We choose a regularized propagator Δ_Λ of the form

$$\Delta_\Lambda(x) = \int \frac{d^d k}{(2\pi)^d} e^{-ikx} \tilde{\Delta}_\Lambda(k) \quad \text{with} \quad \tilde{\Delta}_\Lambda(k) = \frac{C(k^2/\Lambda^2)}{k^2}. \tag{16.30}$$

The function $C(t)$ is regular for $t \geq 0$, positive, decreasing, goes to 1 for $t \to 0$ and goes to zero faster than any power for $t \to \infty$. The field Fourier components corresponding to momenta much higher than Λ thus give a very small contribution to the field integral.

We also introduce the derivative $D_\Lambda(x)$ whose Fourier transform is given by

$$\tilde{D}_\Lambda(k) = -\Lambda \frac{\partial \tilde{\Delta}_\Lambda(k)}{\partial \Lambda} = \frac{2}{\Lambda^2} C'(k^2/\Lambda^2). \tag{16.31}$$

Let us stress an essential property of the function $\tilde{D}_\Lambda$: *it has no pole at $k = 0$ and thus it is not critical*. The function

$$D_\Lambda(x) = -\Lambda \frac{\partial \Delta_\Lambda(x)}{\partial \Lambda} = \int \frac{\mathrm{d}^d k}{(2\pi)^d} \, \mathrm{e}^{ikx} \, \tilde{D}_\Lambda(k) = \Lambda^{d-2} D_{\Lambda=1}(\Lambda x), \tag{16.32}$$

thus decays for $|x| \to \infty$ faster than any power if $C(t)$ is C^∞, exponentially if $C(t)$ is analytical. The same property is verified by the propagator

$$\mathcal{D}(\Lambda_0, \Lambda) = D_{\Lambda_0} - D_\Lambda \quad \text{with} \quad \Lambda_0 > \Lambda,$$

whose inverse now appears in the field integral (16.19).

16.2.1 RGE

With these assumptions and definitions, equation (16.16) becomes

$$\Lambda \frac{\mathrm{d}}{\mathrm{d}\Lambda} \mathcal{V}(\phi, \Lambda) = \frac{1}{2} \int \mathrm{d}^d x \, \mathrm{d}^d y \, D_\Lambda(x-y) \left[\frac{\delta^2 \mathcal{V}}{\delta\phi(x)\delta\phi(y)} - \frac{\delta\mathcal{V}}{\delta\phi(x)} \frac{\delta\mathcal{V}}{\delta\phi(y)} \right], \tag{16.33}$$

where the Fourier transform of $D_\Lambda(x)$ is given by equation (16.31).

Equation (16.33) being exact, one uses also the terminology *exact renormalization group*.

Fundamental remark. Since $D_\Lambda(x)$ decreases faster than any power, if $\mathcal{V}(\phi)$ is initially local, it remains local, a property that will become more apparent when we expand the equation in powers of ϕ. Similarly, in the integrated form (16.19), the decay property of $\mathcal{D}(\Lambda_0, \Lambda)$ ensures the locality of $\mathcal{V}(\phi, \Lambda)$.

Other remarks.

(i) Equations (16.33) and (16.34) differ from the general equation (9.23) introduced in Chapter 9 by the property that the scale parameter Λ appears explicitly through the function D_Λ. We shall later eliminate this explicit dependence.

(ii) To study the existence of fixed points, one starts from an initial critical interaction $\mathcal{V}_0(\phi)$ at scale Λ_0 and uses equation (16.33) to calculate the effective interaction $\mathcal{V}(\phi, \Lambda)$ at scale $\Lambda < \Lambda_0$. A fixed point is defined by the property that $\mathcal{V}(\phi, \Lambda)$, after a suitably chosen renormalization of the field ϕ (which leads to equation (16.29)), has a limit for $\Lambda \ll \Lambda_0$.

(iii) Since equation (16.33) can also be derived from quantum field equations (see Appendix A5.1), in this formulation, partial integration over large-momentum modes does not imply a loss of information, unlike what would happen on a lattice in the case of an integration over short-distance modes.

16.2.2 Fourier representation

In terms of the Fourier components $\tilde{\phi}(k)$ of the field,

$$\phi(x) = \int d^d k \, e^{ikx} \, \tilde{\phi}(k) \iff \tilde{\phi}(k) = \int \frac{d^d x}{(2\pi)^d} e^{-ikx} \phi(x),$$

the functional derivative becomes (using the chain rule)

$$\frac{\delta}{\delta \phi(x)} = \int \frac{d^d k}{(2\pi)^d} e^{-ikx} \frac{\delta}{\delta \tilde{\phi}(k)}.$$

Introducing this representation into equation (16.33), one finds

$$\Lambda \frac{d}{d\Lambda} \mathcal{V}(\phi, \Lambda) = \frac{1}{2} \int \frac{d^d k}{(2\pi)^d} \tilde{D}_\Lambda(k) \left[\frac{\delta^2 \mathcal{V}}{\delta \tilde{\phi}(k) \delta \tilde{\phi}(-k)} - \frac{\delta \mathcal{V}}{\delta \tilde{\phi}(k)} \frac{\delta \mathcal{V}}{\delta \tilde{\phi}(-k)} \right]. \quad (16.34)$$

In the equation, locality translates into regularity of the Fourier components. If the coefficients of the expansion of $\mathcal{V}(\phi, \Lambda)$ in powers of $\tilde{\phi}$, after factorization of the δ function, are regular functions for an initial value of $\Lambda = \Lambda_0$, they remain for $\Lambda < \Lambda_0$ because $\tilde{D}_\Lambda(k)$ is a regular function of k.

Finally, equation (16.18) for the complete Hamiltonian expressed in terms of Fourier components,

$$\mathcal{H}(\phi, \Lambda) = \frac{1}{2}(2\pi)^d \int d^d k \, \tilde{\phi}(k) \tilde{\Delta}_\Lambda^{-1}(k) \tilde{\phi}(-k) + \mathcal{V}(\phi, \Lambda), \quad (16.35)$$

takes the form (omitting the term independent of ϕ)

$$\Lambda \frac{d}{d\Lambda} \mathcal{H}(\phi, \Lambda) = \frac{1}{2} \int \frac{d^d k}{(2\pi)^d} \tilde{D}_\Lambda(k) \left[\frac{\delta^2 \mathcal{H}}{\delta \tilde{\phi}(k) \delta \tilde{\phi}(-k)} - \frac{\delta \mathcal{H}}{\delta \tilde{\phi}(k)} \frac{\delta \mathcal{H}}{\delta \tilde{\phi}(-k)} \right]$$
$$+ \int \frac{d^d k}{(2\pi)^d} \tilde{L}_\Lambda(k) \frac{\delta \mathcal{H}}{\delta \tilde{\phi}(k)} \tilde{\phi}(k) \quad (16.36)$$

with (equation (16.17))

$$\tilde{L}_\Lambda(k) = \tilde{D}_\Lambda(k) / \tilde{\Delta}_\Lambda(k). \quad (16.37)$$

In equation (16.36), we have implicitly subtracted both from $\mathcal{H}(\phi, \Lambda)$ and from the equation their values at $\phi = 0$.

16.2.3 Expansion in powers of the field

We denote by $V^{(n)}$ and $\tilde{V}^{(n)}$ the coefficients of the expansion of $\mathcal{V}(\phi, \Lambda)$ in powers of $\phi(x)$ and of $\tilde{\phi}(p)$, the Fourier transform of the field, respectively:

$$\mathcal{V}(\phi, \Lambda) = \sum_{n=0}^\infty \frac{1}{n!} \int \prod_i d^d x_i \, \phi(x_i) V^{(n)}(x_1, \ldots, x_n) \quad (16.38)$$

$$= \sum_{n=0}^\infty \frac{1}{n!} \int \prod_i d^d p_i \, \tilde{\phi}(p_i) (2\pi)^d \delta^{(d)}(p_1 + \cdots + p_n) \tilde{V}^{(n)}(p_1, \ldots, p_n). \quad (16.39)$$

Equation (16.33) can then be expressed in terms of these components:

$$\Lambda \frac{\mathrm{d}}{\mathrm{d}\Lambda} V^{(n)}(x_1, x_2, \ldots, x_n)$$
$$= \frac{1}{2} \int \mathrm{d}^d x \, \mathrm{d}^d y \, D_\Lambda(x-y) \bigg[V^{(n+2)}(x_1, x_2, \ldots, x_n, x, y)$$
$$- \sum_I V^{(l+1)}(x_{i_1}, \ldots, x_{i_l}, x) V^{(n-l+1)}(x_{i_{l+1}}, \ldots, x_{i_n}, y) \bigg], \quad (16.40)$$

where the set $I \equiv \{i_1, i_2, \ldots, i_l\}$ describes all distinct subsets of $\{1, 2, \ldots, n\}$.

In the Fourier representation, equation (16.34) leads to

$$\Lambda \frac{\mathrm{d}}{\mathrm{d}\Lambda} \tilde{V}^{(n)}(p_1, p_2, \ldots, p_n)$$
$$= \frac{1}{2} \int \frac{\mathrm{d}^d k}{(2\pi)^d} \tilde{D}_\Lambda(k) \tilde{V}^{(n+2)}(p_1, p_2, \ldots, p_n, k, -k)$$
$$- \frac{1}{2} \sum_I \tilde{D}_\Lambda(p_0) \tilde{V}^{(l+1)}(p_{i_1}, \ldots, p_{i_l}, p_0) \tilde{V}^{(n-l+1)}(p_{i_{l+1}}, \ldots, p_{i_n}, -p_0), \quad (16.41)$$

where the momentum p_0 is determined by total momentum conservation.

Let us point out that the integration of these RGE shows that, even if the interaction at the initial scale Λ_0 is proportional to $u\phi^4$, at a generic scale Λ a general local interaction is generated because all functions $V^{(n)}$ are coupled. Nevertheless, in the spirit of the perturbative methods of field theory, it is possible to solve equation (16.33) as an expansion in powers of u with the Ansatz that the terms of $\mathcal{V}(\phi, \Lambda)$ quadratic and quartic in ϕ are of order u and the general term of degree $2n$ of order u^{n-1}. In Section 16.3, we calculate the first terms explicitly.

16.2.4 Correlation functions

With the addition to the interaction of a term corresponding to a linear coupling to an external field,

$$\mathcal{V}(\phi, \Lambda_0) \mapsto \mathcal{V}(\phi, \Lambda_0) - \int \mathrm{d}^d x \, H(x) \phi(x),$$

the partition function becomes a generating functional of correlation functions. However, equation (16.33) shows that $\mathcal{V}(\phi, \Lambda)$ then becomes, in general, a complicated functional of the external field $H(x)$.

Special external field. A first solution to the problem is the following: one chooses an external field H whose Fourier components $\tilde{H}(k)$ vanish for $k^2 \geq \Lambda^2$, together with a function $\tilde{D}_\Lambda(k)$ that vanishes identically for $k^2 \leq \Lambda^2$. This implies that $C'(t)$ vanishes identically for $t \leq 1$. Then $\int \mathrm{d}^d x \, H(x) \phi(x)$ does not contribute to the right-hand side of equation (16.33).

However, we then observe that the RG transformations are such that correlation functions calculated from Hamiltonians $\mathcal{H}(\phi)$ corresponding to different scales Λ, coincide only when all momenta are smaller than the smallest scale Λ. Correlation functions corresponding to different values of Λ then differ by functions that are indefinitely differentiable and thus, after Fourier transformation, by functions decaying, at large distance, faster than any power in a dilatation of space variables, as has been assumed in equation (9.9).

Moreover, the function $C(t)$ then cannot be analytical but only indefinitely differentiable, which affects the critical domain $\xi < \infty$, and the definition of universality becomes there more subtle. Finally, such functions are not well suited to explicit calculations.

RGE modification. It is also possible to choose analytic functions $C(t)$ and generic external fields linearly coupled to ϕ. The RGE obtained in this way then have an obvious problem: for example, in equation (16.36) the right-hand side involves a contribution of the form

$$\int \frac{\mathrm{d}^d k}{(2\pi)^d} \tilde{D}_\Lambda(k) \frac{\delta \mathcal{H}}{\delta \tilde{\phi}(k)} \tilde{H}(-k).$$

By successive iterations, one verifies that the equation generates a general even functional of H, ϕ, that is, invariant under the reflexion $\{H, \phi\} \mapsto \{-H, -\phi\}$.

However, we also note that this additional term is proportional to a functional derivative of $\mathcal{H}$ whose expectation value with the measure $\mathrm{e}^{-\mathcal{H}(\phi)}$ vanishes. This suggests modifying the equation by adding a contribution with zero expectation value. We thus substitute in equation (16.36),

$$\mathcal{H}(\phi) \mapsto \mathcal{H}(\phi, H) = \mathcal{H}(\phi) - (\phi \zeta H) + \tfrac{1}{2}(HUH), \tag{16.42}$$

where ζ and U are two kernels that are determined below. From the relation between derivatives

$$\frac{\delta \mathcal{H}(\phi, H)}{\delta \phi(x)} = \frac{\delta \mathcal{H}(\phi)}{\delta \phi(x)} - [\zeta H](x),$$

one infers

$$\int \mathrm{d}^d x\, \mathrm{d}^d y\, D_\Lambda(x-y) \left[\frac{\delta \mathcal{H}(\phi, H)}{\delta \phi(x)} \frac{\delta \mathcal{H}(\phi, H)}{\delta \phi(y)} - \frac{\delta \mathcal{H}(\phi)}{\delta \phi(x)} \frac{\delta \mathcal{H}(\phi)}{\delta \phi(y)} \right]$$
$$= -2 \int \mathrm{d}^d x\, \mathrm{d}^d y\, D_\Lambda(x-y) [\zeta H](x) \frac{\delta \mathcal{H}(\phi)}{\delta \phi(y)} + (H \zeta D_\Lambda \zeta H)$$

$$\int \mathrm{d}^d x\, \mathrm{d}^d y\, L_\Lambda(x-y) \phi(x) \frac{\delta \mathcal{H}(\phi, H)}{\delta \phi(y)}$$
$$= \int \mathrm{d}^d x\, \mathrm{d}^d y\, L_\Lambda(x-y) \phi(x) \frac{\delta \mathcal{H}(\phi)}{\delta \phi(y)} - (\phi L_\Lambda \zeta H).$$

To eliminate the additional $\mathcal{H}$ dependent term, we use

$$\left\langle \frac{\delta \mathcal{H}(\phi, H)}{\delta \phi(y)} \right\rangle_{\mathrm{e}^{-\mathcal{H}(\phi, H)}} = 0 = \left\langle \frac{\delta \mathcal{H}(\phi)}{\delta \phi(y)} - \int \mathrm{d}^d z\, \zeta(y-z) H(z) \right\rangle_{\mathrm{e}^{-\mathcal{H}(\phi, H)}}.$$

One sees that the form (16.42) is stable if

$$\Lambda \frac{\partial}{\partial \Lambda} \left[-(\phi \zeta H) + \tfrac{1}{2}(HUH) \right] = (H\zeta D_\Lambda \zeta H) - \tfrac{1}{2}(H\zeta D_\Lambda \zeta H) - (\phi L_\Lambda \zeta H).$$

One infers from the two equations, in the Fourier representation,

$$\Lambda \frac{\partial}{\partial \Lambda} \tilde\zeta(k,\Lambda) = \tilde L_\Lambda(k) \tilde\zeta(k,\Lambda), \qquad (16.43)$$

$$\Lambda \frac{\partial}{\partial \Lambda} \tilde U(k,\Lambda) = \tilde D_\Lambda(k) \tilde\zeta^2(k,\Lambda)$$

with the boundary conditions

$$\tilde\zeta(k,\Lambda_0) = 1, \quad \tilde U(k,\Lambda_0) = 0.$$

Using the definitions (16.31), (16.37), one finds the solution of the first equation:

$$\tilde\zeta(k,\Lambda) = \tilde\Delta_{\Lambda_0}(k)/\tilde\Delta_\Lambda(k),$$

which is a regular function of k at $k=0$. Finally,

$$\tilde U(k,\Lambda) = \tilde\Delta^2_{\Lambda_0}(k)/\tilde\Delta_\Lambda(k) - \tilde\Delta_{\Lambda_0}(k) = \tilde\Delta^2_{\Lambda_0}(k)\left[K_\Lambda(k) - K_{\Lambda_0}(k) \right],$$

is also a regular function of k at $k=0$.

One verifies that these results are consistent with identity (16.26).

The factor ζ corresponds to a field renormalization but, since $\tilde\zeta$ depends on k, it is not a simple constant renormalization. However, since the function $\tilde\zeta$ is regular at $k=0$, this renormalization does not affect the leading large-distance behaviour. Similarly, the function U is an addition to the connected two-point function. Again, since $\tilde U$ is a regular function, the addition does not affect the large-distance behaviour.

In the presence of a field renormalization, equation (16.43) gets an additional contribution:

$$\Lambda \frac{\partial}{\partial \Lambda} \tilde\zeta(k,\Lambda) = -\Lambda \frac{\partial \ln \tilde\Delta_\Lambda(k)}{\partial \Lambda} \tilde\zeta(k,\Lambda) + \frac{1}{2}\eta(\Lambda)\tilde\zeta(k,\Lambda).$$

We recall here that the constant η depends on Λ only through the Hamiltonian. In terms of the function $Z(\Lambda)$ the solution of

$$\Lambda \frac{\partial}{\partial \Lambda} Z(\Lambda) = \eta(\Lambda) Z(\Lambda), \quad Z(\Lambda_0) = 1$$

and, thus,

$$\ln Z(\Lambda) = \int_{\Lambda_0}^{\Lambda} \eta(\Lambda') \frac{d\Lambda'}{\Lambda'},$$

the solution can be written as

$$\tilde\zeta(k,\Lambda) = \sqrt{Z(\Lambda)}\,\tilde\Delta_{\Lambda_0}(k)/\tilde\Delta_\Lambda(k). \qquad (16.44)$$

16.3 Perturbative solution: ϕ^4 theory

We first show here that equations (16.40) or (16.41) have a perturbative solution. We consider the example of the ϕ^4 theory and we determine the expansion up to order u^2 since the required algebraic manipulations will be useful later.

Boundary conditions. We assume that the Hamiltonian at scale Λ_0 is of the form

$$\mathcal{V}(\phi, \Lambda_0) = \frac{1}{4!} u \int d^d x\, \phi^4(x) + \frac{1}{2!} r_c(\Lambda_0, u) \int d^d x\, \phi^2(x),$$

where the constant $r_c(\Lambda_0, u)$ is determined by the condition that the initial theory at scale Λ_0 is critical:

$$\tilde{\Gamma}^{(2)}(p=0, \Lambda_0) = 0.$$

In particular, a direct perturbative calculation at order u^2 yields

$$r_c(\Lambda_0, u) = -\frac{u}{2} \int \frac{d^d k}{(2\pi)^d} \tilde{\Delta}_{\Lambda_0}(k)$$
$$+ \frac{u^2}{6} \int \frac{d^d k_1}{(2\pi)^d} \frac{d^d k_2}{(2\pi)^d} \tilde{\Delta}_{\Lambda_0}(k_1) \tilde{\Delta}_{\Lambda_0}(k_2) \tilde{\Delta}_{\Lambda_0}(k_1 + k_2) + O(u^3). \quad (16.45)$$

We take as an Ansatz at order u:

$$\mathcal{V}(\phi, \Lambda) = \frac{1}{4!} u \int d^d x\, \phi^4(x) + \frac{1}{2!} r_{c1}(\Lambda) u \int d^d x\, \phi^2(x) + O(u^2),$$

and use the RGE in the form (16.40).

The equation for the coefficient of $\phi\phi$ at order u is compatible with the Ansatz:

$$\Lambda \frac{d}{d\Lambda} r_{c1}(\Lambda) = \frac{1}{2} D_\Lambda(0).$$

Using (16.31) and the criticality condition at $\Lambda = \Lambda_0$, one can integrate:

$$r_{c1}(\Lambda) = -\tfrac{1}{2} \Delta_\Lambda(0),$$

and the solution indeed corresponds to the criticality condition for a generic scale Λ.

The equation satisfied by $V^{(4)}$ at order u^2 involves the term of order u^2 of $V^{(6)}$, which can be expressed entirely in terms of $V^{(4)}$ at order u ($\mathbf{x} \equiv \{x_1, x_2, \ldots\}$):

$$\Lambda \frac{d}{d\Lambda} \int \prod_i d^d x_i\, \phi(x_i) V^{(6)}(\mathbf{x}, \Lambda) = -u^2 \frac{6!}{2(3!)^2} \int d^d x\, d^d y\, D_\Lambda(x-y) \phi^3(x) \phi^3(y)$$
$$+ O(u^3).$$

The six-point vertex. In Fourier components, denoting by $\tilde{V}_2^{(6)}$ the coefficient of u^2, one finds

$$\Lambda \frac{\mathrm{d}}{\mathrm{d}\Lambda} \tilde{V}_2^{(6)}(p_1,\ldots,p_6) = -\sum_I \tilde{D}_\Lambda(p_1 + p_{i_1} + p_{i_2}),$$

where I represents the 10 subsets $\{i_1, i_2\}$ belonging to $\{2, 3, \ldots, 6\}$. To integrate the equation, we introduce the function

$$\mathcal{D}_\Lambda(x) = \Delta_{\Lambda_0}(x) - \Delta_\Lambda(x). \tag{16.46}$$

It satisfies

$$\Lambda \frac{\mathrm{d}}{\mathrm{d}\Lambda} \mathcal{D}_\Lambda = D_\Lambda, \quad \mathcal{D}_{\Lambda_0} = 0.$$

The essential property of its Fourier transform

$$\tilde{\mathcal{D}}_\Lambda(k) = \frac{1}{k^2}\left[C(k^2/\Lambda_0^2) - C(k^2/\Lambda^2)\right],$$

is to be regular at $k = 0$.

The integration then is simple and yields

$$\int \prod_i \mathrm{d}^d x_i\, \phi(x_i) V^{(6)}(\mathbf{x}, \Lambda) = -\frac{6!}{2(3!)^2} u^2 \int \mathrm{d}^d x\, \mathrm{d}^d y\, \mathcal{D}_\Lambda(x-y) \phi^3(x) \phi^3(y)$$
$$+ O(u^3).$$

In the Fourier representation,

$$\tilde{V}_2^{(6)}(p_i, \Lambda) = -\sum_I \tilde{\mathcal{D}}_\Lambda(p_1 + p_{i_1} + p_{i_2}).$$

The function $\tilde{V}^{(6)}$ thus is also regular at $p_1 = \cdots = p_6 = 0$ and the Hamiltonian remains local.

The four-point vertex. This result allows writing explicitly the equation satisfied by $V_2^{(4)}$, the coefficient of $V^{(4)}$ at order u^2:

$$\Lambda \frac{\mathrm{d}}{\mathrm{d}\Lambda} \frac{1}{4!} \int \prod_i \mathrm{d}^d x_i\, \phi(x_i) V_2^{(4)}(\mathbf{x}, \Lambda)$$
$$= -\int \mathrm{d}^d x\, \mathrm{d}^d y\, D_\Lambda(x-y) \left[\tfrac{1}{8} \mathcal{D}_\Lambda(x-y) \phi^2(x) \phi^2(y) + \tfrac{1}{12} \mathcal{D}_\Lambda(0) \phi(x) \phi^3(y)\right]$$
$$+ \tfrac{1}{12} \Delta_\Lambda(0) \int \mathrm{d}^d x\, \mathrm{d}^d y\, D_\Lambda(x-y) \phi(x) \phi^3(y),$$

where, in the Fourier representation,

$$\Lambda\frac{\mathrm{d}}{\mathrm{d}\Lambda}\tilde{V}_2^{(4)}(\mathbf{p},\Lambda) = \frac{1}{2}\int\frac{\mathrm{d}^d k}{(2\pi)^d}\tilde{D}_\Lambda(k)\tilde{V}_2^{(6)}(p_1,p_2,\ldots,p_4,k,-k)$$
$$+\frac{1}{2}\Delta_\Lambda(0)\sum_i \tilde{D}_\Lambda(p_i)$$
$$= -\int\frac{\mathrm{d}^d k}{(2\pi)^d}\tilde{D}_\Lambda(k)\left[\tilde{D}_\Lambda(k+p_1+p_2)+2\text{ terms}\right]$$
$$+\frac{1}{2}\sum_i\left[-D_\Lambda(0)\tilde{D}_\Lambda(p_i)+\Delta_\Lambda(0)\tilde{D}_\Lambda(p_i)\right].$$

Again, the integration is simple and yields

$$\frac{1}{4!}\int\prod_i \mathrm{d}^d x_i\,\phi(x_i)V_2^{(4)}(\mathbf{x},\Lambda) = -\tfrac{1}{16}\int \mathrm{d}^d x\,\mathrm{d}^d y\,\mathcal{D}_\Lambda^2(x-y)\phi^2(x)\phi^2(y)$$
$$+\tfrac{1}{12}\Delta_\Lambda(0)\int \mathrm{d}^d x\,\mathrm{d}^d y\,\mathcal{D}_\Lambda(x-y)\phi(x)\phi^3(y).$$

In the Fourier representation,

$$\tilde{V}_2^{(4)}(p_i,\Lambda) = -\tfrac{1}{2}\left[\bar{B}_\Lambda(p_1+p_2)+2\text{ terms}\right]+\tfrac{1}{2}\Delta_\Lambda(0)\sum_{i=1}^{4}\tilde{D}_\Lambda(p_i)$$

with

$$\bar{B}_\Lambda(p) = \frac{1}{(2\pi)^d}\int \mathrm{d}^d k\,\tilde{D}_\Lambda(k)\tilde{D}_\Lambda(p-k).$$

One verifies that in the limit $\Lambda_0/\Lambda \to \infty$, $\tilde{V}^{(4)}(p_i,\Lambda)$ diverges logarithmically for $d=4$, and this corresponds to the renormalization of the coefficient of ϕ^4 at one-loop order.

The coefficient $\tilde{V}^{(2)}$ at order u^2. We complete the exercise with a calculation of the coefficient of order u^2 of $\tilde{V}^{(2)}(p,\Lambda)$. It satisfies the equation

$$\Lambda\frac{\mathrm{d}}{\mathrm{d}\Lambda}\tilde{V}_2^{(2)}(p) = \frac{1}{2}\int\frac{\mathrm{d}^d k}{(2\pi)^d}\tilde{D}_\Lambda(k)\tilde{V}_2^{(4)}(p,-p,k,-k) - \frac{1}{4}\tilde{D}_\Lambda(p)\Delta_\Lambda^2(p).$$

The value at $p=0$ yields the criticality condition:

$$\Lambda\frac{\mathrm{d}}{\mathrm{d}\Lambda}\tilde{V}_2^{(2)}(0) = \frac{1}{2}\int\frac{\mathrm{d}^d k}{(2\pi)^d}\tilde{D}_\Lambda(k)\tilde{V}_2^{(4)}(0,0,k,-k) - \frac{1}{4}\tilde{D}_\Lambda(0)\Delta_\Lambda^2(p).$$

We know the value of $\tilde{V}^{(2)}(0) = r_c(\Lambda,u)$ a priori from the usual perturbative expansion. It contains the contribution (16.45) at scale Λ to which the contributions coming from $V_2^{(4)}$ and $V_2^{(6)}$ must be added. One thus finds

$$r_{c2}(\Lambda) = \frac{1}{6}\int \frac{d^d k_1}{(2\pi)^d}\frac{d^d k_2}{(2\pi)^d}\tilde{\Delta}_\Lambda(k_1)\tilde{\Delta}_\Lambda(k_2)\tilde{\Delta}_\Lambda(k_1+k_2)$$
$$- \frac{1}{2}\int \frac{d^d k}{(2\pi)^d}\tilde{\Delta}_\Lambda(k)\tilde{V}_2^{(4)}(0,0,k,-k)$$
$$- \frac{1}{8}\int \frac{d^d k_1}{(2\pi)^d}\frac{d^d k_2}{(2\pi)^d}\tilde{V}_2^{(6)}(0,0,k_1,-k_1,k_2,-k_2)\tilde{\Delta}_\Lambda(k_1)\tilde{\Delta}_\Lambda(k_2). \quad (16.47)$$

One verifies that $r_{c2}(\Lambda)$ indeed satisfies the flow equation. The expression can also be written as

$$r_{c2}(\Lambda) = \frac{1}{6}\int \frac{d^d k_1}{(2\pi)^d}\frac{d^d k_2}{(2\pi)^d}\left[-\tilde{\mathcal{D}}_\Lambda(k_1)\tilde{\mathcal{D}}_\Lambda(k_2)\tilde{\mathcal{D}}_\Lambda(k_1+k_2) + \tilde{\Delta}_\Lambda(k_1)\tilde{\Delta}_\Lambda(k_2)\right.$$
$$\left.\times\tilde{\Delta}_\Lambda(k_1+k_2)\right] - \frac{1}{4}\Delta_\Lambda^2(0)\tilde{\mathcal{D}}_\Lambda(0) + \frac{1}{4}\Delta_\Lambda(0)\bar{B}_\Lambda(0). \quad (16.48)$$

Finally, this boundary condition determines the two-point vertex

$$\tilde{V}_2^{(2)}(p) = \frac{1}{6}\int \frac{d^d k_1}{(2\pi)^d}\frac{d^d k_2}{(2\pi)^d}\left[-\tilde{\mathcal{D}}_\Lambda(k_1)\tilde{\mathcal{D}}_\Lambda(k_2)\tilde{\mathcal{D}}_\Lambda(p+k_1+k_2) + \tilde{\Delta}_\Lambda(k_1)\tilde{\Delta}_\Lambda(k_2)\right.$$
$$\left.\times\tilde{\Delta}_\Lambda(k_1+k_2)\right] - \frac{1}{4}\Delta_\Lambda^2(0)\tilde{\mathcal{D}}_\Lambda(p) + \frac{1}{4}\Delta_\Lambda(0)\bar{B}_\Lambda(0). \quad (16.49)$$

Remark. It results from equation (16.19) in Section 16.1.4, that the same results can be derived from a perturbative calculation of connected correlation functions with the interaction $\mathcal{V}(\phi,\Lambda_0)$ and the propagator $\mathcal{D}(\Lambda_0,\Lambda)$. The purpose of the preceding calculation is to show more explicitly how these exact RGE, with boundary conditions at initial scale, determine the functional $\mathcal{V}(\phi,\Lambda)$. Moreover, one verifies that if the theory is critical for $\Lambda = \Lambda_0$, it remains for generic Λ.

16.4 RG equations: Standard form

After a field renormalization, required to be able to reach non-Gaussian fixed points, the RGE take the form (16.29) with $s = -\ln\Lambda$:

$$\Lambda\frac{\partial}{\partial\Lambda}\mathcal{H}(\phi,\Lambda)$$
$$= \frac{1}{2}\int d^d x\, d^d y\, D_\Lambda(x-y)\left[\frac{\delta^2 \mathcal{H}}{\delta\phi(x)\delta\phi(y)} - \frac{\delta\mathcal{H}}{\delta\phi(x)}\frac{\delta\mathcal{H}}{\delta\phi(y)}\right]$$
$$+ \int d^d x\, d^d y\, \phi(x)L_\Lambda(x-y)\frac{\delta\mathcal{H}}{\delta\phi(y)} + \frac{\eta}{2}\int d^d x\, \phi(x)\frac{\delta\mathcal{H}}{\delta\phi(x)}. \quad (16.50)$$

The function η is *a priori* arbitrary but with one restriction: as we have explained in Chapter 9, the function η must depend on Λ only through $\mathcal{H}(\phi,\Lambda)$. It must be adjusted to ensure the existence of fixed points.

Equation (16.50) does not have a stationary Markovian form since D_Λ and L_Λ depend explicitly on Λ:

$$L_\Lambda(x) = \frac{1}{(2\pi)^d}\int d^d k\, e^{ikx}\, \tilde{L}_\Lambda(k) = \Lambda^d L_1(x), \quad D_\Lambda(x) = \Lambda^{d-2} D_1(\Lambda x),$$

where $\tilde{L}_\Lambda(k)$ is defined by equation (16.37) and the relations (16.32) have been used. To eliminate this dependence, we perform everywhere a dilatation of the position variables of the form

$$x \mapsto x/\Lambda.$$

This has the effect of substituting in the Hamiltonian

$$\int d^d x\, \phi(x) \mapsto \int d^d x\, \Lambda^{-d}\phi(x/\Lambda), \quad \mathcal{H}^{(n)}(x_i, \Lambda) \mapsto \mathcal{H}^{(n)}(x_i/\Lambda, \Lambda).$$

We then define,

$$\phi(x) = \Lambda^{(d-2)/2}\phi'(\Lambda x), \quad \mathcal{H}^{(n)}(x_i, \Lambda) = \Lambda^{n(d+2)/2}\mathcal{H}'^{(n)}(\Lambda x_i, \Lambda).$$

The transformation of ϕ is simply the Gaussian renormalization.

Since $\Lambda^{(d-2)/2}\Lambda^{(d+2)/2} = \Lambda^d$ this change cancels the factor Λ^{-d} of the initial change of variables on x when the coefficient of $\mathcal{H}^{(n)}$ is homogeneous to ϕ^n. Therefore,

$$\int d^d x\, d^d y\, D_\Lambda(x-y) \mapsto \Lambda^{-2-d}\int d^d x\, d^d y\, D_1(x-y),$$

$$\left[\frac{\delta^2 \mathcal{H}}{\delta\phi(x)\delta\phi(y)} - \frac{\delta\mathcal{H}}{\delta\phi(x)}\frac{\delta\mathcal{H}}{\delta\phi(y)}\right] \mapsto \Lambda^{2+d}\left[\frac{\delta^2 \mathcal{H}}{\delta\phi'(x)\delta\phi'(y)} - \frac{\delta\mathcal{H}}{\delta\phi'(x)}\frac{\delta\mathcal{H}}{\delta\phi'(y)}\right],$$

$$\int d^d x\, d^d y\, L_\Lambda(x-y)\phi(x)\frac{\delta\mathcal{H}}{\delta\phi(y)} \mapsto \int \Lambda^{-2d}d^d x\, d^d y\, \Lambda^d L_1(x-y)\Lambda^d \phi'(x)\frac{\delta\mathcal{H}}{\delta\phi'(y)}$$

$$\int d^d x\, \phi(x)\frac{\delta\mathcal{H}}{\delta\phi(x)} \mapsto \int \Lambda^{-d}d^d x\, \Lambda^{(d-2)/2}\phi'(x)\Lambda^{(d+2)/2}\frac{\delta\mathcal{H}}{\delta\phi'(x)}.$$

One verifies that all powers of Λ cancel.

Finally, the relation between ϕ and ϕ' can also be written as

$$\phi'(x) = \Lambda^{(2-d)/2}\phi(x/\Lambda),$$

in such a way that the right-hand side becomes

$$\Lambda\frac{\partial}{\partial\Lambda}\bigg|_\phi = \Lambda\frac{\partial}{\partial\Lambda}\bigg|_{\phi'} + \int d^d x\, \frac{\delta\mathcal{H}}{\delta\phi'(x)}\Lambda\frac{\partial}{\partial\Lambda}\bigg|_\phi \phi'(x)$$

$$= \Lambda\frac{\partial}{\partial\Lambda}\bigg|_{\phi'} + \int d^d x\, \frac{\delta\mathcal{H}}{\delta\phi'(x)}\left(\frac{1}{2}(2-d) - \sum_\mu x^\mu \frac{\partial}{\partial x^\mu}\right)\phi'(x),$$

where we have denoted by x^μ the d coordinates of x.

In what follows, we omit the primes. Moreover, we introduce the dilatation parameter $\lambda = \Lambda_0/\Lambda$ that relates the initial scale Λ_0 to the running scale Λ and thus

$$\Lambda \frac{\mathrm{d}}{\mathrm{d}\Lambda} = -\lambda \frac{\mathrm{d}}{\mathrm{d}\lambda}.$$

Then, the RGE take a form consistent with the general equation (9.23):

$$\lambda \frac{\mathrm{d}}{\mathrm{d}\lambda} \mathcal{H}(\phi, \lambda) = -\frac{1}{2} \int \mathrm{d}^d x\, \mathrm{d}^d y\, D(x-y) \left[\frac{\delta^2 \mathcal{H}}{\delta\phi(x)\delta\phi(y)} - \frac{\delta \mathcal{H}}{\delta\phi(x)} \frac{\delta \mathcal{H}}{\delta\phi(y)} \right]$$
$$- \int \mathrm{d}^d x\, \frac{\delta \mathcal{H}(\phi, \lambda)}{\delta\phi(x)} \left[\frac{1}{2}(d-2+\eta) + \sum_\mu x^\mu \frac{\partial}{\partial x^\mu} \right] \phi(x)$$
$$- \int \mathrm{d}^d x\, \mathrm{d}^d y\, L(x-y) \frac{\delta \mathcal{H}(\phi, \lambda)}{\delta\phi(x)} \phi(y). \tag{16.51}$$

RGE in component form. Expanding equation (16.51) in powers of ϕ, one finds equations in standard form for the components. For $n \neq 2$ ($D_1 \equiv D$),

$$\lambda \frac{\mathrm{d}}{\mathrm{d}\lambda} \mathcal{H}^{(n)}(x_i; \lambda)$$
$$= \left(\frac{1}{2} n(d+2-\eta) + \sum_{j,\mu} x_j^\mu \frac{\partial}{\partial x_j^\mu} \right) \mathcal{H}^{(n)}(x_i; \lambda) - \frac{1}{2} \int \mathrm{d}^d x\, \mathrm{d}^d y\, D(x-y)$$
$$\times \left[\mathcal{H}^{(n+2)}(x_1, x_2, \ldots, x_n, x, y; \lambda) \right.$$
$$\left. - \sum_I \mathcal{H}^{(l+1)}(x_{i_1}, \ldots, x_{i_l}, x; \lambda) \mathcal{H}^{(n-l+1)}(x_{i_{l+1}}, \ldots, x_{i_n}, y; \lambda) \right]. \tag{16.52}$$

In the Fourier representation, the equations take the form

$$\lambda \frac{\mathrm{d}}{\mathrm{d}\lambda} \tilde{\mathcal{H}}^{(n)}(p_i; \lambda)$$
$$= \left(d - \frac{1}{2} n(d-2+\eta) - \sum_{j,\mu} p_j^\mu \frac{\partial}{\partial p_j^\mu} \right) \tilde{\mathcal{H}}^{(n)}(p_i; \lambda)$$
$$- \frac{1}{2} \int \frac{\mathrm{d}^d k}{(2\pi)^d} \tilde{D}(k) \tilde{\mathcal{H}}^{(n+2)}(p_1, p_2, \ldots, p_n, k, -k; \lambda)$$
$$+ \frac{1}{2} \sum_I \tilde{D}(p_0) \tilde{\mathcal{H}}^{(l+1)}(p_{i_1}, \ldots, p_{i_l}, p_0; \lambda) \tilde{\mathcal{H}}^{(n-l+1)}(p_{i_{l+1}}, \ldots, p_{i_n}, -p_0). \tag{16.53}$$

For $n = 2$, it is necessary to distinguish the components of $\mathcal{H}$ and $\mathcal{V}$. In both cases, one finds an additional term in the equation:

$$\lambda \frac{d}{d\lambda} \mathcal{H}^{(2)}(x_1; \lambda)$$
$$= \left(d + 2 - \eta + \sum_\mu x_1^\mu \frac{\partial}{\partial x_1^\mu}\right) \mathcal{H}^{(2)}(x_1; \lambda)$$
$$- \frac{1}{2} \int d^d x \, d^d y \, D(x - y) \left[\mathcal{H}^{(4)}(x_1, 0, , x, y; \lambda) - 2\mathcal{H}^{(2)}(x - x_1; \lambda) \mathcal{H}^{(2)}(y; \lambda)\right]$$
$$- 2 \int d^d y \, L(x_1 - y) \mathcal{H}^{(2)}(y; \lambda). \tag{16.54}$$

In the Fourier representation,

$$\lambda \frac{d}{d\lambda} \tilde{\mathcal{H}}^{(2)}(p; \lambda)$$
$$= \left(2 - \eta - \sum_\mu p^\mu \frac{\partial}{\partial p^\mu}\right) \tilde{\mathcal{H}}^{(2)}(p; \lambda) - 2\tilde{L}(p) \tilde{\mathcal{H}}^{(2)}(p; \lambda)$$
$$- \frac{1}{2} \int \frac{d^d k}{(2\pi)^d} \tilde{D}(k) \tilde{\mathcal{H}}^{(4)}(p, -p, k, -k; \lambda) + \tilde{D}(p) \left(\tilde{\mathcal{H}}^{(2)}(p; \lambda)\right)^2, \tag{16.55}$$

$$\lambda \frac{d}{d\lambda} \tilde{\mathcal{V}}^{(2)}(p; \lambda)$$
$$= \left(2 - \eta - \sum_\mu p^\mu \frac{\partial}{\partial p^\mu}\right) \tilde{\mathcal{V}}^{(2)}(p; \lambda) - \eta \tilde{\Delta}^{-1}(p)$$
$$- \frac{1}{2} \int \frac{d^d k}{(2\pi)^d} \tilde{D}(k) \tilde{\mathcal{H}}^{(4)}(p, -p, k, -k; \lambda) + \tilde{D}(p) \left(\tilde{\mathcal{V}}^{(2)}(p; \lambda)\right)^2. \tag{16.56}$$

16.5 Dimension 4

We have derived general functional RGE. We now apply them to the example of the ϕ^4 theory in the perturbative framework. Our goal is to show explicitly how the FRG allows recovering the results obtained in Chapter 10 in the framework of the asymptotic perturbative RG.

We begin with the dimension 4, where the Gaussian fixed point is marginally stable. The perturbative assumptions and calculations have much in common with those of Section 16.3.

16.5.1 Renormalization conditions. β- and η-functions

Renormalization conditions. In dimension 4, a non-trivial theory can be defined and parametrized, for example, in terms of $g(\lambda)$, the value of $\tilde{\mathcal{H}}^{(4)}(p_i, \lambda)$ at $p_1 = \cdots = p_4 = 0$:

$$g(\lambda) \equiv \tilde{\mathcal{H}}^{(4)}(p_i = 0, \lambda). \tag{16.57}$$

One then solves perturbatively the other flow equations, with appropriate boundary conditions, by the method explained at the end of Section 9.7.2, as has been done in Section 16.3. All other interactions are then determined perturbatively as functions of g, under the assumption that they are at least of order g^2. One suppresses in this way all corrections due to irrelevant operators, keeping only the contributions due to the marginal operator. Then, all interactions become implicit functions of λ through $g(\lambda)$, which satisfies a unique equation (equation (9.62)):

$$\lambda \frac{dg}{d\lambda} = -\beta(g(\lambda)) \tag{16.58}$$

with, as we have already seen, $\beta(g) = O(g^2)$.

The function $\eta(g)$ is determined by the condition

$$\left. \frac{\partial}{\partial p^2} \tilde{\mathcal{H}}^{(2)}(p;g) \right|_{p=0} = 1 \Rightarrow \left. \frac{\partial}{\partial p^2} \tilde{\mathcal{V}}^{(2)}(p;g) \right|_{p=0} = 0, \tag{16.59}$$

which suppresses the redundant operator that corresponds to a change of normalization of the field.

The two conditions (16.57), (16.59) replace the renormalization conditions of the usual renormalization theory (see Section 13.1.2).

RGE: Another form. In equation (16.53), we express the derivative with respect to $\ln \lambda$ in terms of the derivative with respect to g and the function β, using the definition (16.58),

$$\lambda \frac{d}{d\lambda} = -\beta(g) \frac{d}{dg}.$$

For $d = 4$, equation (16.53) then becomes ($n \neq 2$)

$$\beta(g) \frac{d}{dg} \tilde{\mathcal{H}}^{(n)}(p_i; g)$$

$$= \left(n - 4 + \tfrac{1}{2} n \eta(g) + \sum_{j,\mu} p_j^\mu \frac{\partial}{\partial p_j^\mu} \right) \tilde{\mathcal{H}}^{(n)}(p_i; g)$$

$$+ \frac{1}{2} \int \frac{d^4 k}{(2\pi)^4} \tilde{D}(k) \tilde{\mathcal{H}}^{(n+2)}(p_1, p_2, \ldots, p_n, k, -k; g)$$

$$- \frac{1}{2} \sum_I \tilde{D}(p_0) \tilde{\mathcal{H}}^{(l+1)}(p_{i_1}, \ldots, p_{i_l}, p_0; g) \tilde{\mathcal{H}}^{(n-l+1)}(p_{i_{l+1}}, \ldots, p_{i_n}, -p_0; g). \tag{16.60}$$

In the case $n = 2$, equation (16.56) satisfied by $\mathcal{V}^{(2)}$ becomes

$$\beta(g) \frac{d}{dg} \tilde{\mathcal{V}}^{(2)}(p; g)$$

$$= \left(-2 + \eta(g) + \sum_\mu p^\mu \frac{\partial}{\partial p^\mu} \right) \tilde{\mathcal{V}}^{(2)}(p; g) + \eta(g) \tilde{\Delta}^{-1}(p)$$

$$+ \frac{1}{2} \int \frac{d^4 k}{(2\pi)^4} \tilde{D}(k) \tilde{\mathcal{H}}^{(4)}(p, -p, k, -k; g) - \tilde{D}(p) \left(\tilde{\mathcal{V}}^{(2)}(p; g) \right)^2. \tag{16.61}$$

β and η-functions. Using the condition (16.57) in equation (16.60) for $n = 4$, $p_i = 0$, one finds an equation for the function $\beta(g)$:

$$\beta(g) = 2g\eta(g) + \frac{1}{2}\int \frac{\mathrm{d}^4 k}{(2\pi)^4}\tilde{D}(k)\tilde{\mathcal{H}}^{(6)}(0,0,0,0,k,-k;g) - 4g\tilde{D}(0)\tilde{\mathcal{V}}^{(2)}(0;g). \quad (16.62)$$

Equation (16.61) can then be solved recursively in the form

$$\left(2 - \sum_\mu p^\mu \frac{\partial}{\partial p^\mu}\right)\tilde{\mathcal{V}}^{(2)}(p;g)$$
$$= -\beta(g)\frac{\mathrm{d}}{\mathrm{d}g}\tilde{\mathcal{V}}^{(2)}(p;g) - \tilde{D}(p)\left(\tilde{\mathcal{V}}^{(2)}(p;g)\right)^2 + \eta(g)\left(\Delta^{-1}(p) + \tilde{\mathcal{V}}^{(2)}(p;g)\right)$$
$$+ \frac{1}{2}\int \frac{\mathrm{d}^4 k}{(2\pi)^4}\tilde{D}(k)\tilde{\mathcal{H}}^{(4)}(p,-p,k,-k;g). \quad (16.63)$$

For $p = 0$, one obtains the criticality condition that determines

$$r_c(g) = \tilde{\mathcal{V}}^{(2)}(0;g) = \tilde{\mathcal{H}}^{(2)}(0;g).$$

But one also notes that if the right-hand side contains a term proportional to p^2 in its Taylor series expansion at $p = 0$, the solution of the equation cannot be regular. Indeed, the equation

$$\left(2 - \sum_\mu p^\mu \frac{\partial}{\partial p^\mu}\right)f(p^2) = 2\left(1 - p^2 f'(p^2)\right) = p^2$$

has only singular solutions of the form

$$f(p^2) = -p^2 \ln p + \mathrm{const.}\, p^2.$$

Since we look only for regular solutions, we have to impose the condition that the derivative with respect to p^2 of the right-hand side of equation (16.63) vanishes at $p = 0$. This yields the condition

$$\eta(g) = 2C''(0)r_c^2(g) - \frac{1}{2}\frac{\partial}{\partial p^2}\int \frac{\mathrm{d}^4 k}{(2\pi)^4}\tilde{D}(k)\tilde{\mathcal{H}}^{(4)}(p,-p,k,-k;g)\bigg|_{p=0}, \quad (16.64)$$

which determines $\eta(g)$. These arguments justify, in the perturbative framework, the introduction of the function $\eta(g)$ and thus of the renormalization of the field.

16.5.2 Solution of RGE at order g

Since $\eta(g)$ is at least of order g (in fact it is of order g^2) and since we have assumed $\tilde{\mathcal{H}}^{(6)}$ of order g^2, equation (16.62) confirms that $\beta(g)$ is of order g^2.

At order g, since $\tilde{\mathcal{V}}^{(2)}(p;g)$ is of order g, the left-hand side of equation (16.61) can be neglected. The equation reduces to

$$\left(2 - \sum_\mu p^\mu \frac{\partial}{\partial p^\mu}\right) \tilde{\mathcal{V}}^{(2)}(p;g) = \tfrac{1}{2} g D(0) + \eta \tilde{\Delta}^{-1}(p).$$

The left-hand side has no p^2 term in its expansion at $p = 0$ and thus η vanishes at order g. Equation (16.61) thus reduces to

$$\left(2 - \sum_\mu p^\mu \frac{\partial}{\partial p^\mu}\right) \tilde{\mathcal{V}}^{(2)}(p;g) = \tfrac{1}{2} g D(0).$$

We note that the homogeneous equation has regular solutions proportional to p^2 corresponding to the redundant operator that changes the field normalization. The general solution thus is the sum of a solution of the homogeneous equation and a constant, which is a particular solution. The condition (16.59) eliminates the redundant term and selects the constant solution

$$\tilde{\mathcal{V}}^{(2)}(p;g) = \tfrac{1}{4} g D(0) + O(g^2).$$

16.5.3 Solution of RGE at order g^2

For $\tilde{\mathcal{H}}^{(6)}(p_i;g)$, the left-hand side is of order g^3, which can be neglected, and thus

$$\left(2 + \sum_{j,\mu} p_j^\mu \frac{\partial}{\partial p_j^\mu}\right) \tilde{\mathcal{H}}^{(6)}(p_i;g) = g^2 \left[\tilde{D}(p_1 + p_2 + p_3) + 9 \text{ terms}\right] + O(g^3). \quad (16.65)$$

Moreover ($\Delta \equiv \Delta_{\Lambda=1}$),

$$\left(2 + \sum_\mu p^\mu \frac{\partial}{\partial p^\mu}\right) \tilde{\Delta}(p) = \tilde{D}(p).$$

In the left-hand side, one can replace $\tilde{\Delta}(p)$, which is a function singular at $p = 0$, by the function

$$-\tilde{D}_*(p) = \tilde{\Delta}(p) - \frac{1}{p^2} \quad \text{and thus} \quad \left(2 + \sum_\mu p^\mu \frac{\partial}{\partial p^\mu}\right) \tilde{D}_*(p) = -\tilde{D}(p), \quad (16.66)$$

which is regular. We note that the function $\tilde{D}_*(p)$ is the limit of the function (16.46) for $\Lambda_0 \to \infty$.

The regular solution of equation (16.65) can then be written as
$$\mathcal{H}^{(6)}(p_i;g) = -g^2\left[\tilde{\mathcal{D}}_*(p_1+p_2+p_3) + 9 \text{ terms}\right] + O(g^3).$$

β-function. In equation (16.62), since $g\eta(g)$ is of order g^3,
$$\beta(g) = g^2\left[3\tilde{B}_*(0) - 2D(0)\tilde{\mathcal{D}}_*(0)\right] - g^2 D(0)\tilde{D}(0) + O(g^3)$$

with
$$\tilde{B}_*(p) = -\int \frac{d^4k}{(2\pi)^4}\tilde{D}(k)\tilde{\mathcal{D}}_*(p-k). \tag{16.67}$$

The two last terms cancel since
$$\tilde{D}(0) = 2C'(0) = -2\tilde{\mathcal{D}}_*(0), \tag{16.68}$$

and it remains to evaluate
$$\tilde{B}_*(0) = -\int \frac{d^4k}{(2\pi)^4}\tilde{\mathcal{D}}_*(k)\tilde{D}(k).$$

One verifies
$$\frac{d}{dk}\left[\tfrac{1}{2}k^4\tilde{\mathcal{D}}_*^2(k)\right] = 2k^3\tilde{\mathcal{D}}_*^2(k) + k^3\tilde{\mathcal{D}}_*(k)k\frac{d}{dk}\tilde{\mathcal{D}}_*(k)$$
$$= 2k^3\tilde{\mathcal{D}}_*^2(k) - k^3\tilde{\mathcal{D}}_*(k)\left(\tilde{D}(k) + 2\tilde{\mathcal{D}}_*(k)\right) = -k^3\tilde{\mathcal{D}}_*(k)\tilde{D}(k),$$

where relation (16.66) has been used. Thus, the integrand has the form of a total derivative and the integral is determined by the asymptotic behaviour
$$k^2\tilde{\mathcal{D}}_*(k) \to 1 \quad \text{for} \quad k \to \infty.$$

One obtains
$$\tilde{B}_*(0) = \frac{1}{8\pi^2}\int_0^\infty dk \frac{d}{dk}\left[\tfrac{1}{2}k^4\tilde{\mathcal{D}}_*^2(k)\right] = \frac{1}{16\pi^2}. \tag{16.69}$$

One infers
$$\beta(g) = \frac{3}{16\pi^2}g^2 + O(g^3), \tag{16.70}$$

in agreement with the result (10.21) for $\varepsilon = 0$.

η-function. It is then convenient to introduce the function
$$F^{(4)}(p_i;g) = \tilde{\mathcal{H}}^{(4)}(p_i;g) - \tilde{\mathcal{H}}^{(4)}(0;g) = O(g^2). \tag{16.71}$$

At order g^2, $F^{(4)}$ is a solution of

$$\sum_{j,\mu} p_j^\mu \frac{\partial}{\partial p_j^\mu} F^{(4)}(p_i;g) = -\frac{1}{2}\int \frac{\mathrm{d}^4 k}{(2\pi)^4} \tilde{D}(k)\tilde{\mathcal{H}}^{(6)}(p_1,p_2,p_3,p_4,k,-k;g)$$

$$+ g \sum_i \tilde{D}(p_i)\tilde{\mathcal{V}}^{(2)}(p_i;g) - (\mathbf{p}=0).$$

Since the right-hand side of the equation vanishes for $p=0$ (which in fact is a consequence of the flow equation satisfied by $g(\lambda)$), the equation has a regular solution. More explicitly,

$$\sum_{j,\mu} p_j^\mu \frac{\partial}{\partial p_j^\mu} F^{(4)}(p_i;g) = \tfrac{1}{2}g^2 \Big[D(0)\tilde{\mathcal{D}}_*(p_1) + 3 \text{ terms} - 2\tilde{\mathcal{B}}_*(p_1+p_2)$$

$$+ 2 \text{ terms}\Big] + \tfrac{1}{4}D(0)g^2 \sum_i \tilde{D}(p_i) - (\mathbf{p}=0).$$

Using the remark that the equation

$$\sum_\mu p^\mu \frac{\partial}{\partial p^\mu} X(p) = Y(p), \quad Y(0)=0,$$

has the solution

$$X(p) = \int_0^1 \frac{\mathrm{d}\lambda}{\lambda} Y(\lambda p),$$

after a few integrations by parts and algebraic manipulations, one finds

$$F^{(4)}(p_i;g) = -\tfrac{1}{2}g^2 B(p_1+p_2) + 2 \text{ terms} - \tfrac{1}{4}g^2 D(0)\Big[\tilde{\mathcal{D}}_*(p_1) - \tilde{\mathcal{D}}_*(0) + 3 \text{ terms}\Big] \tag{16.72}$$

with

$$B(p) = \int \frac{\mathrm{d}^4 k}{(2\pi)^4} \tilde{\mathcal{D}}_*(k)\Big[\tilde{\mathcal{D}}_*(p+k) - \tilde{\mathcal{D}}_*(k)\Big].$$

From equation (16.64), one then derives, at order g^2,

$$\eta(g) = \frac{1}{2}g^2 \frac{\partial}{\partial p^2} \int \frac{\mathrm{d}^4 k}{(2\pi)^4} \tilde{D}(k)B(p+k)\Big|_{p=0}$$

$$= \frac{1}{2}g^2 \frac{\partial}{\partial p^2} \int \frac{\mathrm{d}^4 k_1}{(2\pi)^4}\frac{\mathrm{d}^4 k_2}{(2\pi)^4} \tilde{D}(k_1)\tilde{\mathcal{D}}_*(k_2)\Big[\tilde{\mathcal{D}}_*(p+k_1+k_2) - \tilde{\mathcal{D}}_*(k_2)\Big]\Big|_{p=0}.$$

For this calculation, it is convenient to return to space variables. The expression then takes the form

$$\eta(g) = \frac{1}{2}g^2 \frac{\partial}{\partial p^2} \int \mathrm{d}^4 x \, e^{ipx} D(x)\mathcal{D}_*^2(x)\Big|_{p=0}.$$

Note that
$$\sum_{\mu=1}^{4}\left(\frac{\partial}{\partial p_\mu}\right)^2 \Phi(p^2) = 8\Phi'(p^2) + 4p^2\Phi''(p^2).$$

Using this relation to evaluate the derivative with respect to p^2, one finds
$$\eta(g) = -\frac{1}{16}g^2 \int d^4x\, x^2 D(x)\mathcal{D}_*^2(x)$$
$$= -\frac{\pi^2}{8}g^2 \int_0^\infty dx\, x^5 D(x)\mathcal{D}_*^2(x).$$

Equation (16.66) implies
$$D(x) = \left(2 + x\frac{\partial}{\partial x}\right)\mathcal{D}_*(x),$$

where, as in the latter integral, x is the radial coordinate. Then,
$$\frac{\partial}{\partial x}x^6 \mathcal{D}_*^3(x) = 6x^5\mathcal{D}_*^3(x) + 3x^5\mathcal{D}_*^2(x)x\frac{\partial}{\partial x}D(x).$$

The integrand thus is an explicit derivative. The contribution at infinity vanishes since all functions are fast decaying. At the origin,
$$\lim_{x\to 0} x^2 \mathcal{D}_*(x) = \lim_{x\to 0} \frac{1}{(2\pi)^4}x^2\int d^4p\, \frac{e^{ipx}}{p^2} = \frac{1}{4\pi^2}.$$

One thus recovers the result (10.39) for $N = 1$:
$$\eta(g) = \frac{1}{6}\frac{1}{(4\pi)^4}g^2 + O(g^3).$$

Remarks.

(i) It is then possible to prove that the RGE can be solved by the same method to all orders in an expansion in powers of g. This also allows proving that the ϕ^4 field theory is renormalizable. Since all functions are regular at zero momentum, it suffices to prove that all integrals converge at large momentum. In particular, this proof of renormalizability avoids the combinatorial considerations about Feynman diagrams associated with other standard proofs.

(ii) One may get the impression that, to eliminate λ, we have used explicitly the property that the field theory involves only one parameter g of ϕ^4 type. Actually, the solution that we have obtained is determined by the conditions on the degree in $g(\lambda)$ of the functions $\mathcal{H}^{(n)}$. For example, in the case of several independent terms of ϕ^4 type, and thus of several coefficients g_i, it suffices to substitute
$$\lambda\frac{d}{d\lambda} = -\sum_i \beta_i(g)\frac{\partial}{\partial g_i}$$

and to use the property $\beta_i(g) = O(g^2)$.

(iii) It is simple to verify that the same method applies to theories of ϕ^n type in the dimensions where these operators are marginal (like the dimension 3 for ϕ^6).

16.6 Fixed point: ε-expansion

To complete the verification of the consistency of the FRG with the asymptotic RG in the form used in the Chapters 10 and 13, we now consider the dimension $d = 4 - \varepsilon$ and expand equations (16.53), (16.56) in powers of g and ε.

Instead of determining the fixed point directly, we follow the strategy of the preceding section, and determine the flow equation satisfied by the effective parameter $g(\lambda)$. The calculations then are very similar.

We start from

$$\beta(g)\frac{\mathrm{d}}{\mathrm{d}g}\tilde{\mathcal{H}}^{(n)}(p_i;g)$$

$$= \left(\frac{1}{2}n(d-2+\eta) - d + \sum_{j,\mu}p_j^\mu\frac{\partial}{\partial p_j^\mu}\right)\tilde{\mathcal{H}}^{(n)}(p_i;g)$$

$$+ \frac{1}{2}\int\frac{\mathrm{d}^d k}{(2\pi)^d}\tilde{D}(k)\tilde{\mathcal{H}}^{(n+2)}(p_1, p_2, \ldots, p_n, k, -k; g)$$

$$- \frac{1}{2}\sum_I \tilde{D}(p_0)\tilde{\mathcal{H}}^{(l+1)}(p_{i_1},\ldots,p_{i_l},p_0)\tilde{\mathcal{H}}^{(n-l+1)}(p_{i_{l+1}},\ldots,p_{i_n},-p_0). \quad (16.73)$$

The equation $n = 2$ reads

$$\beta(g)\frac{\mathrm{d}}{\mathrm{d}g}\tilde{\mathcal{V}}^{(2)}(p;g)$$

$$= \left(\eta - 2 + \sum_\mu p^\mu\frac{\partial}{\partial p^\mu}\right)\tilde{\mathcal{V}}^{(2)}(p;g) + \eta\Delta^{-1}(p)$$

$$+ \frac{1}{2}\int\frac{\mathrm{d}^d k}{(2\pi)^d}\tilde{D}(k)\tilde{\mathcal{H}}^{(4)}(p,-p,k,-k;g) - \tilde{D}(p)\left(\tilde{\mathcal{V}}^{(2)}(p;g)\right)^2. \quad (16.74)$$

Order g. The condition (16.57) implies

$$\beta(g) = -\varepsilon g + 2g\eta(g) + \frac{1}{2}\int\frac{\mathrm{d}^d k}{(2\pi)^d}\tilde{D}(k)\tilde{\mathcal{H}}^{(6)}(0,0,0,0,k,-k;g) - 4g\tilde{D}(0)\tilde{\mathcal{V}}^{(2)}(0;g). \quad (16.75)$$

One verifies that η remains of order g^2 and thus all terms except the first one are at least of order g^2:

$$\beta(g) = -\varepsilon g + O(g^2).$$

By contrast with the case $d = 4$, the left-hand side contributes at leading order.
At order g, $\mathcal{V}^{(2)}(p;g)$ is a constant and thus

$$\mathcal{V}^{(2)}(p;g) \equiv \mathcal{V}^{(2)}(0;g) = r_c(g) = \frac{D(0)}{2(d-2)}g + O(g^2).$$

Note the relation
$$(d-2)\Delta(0) + D(0) = 0, \qquad (16.76)$$
and thus
$$r_c(g) = -\tfrac{1}{2}g\Delta(0) + O(g^2).$$

Order g^2. It is now necessary to determine $\mathcal{H}_*^{(6)}(p_i)$, which is of order g^2, at leading order. At this order, it satisfies

$$\left(2 + \sum_{j,\mu} p_j^\mu \frac{\partial}{\partial p_j^\mu}\right) \mathcal{H}^{(6)}(p_i; g) = g^2 \left[\tilde{D}(p_1 + p_2 + p_3) + 9 \text{ terms}\right] + O(g^3, g^2\varepsilon). \qquad (16.77)$$

Introducing the function (16.66), the regular solution of equation (16.77) can then be written as

$$\mathcal{H}^{(6)}(p_i; g) = -g^2 \left[\tilde{D}_*(p_1 + p_2 + p_3) + 9 \text{ terms}\right] + O(g^3).$$

The function β is then determined up to order g^2:

$$\beta(g) = -\varepsilon g + g^2 \left[3\tilde{B}_*(0) - 2D(0)\tilde{D}_*(0)\right] + 2g^2\Delta(0)\tilde{D}(0) + O(g^3)$$

with

$$\tilde{B}_*(p) = -\int \frac{d^d k}{(2\pi)^d} \tilde{D}(k)\tilde{D}_*(p-k).$$

In the framework of the ε-expansion, one can replace the contributions of order g^2 by their values at $d = 4$, which have already been calculated (equation (16.70)). One infers

$$\beta(g) = -\varepsilon g + \frac{3}{16\pi^2}g^2 + O(g^3, g^2\varepsilon),$$

in agreement with the result (10.21), and one recovers the non-Gaussian fixed point (10.22), $g^* = 16\pi^2\varepsilon/3 + O(\varepsilon^2)$.

As for $d = 4$, one then sets

$$F^{(4)}(p_i, g) = \tilde{\mathcal{H}}^{(4)}(p_i; g) - \tilde{\mathcal{H}}^{(4)}(0; g) = O(g^2). \qquad (16.78)$$

At order g^2, the equation for $F^{(4)}(p_i; g)$ becomes

$$-\left(\varepsilon + \sum_{j,\mu} p_j^\mu \frac{\partial}{\partial p_j^\mu}\right) F^{(4)}(p_i; g)$$
$$= \tfrac{1}{2}g^2 \left[-D(0)\tilde{D}_*(p_1) + 3 \text{ terms} + 2\tilde{B}_*(p_1 + p_2) + 2 \text{ terms}\right]$$
$$+ \tfrac{1}{2}g^2\Delta(0)\left[\tilde{D}(p_1) + 3 \text{ terms}\right] - (\mathbf{p} = 0).$$

One verifies that the regular solution is

$$F^{(4)}(p_i;g) = -\tfrac{1}{2}g^2 B(p_1+p_2) + 2 \text{ terms} + \tfrac{1}{2}g^2 \Delta(0)\left[\tilde{D}_*(p_1) - \tilde{D}_*(0) + 3 \text{ terms}\right] \quad (16.79)$$

with

$$B(p) = \int \frac{d^d k}{(2\pi)^d} \tilde{D}_*(k) \left[\tilde{D}_*(p+k) - \tilde{D}_*(k)\right].$$

It is then possible to calculate $\eta(g)$ at order g^2. At order ε^2, only the value of the function $\eta(g)$ at $d=4$ is needed, which has already been obtained at the end of Section 16.5.1.

16.7 Local stability of the fixed point

Once a fixed point $\mathcal{H}_*$ has been determined, one can expand equation (16.51) in the vicinity of the fixed point:

$$\mathcal{H}(\lambda) = \mathcal{H}_* + E(\lambda).$$

One then obtains the linearized RGE

$$\lambda \frac{d}{d\lambda} E(\lambda) = \mathcal{L}_* E(\lambda),$$

where the linear operator $\mathcal{L}_*$, after an integration by parts, takes the form

$$\mathcal{L}_* = \int d^d x\, \phi(x) \left[\tfrac{1}{2}(d+2-\eta) + \sum_\mu x^\mu \frac{\partial}{\partial x^\mu}\right] \frac{\delta}{\delta\phi(x)}$$

$$+ \int d^d x\, d^d y\, D(x-y) \left[-\frac{1}{2}\frac{\delta^2}{\delta\phi(x)\delta\phi(y)} + \frac{\delta\mathcal{H}_*}{\delta\phi(x)} \frac{\delta}{\delta\phi(y)}\right]$$

$$- \int d^d x\, d^d y\, L(x-y)\phi(x) \frac{\delta}{\delta\phi(y)}. \quad (16.80)$$

We denote by ℓ the eigenvalues and $E_\ell \equiv E_\ell(\lambda=1)$ the eigenvectors of $\mathcal{L}_*$:

$$\mathcal{L}_* E_\ell = \ell E_\ell, \quad (16.81)$$

and thus

$$E_\ell(\lambda) = \lambda^\ell E_\ell(1).$$

Equation (16.81) can be written more explicitly in terms of the components $\tilde{E}_\ell^{(n)}(p_i)$ of E_ℓ as

$$\ell \tilde{E}_\ell^{(n)}(p_i)$$

$$= \left(d - \tfrac{1}{2}n(d-2+\eta) - \sum_j \tilde{D}(p_j)\tilde{\Delta}^{-1}(p_j) - \sum_{j,\mu} p_j^\mu \frac{\partial}{\partial p_j^\mu}\right) \tilde{E}_\ell^{(n)}(p_i)$$

$$- \frac{1}{2}\int \frac{d^d k}{(2\pi)^d} \tilde{D}(k) \tilde{E}_\ell^{(n+2)}(p_1, p_2, \ldots, p_n, k, -k)$$

$$+ \sum_I \tilde{D}(p_0) \tilde{E}_\ell^{(l+1)}(p_{i_1},\ldots,p_{i_l},p_0) \tilde{\mathcal{H}}_*^{(n-l+1)}(p_{i_{l+1}},\ldots,p_{i_n},-p_0). \quad (16.82)$$

16.7.1 Gaussian fixed point

At the Gaussian fixed point, $\eta = 0$ and $\mathcal{H}_*$ reduces to a quadratic Hamiltonian since $\mathcal{V}_*(\phi) = 0$, and the operator $\mathcal{L}_*$ to

$$\mathcal{L}_* = \int d^d x \, \phi(x) \left[\frac{1}{2}(d+2) + \sum_\mu x^\mu \frac{\partial}{\partial x^\mu} \right] \frac{\delta}{\delta \phi(x)}$$

$$- \frac{1}{2} \int d^d x \, d^d y \, D(x-y) \frac{\delta^2}{\delta \phi(x) \delta \phi(y)}.$$

The eigenvectors are obtained by choosing all $\mathcal{H}^{(n)}$ vanishing for $n > N$. The coefficient of the term of degree N in ϕ then satisfies the homogeneous equation

$$\ell \tilde{E}_\ell^{(N)}(p_i) = \left(d - \frac{1}{2} N(d-2) - \sum_{j,\mu} p_j^\mu \frac{\partial}{\partial p_j^\mu} \right) \tilde{E}_\ell^{(N)}(p_i),$$

which is identical to equation (9.56) and which is an eigenvalue equation. The solutions are homogeneous polynomials in the momenta. If r is the degree in the variables p_i, the eigenvalue is given by

$$\ell = d - \tfrac{1}{2} N(d-2) - r.$$

The other coefficients $\mathcal{H}^{(n)}$, $n < N$, are then entirely determined by the equations

$$\left(\frac{1}{2}(N-n)(d-2) + r - \sum_{j,\mu} p_j^\mu \frac{\partial}{\partial p_j^\mu} \right) \tilde{E}_\ell^{(n)}(p_i)$$

$$= \frac{1}{2} \int \frac{d^d k}{(2\pi)^d} \tilde{D}(k) \tilde{E}_\ell^{(n+2)}(p_1, p_2, \ldots, p_n, k, -k). \tag{16.83}$$

One may be surprised by the occurrence of these additional terms. In fact, one can verify that if one sets

$$E(\phi) = \exp\left[-\frac{1}{2} \int d^d x \, d^d y \, \Delta(x-y) \frac{\delta^2}{\delta \phi(x) \delta \phi(y)} \right] \Omega(\phi),$$

the functional $\Omega(\phi)$ satisfies the simpler eigenvalue equation

$$\ell \Omega_\ell(\phi) = \int d^d x \, \phi(x) \left[\frac{1}{2}(d+2) + \sum_\mu x^\mu \frac{\partial}{\partial x^\mu} \right] \frac{\delta}{\delta \phi(x)} \Omega_\ell(\phi),$$

whose solutions are then simple monomials.

The linear operator that transforms $\Omega(\phi)$ into $E(\phi)$,

$$\exp\left[-\frac{1}{2} \int d^d x \, d^d y \, \Delta(x-y) \frac{\delta^2}{\delta \phi(x) \delta \phi(y)} \right],$$

replaces all monomials in ϕ that contribute to Ω by their *normal products*. The normal product $E^{(N)}(\phi)$ of a monomial of degree N in ϕ is such that, for all $n < N$, the Gaussian correlation functions

$$\left\langle \prod_{i=1}^{n} \phi(x_i) E^{(N)}(\phi) \right\rangle$$

with the measure $\mathrm{e}^{-\mathcal{H}_*}$ vanish. Let us point out that the definition of *normal products* depends explicitly on the choice of the Gaussian measure.

For what follows, it is useful to calculate the normal product corresponding to $\int \mathrm{d}^d x\, \phi^3(x)$ which, in the Gaussian approximation, corresponds to the eigenvalue $\frac{1}{2}(6-d)$. One finds

$$\frac{1}{3!}\int \mathrm{d}^d x\, \phi^3(x) \mapsto \frac{1}{3!}\int \mathrm{d}^d x\, \phi^3(x) - \frac{1}{2}\Delta(0)\int \mathrm{d}^d x\, \phi(x).$$

Similarly, for $\int \mathrm{d}^d x\, \phi^4(x)$, one finds

$$\frac{1}{4!}\int \mathrm{d}^d x\, \phi^4(x) \mapsto \frac{1}{4!}\int \mathrm{d}^d x\, \phi^4(x) - \frac{1}{4}\Delta(0)\int \mathrm{d}^d x\, \phi^2(x) + \frac{1}{8}\Delta^2(0).$$

One recognizes the term of order ε in the fixed-point Hamiltonian (Section 16.5.2).

16.7.2 Dimension $d = 4 - \varepsilon$: ϕ^2 and ϕ^4 perturbations

For $n = 2$, at order ε, equation (16.82) reduces to

$$\ell \tilde{E}_\ell^{(2)}(p) = \left(2 + \tfrac{1}{2} g^* D(0)\tilde{D}(p) - \sum_\mu p^\mu \frac{\partial}{\partial p^\mu}\right) \tilde{E}_\ell^{(2)}(p)$$

$$- \frac{1}{2}\int \frac{\mathrm{d}^d k}{(2\pi)^d} \tilde{D}(k) \tilde{E}_\ell^{(4)}(p, -p, , k, -k). \tag{16.84}$$

Since we consider a perturbation that in the Gaussian limit is a constant t, we set

$$\tilde{E}_\ell^{(2)}(p) = t + O(\varepsilon).$$

It is consistent to assume $\tilde{E}_\ell^{(4)}(p)$ of order ε and $\tilde{E}_\ell^{(6)}(p)$ of order ε^2. Then, the eigenvalue ℓ satisfies

$$\ell = 2 + O(\varepsilon),$$

and $\tilde{E}_\ell^{(4)}(p)$ satisfies at leading order

$$2\tilde{E}_\ell^{(4)}(p_i) = -\sum_{j,\mu} p_j^\mu \frac{\partial}{\partial p_j^\mu} \tilde{E}_\ell^{(4)}(p_i) + g^* t \sum_i \tilde{D}(p_i) + O(\varepsilon^2).$$

Equation (16.66) enables integrating:
$$\tilde{E}_\ell^{(4)}(p_i) = -g^* t \sum_i \tilde{D}_*(p_i) + O(\varepsilon^2).$$

Finally, this expression can be inserted into equation (16.84). Then, at leading order,

$$\left(2 - \ell + \tfrac{1}{2}g^* D(0)\tilde{D}(p) + g^* \int \frac{\mathrm{d}^4 k}{(2\pi)^4} \tilde{D}(k)\left(\tilde{D}_*(p) + \tilde{D}_*(k)\right) - \sum_\mu p^\mu \frac{\partial}{\partial p^\mu}\right)$$
$$\times \tilde{E}_\ell^{(2)}(p) = 0.$$

In the limit $p = 0$, using again the condition $\tilde{E}^{(2)}(p=0) \neq 0$ and the regularity at $p = 0$, one finds

$$\ell = 2 + \tfrac{1}{2}g^* D(0)\tilde{D}(0) + g^* D(0)\tilde{D}_*(0) + g^* \int \frac{\mathrm{d}^4 k}{(2\pi)^4} \tilde{D}(k)\tilde{D}_*(k).$$

The two first terms cancel and in the integral one recognizes $\tilde{B}_*(0)$. Thus,

$$\ell = d_t = 2 - \tfrac{1}{3}\varepsilon + O(\varepsilon^2),$$

in agreement with the result (10.32) obtained by perturbative quantum field theory methods.

ϕ^4 *perturbation.* The same method, but now starting from $\tilde{E}^{(4)} = O(1)$, allows calculating the exponent ω, which corresponds to a variation of g and, thus, to the derivative of the β-function as in the field theory RGE. The exponent

$$d_g = -\omega = -\beta'(g^*) = -\varepsilon + O(\varepsilon^2)$$

is the eigenvalue corresponding to a perturbation of ϕ^4 type. One then infers the dimension d_{ϕ^4} of the eigen-operator that is identical to $\phi^4(x)$ at leading order:

$$d_{\phi^4} = d - d_g = 4.$$

Since d_{ϕ^4} is larger than d, this operator is irrelevant.

16.7.3 Perturbations breaking the $\mathbb{Z}_2$ symmetry

We now study the eigenvalues associated to relevant operators breaking the $\mathbb{Z}_2$ reflection symmetry. We want to show that these eigenvalues can be expressed in terms of known exponents.

For this purpose, we start from the fixed-point equation

$$\mathbf{E}_*(\phi) = \int d^d x\, \phi(x) \left[\tfrac{1}{2}(d+2-\eta) + \sum_\mu x^\mu \frac{\partial}{\partial x^\mu} \right] \frac{\delta \mathcal{H}_*(\phi)}{\delta \phi(x)}$$
$$- \frac{1}{2} \int d^d x\, d^d y\, D(x-y) \left[\frac{\delta^2 \mathcal{H}_*}{\delta\phi(x)\delta\phi(y)} - \frac{\delta \mathcal{H}_*}{\delta\phi(x)} \frac{\delta \mathcal{H}_*}{\delta\phi(y)} \right]$$
$$- \int d^d x\, d^d y\, L(x-y) \frac{\delta \mathcal{H}_*}{\delta\phi(x)} \phi(y) = 0. \tag{16.85}$$

By acting on this equation with the operator $\int d^d z\, \delta/\delta\phi(z)$, one infers another equation:

$$\int d^d z\, \frac{\delta \mathbf{E}_*(\phi)}{\delta\phi(z)} = 0. \tag{16.86}$$

We set

$$E^{(3)}(\phi) = \int d^d z\, \frac{\delta \mathcal{H}_*}{\delta\phi(z)}.$$

We note that

$$\int d^d x\, L(x) = \tilde{D}(0)\tilde{\Delta}^{-1}(0) = 0.$$

After an integration by parts, equation (16.86) can then be written as

$$\left[\mathcal{L}_* - \tfrac{1}{2}(d-2+\eta) \right] E^{(3)}(\phi) = 0.$$

We have thus identified an eigenvector, $E^{(3)}(\phi)$, and the corresponding eigenvalue

$$\ell = \tfrac{1}{2}(d-2+\eta).$$

At first order in ε, this relevant eigenvector has the form

$$E^{(3)}(\phi) = r_{c1} g^* \int d^d x\, \phi(x) + \frac{1}{3!} g^* \int d^d x\, \phi^3(x) + O(\varepsilon^2).$$

In the Gaussian limit, it corresponds to the operator $\int d^d x\, \phi^3(x)$. The linear term in ϕ is a correction to ϕ^3 that eliminates a component that is more relevant. Indeed, the action of $\mathcal{L}_*$ on a term linear in ϕ yields

$$\mathcal{L}_* \int d^d z\, \phi(z) = \tfrac{1}{2}(d+2-\eta) \int d^d z\, \phi(z) + \tilde{D}(0) E^{(3)}.$$

One then verifies that

$$\ell = \tfrac{1}{2}(d+2-\eta),$$

is also an eigenvalue, which corresponds to the eigenvector

$$E^{(1)}(\phi) = (1-\eta/2) \int d^d x\, \phi(x) + \tfrac{1}{2}\tilde{D}(0) E^{(3)}(\phi).$$

In the Gaussian approximation, this eigenvector, which is the most relevant, reduces to $\int d^d z\, \phi(z)$ and is associated to the eigenvalue

$$\ell = \tfrac{1}{2}(d+2).$$

From relation (9.64), one also infers the dimension d_ϕ of the operator ϕ:

$$d_\phi = d - \tfrac{1}{2}(d+2-\eta) = \tfrac{1}{2}(d-2+\eta),$$

a result consistent with the initial interpretation of the exponent η as being related to the field renormalization.

Appendix

A1 Technical results

We gather in this section a few useful technical results.

A1.1 Γ, ψ, δ functions

The function $\Gamma(z)$, which interpolates $(z-1)!$ for non-integer arguments, can be defined by

$$\Gamma(z) = \int_0^\infty dt\, t^{z-1} e^{-t}\,. \tag{A1}$$

This integral representation defines an analytic function for $\operatorname{Re} z > 0$. The function satisfies the symmetry relation

$$\Gamma(z)\Gamma(1-z) = \frac{\pi}{\sin(\pi z)}$$

which provides a meromorphic analytic continuation to the whole complex plane, with simple poles for negative integer arguments.

Another identity is the duplication formula

$$\Gamma(2z) = \frac{2^{2z-1}}{\sqrt{\pi}} \Gamma(z)\Gamma(z+1/2).$$

In terms of the logarithmic derivative of Γ,

$$\psi(z) = \Gamma'(z)/\Gamma(z),$$

these relations become

$$\psi(z) - \psi(1-z) + \pi/\tan(\pi z) = 0, \quad 2\psi(2z) = 2\ln 2 + \psi(z) + \psi(z+1/2). \tag{A2}$$

One calls Euler's constant

$$\gamma = -\psi(1). \tag{A3}$$

In particular,

$$\Gamma(1+s) = 1 - \gamma s + O(s^2). \tag{A4}$$

Another useful function is

$$B(\alpha, \beta) = \int_0^1 dt\, t^{\alpha-1}(1-t)^{\beta-1} = \int_0^\infty \frac{t^{\alpha-1} dt}{(1+t)^{\alpha+\beta}} = \frac{\Gamma(\alpha)\Gamma(\beta)}{\Gamma(\alpha+\beta)}. \tag{A5}$$

A1.2 Dirac distribution

We have often referred to Dirac's function or distribution δ. The distribution $\delta(x - x_0)$ is a linear continuous mapping that, to any continuous function $f(x)$, associates the value $f(x_0)$. It is convenient to represent the action of δ by an integral of the form

$$\int dx\, \delta(x - x_0) f(x) = f(x_0).$$

Applying the definition to the function e^{ipx} with p real, one obtains

$$\int dx\, \delta(x)\, e^{ipx} = 1.$$

The Fourier transform of δ thus is the constant function $f \equiv 1$. Conversely,

$$\delta(x) = \frac{1}{2\pi} \int dp\, e^{-ipx}.$$

The distribution δ can be defined as the limit, in the sense of distributions, of ordinary functions, for example,

$$\delta(x) = \lim_{\varepsilon \to 0} \frac{1}{\varepsilon\sqrt{2\pi}}\, e^{-x^2/2\varepsilon^2}.$$

We have also used the sign function:

$$\mathrm{sgn}(x) = +1 \quad \text{for} \quad x > 0, \quad \mathrm{sgn}(x) = -1 \quad \text{for} \quad x < 0.$$

It is simply related to another function, the step or Heaviside function $\theta(x)$ defined by

$$\theta(x) = 1 \quad \text{for} \quad x > 0, \quad \theta(x) = 0 \quad \text{for} \quad x < 0 \;\Rightarrow\; \mathrm{sgn}(x) = \theta(x) - \theta(-x).$$

Finally, in the sense of distributions,

$$\frac{d}{dx}\theta(x) = \delta(x), \qquad \frac{d}{dx}\mathrm{sgn}(x) = 2\delta(x).$$

Moreover, we have also defined the d-dimensional δ-function,

$$\delta^{(d)}(\mathbf{x}) = \prod_{i=1}^{d} \delta(x_i).$$

A1.3 The massive propagator in dimension 2

The massive propagator (8.41) in d dimensions is given by

$$\Delta(x,m) = \frac{1}{(2\pi)^d} \int d^d x \frac{e^{ipx}}{p^2 + m^2} = \frac{2}{(4\pi)^{d/2}} \left(\frac{2m}{x}\right)^{d/2-1} K_{1-d/2}(mx),$$

where the function K_ν is defined by equation (2.54).

For $d > 2$, when $m \to 0$, the propagator has the finite limit (equation (8.40))

$$\Delta(x,0) = \frac{2^{d-2}}{(4\pi)^{d/2}} \Gamma(d/2-1) \frac{1}{x^{d-2}}.$$

For $d \to 2$, the massless propagator diverges as

$$\Delta(x,0) = \frac{1}{2\pi(d-2)} - \frac{1}{4\pi}\left(\gamma + \ln\pi + \ln x^2\right) + O(d-2), \qquad (A6)$$

where the expansion (A4) of the Γ-function has been used. The behaviour of $\Delta(x,m)$ for $d = 2$ when $m \to 0$, can be inferred from the behaviour of $K_0(z)$ for $z \to 0$ (equation (8.41)):

$$K_0(z) = \frac{1}{2} \int_0^\infty \frac{dt}{t} e^{-z(t+1/t)/2} \underset{z \to 0}{=} -\ln(z/2) - \gamma + O(z).$$

(γ is Euler's constant (A3).) Therefore,

$$\Delta(x,m) \underset{m \to 0}{=} -\frac{1}{2\pi}\left(\ln(mx/2) + \gamma\right) + O(m). \qquad (A7)$$

A1.4 Operator determinants

The calculation of Gaussian field integrals generates determinants of integral or differential operators ((6.31), (10.9), (12.26), (14.8)). The evaluation of such determinants can be reduced, if needed after some simple transformations, to the calculation of determinants of operators of the form $\mathbf{M} = \mathbf{1} + \mathbf{K}$. Provided the traces of all powers of $\mathbf{K}$ exist, the identity

$$\ln \det \mathbf{M} \equiv \operatorname{tr} \ln \mathbf{M}, \qquad (A8)$$

expanded in powers of the kernel $\mathbf{K}$:

$$\ln \det(\mathbf{1} + \mathbf{K}) = \sum_{p=1}^\infty \frac{(-1)^{p+1}}{p} \operatorname{tr} \mathbf{K}^p, \qquad (A9)$$

is often useful if the series converges.

When the operators $\mathbf{K}$ can be represented by kernels of the form $K(x,y)$, the product of two operators $\mathbf{K}_1$, $\mathbf{K}_2$ is defined by

$$[\mathbf{K}_1 \mathbf{K}_2](x,y) = \int \mathrm{d}z\, K_1(x,z) K_2(z,y),$$

and the trace by

$$\operatorname{tr} \mathbf{K} = \int \mathrm{d}x\, K(x,x).$$

(One can verify that it obeys the cyclic property.)

The successive traces of powers of $\mathbf{K}$ can then be written more explicitly as

$$\operatorname{tr} \mathbf{K}^p = \int \mathrm{d}x_1 \cdots \mathrm{d}x_p\, K(x_1,x_2) K(x_2,x_3) \cdots K(x_p,x_1).$$

Remark. The determinant of an $n \times n$ matrix, of the form $\mathbf{1} + \lambda \mathbf{K}$, is an n degree polynomial in λ. If one uses the preceding identity, one finds an infinite series. Expressing that all terms beyond order n vanish, one finds identities between the traces of powers of the matrix $\mathbf{K}$.

A1.5 Orthogonal group

The $O(N)$ group is the group of real orthogonal $N \times N$ matrices. An element of $O(N)$ satisfies

$$\mathbf{O}\mathbf{O}^T = \mathbf{1},$$

where T indicates matrix transposition. It is the group of matrices that preserves the scalar product in $\mathbb{R}^N$. Geometrically, it is the group of rotations–reflections in N-dimensional space. The orthogonality relation implies

$$\det \mathbf{O}\mathbf{O}^T = (\det \mathbf{O})^2 = 1$$

and, thus,

$$\det \mathbf{O} = \pm 1.$$

The matrices with determinant $+1$ form a subgroup of $O(N)$ denoted by $SO(N)$, which geometrically corresponds to rotations (transformations that preserve the orientation of the system of coordinates). The group $SO(N)$ is connected to the identity and all matrices of $SO(N)$ can be written as a finite product of matrices belonging to a neighbourhood of the identity. In an infinitesimal neighbourhood of the identity, a matrix belonging to $SO(N)$ can be expanded in the form

$$\mathbf{O} = \mathbf{1} + \mathbf{L} + O(\mathbf{L}^2) \text{ with } \mathbf{L} + \mathbf{L}^T = 0.$$

All antisymmetric matrices form a vector space on $\mathbb{R}$ of dimension $N(N-1)/2$. Let L_α be a basis in this space. The group property of orthogonal matrices leads to the commutation relation

$$[L_\alpha, L_\beta] = \sum_\gamma f_{\alpha\beta\gamma} L_\gamma,$$

where $f_{\alpha\beta\gamma} = -f_{\beta\alpha\gamma}$ are real numbers. These commutation relations, or Lie products, give to the vector space of antisymmetric matrices a Lie algebra structure, $\mathcal{L}(SO(N))$, and the coefficients $f_{\alpha\beta\gamma}$ are called structure constants. Finally, any matrix in $SO(N)$ can be expressed as the exponential of an element of the Lie algebra as

$$\mathbf{O} = \mathrm{e}^{\mathbf{L}}, \quad \mathbf{L} \in \mathcal{L}(SO(N)).$$

A basis of the Lie algebra thus forms a set of generators of $SO(N)$. To generate $O(N)$, one must add one reflection.

In the study of spontaneous symmetry breaking, we have considered the subgroup $O(N-1)$ of $O(N)$ that leaves a given vector $\mathbf{v} \neq 0$ invariant. This condition expressed in a neighbourhood of the identity yields the property of the corresponding elements of $\mathcal{L}(SO(N-1))$:

$$\mathbf{L}\mathbf{v} = 0, \quad \mathbf{L} \in \mathcal{L}(SO(N-1)).$$

The group $SO(N-1)$ has $(N-1)(N-2)/2$ generators. To the complementary set of the other $(N-1)$ remaining generators of $SO(N)$, are associated the $(N-1)$ Goldstone modes.

A2 Fourier transformation: Decay and regularity

In this section, we recall, through a few examples useful for this work, the relation between asymptotic decay for large argument of positive measures and the regularity of their Fourier transform. The reader interested in more details is referred to the relevant mathematical literature.

We first examine the example of discrete measures defined in $\mathbb{Z}$, which is specially simple.

A2.1 Positive discrete measures and Fourier series

Exponential decay. We consider a positive discrete measure ρ_n defined on the lattice of integer points of the real line:

$$\rho_n \geq 0, \quad \sum_{n=-\infty}^{+\infty} \rho_n = 1. \tag{A10}$$

We assume that for $|n| \to \infty$ ρ_n decays exponentially:

$$\rho_n < M \mathrm{e}^{-\mu|n|}, \quad \mu > 0.$$

Therefore, all moments

$$m_p = \sum_{n=-\infty}^{+\infty} n^p \rho_n$$

of ρ_n are finite. The analytic function defined by the Laurent series

$$f(z) = \sum_{n=-\infty}^{+\infty} \rho_n z^{n-1},$$

is analytic and uniform in the ring $e^{-\mu} < |z| < e^{\mu}$ where the Laurent series converges.

The periodic function $\tilde{\rho}(\theta)$, with period 2π,

$$\tilde{\rho}(\theta) = e^{-i\theta} f(e^{-i\theta}) = \sum_{n=-\infty}^{+\infty} e^{-in\theta} \rho_n \quad \Rightarrow \quad \tilde{\rho}^*(\theta) = \tilde{\rho}(-\theta),$$

thus is analytic in the strip $|\operatorname{Im}\theta| < \mu$. Moreover, the positivity of the measure implies $|\tilde{\rho}(\theta)| \leq \tilde{\rho}(0) = 1$.

Conversely, the Fourier coefficients ρ_n are given by

$$\rho_n = \frac{1}{2\pi} \int_0^{2\pi} d\theta \, e^{in\theta} \tilde{\rho}(\theta). \tag{A11}$$

If the function f is analytic and uniform in the ring $e^{-\mu} < |z| < e^{\mu}$, the coefficients of its Laurent series expansion are also given by

$$\rho_n = \frac{1}{2i\pi} \oint_{|z|=1} dz \, z^{-n} f(z).$$

The analyticity domain of f implies the convergence of the Laurent series. Indeed, depending whether n is positive or negative, one can deform the initial circle into circles of radius e^{μ} or $e^{-\mu}$. One infers

$$\forall R, \ 0 < R < e^{\mu} \ : \ \lim_{|n| \to \infty} \rho_n R^n = 0.$$

These results, which establish a relation between the regularity properties of a function and the decay properties of the coefficients of its Fourier expansion, can be generalized in different ways.

Fast decay. Let us assume, for example, that ρ_n decreases, when $n \to \infty$, faster than any power of n:

$$\forall p \in \mathbb{N} \ \exists A_p \text{ such that } \rho_n \leq \frac{A_p}{|n|^p + 1}.$$

All moments

$$m_p = \sum_{n=-\infty}^{+\infty} n^p \rho_n,$$

of ρ_n are still finite. It is actually a necessary and sufficient condition.

It is then simple to verify that the periodic function $\tilde{\rho}(\theta)$ is differentiable, and its derivative is obtained by differentiating term by term the representation (A11):

$$\tilde{\rho}'(\theta) = \sum_{n=-\infty}^{+\infty} (-in)\, e^{-in\theta}\, \rho_n .$$

The series converges for all θ since

$$|\tilde{\rho}'(\theta)| \le |\tilde{\rho}'(0)| \le \sum_n |n|\rho_n < \infty .$$

The measure $n\rho_n$ again being fast decaying, the argument can be iterated and also applies to the successive derivatives:

$$\tilde{\rho}^{(p)}(\theta) = \sum_{n=-\infty}^{+\infty} (-in)^p\, e^{-in\theta}\, \rho_n ,$$

and the function $\tilde{\rho}(\theta)$ thus is indefinitely differentiable. For all θ,

$$\left|\tilde{\rho}^{(p)}(\theta)\right| \le \left|\tilde{\rho}^{(p)}(0)\right| \le \sum_n |n|^p \rho_n < \infty .$$

Conversely, let us assume that $\tilde{\rho}(\theta)$ is indefinitely differentiable. Since $\tilde{\rho}(\theta)$ is differentiable, one can integrate by parts (for $n \ne 0$)

$$\rho_n = \frac{1}{2\pi}\frac{i}{n}\int_0^{2\pi} d\theta\, e^{in\theta}\, \tilde{\rho}'(\theta).$$

More generally, integrating p times, one finds

$$\rho_n = \frac{1}{2\pi}\frac{i^p}{n^p}\int_0^{2\pi} d\theta\, e^{in\theta}\, \tilde{\rho}^{(p)}(\theta)$$

and thus

$$|\rho_n| \le \frac{1}{n^p}\left|\tilde{\rho}^{(p)}(0)\right| \quad \forall p.$$

Algebraic decay. The normalization condition (A10) alone, which implies uniform convergence of the series $\sum_n e^{in\theta}\rho_n$, implies the continuity of $\tilde{\rho}(\theta)$. More generally, the condition

$$\sum_n \rho_n |n|^p < \infty, \quad p \in \mathbb{N},$$

implies the existence and the continuity of the pth derivative of $\tilde{\rho}(\theta)$. This condition is verified if

$$\rho_n \le \frac{R}{|n|^{\sigma+1}+1} \quad \text{with} \quad \sigma > p.$$

Conversely, if $\tilde{\rho}(\theta)$ is a continuous periodic function with period 2π, the Fourier coefficients go to zero for $|n| \to \infty$. Indeed (choosing, e.g., $n > 0$)

$$\rho_n = \frac{1}{2\pi} \int_0^{2\pi} d\theta \, e^{in\theta} \, \tilde{\rho}(\theta) = \frac{1}{2\pi n} \int_0^{2\pi n} d\tau \, e^{i\tau} \, \tilde{\rho}(\tau/n)$$

$$= \frac{1}{2\pi n} \sum_{p=1}^{n} \int_{2\pi(p-1)}^{2\pi p} d\tau \, e^{i\tau} \, \tilde{\rho}(\tau/n)$$

$$= \frac{1}{4\pi^2} \int_0^{2\pi} d\tau \, e^{i\tau} \, \frac{2\pi}{n} \sum_{p=1}^{n} \tilde{\rho}\big(\tau/n + 2\pi(p-1)/n\big).$$

The sum over p is a Riemann sum that converges for $n \to \infty$ toward an integral since $\tilde{\rho}$ is continuous:

$$\lim_{n \to \infty} \frac{2\pi}{n} \sum_{p=1}^{n} \tilde{\rho}\big(\tau/n + 2\pi(p-1)/n\big) = \int_0^{2\pi} ds \, \tilde{\rho}(s),$$

and the convergence is uniform in θ. Subtracting from the sum this limit does not change the integral over τ and, thus,

$$|\rho_n| \le \frac{1}{4\pi^2} \int_0^{2\pi} d\tau \left| \frac{2\pi}{n} \sum_{p=1}^{n} \tilde{\rho}\big(\tau/n + 2\pi(p-1)/n\big) - \int_0^{2\pi} ds \, \tilde{\rho}(s) \right|$$

which goes to zero. Note that only the uniform convergence in θ of the Riemann sum has played a role, and this enables relaxing the continuity condition.

The existence of the pth derivative allows integrating by parts and thus proving a decay at least of order $|n|^{-p}$.

More precise results can sometimes be obtained. For example if $\tilde{\rho}(\theta)$ has an isolated singularity at $\theta = 0$ in $|\theta|^{\sigma}$, $\sigma > 0$, ρ_n decays like $|n|^{-1-\sigma}$

$$\tilde{\rho}(\theta) = A|\theta|^{\sigma} + \text{more regular part} \Rightarrow \rho_n \underset{|n| \to \infty}{\propto} |n|^{-1-\sigma}.$$

This result can be proved, for example, by integrating by parts $[\sigma]+1$ times, $[\sigma]$ being the integer part of σ. One can then calculate the Fourier transform of $|\theta|^{\sigma-[\sigma]-1}$ explicitly.

d-dimensional lattice. The preceding results can be generalized to the example of the transition probabilities considered in Section 3.3.

Let us assume for $\rho(\mathbf{n})$, $\mathbf{n} \in \mathbb{Z}^d$, exponential decay properties,

$$\rho(\mathbf{n}) < M \, e^{-\mu |\mathbf{n}|}, \quad \mu > 0.$$

The function

$$\tilde{\rho}(\mathbf{k}) = \sum_{\mathbf{n} \in \mathbb{Z}^d} e^{-i\mathbf{k}\cdot\mathbf{n}} \, \rho(\mathbf{n}),$$

is analytic in the tube
$$|\operatorname{Im} \mathbf{k}| < \mu.$$

Conversely, if one assumes for $\tilde{\rho}(\mathbf{k})$ the analyticity domain $|\operatorname{Im} \mathbf{k}| \le \mu$, then in the integral over a Brillouin zone,

$$\rho(\mathbf{n}) = \frac{1}{(2\pi)^d} \int d^d k \; e^{i\mathbf{n}\cdot\mathbf{k}} \tilde{\rho}(\mathbf{k}),$$

one can move the contour in the complex domain in the direction

$$\operatorname{Im} \mathbf{k} = \kappa \mathbf{n}, \quad \kappa = \mu/|\mathbf{n}|,$$

and one infers exponential decay.

A2.2 Fourier transformation

The Fourier transforms of positive measures and, more generally, of the functions with exponential or fast decay of the type considered in Section 3.1 also play an important role in various parts of this work. It is thus useful to investigate in which form the properties proved in the preceding section generalize. The main technical difference with the preceding section is that the Fourier transformation is now symmetric. In particular, the initial function and its Fourier transform have both regularity and decay properties and this somewhat complicates the analysis. Since our goal here is to emphasize the duality between regularity and decay, we illustrate this property only with examples where the results can be proved by simple methods.

For the physical problems that are studied in this work, we need as measures not only functions but also distributions in the mathematical sense as Dirac δ 'functions'. In the latter case, the functions $f(q)$ to be measurable, that is, to have a defined expectation value, must be *continuous*. It is then necessary to introduce the theory of distributions, and the interested reader is referred to the corresponding mathematical literature.

Continuous functions. For illustration purpose, we restrict here the discussion to continuous functions $\rho(q)$. Let us assume, for example, that $\rho(q)$ is an exponential decreasing function:

$$|\rho(q)| \le M\, e^{-\mu|q|}, \quad \mu > 0.$$

If $\rho(q)$ is a measure, all moments of the distribution exist and, in addition, they satisfy

$$|\langle q^p \rangle| \le M \int dq \; e^{-\mu|q|} |q|^p \le \frac{2M}{\mu^{p+1}} p!\,.$$

The Fourier transform

$$\tilde{\rho}(k) = \langle e^{-ikq} \rangle \equiv \int dq \; e^{-ikq} \rho(q) \qquad (A12)$$

of $\rho(q)$ is an analytic function for $|\operatorname{Im} k| < \mu$. Indeed,

$$|\tilde{\rho}(k)| \le M \int \mathrm{d}q\; \mathrm{e}^{-(\mu-|\operatorname{Im} k|)|q|} \le \frac{2M}{\mu - |\operatorname{Im} k|}. \tag{A13}$$

Similarly, if $\rho(q)$ decays faster than any power, its Fourier transform is indefinitely differentiable.

Finally, if the integral $\int \mathrm{d}q\, |\rho(q)||q^p|$ converges,

$$\int_{-\infty}^{+\infty} \mathrm{d}q\, |q|^p \rho(q) < \infty,$$

the pth derivative of $\tilde{\rho}(k)$ is continuous.

Reciprocal property. By arguments analogous to those used in the case of Fourier series, one proves that if the integral in the Fourier representation

$$\rho(q) = \frac{1}{2\pi} \int \mathrm{d}k\; \mathrm{e}^{\mathrm{i}kq}\, \tilde{\rho}(k),$$

converges absolutely and if $\tilde{\rho}(k)$ is continuous, $\rho(q)$ goes to zero for $|q| \to \infty$. Indeed,

$$\rho(q) = \frac{1}{2\pi} \int \mathrm{d}k\; \mathrm{e}^{\mathrm{i}kq}\, \tilde{\rho}(k) = \frac{1}{2\pi q} \int \mathrm{d}k\; \mathrm{e}^{\mathrm{i}k}\, \tilde{\rho}(k/q)$$

$$= \frac{1}{2\pi q} \sum_{p=-\infty}^{+\infty} \int_{2\pi(p-1)}^{2\pi p} \mathrm{d}k\; \mathrm{e}^{\mathrm{i}k}\, \tilde{\rho}(k/q)$$

$$= \frac{1}{4\pi^2} \int_0^{2\pi} \mathrm{d}k\; \mathrm{e}^{\mathrm{i}k}\, \frac{2\pi}{q} \sum_{p=-\infty}^{+\infty} \tilde{\rho}(k/q + 2\pi p/q).$$

The assumptions imply that the Riemann sum converges uniformly in k for $|q| \to \infty$ toward an integral that, integrated with $\mathrm{e}^{\mathrm{i}k}$, gives a vanishing contribution.

If $\tilde{\rho}(k)$ has a continuous pth derivative, repeated integrations by parts allow proving that $q^p \rho(q)$ goes to the zero, with the additional assumption that $\int \mathrm{d}k\, |\rho^{(p)}(k)|$ converges.

Similarly, in the case of indefinitely differentiable functions or functions analytic in a strip, one can prove fast or exponential decay of the Fourier transform provided some conditions of convergence at infinity are met.

With more general assumptions, this problem can be discussed in the framework of distributions.

A3 Phase transitions: General remarks

We have gathered here a few additional results relevant for phase transitions.

A3.1 Ground state of the transfer matrix: One dimension

We discuss the problem of the eigenstate with largest eigenvalue of the transfer matrix in the framework of Section 4.1. The notion of transfer matrix has been introduced in Section 4.1.2. Here, we prove that, in one dimension and with the general assumptions of Section 4.1, the eigenfunction $\psi_0(q)$ of the transfer matrix, corresponding to the eigenvalue with largest modulus τ_0, has a constant sign and can thus be chosen positive. Therefore, the eigenvalue τ_0 is *positive and simple*. Using a terminology borrowed from quantum mechanics, we call below the eigenvector of the transfer matrix the *ground state*. Then, the ground state is non-degenerate.

We first show that the function $\psi_0(q)$ has a constant sign and, thus, can be chosen positive.

Any square integrable function $\psi(q)$ satisfies the inequality

$$|\tau_0| \geq \frac{1}{\|\psi\|^2} \left| \int dq\, dq'\, \psi(q') T(q',q) \psi(q) \right|,$$

equality being only possible if $\psi(q)$ belongs to the eigenspace associated to τ_0.

Moreover, since $T(q',q)$ is a symmetric positive function,

$$\left| \int dq\, dq'\, \psi(q') T(q',q) \psi(q) \right| \leq \int dq\, dq'\, |\psi(q')| T(q',q) |\psi(q)|,$$

while

$$\|\psi\| = \| (|\psi|) \|.$$

Thus, if the function $\psi(q)$ has not a constant sign, the function $|\psi(q)|$ has a larger expectation value. We conclude that the function $\psi_0(q)$ has necessarily a constant sign.

A direct consequence of this result is that the eigenvalue τ_0 is positive and simple. Indeed, let us assume that two independent eigenvectors $\psi_0(q)$ and $\tilde\psi_0(q)$ exist corresponding to the eigenvalue τ_0. Since **T** is Hermitian, one can always choose them orthogonal:

$$\int dq\, \psi_0(q) \tilde\psi_0(q) = 0.$$

We then calculate the quantity

$$\int dq'\, dq\, \tilde\psi_0(q') T(q',q) \psi_0(q) = \tau_0 \int dq\, \tilde\psi_0(q) \psi_0(q) = 0,$$

where the condition that $\psi_0(q)$ is an eigenvector with the eigenvalue τ_0 has been used. Since the integrand in the right-hand side is the product of three positive functions and taking into account the special form (4.6) of the kernels T, that is, that $T(q',q)$ is strictly positive for all pairs $\{q,q'\}$, this is impossible.

A3.2 Order parameter and cluster property

When the largest eigenvalue of the transfer matrix is not simple, the proper definition of correlation functions in the infinite volume becomes a subtle question, and depends explicitly on the way the thermodynamic limit is taken; in particular, it can depend on the finite volume boundary conditions. This sensitivity to boundary conditions is another characteristic property of a several-phase region.

Let us examine, in this situation, the decay properties at large distance of correlation functions.

We consider the two-phase region of an Ising type system and call ψ_+, ψ_- the ground state eigenvectors of the transfer matrix that are exchanged by the symmetry operator $\mathbf{P}$ (the reflection that flips the spins),

$$\psi_- = \mathbf{P}\psi_+,$$

and are orthogonal. All vectors $\psi(\alpha)$ of the form

$$\psi(\alpha) = \cos\alpha\,\psi_+ + \sin\alpha\,\psi_-, \tag{A14}$$

are also eigenvectors of the transfer matrix with the same eigenvalue. Let us assume that it is possible to take the infinite volume limit in such a way that the vector $\psi(\alpha)$ is selected, in the sense that $\psi(\alpha)$ is the vector that appears in the expression (4.24) of correlation functions. Correlation functions then involve the action of products of spin matrices at different sites of the transverse lattice on the vector $\psi(\alpha)$. For simplicity reasons, we restrict the analysis to products of spin having the same position on the transverse lattice, and denote by $\mathbf{S}$ the corresponding matrix (defined in (7.50)). In what follows, we thus omit the position subscript σ.

We define

$$(\psi_+ \mathbf{S}\, \psi_+) = m, \tag{A15}$$

where m tend toward 1 in the Ising model in the zero-temperature limit. Then,

$$(\psi_- \mathbf{S}\, \psi_-) = (\psi_+ \mathbf{P}^{-1}\mathbf{S}\mathbf{P}\, \psi_+) = -m. \tag{A16}$$

Also,

$$(\psi_+ \mathbf{S}\psi_-) = -(\psi_- \mathbf{P}^{-1}\mathbf{S}\mathbf{P}\, \psi_+) = -(\psi_- \mathbf{S}\psi_+). \tag{A17}$$

Since $\mathbf{S}$ is a symmetric matrix,

$$(\psi_+ \mathbf{S}\psi_-) = 0. \tag{A18}$$

The matrix $\mathbf{S}$, restricted to the subspace $\{\psi_+, \psi_-\}$, is diagonal in the basis $\{\psi_+, \psi_-\}$. Equations (A15)–(A18) in particular imply:

$$(\psi(\alpha)\mathbf{S}\,\psi(\alpha)) = m\cos 2\alpha, \tag{A19}$$
$$(\psi(\alpha)\mathbf{S}\,\psi(\pi/2 + \alpha)) = m\sin 2\alpha. \tag{A20}$$

Except for $\alpha = \pi/4 \pmod{\pi/2}$, the spin expectation value does not vanish and this characterizes the two-phase region.

We now calculate what naively one would expect to be the connected two-point spin correlation function (see Section 4.3.2), that is, the two-point function of the quantity $\mathbf{S} - \langle \mathbf{S} \rangle$, for two points separated by a distance ℓ in the time direction but at the same transverse position. It is given by

$$W^{(2)}(\ell) = \frac{(\psi(\alpha)\,(\mathbf{S} - m\cos 2\alpha)\,\mathbf{T}^\ell\,(\mathbf{S} - m\cos 2\alpha)\,\psi(\alpha))}{(\psi(\alpha)\mathbf{T}^\ell\psi(\alpha))}. \qquad (A21)$$

The transfer matrix $\mathbf{T}$ projects on the ground states for $\ell \to \infty$:

$$\mathbf{T}^\ell = \tau_0^\ell \left(P_+ + P_- + O\left(e^{-\ell/\xi}\right) \right),$$
$$= \tau_0^\ell \left(P(\alpha) + P(\pi/2 + \alpha) + O\left(e^{-\ell/\xi}\right) \right),$$

where $P_+, P_-, P(\alpha)$ are the projectors on the corresponding vectors. One infers

$$W^{(2)}(\ell) \sim m^2 \sin^2 2\alpha + O\left(e^{-\ell/\xi}\right). \qquad (A22)$$

We note that the correlation functions satisfy the cluster property (see Section 4.3.2, equation (4.26)) only for $\alpha = n\pi$. The corresponding eigenvectors then are ψ_+, ψ_-, which are interchanged by $\mathbf{P}$. The reflection symmetry is spontaneously broken.

Let us point out that the correlation functions calculated by summing over all spin configurations correspond to $\alpha = \pi/4 \pmod{\pi/2}$ and thus do not satisfy the cluster property.

As we have pointed out in the example of Section 7.2, one of the two states $\psi_\pm$ is selected in the thermodynamic limit by the following procedure: one adds to the configuration energy a term coupled linearly to the order parameter (a magnetic field for spin systems). Such a term thus favours one phase. After having taken the thermodynamic limit, one then removes the symmetry breaking term. This procedure chooses one of the state ψ_+ or ψ_- according to the sign of the magnetic field. The correlation functions obtained by this procedure satisfy the cluster property in the infinite volume limit. An alternative method consists in taking the thermodynamic limit with as boundary conditions all spins on the boundary fixed and equal to the same value, $+1$ or -1.

One may wonder about the physical significance of such a procedure: in Section A3.3 we show that such systems in the phase of spontaneous symmetry breaking are no longer ergodic. Once prepared in one phase, they cannot make a transition to another phase. If one then wants to ensure that time expectation values remain equal to ensemble expectation values, one must calculate the partition function by restricting the sum over configurations in which the average spin has the sign of the spontaneous magnetization. A magnetic field selects the relevant configurations automatically.

A3.3 Stochastic dynamics and phase transitions

Ising type model. It is easy to define a stochastic dynamics that converges, asymptotically in time, toward the equilibrium distribution of the Ising model. One can, for example, impose the principle of detailed balance (see Section 3.4.2) and choose as transition probability from a configuration $\{S'_\mathbf{r}\}$ to a configuration $\{S'_\mathbf{r}\}$,

$$\begin{cases} p(S_\mathbf{r}, S'_\mathbf{r}) = e^{-\beta[\mathcal{E}(S'_\mathbf{r}) - \mathcal{E}(S_\mathbf{r})]} & \text{for } \mathcal{E}(S_\mathbf{r}) < \mathcal{E}(S'_\mathbf{r}), \\ p(S_\mathbf{r}, S'_\mathbf{r}) = 1 & \text{otherwise}, \end{cases} \quad (A23)$$

where $\mathbf{r}$ belongs to a set Ω of volume L^d in $\mathbb{Z}^d$ and $\mathcal{E}(S_\mathbf{r})$ is the configuration energy, for example,

$$\mathcal{E}(S) = \sum_{\substack{\text{n.n.} \\ \mathbf{r},\mathbf{r}' \in \Omega \subset \mathbb{Z}^d}} -J S_\mathbf{r} S'_\mathbf{r}. \quad (A24)$$

For the arguments that follow, the precise knowledge of the configurations that are directly connected by the probabilities p is not important provided the configuration space is ergodic in a finite volume, that is, that there exists a non-vanishing probability to relate two arbitrary spin configurations. The important property that is verified by all *local* dynamics is that the probability to go from one configuration to another one with a higher energy is, at low temperature, of the order of $e^{-\beta \Delta \mathcal{E}}$, where $\Delta \mathcal{E}$ is the energy difference.

Therefore, at low temperature, starting from an initial configuration where all spins have, for example, the value $+1$, the probability to generate a bubble of spins with value -1 is proportional to $e^{-\beta JA}$, where A is the measure of the surface of the bubble (as illustrated in Figure 7.3 for the dimension 2). In d dimensions, a bubble with minimal surface is a sphere. If we call L its radius, the measure of the surface is of order L^{d-1} and the probability is of order $e^{-\sigma L^{d-1}}$, where σ is the surface tension. At low temperature, $\sigma \propto \beta = 1/T$.

For $d = 1$, the system thus remains ergodic in the infinite volume limit since the transition probability always remains finite, and thus no phase transition is possible.

For $d > 1$, the transition probability at low temperature goes to zero when $L \to \infty$ and, thus, the same mechanism that leads to the existence of several phases is responsible for a breaking of ergodicity.

Models with continuous symmetries. Again, we take the example of the $O(N)$ symmetry. We now consider a bubble with, at its centre, spin vectors differing by an angle θ from the direction of spontaneous magnetization. In this situation, energy considerations disfavour configurations where, as before, all spins are parallel within the sphere. This would lead to an energy variation proportional to the measure of the surface, thus, of order L^{d-1}. Instead, the energy variation is minimized if one rotates the spins continuously in order to interpolate linearly between θ and zero on a distance of order L. The energy variation between nearest neighbours then is proportional to $1 - \cos(\theta/L) \sim \theta^2/2L^2$. This energy is multiplied by the number of spins in the sphere of radius L. One finds

$$\Delta \mathcal{E} \propto \beta \theta^2 L^{d-2}.$$

One concludes that in dimensions $d \le 2$ this energy remains finite when $L \to \infty$. Thus, a rotation of the spins by an angle θ has a finite probability, and the symmetry $O(N)$ cannot be broken. By contrast, for $d > 2$, a transition is possible.

Conclusion. High- and low-temperature analyses, both static and dynamic, allow proving the existence of a phase transition and determining the nature of the phases. However, they do not reveal the physics at the transition itself, in particular the behaviour of thermodynamic quantities at the critical temperature. Other methods are required to study these problems.

A4 1/N expansion: Calculations

We give here a few additional details about the calculation of the two diagrams that have appeared in the $1/N$ expansion.

A4.1 One-loop Feynman diagrams

The diagram (b) displayed in Figure 2.5 (Section 2.5.1), which is also proportional to the one-loop contribution to the two-point function in the $((\phi)^2)^2$ field theory, has played an important role in the limit $N \to \infty$. We evaluate its behaviour when the mass m goes to zero (with a cut-off Λ).

Regularized diagram. We choose a regularized propagator of the general form (14.9). The diagram in Figure 2.5 then takes the form (14.13),

$$\Omega_d(m) = \int \frac{\mathrm{d}^d k}{(2\pi)^d} \tilde{\Delta}_\Lambda(k) = \frac{1}{(2\pi)^d} \int \frac{\mathrm{d}^d k}{m^2 + k^2 D(k^2/\Lambda^2)} = \Lambda^{d-2} \omega_d(m/\Lambda), \quad (A25)$$

where the function $D(t)$ is positive and regular for $t \ge 0$ and normalized by $D(0) = 1$. By choosing a function $D(k)$ increasing sufficiently fast for $|k| \to \infty$, it is possible to render any diagram convergent and, in particular, the diagram in Figure 2.5.

We need the first terms of the expansion of $\Omega_d(m)$ for $m^2 \to 0$. We present the calculation for the particular function $D(t) = 1 + t$ and, thus,

$$\omega_d(z) = \frac{1}{(2\pi)^d} \int \frac{\mathrm{d}^d k}{z^2 + k^2 + k^4} = N_d \int_0^\infty \frac{k^{d-1} \mathrm{d} k}{z^2 + k^2 + k^4}, \quad (A26)$$

where the constant N_d is the loop factor (14.16a), but the result exhibits most of the properties of the general case. The second integral, where the angular integral has been performed, defines a function analytic for $0 < \operatorname{Re} d < 4$, which is the most interesting situation. The denominator is a second degree polynomial in k^2 which for $z = 0$ has the roots $k^2 = 0$ and $k^2 = -1$. More generally, we can factorize the polynomial:

$$z^2 + k^2 + k^4 = \left(k_1^2 + k^2\right)\left(k_2^2 + k^2\right).$$

The two roots are functions analytic in z^2 in a neighbourhood of $z = 0$. At order z^2, for example,

$$k_1^2 = z^2 + z^4 + O(z^6), \quad k_2^2 = 1 - z^2 + O(z^4).$$

We then proceed by analytic continuation starting from $0 < \operatorname{Re} d < 2$ and separate the integral into two contributions:

$$\omega_d(z) = \frac{N_d}{k_1^2 - k_2^2} \int_0^\infty k^{d-1} dk \left(\frac{1}{k^2 + k_2^2} - \frac{1}{k^2 + k_1^2} \right).$$

Each integral can be calculated explicitly:

$$\int_0^\infty \frac{k^{d-1} dk}{k^2 + \kappa^2} = \kappa^{d-2} \frac{\pi}{2 \sin(\pi d/2)}.$$

Using this result for $\omega_d(z)$, we recognize the constant K_d defined in (14.16b):

$$\omega_d(z) = K(d) \frac{k_2^{d-2} - k_1^{d-2}}{k_2^2 - k_1^2}.$$

This expression now is regular up to $\operatorname{Re} d < 4$. The first terms for $z \to 0$ thus are

$$\omega_d(z) = \omega_d(0) - K(d) z^{d-2} + a(d) z^2 + O(z^4, z^d),$$

with

$$a(d) = (3 - d/2) K(d),$$

an expression that is consistent with expression (14.17) when applied to the particular form of D used in (A26).

More generally, the term proportional to k_2^{d-2} generates a expansion regular in z^2 and the term proportional to k_1^{d-2} generates a regular expansion multiplied by z^{d-2}. One can verify, with a small amount of work, that this structure is general.

A4.2 The two-point function at order $1/N$ for $u \to 0$

At order $1/N$, only one diagram contributes to the two-point function $\langle \sigma \sigma \rangle$ (Figure 14.3). In the limit $\Lambda \to \infty$ and after mass renormalization, one finds

$$\tilde{\Gamma}^{(2)}_{\sigma\sigma}(p) = p^2 + \frac{2}{N(2\pi)^d} \int \frac{d^d q}{(6/u) + b(d) q^{-\varepsilon}} \left(\frac{1}{(p+q)^2} - \frac{1}{q^2} \right) + O\left(\frac{1}{N^2}\right). \quad (A27)$$

To calculate the coefficients of the expansion of $\tilde{\Gamma}^{(2)}_{\sigma\sigma}(p)$ for $u \to 0$, which has the form

$$\sum_{k \geq 1} \alpha_k(\varepsilon) u^k p^{2-k\varepsilon} + \beta_k(\varepsilon) u^{(2+2k)/\varepsilon} p^{-2k}, \quad (A28)$$

it is convenient to introduce the Mellin transform of the integral considered as a function of u. Indeed, if a function $f(u)$ has for $u \to 0$ a behaviour of the form u^t, then the Mellin transform

$$M(s) = \int_0^\infty du\, u^{-1-s} f(u)$$

has a pole at $s = t$. Applying the transformation to the integral and inverting the order between the q and u integrations, one is led to the integral

$$\int_0^\infty du \, \frac{u^{-1-s}}{(6/u) + b(d)q^{-\varepsilon}} = \frac{1}{6} \left(\frac{b(d)q^{-\varepsilon}}{6} \right)^{1-s} \frac{\pi}{\sin \pi s}.$$

The result of the remaining integration over q can be inferred from the generic integral

$$\frac{1}{(2\pi)^d} \int \frac{d^d q}{(p+q)^{2\mu} q^{2\nu}} = p^{d-2\mu-2\nu} \frac{\Gamma(\mu+\nu-d/2)\Gamma(d/2-\mu)\Gamma(d/2-\nu)}{(4\pi)^{d/2} \Gamma(\mu)\Gamma(\nu)\Gamma(d-\mu-\nu)}. \quad (A29)$$

The terms with integer powers of u correspond to the perturbative expansion that exists for ε sufficiently small in dimensional regularization. α_k has a pole at $\varepsilon = (2l+2)/k$ for which the corresponding power of p^2 is $-l$, that is, an integer. One verifies that β_l has a pole for the same value of ε and that the singular contributions cancel in the sum.

A5 Functional flow equations: Additional considerations

In this section, we first recall a general method that allows proving (quantum) field equations for statistical or quantum theories. We then show that the equations of the functional renormalization group (FRG) derived in Chapter 16 provide an example of field equations. Another derivation of the fundamental equation (16.14) follows.

We also show that the Legendre transform of a particular functional, simply related to the Hamiltonian, satisfies a very useful FRG flow equation. Finally, we derive an analogous flow equation for the thermodynamic potential (the generating functional of vertex functions) obtained by a partial integration over the low-momentum modes, which has been widely used, in some approximated form, in numerical calculations.

A5.1 Field equations

We consider the general field integral

$$\mathcal{Z} = \int [d\phi] \, e^{-\mathcal{H}(\phi)}.$$

Field equations can be derived by a simple method, based on the invariance of an integral under an infinitesimal change of variables $\phi \mapsto \varphi$:

$$\phi(x) = \varphi(x) + \varepsilon K(\varphi, x),$$

an identity then being obtained by expanding to first order in the constant parameter ε.

The Jacobian of the transformation is

$$J = \det \frac{\delta\phi(x)}{\delta\varphi(y)} = 1 + \varepsilon \int \mathrm{d}x\, \frac{\delta K(\varphi, x)}{\delta\varphi(x)} + O(\varepsilon^2),$$

where again the identity $\ln\det = \operatorname{tr}\ln$ has been used.

One infers the general equation

$$\left\langle \int \mathrm{d}x \left[\frac{\delta K(\phi, x)}{\delta\phi(x)} - K(\phi, x)\frac{\delta\mathcal{H}(\phi)}{\delta\phi(x)}\right]\right\rangle_{\mathcal{H}} = 0, \qquad (A30)$$

where $\langle\bullet\rangle_{\mathcal{H}}$ denotes an expectation value with the weight $\mathrm{e}^{-\mathcal{H}(\phi)}$.

Equivalently, to derive the equation one notes that the integral of a total derivative vanishes and, thus,

$$\int [\mathrm{d}\phi] \frac{\delta}{\delta\phi(x)} K(\phi)\, \mathrm{e}^{-\mathcal{H}(\phi)} = 0.$$

Calculating the derivative, one obtains the equation

$$\left\langle \frac{\delta K(\phi)}{\delta\phi(x)} - K(\phi)\frac{\delta\mathcal{H}(\phi)}{\delta\phi(x)}\right\rangle_{\mathcal{H}} = 0.$$

An application of this identity to a function $K(\phi, x)$, after integration over x, leads to equation $(A30)$.

One can, of course, wonder whether these formal manipulations are always justified. In some case a limiting procedure starting from an approximation of the continuum by an infinite or even finite lattice can be necessary. Finally, quite often, it is possible to give a formal proof of these identities order by order in some perturbative expansion.

A5.2 Field equations and FRG

One now assumes that the Hamiltonian $\mathcal{H}(\phi)$ depends on a parameter s while the partition function $\mathcal{Z}$ does not depend on it and, thus,

$$\frac{\mathrm{d}\mathcal{Z}}{\mathrm{d}s} = -\left\langle \frac{\mathrm{d}}{\mathrm{d}s}\mathcal{H}(\phi, s)\right\rangle_{\mathcal{H}} = 0.$$

We introduce the notation

$$(\phi \mathcal{K} \phi) \equiv \int \mathrm{d}x\, \mathrm{d}y\, \phi(x) \mathcal{K}(x - y) \phi(y).$$

We decompose $\mathcal{H}(\phi, s)$ into a sum of three terms:

$$\mathcal{H}(\phi, s) = \tfrac{1}{2}\big(\phi \Delta^{-1}(s) \phi\big) + \mathcal{V}(\phi, s) + \tfrac{1}{2}\ln\det \Delta(s),$$

where $\ln\det \Delta = \operatorname{tr}\ln \Delta$ must be regularized.

The equation then becomes

$$\left\langle -\tfrac{1}{2}(\phi\Delta^{-1}(s)D(s)\Delta^{-1}(s)\phi) + \frac{d}{ds}\mathcal{V}(\phi,s)\right\rangle_{\mathcal{H}} + \frac{1}{2}\frac{d}{ds}\ln\det\Delta(s) = 0 \quad (A31)$$

with

$$D(s) = \frac{d\Delta}{ds},$$

an operator with kernel $D(s; x - y)$.

We use the identities

$$\frac{d}{ds}\ln\det\Delta(s) = \frac{d}{ds}\operatorname{tr}\ln\Delta(s) = \operatorname{tr} L(s),$$

with the notation

$$L(s) = D(s)\Delta^{-1}(s).$$

The fundamental equation (16.14) is then obtained with the choice

$$K(\phi, x) = \frac{1}{2}\int dy\, D(s; x - y)\left[[\Delta^{-1}\phi](y) - \frac{\delta\mathcal{V}(\phi, s)}{\delta\phi(y)}\right].$$

The contribution of the Jacobian becomes

$$\int dx\, \frac{\delta K(\phi, x)}{\delta\phi(x)} = \frac{1}{2}\operatorname{tr} L(s) - \frac{1}{2}\int dx\, dy\, D(s; x - y)\frac{\delta^2\mathcal{V}}{\delta\phi(x)\delta\phi(y)}.$$

The second term gives

$$\int dx\, K(\phi, x)\frac{\delta\mathcal{H}(\phi, s)}{\delta\phi(x)}$$

$$= \frac{1}{2}\int dx\, dy\, D(s; x - y)\left\{[\Delta^{-1}\phi](y) - \frac{\delta\mathcal{V}}{\delta\phi(y)}\right\}\left\{[\Delta^{-1}\phi](x) + \frac{\delta\mathcal{V}}{\delta\phi(x)}\right\}$$

$$= \frac{1}{2}\int dx\, dy\, D(s; x - y)\left\{[\Delta^{-1}\phi](y)[\Delta^{-1}\phi](x) - \frac{\delta\mathcal{V}}{\delta\phi(x)}\frac{\delta\mathcal{V}}{\delta\phi(y)}\right\}.$$

Then, eliminating the quadratic term between the two equations (A30) and (A31), one obtains an equation that expresses a sufficient condition for the partition function to be invariant:

$$\frac{d}{ds}\mathcal{V}(\phi, s) = -\frac{1}{2}\int dx\, dy\, D(s; x - y)\left[\frac{\delta^2\mathcal{V}(\phi, s)}{\delta\phi(x)\delta\phi(y)} - \frac{\delta\mathcal{V}(\phi, s)}{\delta\phi(x)}\frac{\delta\mathcal{V}(\phi, s)}{\delta\phi(y)}\right].$$

One recognizes equation (16.14) and by identifying the parameter s with $-\ln\Lambda$, one obtains the renormalization group equations (RGE).

To take into account the field renormalization, one adds a term linear in ϕ: $K(\phi, x) \mapsto K(\phi, x) + \tfrac{1}{2}\eta\phi(x)$. This adds an infinite constant to the Jacobian, which is cancelled by a change of normalization of the partition function, and

$$\int dx\, K(\phi, x)\frac{\delta\mathcal{H}(\phi, s)}{\delta\phi(x)} \mapsto \int dx\, K(\phi, x)\frac{\delta\mathcal{H}(\phi, s)}{\delta\phi(x)} + \frac{\eta}{2}\int dx\, \phi(x)\frac{\delta\mathcal{H}(\phi, s)}{\delta\phi(x)}.$$

More generally, one obtains other RGE by adding to K other local functionals of ϕ. But these algebraic manipulations should not lead us to forget that the goal is to construct a renormalization group that can have local fixed points.

A5.3 FRG: Legendre transformation

We have seen in Section '16.1.4 (equation (16.21)) that the functional

$$\mathcal{W}(H,s) = -\mathcal{V}(\mathcal{D}(s_0,s)H,s) + \tfrac{1}{2}(H\mathcal{D}(s_0,s)H) + \tfrac{1}{2}\ln\det\mathcal{D}(s_0,s), \qquad (A32)$$

where $\mathcal{D}$ is defined in (16.10), has an interpretation as a generating functional of connected correlation functions. It satisfies the flow equation (16.23):

$$\frac{\partial \mathcal{W}}{\partial s} = -\frac{1}{2}\int d^d x\, d^d y\, \frac{\partial \mathcal{K}(s_0,s;x-y)}{\partial s}\left[\frac{\delta^2 \mathcal{W}}{\delta H(x)\delta H(y)} + \frac{\delta \mathcal{W}}{\delta H(x)}\frac{\delta \mathcal{W}}{\delta H(y)}\right], \qquad (A33)$$

where $\mathcal{K}$ is defined in (16.11):

$$\mathcal{D}(s_0,s)\mathcal{K}(s_0,s) = 1.$$

It is then natural to also introduce its Legendre transform:

$$\mathcal{W}(H,s) + \mathcal{G}(\varphi,s) = \int d^d x\, H(x)\varphi(x), \quad H(x) = \frac{\delta\mathcal{G}}{\delta\varphi(x)}.$$

It satisfies a flow equation that is useful for more practical calculations.
The stationarity of the Legendre transformation implies

$$\frac{\partial}{\partial s}\mathcal{W}\Big|_H + \frac{\partial}{\partial s}\mathcal{G}\Big|_\varphi = 0.$$

Moreover, it is convenient to set

$$\Sigma(\varphi,s,x,y) = \int d^d z\, \mathcal{K}(s_0,s;x-z)\frac{\delta^2\mathcal{W}}{\delta H(z)\delta H(y)}.$$

Then, the property (6.19) of the Legendre transformation implies

$$\int d^d z\, d^d z'\, \frac{\delta^2 \mathcal{G}}{\delta\varphi(x)\delta\varphi(z)}\mathcal{D}(s_0,s;z-z')\Sigma(\varphi,s,z',y) = \delta^{(d)}(x-y). \qquad (A34)$$

Equation (A33) becomes

$$\frac{\partial}{\partial s}\mathcal{G}(\varphi,s) = -\frac{1}{2}\int d^d x\, d^d y\, L(s_0,s;x-y)\Sigma(\varphi,s,x,y)$$
$$+ \frac{1}{2}\int d^d x\, d^d y\, \frac{\partial \mathcal{K}(s_0,s;x-y)}{\partial s}\varphi(x)\varphi(y) \qquad (A35)$$

with, in the sense of operators,

$$L(s_0,s) = -\mathcal{K}^{-1}\frac{\partial \mathcal{K}}{\partial s} = \frac{\partial}{\partial s}\ln\mathcal{D}(s_0,s).$$

With these definitions, for $\mathcal{V} = 0$,
$$\mathcal{G}(\varphi, s) = \mathcal{G}(0, s) + \frac{1}{2}\int d^d x \, d^d y \, \varphi(x) \mathcal{K}(s_0, s; x - y) \varphi(y).$$

If one sets
$$\mathcal{G}(\varphi, s) = \frac{1}{2}\int d^d x \, d^d y \, \varphi(x) \mathcal{K}(s_0, s; x - y) \varphi(y) + \mathcal{G}_{\mathrm{I}}(\varphi, s),$$

equation ($A35$) becomes
$$\frac{\partial}{\partial s}\mathcal{G}_{\mathrm{I}}(\varphi, s) = -\frac{1}{2}\int d^d x \, d^d y \, L(s_0, s; x - y) \Sigma(\varphi, s, x, y). \qquad (A36)$$

The kernel $\Sigma(\varphi, s, x, y)$ now is a solution of
$$\Sigma(\varphi, s, x, y) + \int d^d z \, d^d z' \, \frac{\delta^2 \mathcal{G}_{\mathrm{I}}}{\delta\varphi(x)\delta\varphi(z)} \mathcal{D}(s_0, s; z - z') \Sigma(\varphi, s, z', y) = \delta^{(d)}(x - y). \qquad (A37)$$

Equivalently, the Fourier transform
$$\tilde{\Sigma}(\varphi, s, p, q) = \int d^d x \, d^d y \, e^{ipx + iqy} \Sigma(\varphi, s, x, y)$$

satisfies
$$\tilde{\Sigma}(\varphi, s, p, q) + \int \frac{d^d k}{(2\pi)^d} \frac{\delta \mathcal{G}_{\mathrm{I}}}{\delta\tilde{\varphi}(p)\delta\tilde{\varphi}(-k)} \tilde{\mathcal{D}}(s_0, s; k) \tilde{\Sigma}(\varphi, s, k, q) = (2\pi)^d \delta^{(d)}(p + q). \qquad (A38)$$

and equation ($A36$) takes the form
$$\frac{\partial}{\partial s}\mathcal{G}_{\mathrm{I}}(\varphi, s) = -\frac{1}{2}\int \frac{d^d k}{(2\pi)^d} \tilde{L}(s_0, s; k) \tilde{\Sigma}(\varphi, s, k, -k). \qquad (A39)$$

The functional $\mathcal{G}_{\mathrm{I}}(\varphi, s)$ is local and satisfies a relatively simple equation.

To write these equations more explicitly, we expand $\mathcal{G}_{\mathrm{I}}$ and Σ in powers of φ. At order φ^0, in the Fourier representation, $\Sigma = \Sigma_0$ is a solution of
$$\tilde{\Sigma}_0(s, p) + \mathcal{G}_{\mathrm{I}}^{(2)}(p) \tilde{\mathcal{D}}(s_0, s; p) \tilde{\Sigma}_0(s, p) = 1$$

and, thus,
$$\tilde{\Sigma}_0(s, p) = \left[1 + \tilde{\mathcal{D}}(s_0, s; p) \tilde{\mathcal{G}}_{\mathrm{I}}^{(2)}(p)\right]^{-1}.$$

Then, in symbolic notation,
$$\Sigma = \Sigma_0 - \frac{1}{2}\Sigma_0 \frac{\delta^2 \mathcal{G}_{\mathrm{I}}^{(4)}}{\delta\varphi\delta\varphi} \varphi\varphi \mathcal{D}(s_0, s) \Sigma_0 + O(\varphi^4),$$

and thus
$$\frac{\partial}{\partial s}\mathcal{G}_I^{(2)}(p,s) = -\frac{1}{2}\int \frac{d^d k}{(2\pi)^d}\tilde{\mathcal{D}}(s_0,s;k)\tilde{\Sigma}_0^2(k)\tilde{\mathcal{G}}_I^{(4)}(p,-p,k,-k).$$

Application to the ϕ^4 theory. We now set $s = -\ln\Lambda$. We also substitute $\mathcal{D}(s_0,s) \mapsto \mathcal{D}_\Lambda$ and apply these identities to the $u\phi^4$ field theory. At order u^1, one finds
$$\mathcal{V}(\phi) = \frac{1}{2}r_{c1}u\int d^d x\, \phi^2(x) + \frac{1}{4!}u\int d^d x\, \phi^4(x) + O(u^2).$$

One infers
$$\mathcal{W}(H) = \frac{1}{2}\int d^d x\, H\mathcal{K}H - \frac{1}{2}r_{c1}u\int d^d x\,(\mathcal{D}H)^2(x) - \frac{1}{4!}u\int d^d x\,(\mathcal{D}H)^4(x)$$
$$+ O(u^2).$$

Moreover,
$$\varphi(x) = [\mathcal{K}H](x) = \phi(x) + O(u).$$
Using the stationarity of the Legendre transformation that implies
$$\frac{\partial \mathcal{W}}{\partial u} + \frac{\partial \mathcal{G}}{\partial u} = 0,$$
one infers
$$\mathcal{G}_I = \tfrac{1}{2}ur_{c1}\int d^d x\, \varphi^2(x) + \frac{1}{4!}u\int d^d x\, \varphi^4(x) + O(u^2).$$
At this order, one recovers the equation giving r_{c1}, obtained in Section 16.3,
$$\Lambda\frac{d}{d\Lambda}r_{c1} = \tfrac{1}{2}\mathcal{D}_\Lambda(0).$$

More generally, equations (A37) and (A39) can be solved recursively and this generates a perturbative expansion. As a property of the Legendre transformation, only the 1-irreducible diagrams contribute. For example, the term of degree six in φ is now of order u^3.

RGE in standard form. After elimination of the explicit dependence in Λ and field renormalization, one obtains a flow equation analogous to equation (16.51):
$$\lambda\frac{d}{d\lambda}\mathcal{G}_I(\varphi,\lambda) = \int d^d x\, \varphi(x)\left[\frac{1}{2}(d+2+\eta) + \sum_\mu x^\mu \frac{\partial}{\partial x^\mu}\right]\frac{\delta\mathcal{G}_I}{\delta\varphi(x)}$$
$$+ \frac{1}{2}\int d^d x\, d^d y\, \Lambda\frac{d\ln\mathcal{D}_\Lambda}{d\Lambda}(x-y)\Sigma(\varphi,\lambda,y,x)$$
$$+ \tfrac{1}{2}\eta\int d^d x\, d^d y [\Delta_\Lambda]^{-1}(x-y)\varphi(x)\varphi(y).$$

A5.4 Partial mode integration with IR cut-off

We briefly describe a variant of the general formalism of partial mode integration, where low momentum (IR) modes instead are initially cut-off and then added continuously, leading to FRG-type flow equations. The method is intended to deal with problems where physics is dominated by IR modes. The derivation is based on some identities already obtained in Section A5.1.

To simplify and deal with large-momentum divergences, we work in the framework of dimensional regularization where the theory no longer has UV divergences, at least for generic dimensions, but other large-momentum regularizations can be accommodated.

To construct flow equations, following Wetterich (1991), one introduces in the propagator a small-momentum cut-off μ and constructs the flow that corresponds to decreasing μ down to zero. This can be achieved, for example, with the propagator

$$\tilde{\Delta}(\mu; p) = \frac{1 - e^{-p^2/\mu^2}}{p^2},$$

which for $|p| \gg \mu$ has the behaviour of a critical propagator but has no pole at $p = 0$. More generally, one can take as a propagator a regular function of the form

$$\tilde{\Delta}(\mu; p) = \frac{C(p^2/\mu^2)}{p^2}, \quad C(0) = 0, \quad \lim_{t \to \infty} C(t) = 1.$$

We then define

$$\mathcal{K}(\mu; p) = 1/\tilde{\Delta}(\mu; p), \quad \mathcal{S}(\phi) = \tfrac{1}{2}(2\pi)^d \int d^d p \, \tilde{\phi}(p) \mathcal{K}(\mu; p) \tilde{\phi}(-p) + \mathcal{V}(\phi),$$

where $\mathcal{V}$ is independent of μ.

Correlation functions are generated by

$$\mathcal{Z}(J) = \int [d\phi] \exp\left[-\mathcal{S}(\phi) + \int d^d x \, J(x) \phi(x)\right].$$

Then,

$$\mu \frac{d}{d\mu} \mathcal{Z}(J) = -\tfrac{1}{2}(2\pi)^{-d} \int d^d p \, \frac{\delta}{\delta \tilde{J}(-p)} \mu \frac{d}{d\mu} \mathcal{K}(\mu; p) \frac{\delta}{\delta \tilde{J}(p)} \mathcal{Z}(J),$$

where $\tilde{J}(p)$ is the Fourier transform of $J(x)$. In terms of $\mathcal{W}(J) = \ln \mathcal{Z}(J)$, the equation becomes

$$\mu \frac{d}{d\mu} \mathcal{W}(J) = -\tfrac{1}{2}(2\pi)^{-d} \int d^d p \, \frac{\delta \mathcal{W}(J)}{\delta \tilde{J}(-p)} \mu \frac{d}{d\mu} \mathcal{K}(\mu; p) \frac{\delta \mathcal{W}(J)}{\delta \tilde{J}(p)}$$

$$- \tfrac{1}{2}(2\pi)^{-d} \int d^d p \, \frac{\delta}{\delta \tilde{J}(-p)} \mu \frac{d}{d\mu} \mathcal{K}(\mu; p) \frac{\delta}{\delta \tilde{J}(p)} \mathcal{W}(J).$$

Finally, we express the equation in terms of the vertex functional $\Gamma(\phi)$, Legendre transform of $\mathcal{W}(J)$, and find

$$\mu \frac{\mathrm{d}}{\mathrm{d}\mu} \Gamma(\phi) = \tfrac{1}{2}(2\pi)^d \int \mathrm{d}^d p \, \tilde{\phi}(p) \mu \frac{\mathrm{d}}{\mathrm{d}\mu} \mathcal{K}(\mu; p) \tilde{\phi}(-p)$$

$$+ \tfrac{1}{2} \int \frac{\mathrm{d}^d p}{(2\pi)^d} \, \mu \frac{\mathrm{d}}{\mathrm{d}\mu} \mathcal{K}(\mu; p) \left[\frac{\delta^2}{\delta\tilde{\phi}\delta\tilde{\phi}} \Gamma(\phi) \right]^{-1} (p, -p).$$

The equation simplifies if one sets

$$\Gamma(\phi) = \tfrac{1}{2}(2\pi)^d \int \mathrm{d}^d p \, \tilde{\phi}(p) \mathcal{K}(\mu; p) \tilde{\phi}(-p) + \mathcal{G}(\phi).$$

Setting also

$$L(\mu, p) = \mu \frac{\mathrm{d}}{\mathrm{d}\mu} \ln \mathcal{K}(\mu; p),$$

one finds

$$\mu \frac{\mathrm{d}}{\mathrm{d}\mu} \mathcal{G}(\phi) = \tfrac{1}{2}(2\pi)^d \int \mathrm{d}^d p \, L(\mu; p) \Sigma(\phi; p, -p), \qquad (A40)$$

where

$$-\Sigma(\phi; p, q) + \int \frac{\mathrm{d}^d k}{(2\pi)^d} \Sigma(\phi; p, k) \tilde{\Delta}(\mu; k) \frac{\delta^2 \mathcal{G}}{\delta\tilde{\phi}(-k)\delta\tilde{\phi}(q)} = (2\pi)^d \delta^{(d)}(p+q).$$

Equation (A40) is formally identical to equations (A36) or (A39), but has a somewhat different interpretation.

One then studies how the vertex functional $\mathcal{G}(\phi)$ changes when μ goes to zero. For large μ the functional $\mathcal{G}(\phi)$ is local and can be approximated, up to the shift of the kinetic term, by a Hamiltonian or an effective local action, which provide the initial boundary conditions.

In practice, solving the flow equations requires some approximation scheme. The simplest one consists in approximating the functional $\mathcal{G}(\phi)$ by a simple function of ϕ.

Bibliography

[1] J. Zinn-Justin, *Transitions de phase et groupe de renormalisation*, EDP Sciences/ CNRS InterEditions (Les Ulis 2004)

[2] Many interesting details and references about the early history of quantum electrodynamics and the problem of divergences can be found in
S. Weinberg, *The Theory of Quantum Fields*, vol. 1, chap. 1, Cambridge University Press (Cambridge 1995);
A number of original articles are reproduced in
J. Schwinger, ed., *Selected Papers in Electrodynamics*, Dover (New York 1958);
See also
N.N. Bogoliubov and D.V. Shirkov, *Introduction to the Theory of Quantized Fields*, Interscience (New York 1959).

[3] A review of the situation after the construction of the Standard Model of fundamental interactions at the microscopic scale can be found in
Methods in Field Theory, Les Houches 1975, R. Balian and J. Zinn-Justin eds., North-Holland (Amsterdam 1976);
C. Itzykson and J.B. Zuber, *Quantum Field Theory*, McGraw-Hill, (New York 1980).
A selection of original articles is gathered in
Selected Papers on Gauge Theory of Weak and Electromagnetic Interactions, C.H. Lai, ed., World Scientific (Singapore 1981).

[4] In the framework of quantum field theory, the renormalization group ideas have been introduced in
E.C.G. Stueckelberg and A. Peterman, *Helv. Phys. Acta* 26 (1953) 499;
M. Gell-Mann and F.E. Low, *Phys. Rev.* 95 (1954) 1300.
See also
N.N. Bogoliubov and D.V. Shirkov *Introduction to the Theory of Quantized Fields*, Interscience (New York 1959);
K.G. Wilson, *Phys. Rev.* 179 (1969) 1499.

[5] For a presentation of the renormalization group as applied to critical phenomena see
L.P. Kadanoff, *Physics* 2 (1966) 263;
K.G. Wilson, *Phys. Rev.* B4 (1971) 3174, *ibidem* 3184;
K.G. Wilson and J. Kogut, *Phys. Rep.* 12C (1974) 75.

[6] An introduction to path integrals in the spirit of this work can be found in
J. Zinn-Justin, *Path Integrals in Quantum Mechanics*, Oxford University Press (Oxford 2004).

[7] About the origin of Landau's theory see

L.D. Landau, *Phys. Z. Sowjetunion* 11 (1937) 26; reproduced in *Collected Papers of L.D. Landau*, D. ter Haar, ed., Pergamon (New York 1965).

[8] E. Brézin, J.C. Le Guillou and J. Zinn-Justin, *Field Theory Approach to Critical Phenomena* contribution to the work [12],

which describes the application of quantum field theory methods to the calculation of universal quantities.

[9] Additional technical details concerning the topics discussed in this work, presented in the same spirit and, more generally, a unified presentation of quantum field theory adapted to applications both in particle physics and in the theory of critical phenomena is found in

J. Zinn-Justin, *Quantum Field Theory and Critical Phenomena*, Clarendon Press 1989, 4th edn. Oxford University Press (Oxford 2002).

[10] The idea of the ε-expansion is due to

K.G. Wilson and M.E. Fisher, *Phys. Rev. Lett.* 28 (1972) 240.

[11] After Wilson's original articles, several authors have shown that the renormalization group of quantum field theory could be applied to critical phenomena:

C. Di Castro, *Lett. Nuovo Cimento.* 5 (1972) 69;

G. Mack, *Kaiserslautern 1972*, Lecture Notes in Physics vol. 17, W. Ruhl and A. Vancura, eds., Springer-Verlag (Berlin 1972);

E. Brézin, J.C. Le Guillou and J. Zinn-Justin, *Phys. Rev.* D8 (1973) 434, *ibidem* 2418;

P.K. Mitter, *Phys. Rev.* D7 (1973) 2927;

G. Parisi, *Cargèse Lectures 1973*, published in *J. Stat. Phys.* 23 (1980) 49;

B. Schroer, *Phys. Rev.* B8 (1973) 4200;

C. Di Castro, G. Jona-Lasinio and L. Peliti, *Ann. Phys. (NY)* 87 (1974) 327;

F. Jegerlehner and B. Schroer, *Acta Phys. Austr. Suppl.* XI (1973) 389, Springer Verlag (Berlin).

[12] For a pedagogical review, see the volume

Phase Transitions and Critical Phenomena, vol. 6, C. Domb and M.S. Green eds., Academic Press (London 1976).

[13] The first precise estimates of critical exponents, following a suggestion of Parisi [14], obtained by summing the divergent series of [15], have been reported in

J.C. Le Guillou and J. Zinn-Justin, *Phys. Rev. Lett.* 39 (1977) 95; *Phys. Rev.* B21 (1980) 3976.

Improved estimates are found in

R. Guida and J. Zinn-Justin, *J. Phys.* A31 (1998) 8103,

Arxiv: cond-mat/9803240.

[14] The use of fixed dimension perturbative expansions has been advocated by
G. Parisi, *Cargèse Lectures 1973*, published in *J. Stat. Phys.* 23 (1980) 49.

[15] Series at fixed dimension 3 have been reported in
G.A. Baker, B.G. Nickel, M.S. Green and D.I. Meiron, *Phys. Rev. Lett.* 36 (1976) 1351;
B.G. Nickel, D.I. Meiron, G.B. Baker, *University of Guelph Report* 1977, together with first estimates of Ising model exponents. See also
S.A. Antonenko and A.I. Sokolov, *Phys. Rev.* E51 (1995) 1894.

[16] The ε-expansion is summed in
J.C. Le Guillou and J. Zinn-Justin, *J. Physique Lett. (Paris)* 46 (1985) L137; *J. Physique (Paris)* 48 (1987) 19; *ibidem* 50 (1989) 1365.

[17] A numerical determination of the scaling equation of state for the Ising model universality class, using field theory methods, has been reported in
R. Guida and J. Zinn-Justin, *Nucl. Phys.* B489 [FS] (1997) 626.

[18] The example of the N-component model with cubic symmetry is studied in
D.J. Wallace, *J. Phys. C: Solid State Phys.* 6 (1973) 1390;
A. Aharony, *Phys. Rev.* B8 (1973) 3342, 3349, 3358, 3363, 4270, *Phys. Rev. Lett.* 31 (1973) 1494;
The ε-expansion to order ε^5 has been published in
H. Kleinert and V. Schulte-Frohlinde, *Phys. Lett.* B342 (1995) 284;
H. Kleinert, S. Thoms, and V. Schulte-Frohlinde, *Phys. Rev.* B56 (1997) 14428.
Five-loop 3D perturbative expansions are reported in
D.V. Pakhnin and A.I. Sokolov, *Phys. Rev.* B61 (2000) 15130.

[19] E. Vicari and J. Zinn-Justin, *New Journal of Phys.* 8 (2006) 321.
The argument uses a gradient property reported in
D.J. Wallace and R.P.K. Zia, *Phys. Lett.* A48 (1974) 325.

[20] Dimensional regularization has been introduced in
J. Ashmore, *Lett. Nuovo Cimento* 4 (1972) 289;
G. 't Hooft and M. Veltman, *Nucl. Phys.* B44 (1972) 189;
C.G. Bollini and J.J. Giambiagi, *Phys. Lett.* 40B (1972) 566, *Nuovo Cimento* 12B (1972) 20.

[21] Homogeneous renormalization group equations induced by varying the cut-off have been derived in
J. Zinn-Justin, *Cargèse Lectures 1973*, Saclay preprint T73/049, unpublished, a contribution later incorporated into reference [8].

[22] Renormalization group equations in modern form has been first discussed in
C.G. Callan, *Phys. Rev.* D2 (1970) 1541;
K. Symanzik, *Commun. Math. Phys.* 18 (1970) 227.

A pedagogical presentation can be found in

S. Coleman, *Dilatations, Erice Lectures 1971* reproduced in *Aspects of Symmetry*, Cambridge University Press (Cambridge 1985).

[23] Critical or massless theories have been discussed in

K. Symanzik, *Commun. Math. Phys.* 7 (1973) 34.

[24] Homogeneous renormalization group equations, in renormalized form, has been introduced in

S. Weinberg, *Phys. Rev.* D8 (1973) 3497;

G. 't Hooft, *Nucl. Phys.* B61 (1973) 455.

[25] See also the works

M. Le Bellac, *Des Phénomènes Critiques aux Champs de Jauge*, InterEditions (Paris 1988);

C. Itzykson and J.M. Drouffe, *Statistical Field Theory*, 2 vols., Cambridge University Press (Cambridge 1989) 1–403.

[26] As shown by Stanley,

H.E. Stanley, *Phys. Rev.* 176 (1968) 718,

the large-N limit of the classical N-component model coincides with the spherical model solved by

T.H. Berlin and M. Kac, *Phys. Rev.* 86 (1952) 821.

[27] Among the early articles discussing properties of critical phenomena, see

R. Abe, *Prog. Theor. Phys.* 48 (1972) 1414; 49 (1973) 113, 1074, 1877;

S.K. Ma, *Phys. Rev. Lett.* 29 (1972) 1311; *Phys. Rev.* A7 (1973) 2172;

M. Suzuki, *Phys. Lett.* 42A (1972) 5; *Prog. Theor. Phys.* 49 (1973) 424, 1106, 1440;

R.A. Ferrel and D.J. Scalapino, *Phys. Rev. Lett.* 29 (1972) 413;

K.G. Wilson, *Phys. Rev.* D7 (1973) 2911.

[28] The equation of state to order $1/N$ is given in

E. Brézin and D.J. Wallace, *Phys. Rev.* B7 (1973) 1967.

[29] The spin spin correlation function in zero field is obtained in

M.E. Fisher and A. Aharony, *Phys. Rev. Lett.* 31 (1973) 1238;

A. Aharony, *Phys. Rev.* B10 (1974) 2834;

R. Abe and S. Hikami, *Prog. Theor. Phys.* 51 (1974) 1041.

[30] The exponent ω is calculated to order $1/N$ in

S.K. Ma, *Phys. Rev.* A10 (1974) 1818.

[31] See also the contributions of S.K. Ma and E. Brézin, J.C. Le Guillou and J. Zinn-Justin in the review [12].

[32] More recently, large-N techniques have been applied to problems related to weakly interacting Bose gases and Bose–Einstein condensation,

G. Baym, J.-P. Blaizot and J. Zinn-Justin, *Euro. Phys. Lett.* 49 (2000) 150, Arxiv: cond-mat/9907241;

P. Arnold and B. Tomasik, *Phys. Rev.* A 62 (2000) 063604, Arxiv: cond-mat/0005197.

[33] Consistency of the $1/N$ expansion is proved to all orders in
I. Ya Aref'eva, E.R. Nissimov and S.J. Pacheva, *Commun. Math. Phys.* 71 (1980) 213;
A.N. Vasil'ev and M.Yu. Nalimov, *Teor. Mat. Fiz.* 55 (1983) 163.

[34] At present, the longest available series in $1/N$ for exponents and amplitudes are found in
I. Kondor and T. Temesvari, *J. Physique Lett. (Paris)* 39 (1978) L99;
Y. Okabe and M. Oku, *Prog. Theor. Phys.* 60 (1978) 1277, 1287; 61 (1979) 443;
A.N. Vasil'ev, Yu.M. Pis'mak and Yu.R. Honkonen, *Teor. Mat. Fiz.* 46 (1981) 157; 50 (1982) 195.

[35] See also
I. Kondor, T. Temesvari and L. Herenyi, *Phys. Rev.* B22 (1980) 1451.

[36] Long-range forces are discussed in
S.K. Ma, *Phys. Rev.* A7 (1973) 2172.

[37] The renormalization group of the non-linear σ-model is discussed in
A.M. Polyakov, *Phys. Lett.* 59B (1975) 79;
E. Brézin and J. Zinn-Justin, *Phys. Rev. Lett.* 36 (1976) 691; *Phys. Rev.* B14 (1976) 3110;
W.A. Bardeen, B.W. Lee and R.E. Shrock, *Phys. Rev.* D14 (1976) 985.
The results have been generalized to models defined on symmetric spaces
E. Brézin, S. Hikami, J. Zinn-Justin, *Nucl. Phys.* B165, 528-544 (1980).

[38] Critical exponents to higher orders are calculated in
S. Hikami, *Nucl. Phys.* B215[FS7] (1983) 555;
W. Bernreuther and F.J. Wegner, *Phys. Rev. Lett.* 57 (1986) 1383;
F. Wegner, *Nucl. Phys.* B316 (1989) 663.

[39] Speculations about an RG flow near dimension 2, consistent, for $N = 2$, with the Kosterlitz–Thouless transition and, for $N > 2$, with results coming from the non-linear σ-model are presented in
J.L. Cardy and H.W. Hamber, *Phys. Rev. Lett.* 45 (1980) 499.

[40] Long-range forces are discussed in
E. Brézin, J.C. Le Guillou and J. Zinn-Justin, *J. Phys. A: Math. Gen.* 9 (1976) L119.

[41] Various forms of functional renormalization group equations have initially been proposed by

F.J. Wegner and A. Houghton, *Phys. Rev.* A8 (1973) 401;

F.J. Wegner, *J. Physics* C7 (1974) 2098;

K.G. Wilson and J. Kogut, *Phys. Rept.* 12C (1974) 75.

[42] See also the review

F.J. Wegner, in *Phase Transitions and Critical Phenomena* vol. 6, C. Domb and M.S. Green, eds., Academic Press (New York 1976).

[43] The functional renormalization group equations, in the form presented in this work, are inspired from

J. Polchinski, *Nucl. Phys.* B231 (1984) 269,

where they are used to prove renormalizability. See also

G. Keller, C. Kopper, M. Salmhofer, *Helv. Phys. Acta* 65 (1992) 32;

M. Salmhofer, *Renormalization*, Springer (Berlin 1998) and *Comm. Math. Phys.* 194 (1998) 249;

P. Kopietz, *Nucl. Phys.* B595 (2001) 493, Arxiv: hep-th/0007128.

[44] For a few applications and generalizations of FRG equations see, for example,

D.S. Fisher and D.A. Huse, *Phys. Rev.* B32 (1985) 247;

M. Bonini, M. D'Attanasio and G. Marchesini, *Nucl. Phys.* B409 (1993) 441, *ibidem* B444 (1995) 602;

M. D'Attanasio, T. R. Morris, *Phys. Lett.* B409 (1997) 363, Arxiv: hep-th/9704094;

T.R. Morris, *Prog. Theor. Phys. Suppl.* 131 (1998) 395, Arxiv: hep-th/9802039 and references therein;

T.R. Morris and J.F. Tighe, *JHEP* 08 (1999) 007;

J.I. Latorre and T.R. Morris, *JHEP* 11 (2000) 004;

P. Le Doussal, K.J. Wiese and P. Chauve, *Phys. Rev.* E69 (2004) 026112.

[45] Another approach based on an IR cut-off is found in

C. Wetterich, *Nucl. Phys.* B352 (1991) 529; *Phys. Lett.* B 301 (1993) 90;

N. Tetradis, C. Wetterich, *Nucl. Phys.* B422 (1994) 541, Arxiv:hep-ph/9308214;

M. Reuter, C. Wetterich, *Nucl. Phys.* B427 (1994) 291;

J. Berges, N. Tetradis, C. Wetterich, *Phys. Rept.* 363 (2002) 223, Arxiv: hep-ph/0005122;

M. Salmhofer and C. Honerkamp, *Prog. Theor. Phys.* 105 (2001) 1;

J. Berges, N. Tetradis, C. Wetterich, *Phys. Rept.* 363 (2002) 223, Arxiv: hep-ph/0005122.

Index

amplitude ratios, 160, 198, 344.
   numerical results, 265.
annihilation operator, 94.
anomalous dimension, 322.
asymptotic freedom, 355.
asymptotic series, 33.

Bessel function, 42.
   modified, 41, 44.
   third kind, 43.
beta function, 405.
   $N \to \infty$, 347.
   one-loop calculation, 319.
   perturbative calculation, 408, 412.
   σ-model, non-linear, 367, 369.
Borel transformation, 263.
boundary conditions:
   anti-periodic, 167.
   periodic, 128.
   sensitivity to, 147.
Brillouin zone, 67, 181.
Brownian motion, 63.

Callan–Symanzik equations, 327.
central limit theorem, 45, 49.
characteristics: method of, 316.
cluster property, 89, 134, 431.
coexistence curve, 373.
commutation relations: canonical, 94.
configuration energy, 128.
connected contributions, 27.
connected correlation functions, 89.
continuous symmetries, 200.
continuum limit, 69, 88, 98, 193.
correlation functions, 83, 133, 391.
   bare, 312.
   connected, 89, 133, 220, 286.
   critical, power-law behaviour, 320.
   Gaussian, 115.
   renormalized, 312, 316.
   scaling behaviour, 325.
   scaling form, 373.
correlation length, 88, 134, 148, 193.
   divergence, 100, 197.
   quasi-Gaussian approximation, 198.
creation operator, 94.
critical behaviour, 148.
critical domain, 187, 230, 254.
   scaling behaviour, 261.
   scaling property, 230, 261.
   universality, 162.
critical surface, 226.
crossover scale, 346.
cubic anisotropy, 284.
   fixed points, 274.
   linearized flow, 274.
   model with, 272.
cubic symmetry group, 66, 273.
cumulants, 29, 46, 47.
cut-off, 301, 309, 319.
   scale, 218.

decimation, 101.
density matrix, 99.
detailed balance, 71, 432.
determinants: operators, 421.
differential operator: determinant, 333.
diffusion equation: functional, 390.
dimension:
   anomalous, 322.

canonical, 322.
critical, 311.
dimensional analysis, 310.
dimensional continuation, 248, 300.
dimension $d = 4 - \varepsilon$:
  ϕ and ϕ^3 perturbations, 416.
  ϕ^2 and ϕ^4 perturbations, 415.
dimension 4: the role of, 206.
dimension of field, 225, 227, 239.
  $N \to \infty$, 342.
  order ε^2, 253, 259.
  σ-model, non-linear, 374.
Dirac's distribution, function δ, 114, 420.
distribution:
  Gaussian asymptotic, 49.
  mean-spin, 129, 132, 157.
divergences:
  infrared (IR), 313, 364.
  ultraviolet (UV), 244, 290.
divergent series, 32.
Elitzur's conjecture, 376.
epsilon expansion, 239.
  eigen-perturbations, 240.
equation of state, 155.
  $N \to \infty$, 343.
  orthogonal symmetry, 200.
  parametric form, 344.
  quasi-Gaussian, 196.
  RGE, 326.
  scaling form, 160, 372.
  scaling property, 326.
equilibrium distribution, 148.
ergodicity, 148.
  breaking, 148, 431.
eta function, 406.
  perturbative calculation, 410.
Euclidean action, 112, 190, 219, 289.
Euler's constant, 419.
expansion:
  asymptotic, 262.
  ε-, 239.

high-temperature, 182.
in $1/N$, 350.
loop expansion, 296.
low-temperature, 359.
perturbative, 119.
  formal representation, 290.
  one dimension, 121.
exponent:
  correction ω, 239.
    $N \to \infty$, 344.
  correlation ν, definition, 188, 198.
    Gaussian, 188.
    $N \to \infty$, 341.
    order ε, 257, 259.
    scaling, 325.
    σ-model, non-linear, 374.
  δ, definition, 160.
    $N \to \infty$, 343.
    quasi-Gaussian, 160.
  η, definition, 187, 320.
    Gaussian, 187.
    order ε^2, 253, 322.
    order $1/N$, 352.
  magnetic β, 159.
    $N \to \infty$, 340.
    quasi-Gaussian, 159.
  magnetic susceptibility γ, 159.
    order ε, 257.
    quasi-Gaussian, 159.
  specific heat α, 262.
    $N \to \infty$, 344.
    scaling relation, 325.
exponents:
  numerical results, 263.
  scaling relations, 259, 325, 326.
ferromagnetic systems: definition, 127.
Feynman diagrams, 26.
  connected, 291.
  external line, 291.
  faithful representation, 291.
  internal line, 291.

loops, definition, 291.
  one-irreducible, 292.
  one-loop, 246, 297, 335, 433.
  trees, 296.
field dimension: see dimension of field.
field equations, 435.
field integral, 219.
  Gaussian, 190.
    calculation, 190.
  partial mode integration, 384, 441.
  perturbative calculation, 289.
field theory, 218.
  $(\phi^2)^2$: limit $N \to \infty$, 339.
  Gaussian, 287.
  perturbative expansion, 289.
  renormalizable, 307.
field: N-component, 267.
fixed point, 55, 57, 70, 226, 227.
  cubic, 274.
  Gaussian, 231, 414.
    eigen-perturbations, 234.
  infinite dimension, 161.
  local stability, 413.
  non-Gaussian:
    ε-expansion, 237.
    $O(N)$, order ε, 258.
    $O(N)$ symmetric, stability, 277.
    order ε, 252.
  stability, 56, 70, 271.
    and field dimension, 279.
fixed points:
  flow between, 269.
  line of, 379.
flow and potential, 270.
flow between fixed points, 240.
flow of the Hamiltonian, 388.
Fourier representation, 135.
Fourier series, 423.
Fourier transformation, 46, 427.
  decay and regularity, 423.
four-point function, 30.

perturbative calculation, 249, 318.
perturbative expansion, 294.
free energy, 86, 129.
free-field theory, 190.
FRG, 383:
  components, 395.
  correlation functions, 396.
  dimensional regularization, 441.
  ε expansion, 411.
  equations in standard form, 403.
  evolution of the Hamiltonian, 388.
  field equations, 436.
  Fourier representation, 394.
    group, 383.
  interaction flow, 387.
  Legendre transformation, 438.
  locality, 393.
  perturbative ϕ^4 theory, 404.
  perturbative solution, 398, 407.
function β, 252, 258, 314.
function η, 314.
functional derivative, 113.
functional renormalization group
  (FRG), 383.
gamma function, 34, 419.
Gaussian distribution: asymptotic, 70.
Gaussian expectation value, 23.
Gaussian integral, 20, 192.
Gaussian model, 92, 110, 141, 183.
Gaussian path integral, 112.
generating function, 19, 29, 46.
  steepest descent method, 38.
generating functional, 113, 285.
Goldstone modes, 173, 200, 338, 360, 423.
Goldstone phenomenon, 216.
gradient flows, 269.
Hamiltonian, 190, 219.
  critical, 244.
  dimensional analysis, 310.
  effective, 222, 308.

fixed-point, 225.
flow, 392.
general, 289.
general quadratic isotropic, 232.
Landau–Ginzburg–Wilson, 219.
Landau's, 219.
quantum, 96, 100, 103.
harmonic oscillator: perturbed, 119.
Hausdorf dimension, 64.
Heaviside or step function θ, 54.
Hermite polynomials, 35.
Hermitian conjugation, 94.
Hilbert space, 82.

infrared (IR) divergences, 313, 364.
insensitivity to the short-distance
  structure, 307.
integral operator, 71.
interaction vertex, 289.
interaction:
  ferromagnetic, 153, 181.
  short-range, 181.
irrelevant eigenvector, eigen-
  -perturbation, 228.
irrelevant perturbation, 57.
Ising model, 166.
  one dimension, 107.
isotropic fixed point: stability, 278.
Kosterlitz–Thouless phase transition, 172.
Landau–Ginzburg–Wilson Hamiltonian, 219.
Landau's Hamiltonian, 219.
Landau's theory, 199.
Laurent series, 424.
Legendre transformation, 130, 137, 221, 286.
  invertibility, 138.
  stationarity, 131, 138.
  steepest descent method, 142.
local polynomials: dimension, 241.
locality, 60, 66, 79, 220, 289.
loops: number of, topological relation, 291.
magnetic field, 128.
magnetization, 129.

marginal: eigenvector,
  eigen-perturbation, 228.
  perturbation, 57.
Markov chain, 59, 71.
Markovian process, 226.
Markovian stationary process, 59, 229.
mass, 190.
mass operator, 292.
mean random variable: dimension, 57.
mean-field approximation, 153, 208,
  211.
mean-field: corrections, 211.
Mellin transform, 434.
Mermin–Wagner–Coleman theorem,
  173, 338, 364.
metric tensor on the sphere, 362.
model with cubic anisotropy, 284.
model with orthogonal symmetry:
  see $O(N)$ model, 108.
momentum operator, 94.
momentum, 136, 181.

nearest-neighbour interaction, 81, 163.
non-linear σ-model:
  see sigma-model, 355.
non-linear representation of
  the $O(N)$ group, 358.
$O(N)$ model:
  disordered phase, 338.
  ε-expansion, 258.
  limit $N \to \infty$, 337.
  one dimension, 108.
  ordered phase, 338.
$O(N)$ symmetry: general
  Hamiltonian, 331.
one-irreducible functions, 139.
operator σ^2: dimension, 256.
order parameter, 147, 168.
  dimension, 239.
Ornstein–Zernike form, 187, 198.
orthogonal group, 108, 165, 171, 175,
  200, 212, 258, 331, 355, 422.

orthogonal $O(2)$ symmetry, 105.
partial mode integration, 384, 441.
partition function, 80, 121, 128.
   quantum, 99, 103.
path integral, 73.
   Gaussian, 111.
      calculation, 74, 116.
   perturbative calculation, 119.
periodic boundary conditions, 80.
perturbation theory, 24, 119.
perturbative calculations:
   one dimension, 120.
perturbative expansion:
   minimum of the potential, 119.
phase transition, 147, 170.
   first-order, 157.
   helium's superfluid, 264, 265.
   infinite dimension, 153.
   liquid–vapour, 263.
   $O(2)$, 172.
   second-order or continuous, 158.
polymers: statistical properties, 263.
position operator, 83, 94.
propagator, 245, 288.
   massive in dimension 2, 421.
   regularized, 392.
psi function, 419.

quantum field theory: interaction, 289.
quantum ground state, 100.
quantum Hamiltonian, 100.
quantum harmonic oscillator, 96.
quasi-Gaussian approximation, 195, 200.
   corrections, 202.
quotient space, 359.

random variables: integer values, 52.
random walk, 59, 188.
   asymptotic behaviour, 63.
   continuum limit, 63.
   on a circle, 77.
   on a lattice, 65.
redundant eigenvector, eigen-
   -perturbation, 228.
redundant perturbation, 58.
regularization, 290.
   dimensional, 300, 363.
   infrared (IR), 365.
   lattice, 364.
   Pauli–Villars, 363.
relaxation time, 77.
relevant eigenvectors, eigen-
   -perturbations, 227.
relevant perturbation, 58.
renormalizability, 307.
renormalization conditions, 312.
renormalization constants, 312.
renormalization group (RG), 54.
   perturbative, 240.
renormalization group equations
   (RGE), 222.
renormalization theorem, 312.
renormalization, 49, 70.
   field, 222, 392.
   Gaussian, 309.
   mass, 101.
RG, 54, 56.
   exact, 393.
   mass, 244.
   near dimension 4, 259.
   perturbative, 243, 245.
   renormalization group, 54.
RGE:
   asymptotic, 260.
   correlation functions, 229.
   critical domain, 323.
   equation of state, 326.
   field theory, 314.
   functional, 393.
      Hamiltonian flow, 394.
   general solution, 317.
   Hamiltonian flow, 226.
   in a field or below T_c, 262.
   linearized, 227.

N-component field, 268.
$N \to \infty$, 347.
   perturbative solution at $d=4$, 315.
   renormalization group equations, 222.
   renormalized, 316, 326.
σ^4 model, 268.
σ-model, non-linear, 366, 367.
   solution, 372.
scaling behaviour, 225.
   corrections for $N \to \infty$, 348.
scaling property, 64, 70.
Schrödinger equation, 100.
sigma^4 field theory:
   general, RGE, 276.
     β function, 277.
     η function, 277.
     function ν, 277.
   two parameters, 281.
sigma-model, non-linear, 353, 355.
   critical temperature, 370.
   dimension 2, 375.
   Euclidean action, 362.
   $N \to \infty$, 353.
   on the lattice, 356.
   regularization, 363.
   role of dimension 2, 361.
   stability of the Gaussian fixed point, 361.
sign function sgn, 420.
specific heat:
   critical behaviour, 262.
   mean-field, 160.
spin waves, 364.
spins:
   classical, 128.
   independent, 152.
spontaneous magnetization, 147, 157.
spontaneous symmetry breaking, 86, 147, 156.
statistical field theory:
   perturbative expansion, 285.
steepest descent method, 31, 50, 154, 195, 208, 210, 296, 337.
   complex, 34.
   perturbative expansion, 145.
   real, 31.
   several variables, 38.
step or Heaviside function θ, 54, 420.
Stirling's formula, 34.
stochastic dynamics and phase transitions, 432.
super-renormalizable field theories, 313.
symmetry:
   continuous, 171.
   square, 215.

temperature, 128.
thermodynamic limit, 80, 85, 131.
thermodynamic potential, 130.
transfer matrix, 81, 166, 171.
   Gaussian model, 93.
   ground state, 429.
transition scale, 374.
translation invariance, 60, 128, 135.
tricritical point, 211.
two-point function, 28, 122, 197.
   Gaussian, 118, 185, 192.
   order $1/N$, 434.
   perturbative calculation, 122, 248, 252, 254, 293, 369.
   quasi-Gaussian approximation, 201.
   scaling behaviour, 257.
   two-loop calculation, 320.

universality classes, 51, 59, 217, 226.
universality, 49, 57, 69, 226, 322.
   critical domain, 230.
   infinite dimension, 158.
UV divergences, 244.

vacuum diagrams, 293.
vertex functional: one-loop, 298.
vertex functions, 139, 221, 286.
   four-point, 143.

Wick's theorem, 24, 118, 212, 289.
Wiener integral, 73.